From Gene Discovery to Climate-resilient Crops: Omics in Crop Improvement

Edited by

Saurabh Pandey
Department of Molecular Biology and Biotechnology, Indira Gandhi Agricultural University, Chhattisgarh, India

Ashutosh Singh
Centre of Advance Studies on Climate Change, Rajendra Prasad Central Agricultural University, PUSA, Bihar, India

Mithilesh Kumar Singh
Department of Genetics and Plant Breeding, Rajendra Prasad Central Agricultural University, PUSA, Bihar, India

CABI

CABI is a trading name of CAB International

CABI
Nosworthy Way
Wallingford
Oxfordshire OX10 8DE
UK

Tel: +44 (0)1491 832111
E-mail: info@cabi.org
Website: www.cabi.org

CABI
200 Portland Street
Boston
MA 02114
USA

T: +1 (617)682-9015
E-mail: cabi-nao@cabi.org

A catalogue record for this book is available from the British Library, London, UK.

ISBN-13: 9781800628694 (hardback)
9781800628700 (ePDF)
9781800628717 (ePub)

DOI: 10.1079/9781800628717.0000

Commissioning Editor: Ward Cooper
Editorial Assistant: Emma McCann
Production Editor: Rosie Hayden

Typeset by Exeter Premedia Services Pvt Ltd, Chennai, India
Printed and bound by CPI Group (UK) Ltd, Croydon, CR0 4YY

Contents

Contributors

Abhinav, Department of Biotechnology, Motilal Nehru National Institute of Technology, Prayagraj – 211004, UP, India. Email: abhi162416@gmail.com

Aggarwal, Garima, Department of Genetics and Plant Breeding, G. B. Pant University of Agriculture & Technology, Pantnagar, Uttarakhand, India (263145). Email: garimaaggarwalgpb@gmail.com or garimaaggarwal50@gmail.com

Asati, Ruchi, Rani Lakshmi Bai Central Agricultural University, Jhansi-284003, India.

Banshidhar, Department of Genetics and Plant Breeding, ANDUAT, Kumarganj, Ayodhya, Uttar Pradesh, India. Email: banshidhargpb@nduat.org

Bharati, Alka, ICAR–Research Complex for NEH Region, Umiam – 793103, Meghalaya, India. Email: alkabharati.bharati@gmail.com

Chandra, Ajay Kumar, IIT Guwahati, Guwahati, Assam 781039 , India. Email: ajaymbge29@gmail.com

Chaturvedi, S.K., Rani Lakshmi Bai Central Agricultural University, Jhansi-284003, India.

Chaudhary, Divya, ICAR-National Bureau of Plant Genetic Resources, New Delhi (110 012), India. Email: divyachaudhary6767@gmail.com

Chitranshi, Abhishek, ICAR-Central Island Agricultural Research Institute, Sri Vijaya Puram – 744105, Andaman & Nicobar Island, India. Email: chitranshiabhishek@gmail.com

Choudhary, Bhupendra, CPGS-AS, CAU(I), Umiam, Meghalaya 793104, India. Email: bhupendrac910@gmail.com

Choudhary, Girija, Rani Lakshmi Bai Central Agricultural University, Jhansi-284003, India.

Choudhary, Rakesh, Rani Lakshmi Bai Central Agricultural University, Jhansi-284003, India.

Das, Ritwika, Division of Bioinformatics, ICAR-Indian Agricultural Statistics Research Institute (IASRI), New Delhi – 110012, India. Email: ritwika2693@gmail.com

Dinkar, Vishal, ICAR–Indian Institute of Pulses Research (IIPR), Kanpur – 208024, Uttar Pradesh, India. Email: vishaldinkar1991@gmail.com

Dutta, Suman, IRDM Faculty Centre, Ramakrishna Mission Vivekananda Educational and Research Institute, Kolkata – 700103, West Bengal, India. Email: duttasumansiliguri@gmail.com

Edhigalla, Premnath, Department of Genetics and Plant Breeding, Sardar Vallabhbhai Patel University of Agriculture and Technology, Meerut, Uttar Pradesh, India (250110). Email: premnathedhigalla@gmail.com

Gayen, Dipak, Department of Biochemistry, Central University of Rajasthan, Ajmer 305 817, Rajasthan, India. Email: dipak.gayen@curaj.ac.in or gayen.dipak@gmail.com

Irfan, Mohammad, ICAR-Central Institute of Temperate Horticulture (CITH), Srinagar – 191132, J&K, India. Email: irather046@gmail.com

Jaiswal, Priyanka, Department of Genetics and Plant Breeding, ANDUAT, Kumarganj, Ayodhya, Uttar Pradesh, India

Jamir, Temsuramok, CPGS-AS, CAU(I), Umiam, Meghalaya 793104, India. Email:temsuramok@gmail.com

Jeena, Anand Singh, Department of Genetics and Plant Breeding, G. B. Pant University of Agriculture & Technology, Pantnagar, Uttarakhand, India (263145). Email: dr.asjeena@gmail.com

Kampurath, Akshara, Department of Horticulture (Fruit Science) Central University of Tamil Nadu, Thriruvarur – 610005 Tamil Nadu, India Email: akshara.kampurath@gmail.com

Karunakaran, Abinaya, Department of Horticulture (Fruit Science) Central University of Tamil Nadu, Thriruvarur – 610005 Tamil Nadu, India. Email: abinayakarunakaran12@gmail.com

Keerthana, Dhanapati, Rani Lakshmi Bai Central Agricultural University, Jhansi-284003, India.

Kiran, Kadam, Rani Lakshmi Bai Central Agricultural University, Jhansi-284003, India.

Kumar, Amarjeet, Department of Genetics and Plant Breeding, College of Agriculture (CAU, Imphal), Kyrdemkulai – 793104, Meghalaya, India. Email: amarjeetgpb@gmail.com

Kumar, Bishawajit, Department of Genetics and Plant Breeding, G. B. Pant University of Agriculture & Technology, Pantnagar, Uttarakhand, India (263145). Email: bishawajitpbg@gmail.com

Kumar, Narendra, CAU (Imphal) – College of Agriculture, Kyrdemkulai 793105, Meghalaya, India. Email: narendraphysiol@gmail.com

Kumari, Namata, Department of Genetics and Plant Breeding, G. B. Pant University of Agriculture & Technology, Pantnagar, Uttarakhand, India (263145). Email: kumarinamata@gmail.com

Kumari, Shalu, Department of Genetics and Plant Breeding, I.A.S, BHU, Varanasi, 492012, Uttar Pradesh, India. Email: shaluprasad14@bhu.ac.in

Manivannan, Subramanian, Department of Horticulture, North – Eastern Hill University, Shillong – 793022, India. Email: manivannan@cutn.ac.in

Mehra, Kajal, Department of Genetics and Plant Breeding, G. B. Pant University of Agriculture & Technology, Pantnagar, Uttarakhand, India (263145) Email: mehra9505@gmail.com

Mhetre, Vishal Balasaheb, Defence Institute of High Altitude Research, Leh – 194101, UT of Ladakh, India. Email: vishalmhetre8@gmail.com

Mishra, Awdhesh, Department of Biotechnology, Yeungnam University, Gyeongsan 38541, South Korea. Email: awadhesh.biotech07@gmail.com

Nikhil, A.D.S.S. Ajay, Sher-e-Kashmir University of Agricultural Sciences and Technology of Kashmir, Srinagar, 190025, India.

Pandey, Anuradha, Department of Biochemistry, Central University of Rajasthan, Ajmer 305 817, Rajasthan, India. Email: Anuradhapandey678@gmail.com

Pandey, Saurabh, Department of Molecular Biology and Biotechnology, Indira Gandhi Agricultural University, Raipur-492012, Chhattisgarh, India. Email: pandey.saurabh784@gmail.com or saurabh.pandey@igkv.ac.in

Prusty, Reena, ICAR-Central Institute of Temperate Horticulture (CITH), Srinagar – 191132, J&K, India. Email: reenaprusty005@gmail.com

Ramakrishnan P N, Aparna, Department of Biochemistry, Central University of Rajasthan, Ajmer 305 817, Rajasthan, India. Email: aparnak10@gmail.com

Reddy, Basu Sudhakar, Department of Genetics and Plant Breeding, G. B. Pant University of Agriculture and Technology, Pantnagar, Uttrakhand (263145), India. Email: basusudhakarreddy@gmail.com

Rohit, G. B., Pant University of Agriculture & Technology, Pantnagar, Uttarakhand, India, 263145. Email: rohit8zed@gmail.com

Selin, Afiya Rajan, Department of Horticulture (Fruit Science) Central University of Tamil Nadu, Thriruvarur – 610005 Tamil Nadu, India. Email: afiarajan@gmail.com

Shanmugasundram, Senthilkumar, Department of Horticulture (Fruit Science) Central University of Tamil Nadu, Thriruvarur – 610005 Tamil Nadu, India. Email: senthilshanmugam87@gmail.com

Sharma, Soumya, Division of Bioinformatics, ICAR-Indian Agricultural Statistics Research Institute (IASRI), New Delhi – 110012, India. Email: soumya.iasri@gmail.com

Singh, Aditya Pratap, Department of Genetics and Plant Breeding, Bidhan Chandra Krishi Vishwavidyalaya, Mohanpur – 741252, Nadia, West Bengal, India. Email: adityapratapbckv@gmail.com

Singh, Anshuman, Rani Lakshmi Bai Central Agricultural University, Jhansi-284003, India.

Singh, Ashutosh, Centre for Advanced Studies on Climate Change, Dr. Rajendra Prasad Central Agricultural University, Pusa, Samastipur, Bihar, 848125, India.

Singh, Digvijay, Dept of Genetics and Plant Breeding, ANDUAT, Kumarganj, Ayodhya, Uttar Pradesh, India.

Singh, Gurdeep, Baba Farid College, Baba Farid Group of Institutions, Bathinda, Punjab, India, 151001. Email: gurdeepsandhu1997@gmail.com

Singh, Mithilesh Kumar, Department of Genetics and Plant Breeding, Dr. Rajendra Prasad Central Agricultural University, Pusa (Samastipur), 848125, Bihar, India. Email: mithileshgpb@gmail.com

Singh, Roshan Kumar, Department of Botany, Mahishadal Raj College Purba Medinipur, West Bengal 721628, India. Email: rshnsngh8@gmail.com

Suresh, Perinba C., Department of Horticulture, North – Eastern Hill University, Shillong – 793022, India. Email: sureshcp2112@yahoo.com

Tabassum, Baba Farid College, Baba Farid Group of Institutions, Bathinda, Punjab, India, 151001. Email: tabassum12081992@gmail.com

Tiwari, Mayank, Department of Genetics and Plant Breeding, G. B. Pant University of Agriculture & Technology, Pantnagar, Uttarakhand, India (263145). Email: 912mayanktiwari@gmail.com

Unnikrishnan, Neeraj, Department of Biochemistry, Central University of Rajasthan, Ajmer 305 817, Rajasthan, India. Email: neerajneeraj922@gmail.com

Yadav, Vijay Kumar, ICAR-Indian Institute of Grassland and Fodder Research Institute, Jhansi-284003, India.

Zaffar, Aaqif, Division of Genetics & Plant Breeding, SKUAST – Kashmir, Faculty of Agriculture, Wadura – 193201, Sopore, J&K, India. Email: Aaqifzaffar123@gmail.com

Preface

The art of plant breeding dates back to the history of humankind when humans collected seeds of several edible plant species to reap for the next generation. Some intentional or unintended selections were imposed on them to evolve a superior variant. Nevertheless, those variants had some attributes of the increased size of edible parts, taste, and yield. Since then, an array of systematic plant breeding activities continued. New crop varieties were developed using techniques ranging from domestication to selection to exploitation of heterosis and population improvement. This followed precise alteration of traits using backcrossing for improvement of oligogenic traits. Later, genetic gain in several crops reached a plateau, particularly when human population growth was burgeoning. This situation, where the rate of genetic improvement in crop plants through traditional breeding ceased to respond, and population growth exploded, exerted vast pressure on plant breeding to discover novel breeding procedures/techniques to kick-start ceasing breeding practices. Classical plant breeding shifted its gear to modern plant breeding with a parallel emphasis on genotypic assay besides phenotypic assay in the quest for selection of targeted, desirable plant types in less time, thereby accelerating the pace of plant breeding with increased genetic gain for various traits of interest in crop plants to cater to the food, fuel, and fodder need of the humankind.

An enormous amount of genotypic data was generated with the parallel advent of sequencing technologies. This required statistical methods and computer programs for its evaluation, which has given rise to the bioinformatics and omics field. The book *From Gene Discovery to Climate-Resilient Crops: Omics in Crop Improvement* attempts to provide a systematic approach to the theoretical concepts and procedures with a practical inclination for the benefit of students and practitioners of plant breeding. The topic of this book is carefully chosen, keeping in view the recent trends in plant breeding. Various experts on the respective crop-specific omics experience were invited to write chapters for authenticity, increasing the practicality of this book. We assure you that this meticulously crafted book will meet students' and plant breeders' requirements and expectations.

We express our deep sense of appreciation and gratefulness to all the contributors who, besides their other preoccupations, toiled earnestly to make the compilation of this book with vast, recent, and up-to-date information.

We are aware that, despite our sincere efforts, scope prevails for further improvement. We are happy and open to readers in receiving their advice, suggestions, and criticism, which would greatly improve this book in subsequent editions.

Saurabh Pandey
Ashutosh Singh
Mithilesh Kumar Singh

Foreword

The world population is steadily growing and is expected to reach over 10 billion by 2050. Consequently, an increasing pressure to feed a growing population in the future would need the additional production of major crops. On the contrary, the rapidly changing climate is imposing a heavy toll on agriculture, threatening food and nutritional security. The shrinking agricultural resources and the unpredictable climatic variables that mark the baseline for the sustenance of life are some of the most burning challenges to the scientific world in feeding the unprecedented and unbridled population. This has warranted the researchers to deploy classical and advanced tools to improve the climate resilience of cultivated species. Nevertheless, the advancement in molecular biology and computational facilities for data management, computation, and meaningful interpretation of voluminous data arising from genomics, proteomics, and metabolomics is promising new and efficient alternatives to ensure food security for one and all. However, all the other options rely directly or indirectly on exploring or creating genetic variability for the target trait and its integration into an appropriate genetic background.

In this context, the book *From Gene Discovery to Climate-Resilient Crops: Omics in Crop Improvement* is a timely attempt to collate the efforts and outcomes of research on climate-resilient agriculture. Understanding the intricate interactions between climate change and biotic factors is crucial for developing effective mitigation strategies. The book has three parts, where Part A introduces the readers to climate change, and its effect on issues of increased biotic stresses. Part B deals with breeding and molecular approaches used to improve crop species' climate resilience against biotic stresses. Each chapter deals with different strategies of climate change impacts on biotic agents. This part has given a detailed insight into how the various tools and techniques have enhanced the climate resilience of cultivated species. Part C summarizes the advanced tools, including genome editing, that have direct implications for expediting the research on climate-resilient agriculture. This book is an outstanding effort to bring together the knowledge on climate-resilient agriculture, which will serve as a one-stop solution for the researchers as well as students. Also, the book will serve as a guide for researchers aspiring to work in the area of climate-resilient agriculture. Integration of updated knowledge and reliability are the crucial aspects of this book. The efforts of editors to bring out this book are appreciated, and the contributions of authors from different subject domains are acknowledged.

It gives me immense pleasure and a sense of pride while writing these brief words for the illustrative effort of Dr Ashutosh Singh and Dr Mithilesh Kumar Singh, Assistant Professors, RPCAU, and Dr Saurabh Pandey, Assistant Professor, IGKV, Raipur. The book *From Gene Discovery to Climate-Resilient*

Crops: Omics in Crop Improvement is an excellent compendium of information and insights concerning steady transgression from classical to up-to-date molecular breeding approaches for crop improvement. This book's comprehensive information, simplified concepts, and explicit linguistic expertise are expected to be rewarding. Given the available resources and duration, the ideas and methodologies compiled with brevity will undoubtedly prove helpful in fast-pacing the quest to improve economically important traits. This document is a prodigious contribution to the well-being of society and humanity. I commend the authors' endeavors and wish them a bright and prosperous future.

P.K. Singh

1 Crop Genome Sequencing Projects: Basis for Omics

Bishawajit Kumar[1], Premnath Edhigalla[2], Anand Singh Jeena[1], Garima Aggarwal[1*], Kajal Mehra[1], Namata Kumari[1] and Mayank Tiwari[1]
[1]Department of Genetics and Plant Breeding, College of Agriculture, G.B. Pant University of Agriculture and Technology, Pantnagar, Uttarakhand, India; [2]Department of Genetics and Plant Breeding, Sardar Vallabhbhai Patel University of Agriculture and Technology, Meerut (Uttar Pradesh), India

Abstract

The advent of sequencing methodologies has transformed molecular biology, providing sequencing techniques with an unprecedented ability to decode the genetic blueprints of various organisms. The evolution from Sanger sequencing to next-generation sequencing (NGS) and third-generation sequencing technologies has significantly accelerated the speed, accuracy, and affordability of genome sequencing. This advancement has particularly transformed plant genomics, offering deeper insights into gene function, structural variations, and regulatory mechanisms at a genome-wide scale. With the ability to sequence complete genomes, researchers have gained crucial understanding of essential crop traits, such as disease resistance, yield improvement, and adaptation to environmental stresses. Despite these advancements, achieving gapless genomes remains a challenge, particularly in crops with complex polyploidy and large genome sizes. This chapter explores the sequencing efforts in major agricultural crops, horticultural species, bioenergy crops, and oilseeds, including the application of sequencing data in multiomics approaches for improving climate-resilient crops. Additionally, it addresses the current limitations in sequencing technologies and the gaps that hinder fully exploiting genomics for crop improvement. Ultimately, genome sequencing serves as a foundation for advancing precision breeding and gene-editing technologies, highlighting its integral role in shaping future agricultural innovations.

1.1 Introduction

In the face of persistent global challenges like food insecurity and hunger, the world continues to grapple with the immense task of feeding an ever-growing population. The underlying drivers of food insecurity are multifarious and are exacerbated by imbalances in the consumption of agricultural inputs. These include population explosion, the impact of climate change, urbanization, degradation of agricultural land, reliance on carbon-based chemical inputs, and increasing water scarcity (FAO, 2020). Addressing these obstacles requires the adoption of sustainable agricultural practices that ensure increments of crop yields and minimize the environmental footprint. Meeting the rising demand for food in a sustainable manner necessitates improving crop yields even in detrimental environments. This must be accomplished while also reducing the resources required to sustain a burgeoning global population. One of the critical

*Corresponding author: Garimaaggarwalgpb@gmail.com

 From Gene Discovery to Climate-Resilient Crops: Omics in Crop Improvement (eds S. Pandey *et al.*)
DOI: 10.1079/9781800628717.0001

factors in achieving this is understanding how plants tolerate and adapt to uncertain temperatures. Such knowledge is vital for improving plant productivity under changing climatic conditions. Reviewing how plants develop stress tolerance and survival mechanisms will yield new insights, leading to innovative approaches in breeding crops that can withstand the stresses of a changing climate. Success in this global endeavor will require a systems-based approach that integrates new data-driven farming techniques, sustainable agricultural practices, and the development of improved crop cultivars. In this context, genomics offers essential tools and biological insights that can propel the future of 21st-century agriculture.

Genomic technologies could revolutionize crop improvement, providing foundational knowledge and providing the ways to meet those specific challenges of 21st-century agriculture. The year 2025 marks 25 years since the completion of the first-ever plant genome sequence, *Arabidopsis thaliana* (The Arabidopsis Initiative, 2000). During the past two and a half decades, reference genomes were made available for hundreds of plant species ranging from non-vascular plants to flowering plants (Fig. 1.1). A high-quality reference genome sequence provides comprehensive access to the gene catalog of a species, including the regulatory elements controlling gene function, and serves as a framework for understanding genomic variation. Such resources are crucial for fully elucidating the roles of genes in development, efficiently exploiting both natural and induced genetic diversity within organisms, and driving genomics-based approaches to systems biology (Feuillet *et al.*, 2011).

During the last two decades, genomic science has enhanced plant breeding efforts in increasing crop yields, providing greater resilience to environmental stresses and pathogens, and the development of novel crop varieties with enhanced traits (Bevan *et al.*, 2017). Genomics-based tools and knowledge have thus become indispensable for the development of climate-resilient crops, offering a path forward over the daunting agricultural challenges of the 21st century. Sequencing provides a fundamental platform for advancing various omics fields. The introduction of DNA microarray technology (Schena *et al.*, 1995), together with first-generation Sanger sequencing, marked the early steps in the sequencing landscape. This was followed by the development of second-generation or next-generation sequencing (NGS) technologies (Hu *et al.*, 2021), which eventually progressed to third-generation sequencing methods

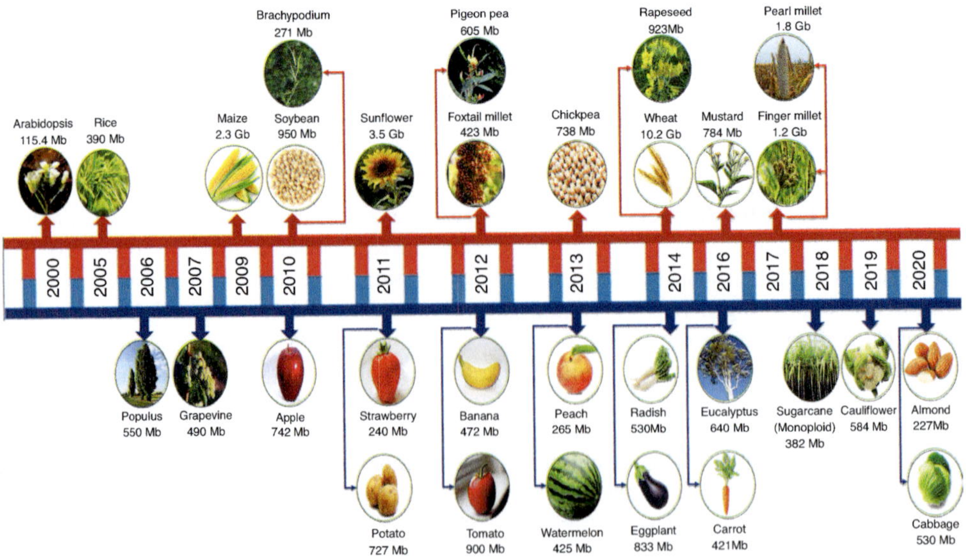

Fig. 1.1. Chronological overview of key crop genome sequencing milestones. Author's own image.

with long-read capabilities (Jain *et al.*, 2015; Xiao and Zhou, 2020). These further enabled whole-genome and whole-exome sequencing to be carried out with the depth required for annotation of the mutational profiles present within samples allowing the creation of molecular markers for trait dissection, selection, and improvement (Stein *et al.*, 2018).

The availability of reference genomes for various crop species has become a key tool in understanding the biological functions of important genes. However, plant species exhibit diverse genome structures, with variations in ploidy levels and repeat content. In response, plant research communities initiated the assembly of reference genomes for different species, a trend that picked up speed as the cost of sequencing plummeted dramatically (Thudi *et al.*, 2021). This coincided with the rise of single-molecule long-range sequencing, advancements in physical mapping, and improvements in genome assembly pipelines. As a result, draft genomes of more than 800 plant species are available to date, with continued improvements in assembly quality, characterized by fewer errors and increased contiguity (Michael and VanBuren, 2020). Interestingly, it was soon realized that an entire genetic information of a species cannot be represented by a single reference genome, leading to the emergence of pan-genomes (Tettelin *et al.*, 2005). Additionally, advancements in genome complexity reduction techniques (Chuang *et al.*, 2003) now enable sequencing even of species with large, complex genomes. Notably, crops such as maize (~2.2 Gb) (Jiao *et al.*, 2017), hexaploid wheat (~17 Gb) (Appels *et al.*, 2018), and even highly complex polyploids like octaploid sugarcane (3 Gb- 10 Gb) (Zhang *et al.*, 2018; Healey *et al.*, 2024) can now be explored for genes and genomics, so extending the applicability of omics in these species. Thus, genome sequencing is instrumental in bolstering global food security globally by allowing the generation of productive, climate-resilient crop varieties. Detailed genetic mapping allows researchers to pinpoint key genes that can be put forth as responsible for crucial traits, including drought tolerance, disease resistance, and nutrient efficiency (Zhang *et al.*, 2004; Francia *et al.*, 2016; Spindel *et al.*, 2018). This, in turn, enables the implementation of precise breeding strategies aimed at enhancing crop performance under unfavorable environmental conditions. Furthermore, by revealing the genetic architecture underlying plant stress responses, sequencing enables the cultivation of crops that can sustain productivity in regions prone to climate variability, thus ensuring consistent food supplies. Genome sequencing provides information related to molecular mechanisms that can be utilized for crop improvement, which is not only essential for meeting the growing demands of an expanding population but also reducing the environmental footprint of agriculture.

This chapter will explore the evolution of sequencing technologies, discuss landmark sequencing projects in model, cereal, bioenergy, horticultural, and oilseed crops, and examine the integration of sequencing in various omics approaches. Additionally, it will address the challenges that persist in sequencing efforts across different crops.

1.2 Evolution of Sequencing Techniques

Genome sequencing plays a central role in molecular breeding, functional genomics, comparative genomics and pangenomics. The most recent developments, especially with NGS technologies, have enabled the identification of a significant number of genetic markers used for detecting desirable traits in crops. The availability of numerous reference genomes developed for various species delivers interesting insights on genes associated with agricultural traits.

1.2.1 First- and second-generation sequencing

The first major successes in DNA sequencing, developed separately in 1977, by Frederick Sanger using the chain termination method, and by Allan Maxam and Walter Gilbert (Maxam and Gilbert, 1977) introducing the chemical degradation method. Although the Maxam–Gilbert method, containing noxious chemicals and not really easy to perform, withered away, Sanger sequencing became widespread. It needed chain-terminating nucleotides to produce fluorescent signals that would determine the presence of nucleotides (Sanger *et al.*, 1977). Due to the size and complexity of plant genomes, first-generation sequencing methods, such as Sanger sequencing, soon became impractical for large-scale genomic studies. However, the introduction of NGS increased the speed of processing and significantly reduced the cost of genome processing involved. NGS systems ended the monopoly of the use of clone-based DNA libraries with large genomes being sequenced much more efficiently. Following the advent of the first NGS platform in 2005 from Roche/454 pyrosequencing, other platforms that had major impacts gained popularity throughout 2008, such as Illumina and SOLiD sequencing (Margulies *et al.*, 2005; Bentley *et al.*, 2008). Roche/454 pyrosequencing was one of the early, widely utilized NGS technologies in plant genomics and was used to sequence crops such as potato, apple, and strawberry (Bevan and Uauy, 2013). Illumina, having high-throughput capacity, was especially useful for the sequencing of large genomes of plants, providing a new power tool in the hands of researchers that allows for fast production of vast genomic data. SOLiD sequencing provides detailed examination of plant genomes and contributes to genome resequencing and transcriptome analysis, thereby allowing researchers to achieve high precision regarding detecting genetic variations in crops that help in the improvement and breeding program (Shulaev *et al.*, 2011). All of these new developments in sequencing technologies have dramatically changed the way the plant genome is approached for gaining insights related to genetic diversity and speedy breeding programs for several crops.

1.2.2 Third-generation sequencing

Recent advances in third-generation sequencing technologies have revolutionized the genome assembly process, making it possible to produce high-quality genomes. These technologies, such as PacBio SMRT and Oxford Nanopore Technology, have longer read lengths than 1 kb; this increases the quality of genome assembles while also reducing the costs and number of cycles for sequencing (Yadav *et al.*, 2018). Unlike short-read sequencing, third-generation techniques do not require DNA amplification; hence, they are more prone to sequencing errors and demand coverage of the genome to a higher order (Hon *et al.*, 2020). PacBio SMRT sequencing is a real-time, single-molecule sequencing method based on the principle of sequencing by synthesis using zero-mode waveguides (ZMWs) to observe the enzymatic activity of an individual DNA polymerase (Hui, 2014). The unmodified DNA polymerase adds fluorescently labeled nucleotides, performing DNA synthesis at the bottom of the ZMWs. The fluorescence emitted during incorporation of the nucleotide recorded as a label is cleaved off at each base addition (Xiao and Zhou, 2020). Although error rates are high, and ligation biases exist, PacBio recently introduced circular consensus sequencing to obtain even higher accuracy reads, with more than 99% accuracy (Dai and Shen, 2022). Oxford Nanopore Technology (ONT) reads native DNA directly without amplification as they pass through the nanoscale pores in lipid membranes. This measure of interruptions in ionic currents as DNA strands pass through the nanopores produces real-time sequence information visualized in a 'squiggle plot' (Deamer *et al.*, 2016). ONT offers ultra-long reads suitable for complex plant genomes with a great number of repetitive sequences, and can also determine native DNA modifications, such as methylation (Schmidt *et al.*, 2017). However, long-range sequencing of large repetitive plant genomes often requires DNA of a good quality, and ONT has a rather reasonable chance of producing long-read data for use in plant genomics and pangenomics (Bolger *et al.*, 2014). Thus, enhancement of accuracy and further reduction in error rates make nanopore sequencing one of the frontiers of innovative tools for plant genomics and epigenomics studies.

1.2.3 Targeted sequencing approaches

Targeted sequencing is a targeted genomic approach to focus selectively on amplifying and determining the sequences of regions-of-interest in a genome, permitting researchers to gain an understanding into genetic variation and functional elements driven by specific traits or diseases. Restriction-site associated DNA (RAD) sequencing takes advantage of the cutting sites of restriction enzymes to reduce complexity in target sequences. It involves breaking genomic DNA at specific recognition sites and then ligating sequencing adapters onto the fragments before quantification and sequencing. RAD sequencing directly provides low-cost tools in population studies, genome-wide polymorphisms, and single nucleotide polymorphism (SNP) discovery and development (Baird *et al.*, 2008). This makes it a powerful method to unravel the genetic variation and evolutionary dynamics in focusing on certain subsets of the genome, and it is very much crucial for research into the genetic diversity of different species (Andrews *et al.*, 2016). Similarly, another sequencing targeted method that has expanded transcriptomic research is RNA sequencing, or RNA-Seq. RNA-Seq allows for thorough gene transcript study and their expression levels, promoting understanding of how a gene is regulated and functional in terms of molecular marker identification. RNA-Seq further clarifies metabolic pathways as well as advancing knowledge of mechanisms underlying plant development, yielding more improved gene annotation and functional genomics (Tyagi *et al.*, 2022). Moreover, RNA-Seq offers the ability to identify new genes and their role in metabolism (Kukurba and Montgomery, 2015). The technique also enables a comparison of gene expression under different conditions. Hybridization capture-based sequencing (Hyb-Seq) utilizes specially designed probes that hybridize to specific target sequences within fragmented genomic DNA. After hybridization, the targeted sequences are enriched through PCR amplification, increasing their abundance for downstream applications. Following amplification, the probes are removed, and the enriched target DNA is prepared for sequencing using NGS technologies. This method is particularly useful for studying target genes, as it allows for focused exploration of regions of interest and enhances understanding of gene function and regulation (Hill *et al.*, 2019). Hyb-Seq is also beneficial in reconstructing phylogenies by comparing genetic variation among different species, thereby advancing our understanding of evolutionary relationships (Mariac *et al.*, 2014). Chromatin immunoprecipitation sequencing (ChIP sequencing) technique is very crucial for determining the protein–DNA interactions within the study area of plant research. The research study identifies how the gene regulatory mechanism works. It identifies transcription-factor binding sites, histone-binding sites, and other regulatory protein–DNA interaction sites that are required to understand the manner in which interactions lead to developmental and environmental responses in plants (Kaufmann *et al.*, 2010). The ChIP-Seq procedure starts with the cross-linking of proteins and DNA in live plants. The chromatin is then fragmented into DNA–protein complexes through sonication or enzymatic digestion. It involves the immunoprecipitation of these complexes with specific antibodies for the formation of protein–DNA complexes. Isolated regions are further processed with reverse cross-linking and DNA purification for sequencing purposes (Kaufmann *et al.*, 2010). ChIP-Seq offers broad potential for regulatory networks across plant species. The binding of key proteins and epigenetic markers provides information related to mechanisms involved in flowering, stress response, and metabolic pathways, and could potentially be key to crop improvement and increased productivity in agriculture (Chen *et al.*, 2018).

1.3 Genome Sequencing in Cereals and Pulses

1.3.1 Major cereals

Bread wheat (*Triticum aestivum* L.) is a significant crop of the world that provides approximately 20% of daily calories and constitutes a primary source of protein. Genome sequencing of wheat has been considered a daunting task because of the large genome size of wheat (~16 Gb) and very repetitive structure, with more than 85% of the genome being repetitive elements (Li *et al.*, 2004). It is more complex due to its polyploid nature, carrying three subgenomes: A, B, and D. This genome was obtained by the hybridization between allotetraploid Emmer wheat *Triticum turgidum* (AABB) and diploid *Aegilops tauschii* (DD) (Marcussen *et al.*, 2014). The first sequencing attempt of the wheat genome utilized Roche 454 pyrosequencing that eventually resulted in a draft fragmented genome. Although very scanty, it was at the time the sole resource on wheat genomics. The International Wheat Genome Sequencing Consortium (IWGSC) later stepped forward and took the lead in genome assembly using chromosome sorting techniques to enhance the assembly, culminating into the IWGSC CSS v1.0 draft genome covering 10.2 Gb (Mayer *et al.*, 2014).

Sequencing the whole chromosome 3B by a BAC-by-BAC approach made a giant step forward and permitted an improved resolution to make genetic mapping much simpler for crop improvement (Choulet *et al.*, 2014). Improved later sequencing technology and assembly methods were able to deliver high-quality genomes on related species, such as *Aegilops tauschii* (Luo *et al.*, 2017), wild emmer (Avni *et al.*, 2017), and *Triticum urartu* (Ling *et al.*, 2018). This draft genome was further refined into CS (TAGC v1.0) by the IWGSC, and the high-quality reference genome CS RefSeq 1.0 (14.5 Gb) was published in 2018 (Appels *et al.*, 2018). CS RefSeq 2.1 is the updated version made based on scaffold orientations and gaps through PacBio sequencing and optical mapping (Zhu *et al.*, 2021). For example, important genomic resources are publicly available in databases (Table 1.1) such as Wheat@Urgi, Graingene, Ensembl Plants, and others for ongoing wheat research and breeding.

Rice is a model organism among cereals due to its diploid nature, high synteny with other cereal genomes, and relatively small genome size of 400–430 Mb (Gale and Devos, 1998). The International Rice Genome Sequencing Project (IRGSP) was initiated in 1997 as a global collaboration to sequence the *Oryza sativa* ssp. japonica (cv. Nipponbare) genome by hierarchical clone-by-clone shotgun sequencing (Eckardt, 2000; Sasaki and Burr, 2000). Assembly contributions from Monsanto and Syngenta resulted in a high-quality assembly of a 390 Mb rice genome that is reported to contain over 32,000 protein-coding genes (Goff *et al.*, 2002; Sasaki, 2005). The Chinese Superhybrid Rice Genome Project sequenced the indica variety Guang Lu Ai 4. The revised IRGSP-1.0 reference genome in 2012 improved assembly with a minimum tiling path validated with optical mapping (Kawahara *et al.*, 2013). Telomere-to-telomere sequence of the Nipponbare cultivar shows some new genes that were not annotated in the previous version using high-throughput sequencing technologies and techniques like Hi-C technology (Shang *et al.*, 2023). The 3K Rice Genomes Project, through the sequencing of 3000 rice accessions, forms a treasure trove for future breeding programs and population-specific genomic selection (3,000 Rice Genomes Project, 2014). Pangenomic research unravels the genetic diversity of rice, including structural variants and presence–absence variants, that are greatly informative to evolutionary biology and speciation (Shang *et al.*, 2022). Genome-wide association studies (GWAS), quantitative trait loci (QTL) mapping, and the identification of stress responsive genes linked to biotic and abiotic stress have paved the way for significant advancements in rice breeding programs (Ahmad, 2022).

Maize is a very crucial cereal crop and also a model organism for studying inheritance, heterosis, gene function, and cytogenetic studies because of its vast diversity and adaptability to environmental stresses. National Science Foundation, U.S. Department of Agriculture, and Department of Energy coordinated a landmark initiative that led to the whole-genome shotgun sequencing of draft B73 maize inbred-line genome using BAC and fosmid libraries (Chandler and Brendel, 2002). In 2009, the B73 RefGen_v1 reference genome consisted of 32,540 protein-coding genes with an 85% proportion of repetitive sequences (Schnable *et al.*, 2009). Later ones, like B73 RefGen_v4, made use of single-molecule sequencing technologies, producing longer contigs and more correct gene annotations with previously concealed sequences revealed (Jiao *et al.*, 2017). Gapless assembly of the Mo17 genome, using hybrid ONT and PacBio HiFi reads, brought the base accuracy to 99.994%, revealing centromeres, transposable element (TE) arrays, and duplicated genes (Chen *et al.*, 2023). The Maize Genetics and Genomics Database (MaizeGDB) serves as a valuable resource for maize research and breeding, offering more than 100 genome assemblies, including key references like B73 (Schnable *et al.*, 2009), PH207 (Hirsch *et al.*, 2016), and Mo17 (Sun *et al.*, 2018). This platform enables an in-depth exploration of maize genomic diversity and is further enhanced by supplemental information from the Nested Association Mapping population (Hufford *et al.*, 2021) and the newly released PanAnd genomes (Stitzer *et al.*, 2024), enriching maize genetics research.

1.3.2 Minor cereals

The genome of the minor millet *Setaria italica* was first to be sequenced, and the result consisted of

Table 1.1. Overview of available crop OMIC databases, detailing resources for genome, epigenome, proteomics, transcriptome, metabolome and multiomics data across various crops. Author's own table.

Database	Omics type	Plant type	Purpose	Reference URL
Phytozome	Genomics	Majority plant species	Integrates plant genomic data for functional and comparative analysis	https://phytozome-next.jgi.doe.gov/
Gramene	Genomics	Cereals, millets, vegetables, pulses, etc.	Provides access to genome data, gene annotations, and comparative genomics tools	https://ensembl.gramene.org/index.html
Graingene	Genomics	Wheat and avena	Identification of QTL, novel genes, molecular markers	https://wheat.pw.usda.gov/GG3/
CerealsDB	Genomics	Wheat and related species	Provides SNPs for wheat to assist molecular breeding	https://www.cerealsdb.uk.net/
PulseDB	Genomics	Pulses	A comprehensive genomic database for pulse crops	https://www.pulsedb.org/
CoGE	Genomics	Majority of crop species	A genomic database useful for conducting comparative genomics	https://genomevolution.org/coge/
Genome Database for Rosaceae	Multiomics	Rosaceae	A curated genomic database of Rosaceae family	https://www.rosaceae.org/
Sol Genomics Network	Multiomics	Solanaceae	A genomic database website of major Solanaceous family members.	https://solgenomics.net/
Brassica database (BRAD)	Multiomics	Brassicaceae	A comprehensive genomic database of Brassica members	http://brassicadb.cn/
MilletDB	Multiomics	Millets	A comprehensive multiomics database for millets	https://milletdb.novogene.com/home
Plant Gene Atlas	Transcriptomics	Model crops	A transcriptome resource database of various crops	https://plantgeneatlas.jgi.doe.gov/
Brassica EDB	Transcriptomics	Brassicaceae members	A gene expression database for transcriptomic studies	https://brassica.biodb.org/
Expression Atlas	Transcriptomics	Major cereals and vegetables	A transcriptome database for studying genes and protein expression	https://www.ebi.ac.uk/
PMHub	Metabolomics	Several plant species	Provides information about plant metabolites, metabolic pathways and their functions.	https://pmhub.org.cn/
RefMeta	Metabolomics	Several plant species	Contains wealth of information pertaining to plant metabolomics.	https://www.biosino.org/RefMetaDB/
Plantcyc	Metabolomics	Several plant species	Provides information about metabolic pathways	https://plantcyc.org/

Continued

Table 1.1. Continued

Database	Omics type	Plant type	Purpose	Reference URL
PPDB	Proteomics	Arabidopsis and maize	Provides proteomic data about Arabidopsis and maize	http://ppdb.tc.cornell.edu/
Croppal	Proteomics	Several plant species	Offers a proteomics resource for mapping protein distributions and enhancing species-specific crop traits	https://croppal.org/
PlantCADB	Epigenomics	Several plant species	Comprehensive insights into genetic regulatory networks aimed at enhancing crop trait development	https://bioinfor.nefu.edu.cn/PlantCADB/
Plant Chromatin State Database (PCSD)	Epigenomics	Several plant species	Provides extensive insights into plant epigenomes	http://systemsbiology.cau.edu.cn/chromstates

two high-quality reference genomes: one, for the cultivar Zhang gu, and another for the cultivar Yugu1. For the Zhang gu genome, Illumina technology resulted in an assembly of 423 Mb, 46% being repetitive sequences, and consisted of 38,801 annotated genes. This assembly enabled research on chromosome evolution and gene families, along with the identification of genes responsible for the C4 pathway, in addition to QTL mapping of herbicide resistance (Zhang *et al.*, 2012). Simultaneously, the Yugu1 genome was assembled based on Sanger capillary sequencing, which covered around 80% of the genome with a total sequence length of 405 Mb. This reference genome contained 24,000 to 29,000 protein-coding genes (Bennetzen *et al.*, 2012). Finger millet (*Eleusine coracana*) is a nutrient-rich crop that thrives under drought conditions and has its genome sequenced extensively for further improvement in breeding. For the Indian cultivar ML 365, the first draft assembly by a combination of both Illumina and SOLiD technologies resulted in having an estimated genome of 1196 Mb with coverage for 82% of the total. This assembly revealed 85,024 genes associated with major traits like drought tolerance and resistance to diseases (Hittalmani *et al.*, 2017). The second draft assembly utilized Illumina sequencing, which reported a genome size of 1.2 Gb and predicted 62,348 genes across the PR-202 cultivar (Hatakeyama *et al.*, 2018). The chromosome-scale reference genome of the KNE796 cultivar, *Eleusine coracana* v1_1, contains 48,836 annotated genes and improved assembly metrics; it was recently published (Devos *et al.*, 2023). A significant milestone in millet breeding was the release of the 1.8 Gb draft genome for the Tift 23 genotype of pearl millet, comprising 38,579 genes. This genome, assembled using Illumina and Sanger sequencing, has been instrumental in comparative genomics, validating candidate genes linked to abiotic stress tolerance, studying heterosis, and analyzing population diversity (Varshney *et al.*, 2017). MilletDB is a pan-genomic database that collates multiomics data for various millet species. These databases are extremely crucial in identifying novel genes and QTLs responsible for the stress tolerance of millets, supporting the improvement of millet resilience and productivity in breeding efforts.

1.3.3 Pulses

Mungbean (*Vigna radiata*) is a self-pollinated, diploid crop with a relatively small genome and its short growth cycle. NGS marked a significant breakthrough in pulse research, allowing chromosome-level draft genome assemblies such as VC1973 to be made. Illumina sequencing of VC1973 resulted in a genome assembly of 431 Mb size and 22,427 protein-coding genes (Kang *et al.*, 2014). Several high-quality genome sequences, such as Tc1966 (Kang *et al.*,

2014), RIL59 (Liu *et al.*, 2016), and Sulv1 (Yan *et al.*, 2020), have been reported. Recently, a large-scale pangenomic study of 217 mungbean accessions improved understanding of the genetic diversity across different species and pointed to genes associated with photoperiodic flowering and other agronomically relevant traits, preparing the way for comprehensive genome-wide association studies (Liu *et al.*, 2022). Chickpea (*Cicer arietinum*) is a nutritionally rich pulse crop, comprising two major types: Desi and Kabuli. A draft genome of the Kabuli cultivar, CDC Frontier, generated through whole-genome shotgun sequencing had approximately 738 Mb and 28,269 protein-coding genes that comprise NBS-LRR genes for resistance to biotic stress (Varshney *et al.*, 2013). The advanced version, CDC Frontier v2.0, was designed by NGS and Hi-C technology for chromosome-length assembly and has helped in the mapping of drought tolerance (Garg *et al.*, 2022). The initially published draft version of Desi cultivar ICC 4958 (Jain *et al.*, 2013) was upgraded improved scaffold lengths and newly annotated genes (Parveen *et al.*, 2024). Advances in chickpea transcriptome sequencing thus found genes associated with resistance to biotic stress (Garg *et al.*, 2023) and abiotic stress (Waqas *et al.*, 2019). Indeed, one milestone pulse breeding has achieved is the availability of two draft genome assemblies for the pigeon pea (*Cajanus cajan*) variety Asha (ICPL87119). The first, by BAC clone-based Sanger sequencing, revealed a genome of 605.7 Mb with 48,680 genes, most of which were linked to drought tolerance (Varshney *et al.*, 2012). The second one with 454 GS-FLX sequencing revealed 12,000 transposable elements and 47,004 genes, of which some were related to stress tolerance (Singh *et al.*, 2012). These drafts were further refined to 82.4 × genome coverage with improved scaffold and contig lengths and identification of an additional 1303 resistance genes (Marla *et al.*, 2020). A de novo assembly of the pigeon pea landrace D30, utilizing Illumina sequencing, produced an 813 Mb genome containing 37,977 genes. This assembly has been pivotal in identifying key agronomic traits and advancing breeding efforts in pigeon pea (Liu *et al.*, 2024). The black gram (*Vigna mungo*) Pant U-31 genotype was recently assembled using the draft genome of Illumina and Oxford Nanopore sequencing. The total size is 475 Mb, and it possesses 18,655 genes (Jegadeesan *et al.*, 2021). Another chromosome-level assembly of high quality included the Chai Nat 80 variety with a genome size of 499 Mb (Pootakham *et al.*, 2021). The Uttara cultivar genome of black gram was sequenced with the help of PacBio SMRT sequencing, which led to the identification of disease-resistance genes and seed storage proteins of utmost importance in accelerating black gram breeding programs (Ambreen *et al.*, 2022). However, these genome assemblies do not cover the large extent of the whole black gram genome, emphasizing the necessity for full pangenomic assembly coverage. A transcriptome level assembly of black gram genome gave insights about the development of molecular markers, enriching genomic resources and advancing breeding initiatives (Raizada and Souframanien, 2019).

1.4 Sequencing of Bio-Energy Crops

Producing biofuels from non-food crops can lead to multiple environmental benefits. Global biofuel demand is projected to increase by 38 billion liters between 2023 and 2028, representing a 30% rise over the previous 5 years, with total demand expected to reach 200 billion liters by 2028 (Renewables, 2023). The cultivation of second-generation, non-food bioenergy crops such as switchgrass, sugarcane, poplar, and Miscanthus is increasingly being considered to meet rising production demands (Happs *et al.*, 2024). Notably, C4 grasses like *Panicum virgatum* (switchgrass), *Miscanthus giganteus* (Miscanthus), and *Sorghum bicolor* L. (sorghum), recognized for their high photosynthetic efficiency, are expected to provide a sustainable source of lignocellulosic biomass for the generation of biofuels (Skaib and Zedan, 2023). Similarly, woody biomass options, such as *Populus* (poplar) and *Eucalyptus* (eucalyptus), demonstrate high productivity per hectare, making them ideal crops for bioenergy production (El Hage *et al.*, 2023). Advancing biofuel crop production requires a detailed description of the genetic and biochemical mechanisms that enable crops to adapt under environmental challenges. (Wasi *et al.*, 2023). Plant-derived secondary metabolites, including terpenoids, phenylpropanoids, alkaloids, cyanogenic glycosides, and glucosinolates, play crucial roles in plant–environment interactions (Chassagne *et al.*, 2019). Among these, terpenoids account for 40% of known plant natural products (Chassagne *et al.*, 2019). Their significance in agriculture and widespread utilization as bioproducts have generated considerable interest in the identification of genes and enzymes involved in plant specialized metabolism. However, progress in this area is frequently slow due to the vast diversity of plant metabolism and constraints in pathway discovery tools, along with the limited distribution of certain genes and enzymes across species, requiring the study of natural producers in lieu of established model organisms (Schilmiller *et al.*, 2012). Advanced techniques for producing transcriptome, metabolome, and genome data have transformed the investigation of biosynthetic and regulatory networks that govern plant metabolism and their influence on crop features in the post-genomic era (Wu *et al.*, 2023). The last few decades have seen the release of many genomes of bioenergy crops, as sequencing and assembly platforms have become more efficient and affordable (Table 1.2).

1.4.1 Woody plants: *Populus* and *Eucalyptus*

Populus, a genus of trees, has become one of the model systems for the study of plant and tree biology (Douglas, 2017). *Populus trichocarpa* (*Nisqually1*) was the first tree species to have its entire genome sequenced in 2006, following Arabidopsis and rice (Tuskan *et al.*, 2006). The genome spans around 550 Mb, organized into 19 chromosomes having a high degree of heterozygosity typical of outcrossing species (Tuskan *et al.*, 2006; Zhang *et al.*, 2020). Substantial resources were committed to polishing the reference genome through genetic and genomic resources like full-length cDNAs, BAC-based physical maps, and thousands of full-length cDNA sequences that aided in contig assembly and anchoring the sequence to genetic and physical maps (Tuskan *et al.*, 2006; Kelleher *et al.*, 2007). Even so, high-quality reference genomes are invariably imperfect due to assembly or gene model prediction errors. Since the initial release, the poplar genome has undergone two significant updates and the current version (v3.0) contains 422.9 Mb of assembled sequence and 41,335 annotated protein-coding loci that are systematically labeled by chromosome and location.

The Phytozome database, maintained by the U.S. Department of Energy's Joint Genome Institute (DOE/JGI), serves as a valuable repository for genomic information of bioenergy crops, including species in the genus *Populus*. Comprehensive gene expression databases for poplar have been generated using whole-genome microarrays and expressed sequence tags (ESTs) (Lee *et al.*, 2005). Reference genomes are available for various *Populus* species, such as *P. euphratica* Oliv. (Zhang *et al.*, 2020), *P. pruinosa* (Yang *et al.*, 2017), *P. tremula* (Lin *et al.*, 2018), *P. alba* (Liu *et al.*, 2019), *P. simonii* (Wu *et al.*, 2020), and *P. deltoides* (Bai *et al.*, 2021). Additionally, genomic data on interspecific hybrids are readily available. A significant feature revealed by the *P. trichocarpa* genome is its highly duplicated nature, a result of a whole genome duplication event in the Salicaceae family about 60 million years ago (Tuskan *et al.*, 2006). Meanwhile, the genome of white poplar revealed a loss of disease-resistance genes (Ma *et al.*, 2019). This has led to annotation of over 45,000 protein-coding genes from the poplar genomes and their functional characterization, offering critical insights into biotic stress resistance, which accelerates breedng efforts. With pan-genome analysis, the comprehension of genetic diversity in poplar has been elevated to the level whereby it may detect structural variations like insertions/deletions (INDELs) and copy number variations (CNVs) across related species such as *P. nigra*, *P. deltoides*, and *P. trichocarpa* (Pinosio *et al.*, 2016). These genomic assemblies paved the way for further tool and resource development, comparative analyses (Brahmia *et al.*, 2020), genome-wide association studies (Nagle *et al.*, 2023), the development of SNP arrays (Faivre-Rampant *et al.*, 2016), and multiomics applications (Yang *et al.*, 2023; Ployet *et al.*, 2024) in bioenergy research. Well-established genomic resources of *Populus* and *Eucalyptus* have significantly shaped the benefits in tree genomics.

Eucalyptus, from the Myrtaceae family, consisting of more than 130 genera present

Table 1.2. Genomic sequencing of key agricultural and horticultural crops. Author's own table.

Crop name	Variety	Genome size	Contig size	Scaffold size	Number of coding genes predicted	Sequencing strategy	Reference
Wheat	Chinese Spring (CS42)	10.2Gb	1.7–8.9 kb		133,090	Whole genome shotgun sequencing (WGS) and Illumina Hiseq	Mayer *et al.*, 2014
Maize	B73	2.3Gb	40 Mb	1.3 Mb	32,450	WGS	Schnable *et al.*, 2009
Rice	Nipponbare	390	6.9 Mb		37,544	WGS	Sasaki, 2005
Foxtail millet	Zhang gu	423Mb	25.4 kb	1 Mb	38,801	WGS + Illumina sequencing	Zhang *et al.*, 2012
Finger millet	ML 365	1196 Mb			85,243	Illumina + SOLiD sequencing	Hittalmani *et al.*, 2017
Pearl millet	Tift 23	1.8 Gb	18 kb	884.95 kb	36,898	WGS + PacBio SMRT sequencing	Varshney *et al.*, 2017
Mungbean	VC1973	579 Mb		1.52 Mb	22,427	Illumina + Roche/454 sequencing	Kang *et al.*, 2014
Chickpea	CDC Frontier	738 Mb	32 kb	645.3 kb	28,269	WGS + Illumina sequencing	Varshney *et al.*, 2013
Pigeon pea	Asha	605.7 Mb	21.95 kb	516.06 kb	48,680	Illumina sequencing	Varshney *et al.*, 2012
Black gram	Pant U-31	475 Mb		1.42 Mb	18,665	Illumina + ONT	Jegadeesan *et al.*, 2021
Soyabean	Williams 82	950 Mb	189.4 kb	47.8 Mb	46,430	WGS	Schmutz *et al.*, 2010
Sunflower	HA412-HO	3.5 Gb			52,232	Illumina + 454/Roche sequencing	Kane *et al.*, 2011
Rapeseed	Darmor-bzh	923.8 Mb	38.9 kb	763 kb	101,040	454/Roche and Sanger sequencing	Chalhoub *et al.*, 2014
Mustard	T84-66	784 Mb	61 kb	855 kb	80,050	Illumina + Pacbio SMRT sequencing	Yang *et al.*, 2016

Continued

Table 1.2. Continued

Crop name	Variety	Genome size	Contig size	Scaffold size	Number of coding genes predicted	Sequencing strategy	Reference
Populus	Nisqually 1	550 Mb	157 kb		45,555	WGS	Tuskan *et al.*, 2006
Eucalyptus	BRASUZ1	640 Mb	67.2 kb	53.9 Mb	36,376	WGS	Myburg *et al.*, 2014
Switch grass	AP13	1.6 Gb	1.1 Mb		40,943	WGS	Sharma *et al.*, 2012
Sorghum	BTx623	730 Mb	187.1 kb	7 Mb	34,496	Sanger sequencing	Paterson *et al.*, 2009
Miscanthus	DH1	2 Gb	33.1 kb	190 kb	67,967	WGS + Illumina sequencing	Mitros *et al.*, 2020
Sugarcane	R570	8.72 Gb		12.6 Mb	194,593	Illumina + Pacbio SMRT sequencing	Healey *et al.*, 2024
Grapevine	Pinot Noir	490 Mb	65.9 kb	2 Mb	30,434	Sanger sequencing	Jaillon *et al.*, 2007
Apple	Golden Delicious	742 Mb	13.4 kb	2 Mb	57,386	Sanger + 454/Roche sequencing	Velasco *et al.*, 2010
Walnut	Chandler	667 Mb	46kb	464 kb	32,498	WGS	Martínez-García *et al.*, 2016
Almond	Texas	227 Mb	103 kb	385 kb	27,969	Illumina + ONT sequencing	Alioto *et al.*, 2020
Banana	DH-Pahang	472 Mb	43 kb	1.3 Mb	36,542	Roche/454 + Sanger + Illumina sequencing	D'Hont *et al.*, 2012
Guava	New Age	443.8 Mb	15.6 Mb	40.4 Mb	25,601	Illumina + PacBio SMRT sequencing	Feng *et al.*, 2021
Watermelon	Inbred line 97,103	425 Mb	26.38 kb	2.38 Mb	23,440	Illumina sequencing	Guo *et al.*, 2013
Potato	Phureja DM1-3 516 R44	727 Mb	31.4 kb	389 kb	39,031	Illumina + 454/ Roche + Sanger sequencing	Xu *et al.*, 2011

Continued

Table 1.2. Continued

Crop name	Variety	Genome size	Contig size	Scaffold size	Number of coding genes predicted	Sequencing strategy	Reference
Tomato	Heinz 1706	900 Mb	55.7 kb	4.4 Mb	34,771	Sanger + Illumina +454/Roche + SOLiD sequencing	The Tomato Genome Consortium, 2012
Melon	DHL92	375 Mb	18.1 kb	4.68 Mb	27,427	454/Roche sequencing	Garcia-Mas *et al.*, 2012
Cabbage	D134	530 Mb	3.68 Mb	8.13 Mb	44,701	Pacbio SMRT sequencing + 10 x Genomics + HiC	Lv *et al.*, 2020
Carrot	DH1	421.5 Mb	64.5 kb	12.7 Mb	32,113	Illumina sequencing	Iorizzo *et al.*, 2016
Cauliflower	C-8	584 Mb	2 Mb		47,772	Illumina + Pacbio SMRT sequencing	Sun *et al.*, 2019

in the family, has become the focus area for genomics studies, and as such, significantly widened the horizon of genomics in the study of trees (Byrne and Hopper, 2008). The most significant advancement came with the sequencing of *Eucalyptus grandis* (flooded gum), the world's most cultivated hardwood species. BRASUZ1 (inbred variety of *Eucalyptus grandis*) was sequenced using whole-genome Sanger shotgun sequencing, paired-end BAC sequencing, and high-density genetic linkage mapping. The genome is approximately 640 Mb in length, and contains 36,376 predicted protein-coding loci (Myburg *et al.*, 2014). A lineage-specific whole-genome duplication (WGD) occurred around 100 million years ago in the evolutionary history of Eucalyptus. Nonetheless, a comparative analysis with the basal rosid lineage, such as *Vitis vinifera*, revealed that more than 95% of paralogs were lost after the WGD event. Interestingly, about 34% of the genes in *Eucalyptus* showed tandem repeats at a much higher rate than was found in *Arabidopsis thaliana* and *Populus trichocarpa* with 18% of the genes. These tandemly duplicated genes, which are frequently linked with stress responses, indicate that they play an integral part in the adaptive evolution of *Eucalyptus* across various environments (Jaillon *et al.*, 2007). The *e.g. randis* genome comprises the largest number of terpene synthase genes reported to date, which contribute to the varied secondary metabolites that improve resistance to diseases and pests (Santos *et al.*, 2019). Another important species, *Eucalyptus camaldulensis* (red river gum), was sequenced using a mix of Sanger sequencing and NGS (Roche 454), with a genome size of 650 Mb (Hirakawa *et al.*, 2011). Databases, including TreeGene (n.d.) and PlantGenIE (n.d.), were essential to the study of both model and non-model tree systems. Both of these platforms also include existing resources such as Phytozome (n.d.) and PLAZA (n.d.), which are used for comparative genomics for *Eucalyptus* species. RNA-seq technologies have turned out to be essential in unraveling the molecular mechanisms behind *Eucalyptus* species, particularly for identifying genes related to pathogenesis (Tang and Tang, 2019). As many as 1400 genes and a host of pathogenicity factors were identified in *Eucalyptus nitens* (Meyer *et al.*, 2016). An additional application of RNA-seq has been in the study of genes involved in wood formation, such as NAC domain genes (Hussey *et al.*, 2015), lignin biosynthesis (Carocha *et al.*, 2015), and responses to stresses (Ulloa *et al.*, 2022). The genomic information from *Eucalyptus* has supported various omics approaches, including haplotype and SNP-based GWAS (Sekiya *et al.*, 2021; Valenzuela *et al.*, 2021), furthering our understanding of *Eucalyptus* genetics.

1.4.2 C4 Grasses: switchgrass, sorghum, Miscanthus, and sugarcane

Switchgrass (*Panicum virgatum* L.), a perennial C4 grass of the family Panicoideae, holds immense potential as a biofuel crop due to its ability to yield high energy returns—producing 540% more energy than what is required for its cultivation and it can be easily grown on marginal cropland with very low fertilizer and management requirements (Margaritopoulou *et al.*, 2016). Tetraploid habit and outcrossing reproductive strategy may explain its adaptation to so many different environments. However, the genetic complexity of switchgrass, characterized by high ploidy levels and self-incompatibility among diverse ecotypes, presents significant challenges for genetic studies and crop improvement (Hestrin *et al.*, 2021). A notable breakthrough came with the release of a highly contiguous genome assembly for the lowland genotype AP13 and has provided important information about the genes underlying biomass productivity (Sharma *et al.*, 2012). Genome sequencing consisted of generating 330,297 high-quality BAC-end sequences (BES) of both genomic libraries, thereby allowing intensive survey of the switchgrass genome structure and organization that revealed a genome size of around 1372–1666 Mb (Sharma *et al.*, 2012). Researchers have begun to define the transcriptional landscape of switchgrass using NGS technologies. This approach generated 980,000 reads possessing an average length of 367 bp using Roche 454 GS-FLX Titanium sequencing and assembled these reads into 243,600 contigs containing an average length of 535 bp in a de novo approach (Wang *et al.*, 2012). The reference genome of switchgrass has been utilized in different omics studies to improve lignocellulosic biomass (Poudel *et al.*, 2017). While switchgrass has been subjected to massive resequencing and genotyping, GWAS have not yet been conducted.

Sorghum, the fifth most important cereal crop in the world and a valuable source of fermentable free sugars and lignocellulosics, is another crop with high potential for biofuel. As a monocot model crop for biofuel research, sorghum is versatile, being produced for grain, forage, sugar, and biomass. Biomass sorghum varieties, in particular, are cultivated for cellulosic biofuels (Rooney *et al.*, 2007). The first sorghum reference genome was produced by whole-genome shotgun sequencing from the grain sorghum variety BTx623 (Paterson *et al.*, 2009). The ideal model system for studying closely related difficult-to-handle polyploid bioenergy crops like sugarcane and Miscanthus is sorghum due to a relatively smaller (730 Mb) and less complex genome. The BTx623 genome has already been improved, with the version 3.1.1 containing improved assembly and annotation (McCormick *et al.*, 2018). Subsequently the high-quality reference genome of the sweet sorghum variety 'Rio' was sequenced using Pacific Biosciences long-read sequencing (Cooper *et al.*, 2019). Bioinformatics resources like Phytozome, Sorghum FDB (Functional Genomics Database) (Tian *et al.*, 2016), SorghumBase (Gladman *et al.*, 2022) and MilletDB (Sun *et al.*, 2023) support sorghum research and genome data access. Genotyping-by-sequencing (GBS) (Maulana *et al.*, 2023) has improved precision in genetic studies of sorghum. Moreover, high-density genetic linkage maps (Guden *et al.*, 2023; Tu *et al.*, 2023), GWAS (Kebede *et al.*, 2024), and integrated multiomics approaches (Zhao *et al.*, 2024) have been used for studying stress responses and improving the bioenergy potential.

Interestingly, the genomes of sorghum and sugarcane share high collinearity (Wang *et al.*, 2021), making sorghum a strong genetic model for sugarcane. Recent pan-genome analyses in sorghum have identified genes that confer resistance to sugarcane aphids (Wang *et al.*, 2021). Sugarcane, a member of the Poaceae family, is a tall grass that is propagated vegetatively and well known for accumulating sugar inside its stem. The concern over global warming, combined with the interest of finding sustainable alternatives to fossil fuels, has made sugarcane a leading crop for generating bioenergy and serving as a biomaterial crop, accounting for approximately 60% of global biofuel production (Yang *et al.*, 2020). Though sugarcane presents excellent potential, breeding relied much on conventional techniques due to the supposed complexity of its genetics. The sugarcane genome is interspecific, polyploid, and aneuploid, which makes it one of the most complex crop genomes. Genome size for *Saccharum officinarum* ranges between 7.50 and 8.55 Gb averaging 7.88 Gb, while *S. robustum* has a genome size between 7.65 and 11.78 Gb (D'Hont *et al.*, 1996). *S. spontaneum* exhibits wide variability in its genome, which ranges from 3.36 to 12.64 Gb (Daniels and Roach, 1987; Sreenivasan *et al.*, 1987). Modern sugarcane (*Saccharum* spp. hybrids) hosts two sub-genomes originating from a high-sugar, thick-stalked species, *S. officinarum*, and a low-sugar wild thin-stalked species, *S. spontaneum* (Daniels and Roach, 1987; D'Hont *et al.*, 1996). Currently, three reference genomes for *Saccharum* hybrids are available. These include the monoploid genome of the sugarcane hybrid R570 (Garsmeur *et al.*, 2018), the unscaffolded genome of the Brazilian cultivar SP80-3280 (Souza *et al.*, 2019), and the first polyploid reference genome for R570 (Healey *et al.*, 2024). This polyploid genome is an 8.7 billion base pair assembly and represents unique DNA sequences across approximately 12 chromosome copies. The assembly yielded a 5.04 Gb primary genome, making up around half of the 10 Gb sequence. Notably, *S. spontaneum* is the only sugarcane genome assembled at the chromosome and allele-defined levels (Zhang *et al.*, 2018). The high gene duplication and segmental duplications in sugarcane make it likely to have misidentified large genomic fragments; there is a higher risk of misidentifying large genomic fragments, especially with multiple homologous and homoeologous chromosomes. A WGD event took place in the *Saccharum* lineage following divergence from sorghum around 10 million years ago (Paterson *et al.*, 2009). Nevertheless, the available reference genome is currently being applied in numerous biotechnological applications and integrated multiomics approaches (Perlo *et al.*, 2022; Li *et al.*, 2023).

Miscanthus, much like sugarcane, belongs to the Poaceae family, specifically within the saccharine subtribe. It is unpretentious, easy to grow, and can thrive in marginal lands, requiring minimal fertilizer. Additionally, it can even survive drought conditions and low temperatures efficiently by employing the C4 photosynthesis pathway (Weijde *et al.*, 2013). The most commercially grown *M. x giganteus* is a sterile allotriploid, so does not produce viable seeds, minimizing its risk of becoming invasive (Hodkinson *et al.*, 2002). *M. x giganteus* ($3n=57$ chromosomes) is considered as the hybrid between diploid *Miscanthus sinensis* ($2n=38$) and tetraploid *Miscanthus sacchariflorus* ($4n=76$), species native to South-east Asia and Japan (Hirayoshi and Nishikawa, 1960). *Miscanthus sinensis* is a diploid ($2n=38$) with a haploid genome size of 2.4–2.6 Gb (Rayburn *et al.*, 2009), whereas *M. sacchariflorus* occurs in both diploid and tetraploid forms. *Miscanthus* is genetically linked to sorghum and sugarcane, sharing a close evolutionary relationship (Mitros *et al.*, 2020). The genome of *Miscanthus sinensis* was put together on to 19 chromosomes using WGS data (Mitros *et al.*, 2020). Such an assembly anchored 1.68 Gb of contigs to the chromosomes. A draft genome assembly of diploid *M. sacchariflorus* cv. 'Robustus 297' of more recent origin has been generated; it spans to a size of 2.54 Gb, accounting for approximately 59% of the expected total genome size (De Vega *et al.*, 2021). Similarly, for *Miscanthus floridulus* and *Miscanthus lutarioriparius*, high-quality reference genomes were generated using third-generation sequencing and Hi-C mapping technologies (Miao *et al.*, 2021; Zhang *et al.*, 2021). Comparative analyses of Saccharinae genomes suggest that sorghum diverged first from the common ancestor of *Saccharum* and *Miscanthus* and, subsequently that there are independent whole-genome duplication events in the genomes of *Saccharum* and *Miscanthus* (Zhang *et al.*, 2021). Strong collinearity amongst the *Miscanthus* and sorghum genomes suggests a conserved evolutionary relationship between them (Mitros *et al.*, 2020). These genomes have been comprehensively annotated and re-sequenced so as to offer refined genetic blueprints amenable to omics studies and other advanced sequence-driven analyses.

1.5 Sequencing of Horticultural Crops

Horticultural crops, comprising a diverse array of fruits, vegetables, nuts, and ornamental species, have garnered significant attention in genomic research due to their economic value and role in human health. Recent breakthroughs in sequencing throw light on the complex genetic architecture of key traits like fruit ripening, flavor, color, nutritional contents, and resistance to biotic as well as abiotic stresses.

1.5.1 Fruit and nut crops

Ultimately, genomic research in fruit and nut crops is to sequence entire genomes and understand the functions of individual genes, both critical steps toward both functional genomics and breeding programs. However, for most perennial fruit tree crops, great levels of heterozygosity and large repetitive sequences were important barriers to assembly of the genome (Linsmith *et al.*, 2019). Improvement in techniques, such as the establishment of doubled haploid lines, might have overcome these impediments, thereby making high-fidelity genome assemblies possible. One of the first breakthroughs came with the sequencing of the grapevine (*Vitis vinifera*) genome through the use of nearly homozygous genotype (cv. PN40024) derived from Pinot Noir. This pioneering effort used Sanger sequencing and provided an 8× coverage of the genome, comprising 490 Mb with around 30,000 protein-coding genes, and has since been a cornerstone for *Vitis* genetic research; it was one of the first fruit crop genomes that was sequenced (Jaillon *et al.*, 2007).

Equally significant was the completion of the sequence of apple (*Malus* ×*domestica*) in 2010; the variety Golden Delicious, with a genome size of 742 Mb, became the first published apple genome (Velasco *et al.*, 2010). Subsequent resequencing has produced a pan-genome including 91 accessions of domesticated apples and their wild relatives. This revealed significant evolutionary traits, including selective sweeps related to domestication (Sun *et al.*, 2020). In the case of peach (*Prunus persica*), its compact genome (230 Mb) was particularly suited for rapid genomic development credited to its short juvenile period and diploid structure. This allowed the sequencing of the peach genome to propel functional genomics and breeding efforts (Arús *et al.*, 2012). Other fruit crops that have garnered specific interest are the woodland strawberry (*Fragaria vesca*), which, despite its small genome size of 240Mb, has been instrumental in identifying nearly 34,809 genes that influence vital horticultural traits like flavor and flowering time (Shulaev *et al.*, 2011). Similarly, the sequencing of walnut (*Juglans regia*) in the cultivar Chandler marked a significant advancement in nut crop genomics. The assembly of the 667Mb genome size featured important information for complex trait inheritance and genetic improvement (Martínez-García *et al.*, 2016). More recently, an upgraded version of the Chandler genome v2.0 was produced using advanced sequencing technologies like Oxford Nanopore long-read sequencing combined with Hi-C chromosome conformation capture (Marrano *et al.*, 2020). Furthermore, in addition to these developments, sequencing of the almond (*Prunus dulcis*) cultivar Texas revealed a genome assembly of 227.6Mb. Phylogenomic analysis has also revealed that almond diverged from peach some 5.88million years ago (Alioto *et al.*, 2020).

Genomic advances in tropical fruit crops such as mango (*Mangifera indica*), banana (*Musa acuminata*), and guava (*Psidium guajava*) have also been achieved (D'Hont *et al.*, 2012; Singh, 2016; Feng *et al.*, 2021; Banana genome hub, n.d.) A more productive alternative to WGS has been to focus research more on the parts of the genome encoding proteins or RNAs, which is called transcriptome analysis. This has been utilized in various crops of fruits such as citrus species (Zhang *et al.*, 2017), persimmon (Chen *et al.*, 2024), and litchi (Hu *et al.*, 2018). These genomic sequences have, therefore, led to the setting up of several databases including the Rosaceae Genome Database (Jung *et al.*, 2014), the Citrus Genome Database (www.citrusgenomedb.org), and the Banana Genome Hub (banana-genome-hub.southgreen.fr). These resources will prove crucial in high-density genetic markers development, discovery of genetic variation, and amelioration of economically important traits in fruit and nut crops. The genomic revolution, supported by large-scale initiatives like the Earth Genome Project, will undoubtedly be fruitful for covering previously unsequenced fruit tree species, paving the way for future advancements in crop genomics. Fruits and nut crops improvement innovations are further propelled by the multiple-omics approaches like transcriptomics (Hou *et al.*, 2022), proteomics (Bhuiyan *et al.*, 2020), and comparative genomics through stress responses (Borredá *et al.*, 2022; Zheng *et al.*, 2024).

1.5.2 Vegetable crops

The availability of high-quality sequencing data for vegetable crops has advanced our ability to identify the genetic basis of disease resistance, thereby improving our understanding of plant–pathogen interactions at a molecular level. For example, the *Cucumis sativus* (cucumber) draft genome was sequenced using a combination of Sanger and Illumina GA technologies, achieving 72.2-fold genome coverage and revealing the involvement of 137 genes linked to ethylene biosynthesis and signaling pathways (Huang *et al.*, 2009). A subsequent study further refined the cucumber reference genome ('9930') to 367 Mb and contributed to our knowledge of this species. Melon (*Cucumis melo*), although belonging to the same genus as the cucumber, has a larger genome size (450 Mb), which contributes to key differences in stress-related and flavor-related genes (Garcia-Mas *et al.*, 2012). Similarly, the sequencing of watermelon (*Citrullus lanatus*) achieved a high-quality draft genome, providing insights into the implications for breeding and stress resistance (Guo *et al.*, 2013).

Within the Solanaceae family, key vegetables such as tomato, potato, and eggplant have seen significant advancements. The tomato genome was assembled using Sanger and NGS technologies, yielding a high-quality genome sequence of the cultivated 'Heinz 1706' variety, amounting to approximately 900 Mb, together with very important nucleotide differences between wild and cultivated tomatoes (Tomato Genome Consortium, 2012). The potato genome assembly covered 86% of its 844 Mb genome, a crucial step toward addressing susceptibility to abiotic stress through breeding programs (Xu *et al.*, 2011). Comparative genomic studies between tomato and eggplant have discovered genes associated with crop-specific traits, such as anthocyanin production and fruit size, thus underlining the scope for the transfer of genes across species to be exploited to improve crop traits (Hirakawa *et al.*, 2014). In parallel, genome sequencing of the pepper (*Capsicum annuum*) revealed that the cultivated lineage had a more reduced level of genome-wide polymorphism when compared to its wild ancestors that served as its progenitors, thus throwing light on how long-term breeding has decreased the level of polymorphism in a crop species (Qin *et al.*, 2014).

Efforts to sequence members of the *Brassicaceae* family, including Chinese cabbage (*Brassica rapa*), cabbage (*Brassica oleracea*), and cauliflower (*Brassica oleracea var. botrytis*), have similarly enriched genomic resources for vegetable crops (Wang *et al.*, 2011; Liu *et al.*, 2014; Sun *et al.*, 2019). For instance, the Chinese cabbage genome (485 Mb) revealed an abundance of genes associated with environmental adaptability, contributing to its ability to tolerate various stresses such as salt, cold, and osmotic conditions (Wang *et al.*, 2011). Radish genome sequencing (529 Mb) indicated shared ancestry with Chinese cabbage, and led to the disclosure of gene duplication events that occurred post-genome triplication (Kitashiba *et al.*, 2014). Sequencing efforts in carrot (*Daucus carota*) have unraveled the genetic basis of high carotenoid accumulation, shedding light on the evolution of this trait and guiding future breeding programs for improved nutrient content (Iorizzo *et al.*, 2016). The completion of genome sequencing across a range of vegetable crops not only serves as a foundation for future genome resequencing and transcriptome analysis but also facilitates the application of omics technologies and novel marker identification strategies for crop improvement.

1.5.3 Genome sequencing in oilseed crops

Soybean, an important oilseed crop, was the first legume crop to have its genome sequenced. The draft genome covers an area of about 950 Mb and houses some 46,000 protein-coding genes, with around 60% of its composition comprising repetitive elements. The whole-genome sequencing effort provided great insights in advancing knowledge of whole-genome duplication, including critical information about the phylogenetic relationships among legume crops and helping to identify genes involved in nodulation and oil biosynthesis (Schmutz *et al.*, 2010). Following the first studies, several high-quality soybean genome assemblies based on long-read sequencing have appeared, such as Zhonghuang 13 (Shen *et al.*, 2019) and wild soybean W05 (Xie *et al.*, 2019). Pangenomic studies of wild soybeans (Li *et al.*, 2014) and cultivated soybeans (Liu *et al.*, 2020) supported functional genomic research. Soybase is an online resource for soybean research that contains genetic, genomic, and phenotypic data (Grant *et al.*, 2010).

Recent developments in genomics sequencing have greatly expanded the knowledge of the *Brassica napus* genome. Developments of multiple genome assemblies (Sun *et al.*, 2017) and a detailed pan-genome (Song *et al.*, 2020) have furnished very important resources for the better study of the structural relationship of the genome and on the acceleration of improvement genetics in the crop. The draft genome of the *Brassica juncea* variety T84-66 was assembled through whole-genome sequencing, with a genome size of 784 Mb containing 80,050 protein-coding genes (Yang *et al.*, 2016). The chromosome-scale assembly of *B. juncea* variety Varuna produced a genome size of about 850 Mb by Pacbio SMRT sequencing (Paritosh *et al.*, 2021). Previous studies focused on identifying salt-tolerant *B. juncea* varieties through morphological traits (Aggarwal *et al.*, 2024), can be enhanced by incorporating genomic information, particularly through the development of trait-specific markers. BrassicaDB is the primary repository of genomic and genetic data of Brassica species, which gives researchers an outstanding tool with which to explore the diversity and phylogenetic relationship of these crops (Chen *et al.*, 2022).

The sunflower genome project began through sequencing HA412-HO by a hybrid approach combining Solexa and 454/Roche pyrosequencing. The result was a genome size of 3.5 Gb at 80 × genome coverage (Kane *et al.*, 2011). Since then, several high-quality reference genomes including XRQ, HA412, and PSC8 have been added that offer valuable resources in the mapping of genes related to desirable agronomic traits (Terzić *et al.*, 2020). Cultivated sunflowers have unfortunately lost many genes that are responsible for the resistance of various biotic and abiotic stresses. This has necessitated the exploration of wild sunflower varieties as a source of these valuable traits. Pangenome sequencing of wild relatives does hold much promise to identify those traits and introgress them in cultivated sunflowers. The comprehensive pangenome analysis of sunflower identified 61,205 genes that include many genes for resistance to downy mildew (Hübner *et al.*, 2019).

1.6 Applications of Sequencing Technologies Across Multiomics

From early Sanger sequencing to NGS and now third-generation sequencing, the evolving sequencing technologies have revolutionized agriculture and omics research. The development of NGS in the 2000s brought a major breakthrough, dramatically increasing sequencing speed and reducing costs. This allowed for the high-throughput analysis of whole genomes, in which formerly inaccessible genetic variation was uncovered, thereby accelerating crop improvement programs and the precision of molecular breeding techniques. NGS integrated genomics, transcriptomics, proteomics, and epigenomics and therefore enabled gene expression, protein function, and epigenetic modification studies at a genome-wide scale. More recently, third-generation sequencing technologies, such as single-molecule real-time sequencing (SMRT) and nanopore sequencing, have pushed the boundaries further by providing longer read lengths and real-time sequencing capabilities. Taken together, the several leaps in technology have, therefore, accelerated the pace of agricultural genomics such that there is more precise, targeted intervention to improve crop productivity and resilience under prevailing climatic changes and ever-increasing demands for food everywhere in the world.

Advances in sequencing technologies have significantly shortened project timelines, from eight or ten years to just a few years with reduced effort. These technologies reveal key genetic variations, SNPs, structural variants, and a broader understanding of genome architecture, which has been instrumental across many biological fields. In particular, NGS has become a crucial tool for cost-effective transcriptome and interactome profiling, as well as for detailed analyses of epigenomic modifications (Varshney *et al.*, 2009). Re-sequencing has greatly improved the availability of genetic information by providing access to genome-wide sequence variation. Such information can now be translated into the generation of molecular markers that improve precision and depth in genetic mapping of key agronomic traits. These data from high-density SNPs have been really useful in a variety of systems-genetics, including QTL mapping in bi-parental crosses, recombinant inbred lines, and advanced populations such as MAGIC (multi-parent advanced generation inter-cross) and NAM (nested association mapping) (Kover *et al.*, 2009; McMullen *et al.*, 2009). Such markers are subsequently used in crop-breeding programs in which genomic segments carrying desirable alleles are combined into a single genetic background through the application of marker-assisted selection (MAS). Additionally, the sequencing of parental lines influences genomic selection (GS) by identifying polymorphisms in both genic and inter-genic regions, estimating gene expression levels, and providing data on epigenetic states. The expanding genomic resources offer a platform for models like Encyclopedia of DNA Elements (ENCODE) (Luo *et al.*, 2020) and plant pENCODE (Lane *et al.*, 2014) that could be used in predictive models based on genomic information derived from NGS. Resources and insights expected from structural genomics are meant to enhance functional genomics, an important crossing point with technologies such as CRISPR/Cas9 genome editing (Cao *et al.*, 2022).

RNA sequencing (RNA-seq) is the newest in a line of studies generating information on mRNA molecules, which has become indispensable in tracking gene expression and identifying regulatory pathways that link transcriptional activity with phenotypic traits. Several model crop plants such as Arabidopsis, maize and rice were examined through RNA-seq to reveal the genes that contribute to climate resilience (Pan *et al.*, 2020; Schaarschmidt *et al.*, 2020; Tao *et al.*, 2022). This technique forms part of transcriptomics, which deals with the total number of transcripts in an organism, their abundance in specific cells, and post-transcriptional modifications (Tyagi *et al.*, 2022). Through transcriptomics, researchers gain critical insights into gene structure, expression, and regulation, providing valuable information that enhances our knowledge of plant responses to various environmental conditions. A very important mechanism for the regulation of gene expression at the epigenetic level includes DNA methylation and histone modifications that are crucial in maintaining the genomic integrity. Sequencing technologies such as bisulfite sequencing have been pivotal in detecting these epigenetic markers and thus providing genome-wide methylation profiling at single-nucleotide resolution (Boquete *et al.*, 2021). Methylation Capture Sequencing (MC-seq) is an example of how bisulfite sequencing is used to gather methylation data at this high resolution, giving researchers a scientific understanding of how epigenetic modifications impact gene expression (Morselli *et al.*, 2021). These insights have broadened our comprehension of plant adaptability through phenotypic plasticity, highlighting how epigenetic regulation contributes to the ability of species to cope with environmental changes (Cui *et al.*, 2021).

Further complementing genomic data is proteomics, which focuses on understanding protein roles, structures, functions, and interactions within an organism. NGS provides information about the genetic sequences that encode proteins, allowing researchers to map the protein landscape of an organism and analyze protein interactions, post-translational modifications, and their roles in cellular functions. Proteomics thus helps decipher the molecular mechanisms behind stress responses and other physiological processes, which is particularly relevant in improving crop stress tolerance (Mustafa and Komatsu, 2021). Proteomics continues to shape functional genomics by facilitating the development of biotechnological tools aimed at improving plant stress resilience. The integration of metabolomics with NGS has improved the way for understanding how genetic sequences influence metabolic networks. Metabolomics, like other omics approaches, fills its niche in systems biology, where the genetic, transcriptomic, and proteomic data all become related to physiological traits. This map of biological processes along the way enables predicting and analyzing pathways of metabolism, aiding in a more holistic approach toward the study of organismal functions (Pandey *et al.*, 2016). Such integrated multiomics approaches are necessary for advancing our understanding of gene function, molecular interactions, and how plants respond to biotic and abiotic stressors.

1.7 Challenges and Gaps in High-throughput Sequencing

While undoubtedly revolutionary in changing the landscape of genome study, for many applications and organisms, sequencing technologies still pose huge challenges. One major limitation involves the accuracy of sequencing, particularly in repetitive or complex regions of the genome. Even with third-generation sequencing technologies, achieving complete, error-free assemblies remains difficult for many species, especially those with large or polyploid genomes. Although advancements have been made, such as the successful assembly of gapless genomes in crops like rice and maize (Li *et al.*, 2021; Chen *et al.*, 2023), other plant species continue to present hurdles due to their intricate genomic features. Draft genomes are helpful but generally insufficient when solving questions of speciation, the dynamics of evolution, and in the correct association of phenotypes with genetic variations. Assembly errors also can introduce false variation, complicating downstream analyses, and limit further ability to understand complex traits. High sequencing costs, especially for larger genomes, is another challenge. Although the rate of decline in sequencing cost has been relatively rapid, this still forms a significant barrier to many researchers, particularly those in the developing world. Moreover, a single reference genome is rarely sufficient to capture the full genetic diversity within a species. This has paved the way to the concept of pan-genomes, which aims to encompass the full range of DNA sequences within a species from the presence–absence variations, all the way to copy-number variations. However, constructing these pan-genomes is still in its infancy, and the tools used are often tailored to specific species, makes it difficult to generalize across different organisms (Bayer *et al.*, 2020). The functional interpretation of sequencing data poses another significant challenge. Sequencing alone is often insufficient to fully understand the complex relationship between genes, proteins, and environmental factors in agricultural species. Epigenetic modifications such as DNA methylation represent crucial means to regulate gene expression, but existing sequencing technologies lack the depth to capture these modifications. Although improvements such as methylation sequencing are generating results in genome-wide regulation, the precise understanding of control mechanisms of gene activity is still not achieved (Cui *et al.*, 2021). Lastly, high-throughput sequencing generates a very large amount of data that may inundate a researcher without the computational resources to uncover meaningful biological findings. What seems to keep afloat in these data deluges are only ongoing advancements in sequencing technology and bioinformatics tools that continue to hone the accuracy, accessibility, and utility of genome sequencing in agriculture and beyond.

1.8 Conclusion

Over the past 25 years, sequencing technologies have undergone major improvements. Beginning with Sanger sequencing, the introduction of next-generation sequencing and, more recently, third-generation sequencing technologies has improved our ability to map genomes, detect structural variations, and study the intricacies of gene regulation and epigenetics. These innovations have increased the availability of reference genomes and, more recently, pan-genome studies, both of which have expanded our understanding of plant genetic diversity. However, challenges remain, such as generating gapless reference genomes, particularly due to polyploidy and heterozygosity, which complicate the assembly process. Despite these challenges, the future of plant genome sequencing looks promising, with expectations that the field will continue to grow and evolve over the next few decades, transforming plant biology in the process. Flowering plants alone, with their approximately 300,000 species, demonstrate enormous genomic diversity. This diversity arises from processes such as polyploidy, transposable element proliferation, and small RNA-mediated regulatory events, all influenced by complex biotic and abiotic environmental interactions. As we advance, the use of synthetic polyploids and other experimental models will allow scientists to explore the evolutionary and ecological dynamics of these phenomena. The increasing availability of large-scale multiomics datasets, spanning genomes, transcriptomes, and epigenomes, will further empower researchers to understand plant responses to environmental changes and their long-term adaptation strategies. While estimating the impact of climate change on crop production remains uncertain, continuous progress in genomics and genomics-assisted breeding will undoubtedly play a critical role in mitigating its negative effects on crop plants.Banana genome hub, n.d.

References

3,000 Rice Genomes Project (2014) The 3,000 rice genomes project. *GigaScience* 3(1), 2047–217X.

Aggarwal, G., Edhigalla, P., Walia, P., Jindal, S. and Sandal, S.S. (2024) A method for screening salt stress tolerance in Indian mustard (*Brassica juncea*) (L.) Czern & Coss at seedling stage. *Scientific Reports* 14(1), 12705.

Ahmad, M. (2022) Genomics and transcriptomics to protect rice (*Oryza sativa*. L.) from abiotic stressors: -Pathways to achieving zero hunger. *Frontiers in Plant Science* 13, 1002596.

Alioto, T., Alexiou, K.G., Bardil, A., Barteri, F., Castanera, R. *et al.* (2020) Transposons played a major role in the diversification between the closely related almond and peach genomes: Results from the almond genome sequence. *The Plant Journal* 101(2), 455–472.

Ambreen, H., Oraon, P.K., Wahlang, D.R., Satyawada, R.R., Katiyar-Agarwal, S. *et al.* (2022) Long-read-based draft genome sequence of Indian black gram IPU-94-1 'Uttara': Insights into disease resistance and seed storage protein genes. *The Plant Genome* 15(3), e20234.

Andrews, K.R., Good, J.M., Miller, M.R., Luikart, G. and Hohenlohe, P.A. (2016) Harnessing the power of RADseq for ecological and evolutionary genomics. *Nature Reviews Genetics* 17(2), 81–92.

Arús, P., Verde, I., Sosinski, B., Zhebentyayeva, T. and Abbott, A.G. (2012) The peach genome. *Tree Genetics & Genomes* 8, 531–547.

Avni, R., Nave, M., Barad, O., Baruch, K., Twardziok, S.O. *et al.* (2017) Wild emmer genome architecture and diversity elucidate wheat evolution and domestication. *Science* 357(6346), 93–97.

Bai, S., Wu, H., Zhang, J., Pan, Z., Zhao, W. *et al.* (2021) Genome assembly of Salicaceae *Populus deltoides* (Eastern Cottonwood) I-69 based on nanopore sequencing and Hi-C technologies. *Journal of Heredity* 112(3), 303–310.

Baird, N.A., Etter, P.D., Atwood, T.S., Currey, M.C., Shiver, A.L. *et al.* (2008) Rapid SNP discovery and genetic mapping using sequenced RAD markers. *PloS One* 3(10), e3376.

Banana genome hub (n.d.). Available at: https://banana-genome-hub.southgreen.fr (accessed 23 September 2024).

Bayer, P.E., Golicz, A.A., Scheben, A., Batley, J. and Edwards, D. (2020) Plant pan-genomes are the new reference. *Nature Plants* 6(8), 914–920.

Bennetzen, J.L., Schmutz, J., Wang, H., Percifield, R., Hawkins, J. *et al.* (2012) Reference genome sequence of the model plant Setaria. *Nature Biotechnology* 30(6), 555–561.

Bentley, D.R., Balasubramanian, S., Swerdlow, H.P., Smith, G.P., Milton, J. *et al.* (2008) Accurate whole human genome sequencing using reversible terminator chemistry. *Nature* 456(7218), 53–59.

Bevan, M.W. and Uauy, C. (2013) Genomics reveals new landscapes for crop improvement. *Genome Biology* 14, 1–11.

Bevan, M.W., Uauy, C., Wulff, B.B., Zhou, J., Krasileva, K. *et al.* (2017) Genomic innovation for crop improvement. *Nature* 543(7645), 346–354.

Bhuiyan, F., Campos, N.A., Swennen, R. and Carpentier, S. (2020) Characterizing fruit ripening in plantain and Cavendish bananas: A proteomics approach. *Journal of Proteomics* 214, 103632.

Bolger, M.E., Weisshaar, B., Scholz, U., Stein, N., Usadel, B. *et al.* (2014) Plant genome sequencing—applications for crop improvement. *Current Opinion in Biotechnology* 26, 31–37.

Boquete, M.T., Muyle, A. and Alonso, C. (2021) Plant epigenetics: Phenotypic and functional diversity beyond the DNA sequence. *American Journal of Botany* 108(4), 553–558.

Borredá, C., Perez-Roman, E., Talon, M. and Terol, J. (2022) Comparative transcriptomics of wild and commercial citrus during early ripening reveals how domestication shaped fruit gene expression. *BMC Plant Biology* 22(1), 123.

Brahmia, F.Z., Alpar, T., Horváth, P.G. and Csiha, C. (2020) Comparative analysis of wettability with fire retardants of poplar (*Populus cv. euramericana* I214) and Scots pine (*Pinus sylvestris*). *Surfaces and Interfaces* 18, 100405.

Byrne, M. and Hopper, S.D. (2008) Granite outcrops as ancient islands in old landscapes: Evidence from the phylogeography and population genetics of *Eucalyptus caesia* (Myrtaceae) in Western Australia. *Biological Journal of the Linnean Society* 93(1), 177–188.

Cao, H.X., Vu, G.T.H. and Gailing, O. (2022) From genome sequencing to CRISPR-based genome editing for climate-resilient forest trees. *International Journal of Molecular Sciences* 23(2), 966.

Carocha, V., Soler, M., Hefer, C., Cassan-Wang, H., Fevereiro, P. *et al.* (2015) Genome-wide analysis of the lignin toolbox of *Eucalyptus grandis*. *New Phytologist* 206(4), 1297–1313.

Chalhoub, B., Denoeud, F., Liu, S., Parkin, I.A., Tang, H. *et al.* (2014) Early allopolyploid evolution in the post-neolithic *Brassica napus* oilseed genome. *Science* 345(6199), 950–953.

Chandler, V.L. and Brendel, V. (2002) The maize genome sequencing project. *Plant Physiology* 130(4), 1594–1597.

Chassagne, F., Cabanac, G., Hubert, G., David, B. and Marti, G. (2019) The landscape of natural product diversity and their pharmacological relevance from a focus on the Dictionary of natural products. *Phytochemistry Reviews* 18, 601–622.

Chen, H., Wang, T., He, X., Cai, X., Lin, R. *et al.* (2022) BRAD V3. 0: An upgraded *Brassicaceae* database. *Nucleic Acids Research* 50(D1), D1432–D1441.

Chen, J., Wang, Z., Tan, K., Huang, W., Shi, J. *et al.* (2023) A complete telomere-to-telomere assembly of the maize genome. *Nature Genetics* 55(7), 1221–1231.

Chen, X., Bhadauria, V. and Ma, B. (2018) ChIP-seq: A powerful tool for studying protein–DNA interactions in plants. *Current Issues in Molecular Biology* 27(1), 171–180.

Chen, Y.T., Su, W.Y., Wu, W., Zhu, Q.G., Yin, X.R. *et al.* (2024) Transcriptional responses of postharvest ABA treatment in persimmon fruit. *Postharvest Biology and Technology* 217, 113097.

Choulet, F., Alberti, A., Theil, S., Glover, N., Barbe, V. *et al.* (2014) Structural and functional partitioning of bread wheat chromosome 3B. *Science* 345(6194), 1249721.

Chuang, T.J., Lin, W.C., Lee, H.C., Wang, C.W., Hsiao, K.L. *et al.* (2003) A complexity reduction algorithm for analysis and annotation of large genomic sequences. *Genome Research* 13(2), 313–322.

Citrus genome datab (n.d.). Available at: https://www.citrusgenomedb.org (accessed 23 September 2024).

Cooper, E.A., Brenton, Z.W., Flinn, B.S., Jenkins, J., Shu, S. *et al.* (2019) A new reference genome for *Sorghum bicolor* reveals high levels of sequence similarity between sweet and grain genotypes: Implications for the genetics of sugar metabolism. *BMC Genomics* 20, 1–13.

Cui, N., Chen, X., Shi, Y., Chi, M., Hu, J. *et al.* (2021) Changes in the epigenome and transcriptome of rice in response to *Magnaporthe oryzae* infection. *The Crop Journal* 9(4), 843–853.

Dai, X. and Shen, L. (2022) Advances and trends in omics technology development. *Frontiers in Medicine* 9, 911861.

Daniels, J. and Roach, B. (1987) Taxonomy and evolution. In: Heinz, D. (ed.) *Sugarcane Improvement Through Breeding*. Elsevier Press, The Netherlands, pp. 1–14.

Deamer, D., Akeson, M. and Branton, D. (2016) Three decades of nanopore sequencing. *Nature Biotechnology* 34(5), 518–524.

De Vega, J., Donnison, I., Dyer, S. and Farrar, K. (2021) Draft genome assembly of the biofuel grass crop *Miscanthus sacchariflorus*. *F1000Research* 10.

Devos, K.M., Qi, P., Bahri, B.A., Gimode, D.M., Jenike, K. *et al.* (2023) Genome analyses reveal population structure and a purple stigma color gene candidate in finger millet. *Nature Communications* 14(1), 3694.

D'Hont, A., Grivet, L., Feldmann, P., Glaszmann, J.C., Rao, S. *et al.* (1996) Characterisation of the double genome structure of modern sugarcane cultivars (*Saccharum* spp.) by molecular cytogenetics. *Molecular and General Genetics MGG* 250, 405–413.

D'Hont, A., Denoeud, F., Aury, J.M., Baurens, F.C., Carreel, F. *et al.* (2012) The banana (*Musa acuminata*) genome and the evolution of monocotyledonous plants. *Nature* 488(7410), 213–217.

Douglas, C.J. (2017) Populus as a model tree. In: Groover, A. and Kronk, Q. (eds) *Comparative and Evolutionary Genomics of Angiosperm Trees*. Springer, pp. 61–84.

Eckardt, N.A. (2000) Sequencing the rice genome. *The Plant Cell* 12(11), 2011–2017. DOI: 10.1105/tpc.12.11.2011.

El Hage, M., Louka, N., Rezzoug, S.A., Maugard, T., Sablé, S. *et al.* (2023) Bioethanol production from woody biomass: Recent advances on the effect of pretreatments on the bioconversion process and energy yield aspects. *Energies* 16(13), 5052.

Faivre-Rampant, P., Zaina, G., Jorge, V., Giacomello, S., Segura, V. *et al.* (2016) New resources for genetic studies in *Populus nigra*: Genome-wide SNP discovery and development of a 12k Infinium array. *Molecular Ecology Resources* 16(4), 1023–1036.

Feng, C., Feng, C., Lin, X., Liu, S., Li, Y. *et al.* (2021) A chromosome-level genome assembly provides insights into ascorbic acid accumulation and fruit softening in guava (*Psidium guajava*). *Plant Biotechnology Journal* 19(4), 717–730.

Feuillet, C., Leach, J.E., Rogers, J., Schnable, P.S. and Eversole, K. (2011) Crop genome sequencing: Lessons and rationales. *Trends in Plant Science* 16(2), 77–88.

Food and Agriculture Organization (2020) The State of Food Security and Nutrition in the World 2020. FAO, Rome.

Francia, E., Morcia, C., Pasquariello, M., Mazzamurro, V., Milc, J.A. *et al.* (2016) Copy number variation at the HvCBF4–HvCBF2 genomic segment is a major component of frost resistance in barley. *Plant Molecular Biology* 92, 161–175.

Gale, M.D. and Devos, K.M. (1998) Comparative genetics in the grasses. *Proceedings of the National Academy of Sciences* 95(5), 1971–1974.

Garcia-Mas, J., Benjak, A., Sanseverino, W., Bourgeois, M., Mir, G. *et al.* (2012) The genome of melon (*Cucumis melo* L.). *Proceedings of the National Academy of Sciences* 109(29), 11872–11877.

Garg, V., Dudchenko, O., Wang, J., Khan, A.W., Gupta, S. *et al.* (2022) Chromosome-length genome assemblies of six legume species provide insights into genome organization, evolution, and agronomic traits for crop improvement. *Journal of Advanced Research* 42, 315–329.

Garg, V., Chitikineni, A., Sharma, M., Ghosh, R., Samineni, S. *et al.* (2023) Transcriptome profiling reveals the expression and regulation of genes associated with Fusarium wilt resistance in chickpea (*Cicer arietinum* L.). *The Plant Genome* 16(3), e20340.

Garsmeur, O., Droc, G., Antonise, R., Grimwood, J., Potier, B. *et al.* (2018) A mosaic monoploid reference sequence for the highly complex genome of sugarcane. *Nature Communications* 9(1), 2638.

Gladman, N., Olson, A., Wei, S., Chougule, K., Lu, Z. *et al.* (2022) SorghumBase: A web-based portal for sorghum genetic information and community advancement. *Planta* 255(2), 35.

Goff, S.A., Ricke, D., Lan, T.H., Presting, G., Wang, R. *et al.* (2002) A draft sequence of the rice genome (*Oryza sativa* L.ssp. *japonica*). *Science* 296(5565).

Grant, D., Nelson, R.T., Cannon, S.B. and Shoemaker, R.C. (2010) SoyBase, the USDA-ARS soybean genetics and genomics database. *Nucleic Acids Research* 38(suppl_1), D843–D846.

Guden, B., Yol, E., Erdurmus, C., Lucas, S.J. and Uzun, B. (2023) Construction of a high-density genetic linkage map and QTL mapping for bioenergy-related traits in sweet sorghum [*Sorghum bicolor* (L.) Moench]. *Frontiers in Plant Science* 14, 1081931.

Guo, S., Zhang, J., Sun, H., Salse, J., Lucas, W.J. *et al.* (2013) The draft genome of watermelon (*Citrullus lanatus*) and resequencing of 20 diverse accessions. *Nature Genetics* 45(1), 51–58.

Happs, R.M., Hanes, R.J., Bartling, A.W., Field, J.L., Harman-Ware, A.E. *et al.* (2024) Economic and sustainability impacts of yield and composition variation in bioenergy crops: Switchgrass (*Panicum virgatum* L.). *ACS Sustainable Chemistry & Engineering* 12(5), 1897–1910.

Hatakeyama, M., Aluri, S., Balachadran, M.T., Sivarajan, S.R., Patrignani, A. *et al.* (2018) Multiple hybrid de novo genome assembly of finger millet, an orphan allotetraploid crop. *DNA Research* 25(1), 39–47.

Healey, A.L., Garsmeur, O., Lovell, J.T., Shengquiang, S., Sreedasyam, A. *et al.* (2024) The complex polyploid genome architecture of sugarcane. *Nature* 628(8009), 804–810.

Hestrin, R., Lee, M.R., Whitaker, B.K. and Pett-Ridge, J. (2021) The switchgrass microbiome: A review of structure, function, and taxonomic distribution. *Phytobiomes Journal* 5(1), 14–28.

Hill, C.B., Wong, D., Tibbits, J., Forrest, K., Hayden, M. *et al.* (2019) Targeted enrichment by solution-based hybrid capture to identify genetic sequence variants in barley. *Scientific Data* 6(1), 12.

Hirakawa, H., Nakamura, Y., Kaneko, T., Isobe, S., Sakai, H. *et al.* (2011) Survey of the genetic information carried in the genome of *Eucalyptus camaldulensis*. *Plant Biotechnology* 28(5), 471–480.

Hirakawa, H., Shirasawa, K., Miyatake, K.O.J.I., Nunome, T., Negoro, S. *et al.* (2014) Draft genome sequence of eggplant (*Solanum melongena* L.): The representative solanum species indigenous to the old world. *DNA Research* 21(6), 649–660.

Hirayoshi, I. and Nishikawa, K. (1960) Hakura. Cytogenetical forage studies on forage plants (VIII): 3x-and 4x-hybrids raized from the cross, *Miscanthus sinensis* var. condensatus x *M. sacchariflorus*. *Research Bulletin Faculty Agriculture Gifu University* 12, 82–88.

Hirsch, C.N., Hirsch, C.D., Brohammer, A.B., Bowman, M.J., Soifer, I. *et al.* (2016) Draft assembly of elite inbred line PH207 provides insights into genomic and transcriptome diversity in maize. *The Plant Cell* 28(11), 2700–2714.

Hittalmani, S., Mahesh, H.B., Shirke, M.D., Biradar, H., Uday, G. *et al.* (2017) Genome and transcriptome sequence of finger millet (*Eleusine coracana* (L.) Gaertn.) provides insights into drought tolerance and nutraceutical properties. *BMC Genomics* 18, 1–16.

Hodkinson, T.R., Chase, M.W. and Renvoize, S.A. (2002) Characterization of a genetic resource collection for *Miscanthus* (Saccharinae, Andropogoneae, Poaceae) using AFLP and ISSR PCR. *Annals of Botany* 89(5), 627–636.

Hon, T., Mars, K., Young, G., Tsai, Y.C., Karalius, J.W. *et al.* (2020) Highly accurate long-read HiFi sequencing data for five complex genomes. *Scientific Data* 7(1), 399.

Hou, L., Li, M., Zhang, C., Liu, N., Liu, X. *et al.* (2022) Comparative transcriptomic analyses of different jujube cultivars reveal the co-regulation of multiple pathways during fruit cracking. *Genes* 13(1), 105.

Hu, F., Chen, Z., Zhao, J., Wang, X., Su, W. *et al.* (2018) Differential gene expression between the vigorous and dwarf litchi cultivars based on RNA-seq transcriptome analysis. *PloS One* 13(12), e0208771.

Hu, T., Chitnis, N., Monos, D. and Dinh, A. (2021) Next-generation sequencing technologies: An overview. *Human Immunology* 82(11), 801–811.

Huang, S., Li, R., Zhang, Z., Li, L.I., Gu, X. *et al.* (2009) The genome of the cucumber, *Cucumis sativus* L. *Nature Genetics* 41(12), 1275–1281.

Hübner, S., Bercovich, N., Todesco, M., Mandel, J.R., Odenheimer, J. *et al.* (2019) Sunflower pan-genome analysis shows that hybridization altered gene content and disease resistance. *Nature Plants* 5(1), 54–62.

Hufford, M.B., Seetharam, A.S., Woodhouse, M.R., Chougule, K.M., Ou, S. *et al.* (2021) De novo assembly, annotation, and comparative analysis of 26 diverse maize genomes. *Science* 373(6555), 655–662.

Hui, P. (2014) Next generation sequencing: Chemistry, technology and applications. In: Tang, N.L.S. and Poon, T. (eds) *Chemical Diagnostics: From Bench to Bedside*. Springer, Berlin, Heidelberg, Germany, pp. 1–18.

Hussey, S.G., Saïdi, M.N., Hefer, C.A., Myburg, A.A. and Grima-Pettenati, J. (2015) Structural, evolutionary and functional analysis of the NAC domain protein family in Eucalyptus. *New Phytologist* 206(4), 1337–1350.

International Rice Genome Sequencing Project and Sasaki, T. (2005) The map-based sequence of the rice genome. *Nature* 436(7052), 793–800.

International Wheat Genome Sequencing Consortium (IWGSC), Mayer, K.F., Rogers, J., Doležel, J., Pozniak, C. *et al.* (2014) A chromosome-based draft sequence of the hexaploid bread wheat (*Triticum aestivum*) genome. *Science* 345(6194), 1251788.

International Wheat Genome Sequencing Consortium (IWGSC), Appels, R., Eversole, K., Stein, N., Feuillet, C. *et al.* (2018) Shifting the limits in wheat research and breeding using a fully annotated reference genome. *Science* 361(6403), eaar7191.

Iorizzo, M., Ellison, S., Senalik, D., Zeng, P., Satapoomin, P. *et al.* (2016) A high-quality carrot genome assembly provides new insights into carotenoid accumulation and asterid genome evolution. *Nature Genetics* 48(6), 657–666.

Jaillon, O., Aury, J.M., Noel, B., Policriti, A., Clepet, C. *et al.* (2007) The grapevine genome sequence suggests ancestral hexaploidization in major angiosperm phyla. *Nature* 449(7161), 463–467.

Jain, M., Misra, G., Patel, R.K., Priya, P., Jhanwar, S. *et al.* (2013) A draft genome sequence of the pulse crop chickpea (*Cicer arietinum* L.). *The Plant Journal* 74(5), 715–729.

Jain, M., Fiddes, I.T., Miga, K.H., Olsen, H.E., Paten, B. *et al.* (2015) Improved data analysis for the MinION nanopore sequencer. *Nature Methods* 12(4), 351–356.

Jegadeesan, S., Raizada, A., Dhanasekar, P. and Suprasanna, P. (2021) Draft genome sequence of the pulse crop blackgram [*Vigna mungo* (L.) Hepper] reveals potential R-genes. *Scientific Reports* 11(1), 11247.

Jiao, Y., Peluso, P., Shi, J., Liang, T., Stitzer, M.C. *et al.* (2017) Improved maize reference genome with single-molecule technologies. *Nature* 546(7659), 524–527.

Jung, S., Ficklin, S.P., Lee, T., Cheng, C.H., Blenda, A. *et al.* (2014) The genome database for Rosaceae (GDR): Year 10 update. *Nucleic Acids Research* 42(D1), D1237–D1244.

Kane, N.C., Gill, N., King, M.G., Bowers, J.E., Berges, H. *et al.* (2011) Progress towards a reference genome for sunflower. *Botany* 89(7), 429–437.

Kang, Y.J., Kim, S.K., Kim, M.Y., Lestari, P., Kim, K.H. *et al.* (2014) Genome sequence of mungbean and insights into evolution within Vigna species. *Nature Communications* 5(1), 5443.

Kaufmann, K., Muino, J.M., Østerås, M., Farinelli, L., Krajewski, P. *et al.* (2010) Chromatin immunoprecipitation (ChIP) of plant transcription factors followed by sequencing (ChIP-SEQ) or hybridization to whole genome arrays (ChIP-CHIP). *Nature Protocols* 5(3), 457–472.

Kawahara, Y., de la Bastide, M., Hamilton, J.P., Kanamori, H., McCombie, W.R. *et al.* (2013) Improvement of the *Oryza sativa* Nipponbare reference genome using next generation sequence and optical map data. *Rice* 6, 1–10.

Kebede, A., Barka, G.D., Kebede, M., Tadesse, T., Girma, G. *et al.* (2024) Multi-locus genome-wide association analysis for root and shoot traits at seedling stage in Ethiopian sorghum (*Sorghum bicolor* (L.) Moench) accessions. *Genetic Resources and Crop Evolution* 1–23.

Kelleher, C.T., Chiu, R., Shin, H., Bosdet, I.E., Krzywinski, M.I. *et al.* (2007) A physical map of the highly heterozygous *Populus* genome: Integration with the genome sequence and genetic map and analysis of haplotype variation. *The Plant Journal* 50(6), 1063–1078.

Kitashiba, H., Li, F., Hirakawa, H., Kawanabe, T., Zou, Z. *et al.* (2014) Draft sequences of the radish (*Raphanus sativus* L.) genome. *DNA Research: An International Journal for Rapid Publication of Reports on Genes and Genomes* 21(5), 481–490.

Kover, P.X., Valdar, W., Trakalo, J., Scarcelli, N., Ehrenreich, I.M. *et al.* (2009) A multiparent advanced generation inter-cross to fine-map quantitative traits in *Arabidopsis thaliana*. *PLoS Genetics* 5(7), e1000551.

Kukurba, K.R. and Montgomery, S.B. (2015) RNA sequencing and analysis. *Cold Spring Harbor Protocols* 2015(11), pdbtop084970.

Lane, A.K., Niederhuth, C.E., Ji, L. and Schmitz, R.J. (2014) pENCODE: A plant encyclopedia of DNA elements. *Annual Review of Genetics* 48(1), 49–70.

Lee, H., Lee, J.S., Noh, E.W., Bae, E.K., Choi, Y.I. *et al.* (2005) Generation and analysis of expressed sequence tags from poplar (*Populus alba*× *P. tremula* var. Glandulosa) suspension cells. *Plant Science* 169(6), 1118–1124.

Li, A.M., Liao, F., Wang, M., Chen, Z.L., Qin, C.X. *et al.* (2023) Transcriptomic and proteomic landscape of sugarcane response to biotic and abiotic stressors. *International Journal of Molecular Sciences* 24(10), 8913.

Li, K., Jiang, W., Hui, Y., Kong, M., Feng, L.Y. *et al.* (2021) Gapless indica rice genome reveals synergistic contributions of active transposable elements and segmental duplications to rice genome evolution. *Molecular Plant* 14(10), 1745–1756.

Li, W., Zhang, P., Fellers, J.P., Friebe, B. and Gill, B.S. (2004) Sequence composition, organization, and evolution of the core *Triticeae* genome. *The Plant Journal* 40(4), 500–511.

Li, Y.H., Zhou, G., Ma, J., Jiang, W., Jin, L.G. *et al.* (2014) De novo assembly of soybean wild relatives for pan-genome analysis of diversity and agronomic traits. *Nature Biotechnology* 32(10), 1045–1052.

Luo, M.C., Gu, Y.Q., Puiu, D., Wang, H., Twardziok, S.O. *et al.* (2017) Genome sequence of the progenitor of the wheat D genome *Aegilops tauschii*. *Nature* 551(7681), 498–502.

Lin, Y.C., Wang, J., Delhomme, N., Schiffthaler, B., Sundström, G. *et al.* (2018) Functional and evolutionary genomic inferences in *Populus* through genome and population sequencing of American and European aspen. *Proceedings of the National Academy of Sciences* 115(46), E10970–E10978.

Ling, H.Q., Ma, B., Shi, X., Liu, H., Dong, L. *et al.* (2018) Genome sequence of the progenitor of wheat A subgenome *Triticum urartu*. *Nature* 557(7705), 424–428.

Linsmith, G., Rombauts, S., Montanari, S., Deng, C.H., Celton, J.M. *et al.* (2019) Pseudo-chromosome–length genome assembly of a double haploid “Bartlett” pear (*Pyrus communis* L.). *Gigascience* 8(12), giz138.

Liu, C., Wang, Y., Peng, J., Fan, B., Xu, D. *et al.* (2022) High-quality genome assembly and pan-genome studies facilitate genetic discovery in mung bean and its improvement. *Plant Communications* 3(6).

Liu, C., Ding, X., Wu, Y., Zhang, J., Huang, R. *et al.* (2024) Chromosome-scale reference genome of an ancient landrace: Unveiling the genetic basis of seed weight in the food legume crop pigeonpea (*Cajanus cajan*). *Horticulture Research* 11(9), uhae201.

Liu, M.S., Kuo, T.C.Y., Ko, C.Y., Wu, D.C., Li, K.Y. *et al.* (2016) Genomic and transcriptomic comparison of nucleotide variations for insights into bruchid resistance of mungbean *Vigna radiata* [L.] R. Wilczek). *BMC Plant Biology* 16, 1–16.

Liu, S., Liu, Y., Yang, X., Tong, C., Edwards, D. *et al.* (2014) The *Brassica oleracea* genome reveals the asymmetrical evolution of polyploid genomes. *Nature Communications* 5(1), 3930.

Liu, Y., Du, H., Li, P., Shen, Y., Peng, H. *et al.* (2020) Pan-genome of wild and cultivated soybeans. *Cell* 182(1), 162–176.

Liu, Y.J., Wang, X.R. and Zeng, Q.Y. (2019) De novo assembly of white poplar genome and genetic diversity of white poplar population in Irtysh River Basin in China. *Science China Life Sciences* 62, 609–618.

Luo, Y., Hitz, B.C., Gabdank, I., Hilton, J.A., Kagda, M.S. *et al.* (2020) New developments on the encyclopedia of DNA elements (ENCODE) data portal. *Nucleic Acids Research* 48(D1), D882–D889.

Lv, H., Wang, Y., Han, F., Ji, J., Fang, Z. *et al.* (2020) A high-quality reference genome for cabbage obtained with SMRT reveals novel genomic features and evolutionary characteristics. *Scientific Reports* 10(1), 12394.

Ma, J., Wan, D., Duan, B., Bai, X., Bai, Q. *et al.* (2019) Genome sequence and genetic transformation of a widely distributed and cultivated poplar. *Plant Biotechnology Journal* 17(2), 451–460.

Marcussen, T., Sandve, S.R., Heier, L., Spannagl, M., Pfeifer, M. *et al.* (2014) Ancient hybridizations among the ancestral genomes of bread wheat. *Science* 345(6194), 1250092.

Margaritopoulou, T., Roka, L., Alexopoulou, E., Christou, M., Rigas, S. *et al.* (2016) Biotechnology towards energy crops. *Molecular Biotechnology* 58, 149–158.

Margulies, M., Egholm, M., Altman, W.E., Attiya, S., Bader, J.S. *et al.* (2005) Genome sequencing in microfabricated high-density picolitre reactors. *Nature* 437(7057), 376–380.

Mariac, C., Scarcelli, N., Pouzadou, J., Barnaud, A., Billot, C. *et al.* (2014) Cost-effective enrichment hybridization capture of chloroplast genomes at deep multiplexing levels for population genetics and phylogeography studies. *Molecular Ecology Resources* 14(6), 1103–1113.

Marla, S.S., Mishra, P., Maurya, R., Singh, M., Wankhede, D.P. *et al.* (2020) Refinement of draft genome assemblies of pigeonpea (*Cajanus cajan*). *Frontiers in Genetics* 11, 607432.

Marrano, A., Britton, M., Zaini, P.A., Zimin, A.V., Workman, R.E. *et al.* (2020) High-quality chromosome-scale assembly of the walnut (*Juglans regia* L.) reference genome. *Gigascience* 9(5), giaa050.

Martínez-García, P.J., Crepeau, M.W., Puiu, D., Gonzalez-Ibeas, D., Whalen, J. *et al.* (2016) The walnut (*Juglans regia*) genome sequence reveals diversity in genes coding for the biosynthesis of non-structural polyphenols. *The Plant Journal* 87(5), 507–532.

Maulana, F., Perumal, R., Serba, D.D. and Tesso, T. (2023) Genomic prediction of hybrid performance in grain sorghum (*Sorghum bicolor* L.). *Frontiers in Plant Science* 14, 1139896.
Maxam, A.M. and Gilbert, W. (1977) A new method for sequencing DNA. *Proceedings of the National Academy of Sciences* 74(2), 560–564.
McCormick, R.F., Truong, S.K., Sreedasyam, A., Jenkins, J., Shu, S. *et al.* (2018) The *Sorghum bicolor* reference genome: Improved assembly, gene annotations, a transcriptome atlas, and signatures of genome organization. *The Plant Journal* 93(2), 338–354.
McMullen, M.D., Kresovich, S., Villeda, H.S., Bradbury, P., Li, H. *et al.* (2009) Genetic properties of the maize nested association mapping population. *Science* 325(5941), 737–740.
Meyer, F.E., Shuey, L.S., Naidoo, S., Mamni, T., Berger, D.K. *et al.* (2016) Dual RNA-sequencing of *Eucalyptus nitens* during *Phytophthora cinnamomi* challenge reveals pathogen and host factors influencing compatibility. *Frontiers in Plant Science* 7, 191.
Miao, J., Feng, Q., Li, Y., Zhao, Q., Zhou, C. *et al.* (2021) Chromosome-scale assembly and analysis of biomass crop *Miscanthus lutarioriparius* genome. *Nature Communications* 12(1), 2458.
Michael, T.P. and VanBuren, R. (2020) Building near-complete plant genomes. *Current Opinion in Plant Biology* 54, 26–33.
Mitros, T., Session, A.M., James, B.T., Wu, G.A., Belaffif, M.B. *et al.* (2020) Genome biology of the paleo-tetraploid perennial biomass crop *Miscanthus*. *Nature Communications* 11(1), 5442.
Morselli, M., Farrell, C., Rubbi, L., Fehling, H.L., Henkhaus, R. *et al.* (2021) Targeted bisulfite sequencing for biomarker discovery. *Methods* 187, 13–27.
Mustafa, G. and Komatsu, S. (2021) Plant proteomic research for improvement of food crops under stresses: A review. *Molecular Omics* 17(6), 860–880.
Myburg, A.A., Grattapaglia, D., Tuskan, G.A., Hellsten, U., Hayes, R.D. *et al.* (2014) The genome of *Eucalyptus grandis*. *Nature* 510(7505), 356–362.
Nagle, M.F., Yuan, J., Kaur, D., Ma, C., Peremyslova, E. *et al.* (2023) GWAS identifies candidate genes controlling adventitious rooting in *Populus trichocarpa*. *Horticulture Research* 10(8), uhad125.
Pan, Y., Liang, H., Gao, L., Dai, G., Chen, W. *et al.* (2020) Transcriptomic profiling of germinating seeds under cold stress and characterization of the cold-tolerant gene LTG5 in rice. *BMC Plant Biology* 20, 1–17.
Paritosh, K., Yadava, S.K., Singh, P., Bhayana, L., Mukhopadhyay, A., *et al.* (2021) A chromosome-scale assembly of allotetraploid *Brassica juncea* (AABB) elucidates comparative architecture of the A and B genomes. *Plant Biotechnology Journal* 19(3), 602–614.
Pandey, M.K., Roorkiwal, M., Singh, V.K., Ramalingam, A., Kudapa, H. *et al.* (2016) Emerging genomic tools for legume breeding: Current status and future prospects. *Frontiers in Plant Science* 7, 455.
Parveen, K., Saddique, M.A.B., Waqas, M.U., Attia, K.A., Rizwan, M. *et al.* (2024) Genome-wide analysis and expression divergence of protein disulfide isomerase (PDI) gene family members in chickpea (*Cicer arietinum*) under salt stress. *Functional Plant Biology* 51(2).
Paterson, A.H., Bowers, J.E., Bruggmann, R., Dubchak, I., Grimwood, J. *et al.* (2009) The *Sorghum bicolor* genome and the diversification of grasses. *Nature* 457(7229), 551–556.
Perlo, V., Furtado, A., Botha, F.C., Margarido, G.R., Hodgson-Kratky, K. *et al.* (2022) Transcriptome and metabolome integration in sugarcane through culm development. *Food and Energy Security* 11(4), e421.
Phytozome (n.d.). Available at: https://phytozome-next.jgi.doe.gov/ (accessed 20 September 2024).
Pinosio, S., Giacomello, S., Faivre-Rampant, P., Taylor, G., Jorge, V. *et al.* (2016) Characterization of the poplar pan-genome by genome-wide identification of structural variation. *Molecular Biology and Evolution* 33(10), 2706–2719.
PlantGenIE Platform (n.d.). Available at: https://plantgenie.org/ (accessed 20 September 2024).
PLAZA (n.d.). Available at: https://bioinformatics.psb.ugent.be/plaza/ (accessed 20 September 2024).
Ployet, R., Feng, K., Zhang, J., Baxter, I., Glasgow, D.C. *et al.* (2024) Elemental profiling and genome-wide association mapping reveal genomic variants modulating ionomic composition in *Populus trichocarpa* leaves. *bioRxiv* 2024–05.
Pootakham, W., Nawae, W., Naktang, C., Sonthirod, C., Yoocha, T. *et al.* (2021) A chromosome-scale assembly of the black gram (*Vigna mungo*) genome. *Molecular Ecology Resources* 21(1), 238–250.
Poudel, S., Giannone, R.J., Rodriguez, M., Raman, B., Martin, M.Z. *et al.* (2017) Integrated omics analyses reveal the details of metabolic adaptation of *Clostridium thermocellum* to lignocellulose-derived growth inhibitors released during the deconstruction of switchgrass. *Biotechnology for Biofuels* 10, 1–14.

Qin, C., Yu, C., Shen, Y., Fang, X., Chen, L. *et al.* (2014) Whole-genome sequencing of cultivated and wild peppers provides insights into capsicum domestication and specialization. *Proceedings of the National Academy of Sciences* 111(14), 5135–5140.

Raizada, A. and Souframanien, J. (2019) Transcriptome sequencing, de novo assembly, characterisation of wild accession of blackgram (*Vigna mungo* var. *silvestris*) as a rich resource for development of molecular markers and validation of SNPs by high resolution melting (HRM) analysis. *BMC Plant Biology* 19, 1–16.

Rayburn, A.L., Crawford, J., Rayburn, C.M. and Juvik, J.A. (2009) Genome size of three *Miscanthus* species. *Plant Molecular Biology Reporter* 27, 184–188.

Renewables (2023) Analysis and forecasts to 2028, International Energy Agency, Paris. Available at: https://www.iea.org/reports/renewables-2023 (accessed 20 September 2024).

Rooney, W.L., Blumenthal, J., Bean, B. and Mullet, J.E. (2007) Designing sorghum as a dedicated bioenergy feedstock. *Biofuels, Bioproducts and Biorefining* 1(2), 147–157.

Sanger, F., Nicklen, S. and Coulson, A.R. (1977) DNA sequencing with chain-terminating inhibitors. *Proceedings of the National Academy of Sciences* 74(12), 5463–5467.

Santos, B.M.D., Zibrandtsen, J.F., Gunbilig, D., Sørensen, M., Cozzi, F. *et al.* (2019) Quantification and localization of formylated phloroglucinol compounds (FPCs) in *Eucalyptus* species. *Frontiers in Plant Science* 10, 186.

Sasaki, T. and Burr, B. (2000) International rice genome sequencing project: The effort to completely sequence the rice genome. *Current Opinion in Plant Biology* 3(2), 138–142.

Schaarschmidt, S., Fischer, A., Zuther, E. and Hincha, D.K. (2020) Evaluation of seven different RNA-seq alignment tools based on experimental data from the model plant *Arabidopsis thaliana*. *International Journal of Molecular Sciences* 21(5), 1720.

Schena, M., Shalon, D., Davis, R.W. and Brown, P.O. (1995) Quantitative monitoring of gene expression patterns with a complementary DNA microarray. *Science* 270(5235), 467–470.

Schilmiller, A.L., Pichersky, E. and Last, R.L. (2012) Taming the hydra of specialized metabolism: How systems biology and comparative approaches are revolutionizing plant biochemistry. *Current Opinion in Plant Biology* 15(3), 338–344.

Schmidt, M.H.W., Vogel, A., Denton, A.K., Istace, B., Wormit, A. *et al.* (2017) De novo assembly of a new *Solanum pennellii* accession using nanopore sequencing. *The Plant Cell* 29(10), 2336–2348.

Schmutz, J., Cannon, S.B., Schlueter, J., Ma, J., Mitros, T. *et al.* (2010) Genome sequence of the palaeopolyploid soybean. *Nature* 463(7278), 178–183.

Schnable, P.S., Ware, D., Fulton, R.S., Stein, J.C., Wei, F. *et al.* (2009) The B73 maize genome: Complexity, diversity, and dynamics. *Science* 326(5956), 1112–1115.

Sekiya, A., Marques, F.G., Leite, T.F., Cataldi, T.R., De Moraes, F.E. *et al.* (2021) Network analysis combining proteomics and metabolomics reveals new insights into early responses of *Eucalyptus grandis* during rust infection. *Frontiers in Plant Science* 11, 604849.

Shang, L., Li, X., He, H., Yuan, Q., Song, Y. *et al.* (2022) A super pan-genomic landscape of rice. *Cell Research* 32(10), 878–896.

Shang, L., He, W., Wang, T., Yang, Y., Xu, Q. *et al.* (2023) A complete assembly of the rice *Nipponbare* reference genome. *Molecular Plant* 16(8), 1232–1236.

Sharma, M.K., Sharma, R., Cao, P., Jenkins, J., Bartley, L.E. *et al.* (2012) A genome-wide survey of switchgrass genome structure and organization. *PloS One* 7(4), e33892.

Shen, Y., Du, H., Liu, Y., Ni, L., Wang, Z. *et al.* (2019) Update soybean Zhonghuang 13 genome to a golden reference. *Science China Life Sciences* 62(9), 1257–1260.

Shulaev, V., Sargent, D.J., Crowhurst, R.N., Mockler, T.C., Folkerts, O. *et al.* (2011) The genome of woodland strawberry (*Fragaria vesca*). *Nature Genetics* 43(2), 109–116.

Singh, N.K. (2016) Origin, diversity and genome sequence of mango (*Mangifera indica* L.). *Indian Journal of History of Science* 51(2.2), 355.

Singh, N.K., Gupta, D.K., Jayaswal, P.K., Mahato, A.K., Dutta, S. *et al.* (2012) The first draft of the pigeonpea genome sequence. *Journal of Plant Biochemistry and Biotechnology* 21, 98–112.

Skaib, A. and Zedan, R. (2023) Bioenergy: Socio-economic and environmental implications and future of biofuels. *African Journal of Advanced Pure and Applied Sciences (AJAPAS)* 19–35.

Song, J.M., Guan, Z., Hu, J., Guo, C., Yang, Z. *et al.* (2020) Eight high-quality genomes reveal pan-genome architecture and ecotype differentiation of *Brassica napus*. *Nature Plants* 6(1), 34–45.

Souza, G.M., Sluys, M.A., Lembke, C.G., Lee, H., Margarido, G.R.A. *et al.* (2019) Assembly of the 373k gene space of the polyploid sugarcane genome reveals reservoirs of functional diversity in the world's leading biomass crop. *GigaScience* 8(12), giz129.

Spindel, J.E., Dahlberg, J., Colgan, M., Hollingsworth, J., Sievert, J. *et al.* (2018) Association mapping by aerial drone reveals 213 genetic associations for *Sorghum bicolor* biomass traits under drought. *BMC Genomics* 19, 1–18.

Sreenivasan, T.V., Ahloowalia, B.S. and Heinz, D.J. (1987) Chapter 5 Cytogenetics. In: Don, J.H. (ed.) *Developments in Crop Science*. Elsevier, Amsterdam, pp. 211–253.

Stein, J.C., Yu, Y., Copetti, D., Zwickl, D.J., Zhang, L. *et al.* (2018) Genomes of 13 domesticated and wild rice relatives highlight genetic conservation, turnover and innovation across the genus *Oryza*. *Nature Genetics* 50(2), 285–296.

Stitzer, M., Seetharam, A., Armin, S., Hsu, S.K., Schulz, A. *et al.* (2024) Extensive genome evolution distinguishes maize within a stable tribe of grasses. *Zenodo. Doi* 10.

Sun, D., Wang, C., Zhang, X., Zhang, W., Jiang, H. *et al.* (2019) Draft genome sequence of cauliflower (*Brassica oleracea* L. var. Botrytis) provides new insights into the c genome in *Brassica* species. *Horticulture Research* 6.

Sun, F., Fan, G., Hu, Q., Zhou, Y., Guan, M. *et al.* (2017) The high-quality genome of *Brassica napus* cultivar 'ZS 11'reveals the introgression history in semi-winter morphotype. *The Plant Journal* 92(3), 452–468.

Sun, M., Yan, H., Zhang, A., Jin, Y., Lin, C. *et al.* (2023) Milletdb: A multi-omics database to accelerate the research of functional genomics and molecular breeding of millets. *Plant Biotechnology Journal* 21(11), 2348–2357.

Sun, S., Zhou, Y., Chen, J., Shi, J., Zhao, H. *et al.* (2018) Extensive intraspecific gene order and gene structural variations between Mo17 and other maize genomes. *Nature Genetics* 50(9), 1289–1295.

Sun, X., Jiao, C., Schwaninger, H., Chao, C.T., Ma, Y. *et al.* (2020) Phased diploid genome assemblies and pan-genomes provide insights into the genetic history of apple domestication. *Nature Genetics* 52(12), 1423–1432.

Tang, W. and Tang, A.Y. (2019) Biological significance of RNA-seq and single-cell genomic research in woody plants. *Journal of Forestry Research* 30, 1555–1568.

Tao, S., Liu, P., Shi, Y., Feng, Y., Gao, J. *et al.* (2022) Single-cell transcriptome and network analyses unveil key transcription factors regulating mesophyll cell development in maize. *Genes* 13(2), 374.

Terzić, S., Boniface, M.C., Marek, L.F., Alvarez, D., Baumann, K. *et al.* (2020) Gene banks for wild and cultivated sunflower genetic resources. *OCL Oilseeds and Fats, Crops and Lipids* 27(9), 1–14.

Tettelin, H., Masignani, V., Cieslewicz, M.J., Donati, C., Medini, D. *et al.* (2005) Genome analysis of multiple pathogenic isolates of *Streptococcus agalactiae*: Implications for the microbial "pan-genome". *Proceedings of the National Academy of Sciences* 102(39), 13950–13955.

The Arabidopsis Initiative (2000) Analysis of the genome sequence of the flowering plant *Arabidopsis thaliana*. *Nature* 408, 796–815.

Thudi, M., Palakurthi, R., Schnable, J.C., Chitikineni, A., Dreisigacker, S. *et al.* (2021) Genomic resources in plant breeding for sustainable agriculture. *Journal of Plant Physiology* 257, 153351.

Tian, T., You, Q., Zhang, L., Yi, X., Yan, H. *et al.* (2016) SorghumFDB: Sorghum functional genomics database with multidimensional network analysis. *Database*2016, baw099.

Tomato Genome Consortium, X. (2012) The tomato genome sequence provides insights into fleshy fruit evolution. *Nature* 485(7400), 635.

TreeGene (n.d.). Available at: https://treegenesdb.org/ (accessed 20 September 2024).

Tu, W., Wang, L., Jin, P., Meng, R., Zheng, J. *et al.* (2023) Construction of high-density genetic map and quantitative trait locus mapping in a sorghum Tx623A× sudangrass S722 population. *Grassland Science* 69(2), 132–141.

Tuskan, G.A., Difazio, S., Jansson, S., Bohlmann, J., Grigoriev, I. *et al.* (2006) The genome of black cottonwood, *Populus trichocarpa* (Torr. & Gray). *Science* 313(5793), 1596–1604.

Tyagi, P., Singh, D., Mathur, S., Singh, A. and Ranjan, R. (2022) Upcoming progress of transcriptomics studies on plants: An overview. *Frontiers in Plant Science* 13, 1030890.

Ulloa, J.L., Aguayo, P., Conejera, D., Rubilar, R., Balocchi, C. *et al.* (2022) Transcriptomic response in foliar and root tissues of a drought-tolerant *Eucalyptus globulus* genotype under drought stress. *Trees* 1–13.

Valenzuela, C.E., Ballesta, P., Ahmar, S., Fiaz, S., Heidari, P. *et al.* (2021) Haplotype-and SNP-based GWAS for growth and wood quality traits in *Eucalyptus cladocalyx* trees under arid conditions. *Plants* 10(1), 148.

Varshney, R.K., Nayak, S.N., May, G.D. and Jackson, S.A. (2009) Next-generation sequencing technologies and their implications for crop genetics and breeding. *Trends in Biotechnology* 27(9), 522–530.

Varshney, R.K., Chen, W., Li, Y., Bharti, A.K., Saxena, R.K. *et al.* (2012) Draft genome sequence of pigeonpea (*Cajanus cajan*), an orphan legume crop of resource-poor farmers. *Nature Biotechnology* 30(1), 83.

Varshney, R.K., Song, C., Saxena, R.K., Azam, S., Yu, S. *et al.* (2013) Draft genome sequence of chickpea (*Cicer arietinum*) provides a resource for trait improvement. *Nature Biotechnology* 31(3), 240–246.

Varshney, R.K., Shi, C., Thudi, M., Mariac, C. and Wallace, J. (2017) Pearl millet genome sequence provides a resource to improve agronomic traits in arid environments. *Nature Biotechnology* 35(10), 969–976.

Velasco, R., Zharkikh, A., Affourtit, J., Dhingra, A., Cestaro, A. *et al.* (2010) The genome of the domesticated apple (*Malus × domestica* Borkh.). *Nature Genetics* 42(10), 833–839.

Wang, B., Jiao, Y., Chougule, K., Olson, A., Huang, J. *et al.* (2021) Pan-genome analysis in sorghum highlights the extent of genomic variation and sugarcane aphid resistance genes. *BioRxiv* 2021-01.

Wang, X., Wang, H., Wang, J., Sun, R., Wu, J. *et al.* (2011) The genome of the mesopolyploid crop species *Brassica rapa*. *Nature Genetics* 43(10), 1035–1040.

Wang, Y., Zeng, X., Iyer, N.J., Bryant, D.W., Mockler, T.C. *et al.* (2012) Exploring the switchgrass transcriptome using second-generation sequencing technology. *PloS One* 7(3), e34225.

Waqas, M., Azhar, M.T., Rana, I.A., Azeem, F., Ali, M.A. *et al.* (2019) Genome-wide identification and expression analyses of WRKY transcription factor family members from chickpea (*Cicer arietinum* L.) reveal their role in abiotic stress-responses. *Genes & Genomics* 41, 467–481.

Wasi, A., Tahseen, S., Arishakausar, a, Bhatt, A.Y. and Shahzad, A. (2023) Current status and future prospectus of bioenergy crops. In: *Biotechnology and Omics Approaches for Bioenergy Crops*. Springer Nature, Singapore, pp. 271–288.

Weijde, T., Alvim, K.C., Torres, A., Vermerris, W., Dolstra, O. *et al.* (2013) The potential of C4 grasses for cellulosic biofuel production. *Frontiers in Plant Science* 4, 107.

Wu, H., Yao, D., Chen, Y., Yang, W., Zhao, W. *et al.* (2020) De novo genome assembly of *Populus simonii* further supports that *Populus simonii* and *Populus trichocarpa* belong to different sections. *G3: Genes, Genomes, Genetics* 10(2), 455–466.

Wu, M., Northen, T.R. and Ding, Y. (2023) Stressing the importance of plant specialized metabolites: Omics-based approaches for discovering specialized metabolism in plant stress responses. *Frontiers in Plant Science* 14, 1272363.

Xiao, T. and Zhou, W. (2020) The third-generation sequencing: The advanced approach to genetic diseases. *Translational Pediatrics* 9(2), 163.

Xie, M., Chung, C.Y.L., Li, M.W., Wong, F.L., Wang, X. *et al.* (2019) A reference-grade wild soybean genome. *Nature Communications* 10(1), 1216.

Xu, X., Pan, S., Cheng, S., Zhang, B., Mu, D. *et al.* (2011) Genome sequence and analysis of the tuber crop potato. *Nature* 475(7355), 189–195.

Yadav, P., Vaidya, E., Rani, R., Yadav, N.K., Singh, B.K. *et al.* (2018) Recent perspective of next generation sequencing: Applications in molecular plant biology and crop improvement. *Proceedings of the National Academy of Sciences, India Section B: Biological Sciences* 88, 435–449.

Yan, Q., Wang, Q., Xuzhen, C., Wang, L., Somta, P. *et al.* (2020) High-quality genome assembly, annotation and evolutionary analysis of the mungbean (*Vigna radiata*) genome. *Authorea Preprints*.

Yang, J., Liu, D., Wang, X., Ji, C., Cheng, F. *et al.* (2016) The genome sequence of allopolyploid *Brassica juncea* and analysis of differential homoeolog gene expression influencing selection. *Nature Genetics* 48(10), 1225–1232.

Yang, W., Wang, K., Zhang, J., Ma, J., Liu, J. *et al.* (2017) The draft genome sequence of a desert tree *Populus pruinosa*. *Gigascience* 6(9), gix075.

Yang, X., Luo, Z., Todd, J., Sood, S. and Wang, J. (2020) Genome-wide association study of multiple yield traits in a diversity panel of polyploid sugarcane (*Saccharum* spp.). *The Plant Genome* 13(1), e20006.

Yang, X., Li, Y., Yu, R., Zhang, L., Yang, Y. *et al.* (2023) Integrated transcriptomic and metabolomic profiles reveal adaptive responses of three poplar varieties against the bacterial pathogen *Lonsdalea populi*. *Plant, Cell & Environment* 46(1), 306–321.

Zhang, G., Liu, X., Quan, Z., Cheng, S., Xu, X. *et al*. (2012) Genome sequence of foxtail millet (*Setaria italica*) provides insights into grass evolution and biofuel potential. *Nature Biotechnology* 30(6), 549–554.

Zhang, G., Ge, C., Xu, P., Wang, S., Cheng, S. *et al*. (2021) The reference genome of *Miscanthus floridulus* illuminates the evolution of *Saccharinae*. *Nature Plants* 7(5), 608–618.

Zhang, J., Huguet-Tapia, J.C., Hu, Y., Jones, J., Wang, N. *et al*. (2017) Homologues of CsLOB1 in citrus function as disease susceptibility genes in citrus canker. *Molecular Plant Pathology* 18(6), 798–810.

Zhang, J., Zhang, X., Tang, H., Zhang, Q., Hua, X. *et al*. (2018) Allele-defined genome of the autopolyploid sugarcane *Saccharum spontaneum* L. *Nature Genetics* 50(11), 1565–1573.

Zhang, J.Z., Creelman, R.A. and Zhu, J.K. (2004) From laboratory to field. Using information from Arabidopsis to engineer salt, cold, and drought tolerance in crops. *Plant Physiology* 135(2), 615–621.

Zhang, Z., Chen, Y., Zhang, J., Ma, X., Li, Y. *et al*. (2020) Improved genome assembly provides new insights into genome evolution in a desert poplar (*Populus euphratica*). *Molecular Ecology Resources* 20(3), 781–794.

Zhao, X.R., Zhao, D.T., Zhang, L.Y., Chang, J.H. and Cui, J.H. (2024) Combining transcriptome and metabolome analysis to understand the response of sorghum to *Melanaphis sacchari*. *BMC Plant Biology* 24(1), 529.

Zheng, Y.Y., Chen, L.H., Fan, B.L., Xu, Z., Wang, Q. *et al*. (2024) Integrative multiomics profiling of passion fruit reveals the genetic basis for fruit color and aroma. *Plant Physiology* 194(4), 2491–2510.

Zhu, T., Wang, L., Rimbert, H., Rodriguez, J.C., Deal, K.R. *et al*. (2021) Optical maps refine the bread wheat *Triticum aestivum* cv Chinese spring genome assembly. *The Plant Journal* 107(1), 303–314.

2 Crop Genomics: A Tool for Crop Improvement

Awdhesh Mishra[1] and Roshan Kumar Singh[2*]
[1]Department of Biotechnology, Yeungnam University, Gyeongsan, South Korea;
[2]Department of Botany, Mahishadal Raj College Purba Medinipur, West Bengal, India

Abstract
Plant genomics has revolutionized the field of crop improvement by enabling the investigation of genetic traits at the molecular level. Recent advancements in nucleic acid sequencing technologies have greatly facilitated the in-depth investigation of syntenic relationships between related crop genomes, identification of novel gene families, and portrayal of the molecular function of newly identified gene pools. In the previous decade, integrating crop genomics with other omics facilities has proven a remarkable approach for understanding the genetic complexity of various agronomic traits. Genetic determinants that confer adaptation to adverse environmental conditions and enhanced crop yields are being identified using various genomic tools. These genetic determinants are being used in marker-assisted breeding and genetic engineering to introgress the superior trait into popular crops. Cutting-edge technologies like advanced genome sequencing platforms, high-throughput genotyping, precise genome editing, and space technology-enabled accelerated crop growth have transformed the crop improvement programs. The present chapter provides a comprehensive overview of recent advances in functional and structural genomics, application of comparative genomics, pangenomics, and recent tools and technologies in crop genomics. The chapter also discusses the applications, advantages, limitations, and scope for improvement of these approaches, offering a systematic understanding of the role of genomics in meeting the increasing demands for agricultural produce amidst a rapidly growing global population.

2.1 Introduction

The world is facing an unprecedented surge in demand for food crops, driven by a rapidly growing population and dwindling natural resources. In terms of hunger, the situation is alarming, with around 830 million people expected to experience hunger by 2030. This highlights the need for concerted efforts to address food insecurity, improve agricultural productivity, and reduce poverty and inequality (FAO, 2020). To ensure global food and agriculture security amidst climate change, it is imperative that we develop high-yielding and climate-smart crops. Genomic analysis using computational tools plays a pivotal role in crop improvement. The completion of the Arabidopsis genome marked a significant milestone, and since then, more than 800 crop genomes have been sequenced. Recent advancements in sequencing technologies have revolutionized the field of genomics, making it more accessible and affordable. The creation of high-quality genetic maps has bridged the gap between laboratory research and practical

*Corresponding author: rshnsngh8@gmail.com

 From Gene Discovery to Climate-Resilient Crops: Omics in Crop Improvement (eds S. Pandey *et al.*)
DOI: 10.1079/9781800628717.0002

field applications, paving the way for more effective crop improvement strategies. Molecular breeding and transgenic approaches have collectively contributed to the development of high-yielding crops, characterized by enhanced productivity and improved nutritional profiles (Singh *et al.*, 2020).

Recent breakthroughs in cutting-edge technologies have revolutionized crop genomics. Advances in complex genome sequencing, high-throughput phenotyping, epigenomic attributes, and pangenome analysis have significantly enhanced our understanding of genetic determinants underlying a particular trait. The 3000 Rice Genome Project and 3000 Chickpea Genome Sequencing initiatives have yielded a treasure trove of valuable genetic information (Varshney, 2016; Wang *et al.*, 2018). This wealth of data has significantly expanded the gene pool, empowering breeders with the necessary tools to drive genetic gains and accelerate crop improvement. Comprehensive exploration of pangenomes, encompassing both the core and accessory genomes, offers a valuable opportunity to identify and exploit genes that have driven evolutionary advantages. The surge in quantitative trait loci (QTL) and genome-wide association studies (GWAS) has significantly enhanced our understanding of the genetic architecture underlying key plant traits. Specifically, these studies have shed light on the genetic factors influencing grain size, growth patterns, and disease resistance, ultimately informing more effective crop improvement strategies. Genomics-assisted breeding (GAB) has emerged as a powerful tool for cultivar development, enabling breeders to harness the power of allelic variation to create improved crop varieties with desirable traits. Furthermore, the emergence of biotechnological tools like CRISPR-Cas genome editing, long-read single molecule sequencing, and artificial intelligence have enabled precise and efficient crop improvement (Harfouche *et al.*, 2019; Singh *et al.*, 2021). Translational genomics leverages comparative genomics to investigate the similarities and differences among various plant species. Notably, the accuracy of the functional annotation of novel sequences improves significantly when species are evolutionarily closely related. Consequently, plant species with well-studied model relatives become prime targets for translational genomic approaches, facilitating the discovery of new genes, functions, and traits.

2.2 Advancement in Genome Sequencing Technology

The past few years have witnessed tremendous advancements in genome sequencing technologies. Consequently, numerous genome sequencing projects have been successfully completed for a diverse range of plant species. High-throughput sequencing technologies have revolutionized the field of genomics, enabling rapid, accurate, and cost-effective sequencing of large volumes of data. Prior to 2004, genome sequencing relied heavily on Sanger sequencing technology, which, although renowned for its accuracy, had significant limitations in terms of throughput and scalability. The groundbreaking sequencing of the Arabidopsis and rice genomes served as a catalyst, inspiring the sequencing of numerous other crop and model plant species. Since then, sequencing technologies have undergone a transformative evolution, driven by the integration of high-performance computing system. This synergy has given rise to innovative, advanced, and cost-effective next-generation sequencing (NGS) technologies. NGS technology has revolutionized the field of genomics, enabling the rapid generation of reference genomes and re-sequencing of related species to elucidate genetic diversity. Additionally, NGS facilitates transcriptome sequencing, providing valuable insights into complex gene networks, as well as metagenomics and high-throughput genotyping methods such as genotyping by sequencing (GBS) and QTL mapping. Several NGS platforms have been established including Illumina, Pacific Biosciences, and Oxford Nanopore, that can enable the simultaneous sequencing of millions to billions of DNA fragments, dramatically accelerating genomic research and novel allele discovery (Satam *et al.*, 2023). Through the sequencing of related genotypes, either individually or in pooled sets, within a species, and comparing them to a reference genome, genome-wide variations can be identified. These variations include microsatellites or simple sequence repeats (SSRs), single nucleotide

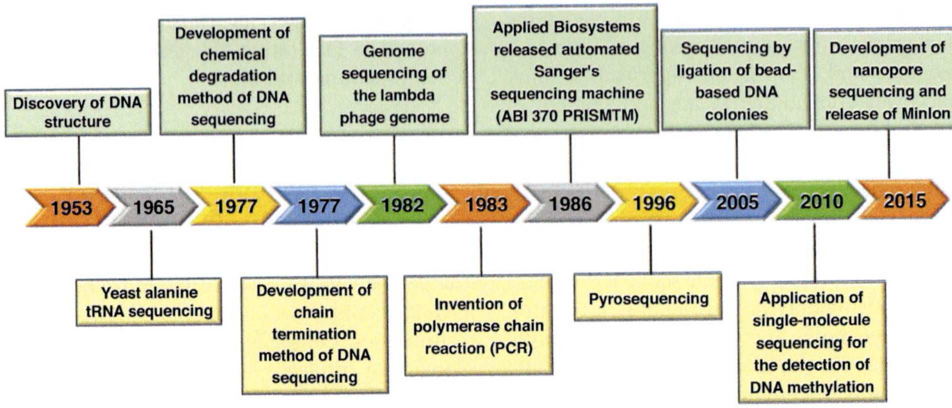

Fig. 2.1. A timeline of key milestones, discoveries, and innovations in DNA sequencing technology. Author's own image.

polymorphisms (SNPs), insertions/deletions (InDels), and structural variations such as translocations, chromosome fusions, and copy number variations (CNVs). Analyzing these variations is a crucial factor influencing plant evolution, adaptation, and diversity. The evolution and milestone achievements in DNA sequencing technology through different time periods are briefly illustrated in Fig. 2.1.

2.2.1 The Pangenome concept

A pangenome refers to the comprehensive analysis of the entire genetic information present within a specific biological clade, encompassing all genes found in a particular species, genus, or other taxonomic group. This concept is analogous to metagenomics, which involves the study of multiple genomes from the same environment, but at the ecosystem level. Within a pangenome, genes can be categorized into two main groups: core genes and dispensable genes. Core genes are universally present among all studied individuals, representing a shared genetic foundation. In contrast, dispensable genes are found in specific subgroups or are unique to individual organisms, contributing to genetic diversity and adaptability (Yocca and Edger, 2022). In 2005, a landmark study on genome sequencing of multiple *Streptococcus agalactiae* isolates gave rise to the pangenome concept. The data demonstrated that while 80% of the genes were conserved across all *Streptococcus agalactiae* isolates (core genes), 20% were variably present, highlighting the dynamic nature of bacterial genomes (Tettelin *et al.*, 2005). Since then, pangenome studies have been conducted in several crops including rice, wheat, brassica, maize (corn), soybean, pigeonpea, and others (Table 2.1).

Pangenome analysis of crops may be influenced by the phylogenetic resolution of the study population, as analyzing a larger population may reveal distinct patterns compared to a smaller one (Vernikos *et al.*, 2015). Pangenomic analysis offers a dual benefit, allowing researchers to gauge the genomic diversity within a dataset while also providing a predictive framework to determine how many more genomes are necessary to exhaustively characterize the more efficient pangenome. Two distinct categories of methods are utilized in pangenomics research. The first set of methods is designed for building pangenomes from datasets, whereas the second set is aimed at analyzing existing pangenomes to identify core, dispensable, and strain-specific genes. Comparative genomic analysis, which involves comparing two or more genomes within a group, is essential for identifying shared and non-shared sequences. Nevertheless, inconsistencies may occur between studies focusing on the same genomic dataset, primarily due to variations in bioinformatic pipelines and sequence alignment techniques. Each species possesses unique genetic and demographic characteristics, such

Table 2.1. A summary of major crops with pangenome analysis. Author's own table.

Crop species	No. of accessions used	Sequencing method/ platform	No. of genes identified in pangenome	Reference
Zea mays	503	Illumina HiSeq	41,903	Hirsch *et al.*, 2014
Brassica oleracea	10	Illumina HiSeq	61,379	Golicz *et al.*, 2016
Triticum aestivum	19	Illumina HiSeq	139,747	Montenegro *et al.*, 2017
Brachypodium distachyon	54	Illumina HiSeq	37,886	Gordon *et al.*, 2017
Brassica napus	53	Illumina HiSeq	94,013	Hurgobin *et al.*, 2018
Oryza sativa	3010	Illumina HiSeq PacBio	48,098	Wang *et al.*, 2018
Solanum lycopersicum	725	Illumina NextSeq	40,369	Gao *et al.*, 2019
Cajanus cajan	206	Illumina HiSeq	55,512	Zhao *et al.*, 2020
Glycine max	29	Illumina HiSeq PacBio	57,492	Liu *et al.*, 2020
Sorghum bicolor	354	Illumina HiSeq	35,719	Ruperao *et al.*, 2021
Oryza sativa and *O. glaberrima*	34	PacBio SMRT long-read sequencing	66,636	Qin *et al.*, 2021
Cicer arietinum	3366	Illumina HiSeq	29,870	Varshney *et al.*, 2021
Oryza sativa, O. glaberrima, O. barthii, and *O. rufipogon*	251	Oxford Nanopore Technology and Illumina NovaSeq	51,359	Shang *et al.*, 2022
Panicum miliaceum	516	Hi-C sequencing	66,042	Chen *et al.*, 2023
Setaria italica and *S. viridis*	110	Illumina NovaSeq	73,528	He *et al.*, 2023
Cicer sp.	8	NA	24,827	Khan *et al.*, 2024

as effective population size, genetic diversity, and demographic parameters, which can significantly impact the performance and interpretation of various methods in genomics. The emergence of pan-genomic research has significantly advanced our understanding of crop diversity and improvement. Nevertheless, the applications of crop pangenomics in crop improvement are only beginning to be explored, with substantial potential remaining to be harnessed. To effectively leverage pangenomics insights for crop improvement, it is essential to integrate pangenomic insights with QTL/GWAS findings and resequencing data. This integrative strategy facilitates the discovery of advantageous genes and alleles, ultimately informing the development of effective breeding programs tailored to specific crop improvement objectives.

2.2.2 Genotyping by sequencing (GBS) and genome-wide association analysis (GWAS)

GBS represents a robust method for SNP identification within genomes and populations, offering a unique advantage by eliminating the need for prior genome sequence information (Pootakham, 2023). The GBS technique involves a series of steps, including restriction digestion of genomic DNA, ligation of barcode adaptors, PCR amplification, and pooling to generate multiplex libraries for subsequent sequencing. The simplicity of library preparation, low cost, and fewer PCR and purification steps in GBS enables researchers to efficiently process and analyze large collections of samples (Favre *et al.*, 2021). GBS has proven to be a valuable tool in plant breeding, with applications in genomic selection, assessment of genomic diversity, genome-wide association mapping, linkage analysis, and the identification of novel molecular markers. The precise location of the semi-dwarfing gene (*ari-e*) on barley chromosome 5 hour was confirmed through the integration of GBS and high-density SNP mapping (Liu *et al.*, 2014).

GWAS involves scanning the entire genome to uncover associations between specific SNPs and particular phenotypic traits, offering a comprehensive understanding of the genetic factors influencing trait variation. GWAS has undergone remarkable progress, from its inception using microarray-based analyses to its current status as a high-resolution tool, enabled by next-generation sequencing technologies. This evolution has enabled GWAS to associate natural genomic variations underlying complex traits in crops. The quantitative framework of GWAS in crops is rooted in the analysis of linkage disequilibrium (LD) patterns. This involves genotyping and phenotyping a diverse set of accessions or varieties, which are evaluated repeatedly for several traits and genotyped once. By analyzing the associations between the phenotypic data and genomic variations, genetic factors underlying complex traits can be identified. Various population types, including biparental populations, multiparental populations [such as nested association mapping (NAM), multiparent advanced generation intercross (MAGIC), and random-open-parent association mapping (ROAM) populations], and breeding populations, are currently being utilized for GWAS and linkage analyses (Singh *et al.*, 2022a). GWAS have been performed in many crops, such as rice, maize, wheat, millets, and others (Xiao *et al.*, 2021; Demirjian *et al.*, 2023). A GWAS analysis was conducted with a set of 104 accessions of foxtail millet for 11 micro- and macronutrient related traits (Jaiswal *et al.*, 2023). A total of 24 candidate genes were identified for 28 marker-trait associations (MTAs) involving seven traits. Similarly, a comparable study conducted in rice successfully identified alleles linked to ten grain-related traits and flowering time, leveraging a comprehensive dataset comprising approximately 1.3 million SNPs (Huang *et al.*, 2012). The molecular function of candidate gene(s) discovered through GWAS and MTAs is subsequently validated using a reverse genetics approach, providing further insight into the genetic mechanisms underlying the trait of interest.

2.2.3 Genomic-assisted breeding

The integration of genomic information into crop breeding programs has become a standard practice in the NGS era. This has led to the development of elite cultivars with enhanced yield, improved stress tolerance, and superior nutritional content (Singh *et al.*, 2020). Two prominent approaches have emerged in genomics-assisted breeding: marker-assisted selection (MAS) and genomic selection (GS). MAS involves the use of molecular markers that are linked to a specific gene or QTLs associated with a desired trait or phenotype. These markers enable breeders to selectively identify and retain progeny that possess favorable alleles for the target trait. GS involves a holistic strategy, where all molecular markers present throughout the genome are used to predict the genetic merit or breeding value of an individual, enabling a more accurate and comprehensive selection process (Mohanavel *et al.*, 2024).

MAS has been successfully employed in several key areas of plant breeding, including the incorporation of recessive alleles, stacking multiple genes (gene pyramiding), and selecting for traits that manifest at later stages of plant growth or are difficult and expensive to evaluate phenotypically. The use of DNA markers allows breeders to identify the genotype of traits that are typically expressed only at flowering or harvest stages. This early identification facilitates the planning of targeted crosses, streamlining the breeding process. Certain traits require specific environmental conditions to manifest, making genotype selection challenging. MAS offers a solution by enabling breeders to select desirable genotypes regardless of the environment. The effective application of MAS in crop improvement relies on two critical conditions. The molecular marker used must be closely linked to the gene of interest, and the target gene should have a high heritability value, indicating a significant genetic component. The application of MAS has yielded significant improvements in elite cultivars of numerous crops, demonstrating its effectiveness in modern plant breeding. For example, the generation of a drought-, bacterial-leaf-blight-, and blast-resistant rice cultivar (Singh *et al.*, 2022b), heat tolerance in wheat (Bellundagi *et al.*, 2022), climate smart coffee (Ferrão *et al.*, 2024), nutritional quality in maize (Hossain *et al.*, 2023), and many others. Marker-assisted gene pyramiding is a powerful breeding approach that enables the simultaneous combination of multiple genes from different donor parents into a single genotype. This is achieved by using linked molecular markers to track the genes of interest. To combat blast disease in rice, this study employed a marker-assisted backcrossing (MABC) approach, coupled with foreground and background selection using KASP marker assays, to pyramid four blast-resistance (Pi) genes (*Piz*, *Pib*, *Pita*, and *Pik*) into a prominent, yet susceptible, Italian *japonica* rice variety (Zampieri *et al.*, 2023). This strategy aims to confer broad-spectrum and durable resistance while preserving the desirable traits of the elite cultivar.

2.2.4 Speed breeding

Conventional breeding approaches have been instrumental in shaping modern agriculture since its inception. The longer generation cycle of many crop plants has long been a constraint in crop breeding. This bottleneck has made it difficult for breeders to rapidly achieve substantial genetic improvements. Speed breeding is a groundbreaking approach that expedites the breeding process of crop plants by cultivating them under controlled artificial light and temperature conditions, thereby reducing the time required for breeding cycles. The speed breeding technique primarily relies on the strategic modification of light parameters, including quality, intensity, and duration. By optimizing these factors, plants perceive the signals through photoreceptors, which in turn stimulates accelerated reproductive development, enabling faster breeding cycles (Ghosh *et al.*, 2018; Ahmar *et al.*, 2020). Speed breeding protocols have been developed by grouping plants into three primary categories: short-day plants (SDP), which flower in response to shorter daylight periods; long-day plants (LDP), which require longer daylight periods to induce flowering; and day-neutral plants (DNP), the flowering of which is unaffected by daylight duration. The convergence of speed breeding with omics technologies has the potential to transform the field of plant breeding (Pandey *et al.*, 2022). By leveraging genomics, transcriptomics, proteomics, phenotyping, and metabolomics, breeders can now gather and analyze vast amounts of data, leading to unprecedented advances in crop improvement programs. A significant advancement in speed breeding has already been achieved, with researchers successfully generating 4–6 generations of key crop species, including brassica, pea, wheat, barley, rice, soybean, and chickpea within a single year (Rai, 2022). This accelerated breeding approach far surpasses the two generations per year typically achieved through conventional breeding methods. Apart from its potential application, a significant constraint is to implementing speed breeding on a larger scale due to limited availability of trained breeders with expertise in operating and maintaining controlled growth environments. Several other limitations hinder the widespread adoption of speed breeding, including inadequate infrastructure, the need for genotype-specific protocols, insufficient phenotyping capabilities, and the risk of losing superior genotypes during field transfer (Gudi *et al.*, 2022).

2.3 Advancement in Epigenomics Tools

Epigenetics refers to the heritable modifications in chromatin structure that significantly influence gene expression and cellular function without altering the underlying DNA sequence. This complex regulatory process involves various molecular mechanisms, including histone modifications (acetylation and methylation), DNA methylation, and non-coding RNA interactions, which collectively control chromatin accessibility, gene activation, and silencing. The advent of NGS technology has enabled the development of various approaches to study epigenetic changes, including whole-genome bisulfite sequencing (WGBS), reduced representation bisulfite sequencing (RRBS), and chromatin immunoprecipitation sequencing (ChIP-Seq). These techniques have been successfully applied to numerous crop plants. WGBS, also referred to as MethylC-Seq, is employed to map cytosine methylation patterns comprehensively across the entire genome. This method was first applied to plants to decipher the methylome of Arabidopsis (Gu *et al.*, 2011), and subsequently, this approach has been expanded to investigate the epigenetic landscapes of other crops including rice (Li *et al.*, 2012), brassica (Gao *et al.*, 2014), foxtail millet (Singh *et al.*, 2016), and others. WGBS analysis of pigeonpea cytoplasmic male sterile, restorer, and F1 hybrid lines revealed potential epigenetic mechanisms regulating pollen fertility-related genes (Junaid *et al.*, 2018). ChIP is a powerful technique for studying DNA–protein interactions *in vivo*. The process involves formaldehyde fixation of genomic DNA and protein interactions, followed by restriction digestion to produce 200–800 bp fragments. These fragments are then immunoprecipitated using antibodies specific to the protein of interest. The immunoprecipitated DNA is released from the protein complex through acid treatment and amplified by PCR. Similarly, histone modification landscapes are captured using antibodies specific to particular histone proteins (Ranawaka *et al.*, 2020). The further application of NGS technologies in combination with ChIP led to the development of ChIP-seq. In this technology, the immunoprecipitated DNA fragments are finally sequenced using NGS platforms and the resulting epigenetic profiles are compared to corresponding transcriptome data to assess the epigenetic regulation of gene expression (Kumar *et al.*, 2024).

2.4 Biotechnological Advancements for Crop Improvement

Biotechnology is an interdisciplinary field that integrates concepts and techniques from molecular biology, genetics, biochemistry, microbiology, and chemical and computer science. Biotechnological strategies involve modifying genes and genomic regions associated with desirable traits either through overexpression, RNA interference, or genome editing, allowing for precise control over genetic changes, facilitating the development of improved crop varieties (Sood *et al.*, 2021). The creation of transgenic plants involves the use of recombinant DNA technology to transfer foreign genes into the plant genome. These genes, which originate from same or unrelated plants, are stably integrated into the plant's genome. Transgenic technology has the advantage over conventional breeding as it eliminates the cross genera barriers, has a low labor requirement, is efficient, and less time consuming. The construction of transgenic plants involves the following steps including cloning of the desired gene, designing the vector construct for plant transformation, the transformation process, and finally, regeneration and agronomic evaluation of genetically modified plants. During the past two decades, this approach has been extensively utilized to modify plant genomes and introduce desirable traits in several crops (Singh *et al.*, 2020). Some commercially available genetically modified crops are insect-resistant Bt Shanyou 63 Rice, insect-resistant Bt Xtra™ Maize, herbicide-tolerant Roundup Ready™ Wheat, provitamin-A-rich golden rice, low indole glucosinolate canola, etc.

The advent of precise genome editing has transformed the field of biotechnology. This breakthrough began in 1996, when researchers demonstrated that DNA-binding zinc finger domains, combined with Fok1 endonuclease domains, could cleave DNA at specific locations, functioning as site-specific nucleases (Kim *et al.*, 1996). In due course, several such editing tools have been invented including transcription activator-like effector nucleases (TALENs), meganucleases (MegaNs), and clustered regularly interspaced short palindrome repeats (CRISPR)/CRISPR-associated protein (Cas).

The process of precise genome editing involves three key stages. Initially, a custom-designed nuclease recognizes the target DNA sequence and generates double-strand breaks (DSBs) at or near the desired location. Subsequently, the cell's repair machinery is activated, utilizing either nonhomologous end-joining (NHEJ) or homologous recombination (HR) to repair the DSBs. By harnessing the susceptibility of NHEJ, precise modifications can be introduced to the genome. CRISPR was first identified in *Escherichia coli* in 1987 as a novel immune system defense mechanism that protected the bacteria against foreign plasmid and viral DNA invasion (Ishino *et al.*, 1987). Since its discovery, CRISPR/Cas techniques have undergone significant development and have emerged as the most widely used gene editing techniques in recent years. CRISPR-Cas9 is regarded as the most impactful gene editing tool due to its reliance on guide RNA (gRNA) for target recognition. This eliminates the need to design new proteins for each novel target site, streamlining the gene editing process. One of the technical limitations of CRISPR-Cas9 is the necessity for the protospacer adjacent motif (PAM) sequence to be situated immediately downstream of the gRNA target site, which is a prerequisite for Cas9 protein recognition and subsequent DNA cleavage. Different variants of Cas9, which differ in biological origin, are based on varied PAM recognition sequences as summarized in Table 2.2. Another major concern associated with the CRISPR/Cas9 system is the potential for off-target cleavage by the Cas9 enzyme (Liu *et al.*, 2022), although its effect can be minimized by performing backcrossing repeatedly to eliminate undesirable mutations. However, backcrossing takes a lot of time and slows down crop improvement. Through the application of the CRISPR/Cas technique, significant genetic modifications have been achieved in plants, resulting in enhanced crop yield, improved tolerance to biotic and abiotic stresses, enhanced nutrient quality and production of monoploid seeds (Zhu *et al.*, 2020; Mishra *et al.*, 2024). Notably, genome-edited lines have largely avoided the intense resistance and rejection that genetically modified organisms (GMOs) faced in the past.

2.5 Application of Single Cell Genomics in Crop Improvement

The complexity of crop plants lies in their multicellular nature, consisting of various cell types that contribute distinctively to plant development. Interestingly, these diverse cell types share a common genome. Simultaneous analysis of multiple omics domains, including the genome, transcriptome, epigenome, proteome, and metabolome, at a single cell level can provide a holistic understanding of cellular biology that maximizes the impact of crop improvement. Single cell transcriptome analysis has been carried out in several crops including soybean (Libault *et al.*, 2010), Arabidopsis (Shulse *et al.*, 2019), rice (Zhang *et al.*, 2021), maize (Sun *et al.*, 2022), and others. Fluorescence-activated cell sorting (FACS)

Table 2.2. PAM recognition sequence of different Cas9 orthologs. Author's own table.

Cas9	Source	PAM sequence (5′–3′)
SpCas9	*Streptococcus pyrogenes*	NGG
dCas9	Deactivated SpCas9	NGG
Cas9n	SpCas9 nickase	NGG
dCas9-cm	dCas9 fused with cytidine deaminase	NGG
StCas9	*Streptococcus thermophilus*	NNAGAAW
NmCas9	*Neisseria meningitidis*	NNNNGATT
SaCas9	*Staphylococcus aureus*	NNGRRT NNGRR(N)
Cpf1	*Prevoltella and Francisella1*	TTN
dCpf1	Deactivated Cpf1	TTN

is the most widely used method for isolating individual cells. This rapid cell-sorting technique involves staining live cells with fluorescent markers, analyzing them simultaneously, and then separating them. FACS can efficiently distribute single cells into 96- or 384-well plates, enabling pooled sorting of plant protoplasts. Moreover, plant nuclei can be isolated quickly using FACS combined with cell lysis (Borges *et al.*, 2012). Single-cell genomics approaches in plants hold great promise for enhancing our understanding of plant biology in two key areas. First, these approaches can provide insights into the developmental dynamics of plant tissues, enabling the identification of non-anatomical markers for crucial cell populations. Second, they can shed light on plant stress signaling, responses, and adaptation mechanisms.

2.6 Conclusion and Way Forward

Genomics is a multidisciplinary field that employs a comprehensive approach to characterize and quantify key genes within an organism. This involves integrating mapping, editing, functional, and structural studies to elucidate the intricate relationships between genes and their functions. This genomics-driven approach enables the precise identification and manipulation of key genetic factors, ultimately leading to more effective and sustainable crop genetic enhancement strategies. Recent advancements in genomic technologies and molecular marker development have significantly enhanced the efficiency of plant breeding programs. By employing genome-wide association studies and QTL mapping, researchers can accurately identify and tag genes or markers associated with important agronomic and physiological traits. Identification and characterization of genes, along with knowledge of molecular markers associated with them and their physical location on chromosome, offer a powerful means of enhancing crop yields and tailoring stress responses. This can be achieved through the strategic transfer of beneficial genes into popular cultivars using traditional breeding, marker-assisted selection, or genetic engineering techniques coupled with speed breeding. Targeted genome editing has far-reaching implications, allowing investigators to delve into the fundamental principles governing system biology and address various objectives aimed at enhancing agricultural productivity and quality. The integration of next-generation artificial intelligence (AI) in plant biology has facilitated a deeper understanding of the complex mechanisms underlying plant function and response to environmental cues. Additionally, the ability of AI to assimilate and analyze large-scale omics data is instrumental in optimizing speed breeding protocols, enabling the rapid development of crop varieties with improved yield and climate resilience.

Acknowledgements

The author's work in this area is supported by the grants from Rashtriya Uchchatar Shiksha Abhiyan (RUSA), Govt of India and funds from Mahishadal Raj College, West Bengal.

References

Ahmar, S., Gill, R.A., Jung, K.H., Faheem, A., Qasim, M.U. *et al.* (2020) Conventional and molecular techniques from simple breeding to speed breeding in crop plants: Recent advances and future outlook. *International Journal of Molecular Sciences* 21, 2590.

Bellundagi, A., Ramya, K.T., Krishna, H., Jain, N., Shashikumara, P. *et al.* (2022) Marker-assisted backcross breeding for heat tolerance in bread wheat (*Triticum aestivum* L.). *Frontiers in Genetics* 13, 1056783.

Borges, F., Gardner, R., Lopes, T., Carlarco, J.P., Boavida, L.C. *et al.* (2012) FACS-based purification of Arabidopsis microspores, sperm cells and vegetative nuclei. *Plant Methods* 8, 44.

Chen, J., Liu, Y., Liu, M., Guo, W., Wang, Y. *et al.* (2023) Pangenome analysis reveals genomic variations associated with domestication traits in broomcorn millet. *Nature Genetics* 55, 2243–2254.

Demirjian, C., Vailleau, F., Berthomé, R. and Roux, F. (2023) Genome-wide association studies in plant pathosystems: Success or failure? *Trends in Plant Science* 28, 471–485.

Favre, F., Jourda, C., Besse, P. and Charron, C. (2021) Genotyping-by-sequencing technology in plant taxonomy and phylogeny. *Methods in Molecular Biology* 2222, 167–178.

Ferrão, M.A.G., Fonseca, A.F.A., Volpi, P.S., Souza, L.C., Comério, M. *et al.* (2024) Genomic-assisted breeding for climate-smart coffee. *Plant Genome* 17, e20321.

Food and Agriculture Organization (2020) *The state of food security and nutrition in the world 2020*. FAO, Rome.

Gao, G., Li, J., Li, H., Li, F., Xu, K. *et al.* (2014) Comparison of the heat stress induced variations in DNA methylation between heat-tolerant and heat-sensitive rapeseed seedlings. *Breeding Science* 64, 125–133.

Gao, L., Gonda, I., Sun, H., Ma, Q., Bao, K. *et al.* (2019) The tomato pan-genome uncovers new genes and a rare allele regulating fruit flavor. *Nature Genetics* 51, 1044.

Ghosh, S., Watson, A., Gonzalez-Navarro, O.E., Ramirez-Gonzalez, R.H., Yanes, L. *et al.* (2018) Speed breeding in growth chambers and glasshouses for crop breeding and model plant research. *Nature Protocols* 13, 2944–2963.

Golicz, A.A., Bayer, P.E., Barker, G.C., Edger, P.P., Kim, H. *et al.* (2016) The pangenome of an agronomically important crop plant *Brassica oleracea*. *Nature Communication* 7, 13390.

Gordon, S.P., Contreras-Moreira, B., Woods, D.P., Des Marais, D.L., Burgess, D. *et al.* (2017) Extensive gene content variation in the *Brachypodium distachyon* pan-genome correlates with population structure. *Nature Communication* 8, 2184.

Gu, H., Smith, Z.D., Bock, C., Boyle, P., Gnirke, A. *et al.* (2011) Preparation of reduced representation bisulfite sequencing libraries for genome-scale DNA methylation profiling. *Nature Protocols* 6, 468–481.

Gudi, S., Kumar, P., Singh, S., Tanin, M.J. and Sharma, A. (2022) Strategies for accelerating genetic gains in crop plants: Special focus on speed breeding. *Physiology and Molecular Biology of Plants* 28, 1921–1938.

Harfouche, A.L., Jacobson, D.A., Kainer, D., Romero, J.C., Harfouche, A.H. *et al.* (2019) Accelerating climate resilient plant breeding by applying next-generation artificial intelligence. *Trends in Biotechnology* 37, 1217–1235.

He, Q., Tang, S., Zhi, H., Chen, J., Zhang, J. *et al.* (2023) A graph-based genome and pan-genome variation of the model plant *Setaria*. *Nature Genetics* 55, 1232–1242.

Hirsch, C.N., Foerster, J.M., Johnson, J.M., Sekhon, R.S., Muttoni, G. *et al.* (2014) Insights into the maize pan-genome and pan transcriptome. *The Plant Cell* 26, 121–135.

Hossain, F., Jaiswal, S.K., Muthusamy, V., Zunjare, R.U., Mishra, S.J. *et al.* (2023) Enhancement of nutritional quality in maize kernel through marker-assisted breeding for vte4, crtRB1, and opaque2 genes. *Journal of Applied Genetics* 64, 431–443.

Huang, X., Zhao, Y., Wei, X., Li, C., Wang, A. *et al.* (2012) Genome-wide association study of flowering time and grain yield traits in a worldwide collection of rice germplasm. *Nature Genetics* 44, 32–39.

Hurgobin, B., Golicz, A.A., Bayer, P.E., Chan, C.-K.K., Tirnaz, S. *et al.* (2018) Homoeologous exchange is a major cause of gene presence/absence variation in the amphidiploid *Brassica napus*. *Plant Biotechnology Journal* 16, 1265–1274.

Ishino, Y., Shinagawa, H., Makino, K., Amemura, M. and Nakata, A. (1987) Nucleotide sequence of the iap gene, responsible for alkaline phosphatase isozyme conversion in *Escherichia coli*, and identification of the gene product. *Journal of Bacteriology* 169, 5429–5433.

Jaiswal, V., Bandyopadhyay, T., Singh, R.K., Gahlaut, V., Muthamilarasan, M. *et al.* (2023) Multi-environment GWAS identifies genomic regions underlying grain nutrient traits in foxtail millet (*Setaria italica*). *Plant Cell Reports* 43, 6.

Junaid, A., Kumar, H., Rao, A.R., Patil, A.N., Singh, N.K. *et al.* (2018) Unravelling the epigenomic interactions between parental inbreds resulting in an altered hybrid methylome in pigeonpea. *DNA Research* 25, 361–373.

Khan, A.W., Garg, V., Sun, S., Gupta, S., Dudchenko, O. *et al.* (2024) Cicer super-pangenome provides insights into species evolution and agronomic trait loci for crop improvement in chickpea. *Nature Genetics* 56, 1225–1234.

Kim, Y.G., Cha, J. and Chandrasegaran, S. (1996) Hybrid restriction enzymes: Zinc finger fusions to Fok I cleavage domain. *Proceedings of the National Academy of Sciences USA* 93, 1156–1160.

Kumar, V., Singh, B., Kumar Singh, R., Sharma, N., Muthamilarasan, M. *et al.* (2024) Histone deacetylase 9 interacts with SiHAT3.1 and SiHDA19 to repress dehydration responses through H3K9 deacetylation in foxtail millet. *Journal of Experimental Botany* 75, 1098–1111.

Li, X., Zhu, J., Hu, F., Ge, S., Ye, M. *et al.* (2012) Single-base resolution maps of cultivated and wild rice methylomes and regulatory roles of DNA methylation in plant gene expression. *BMC Genomics* 13, 300.

Libault, M., Farmer, A., Brechenmacher, L., Drnevich, J., Langley, R.J. *et al.* (2010) Complete transcriptome of the soybean root hair cell, a single-cell model, and its alteration in response to *Bradyrhizobium japonicum* infection. *Plant Physiology* 152, 541–552.

Liu, G., Lin, Q., Jin, S. and Gao, C. (2022) The CRISPR-Cas toolbox and gene editing technologies. *Molecular Cell* 82, 333–347.

Liu, H., Bayer, M., Druka, A., Russell, J.R., Hackett, C.A. *et al.* (2014) An evaluation of genotyping by sequencing (GBS) to map the *Breviaristatum-e* (*ari-e*) locus in cultivated barley. *BMC Genomics* 15, 104.

Liu, Y., Du, H., Li, P., Shen, Y., Peng, H. *et al.* (2020) Pan-genome of wild and cultivated soybeans. *Cell* 182, 162–176.

Mishra, S., Nayak, S., Tuteja, N., Poosapati, S., Swain, D.M. *et al.* (2024) CRISPR/Cas-mediated genome engineering in plants: Application and prospectives. *Plants (Basel)* 13, 1884.

Mohanavel, V., Muthu, V., Kambale, R., Palaniswamy, R., Seeli, P. *et al.* (2024) Marker-assisted breeding accelerates the development of multiple-stress-tolerant rice genotypes adapted to wider environments. *Frontiers in Plant Science* 15, 1402368.

Montenegro, J.D., Golicz, A.A., Bayer, P.E., Hurgobin, B., Lee, H.T. *et al.* (2017) The pangenome of hexaploid bread wheat. *Plant Journal* 90, 1007–1013.

Pandey, S., Singh, A., Parida, S.K. and Prasad, M. (2022) Combining speed breeding with traditional and genomics-assisted breeding for crop improvement. *Plant Breeding* 141(3), 301–313.

Pootakham, W. (2023) Genotyping by sequencing (GBS) for genome-wide SNP identification in plants. *Methods in Molecular Biology* 2638, 1–8.

Qin, P., Lu, H., Du, H., Wang, H., Chen, W. *et al.* (2021) Pan-genome analysis of 33 genetically diverse rice accessions reveals hidden genomic variations. *Cell* 184, 3542–3558.

Rai, K.K. (2022) Integrating speed breeding with artificial intelligence for developing climate-smart crops. *Molecular Biology Reports* 49, 11385–11402.

Ranawaka, B., Tanurdzic, M., Waterhouse, P. and Naim, F. (2020) An optimised chromatin immunoprecipitation (ChIP) method for starchy leaves of *Nicotiana benthamiana* to study histone modifications of an allotetraploid plant. *Molecular Biology Reports* 47, 9499–9509.

Ruperao, P., Thirunavukkarasu, N., Gandham, P., Selvanayagam, S., Govindaraj, M. *et al.* (2021) Sorghum pan-genome explores the functional utility for genomic-assisted breeding to accelerate the genetic gain. *Frontiers in Plant Science* 12, 963.

Satam, H., Joshi, K., Mangrolia, U., Waghoo, S., Zaidi, G. *et al.* (2023) Next-generation sequencing technology: Current trends and advancements. *Biology (Basel)* 12, 997.

Shang, L., Li, X., He, H., Yuan, Q., Song, Y. *et al.* (2022) A super pan-genomic landscape of rice. *Cell Research* 32, 878–896.

Shulse, C.N., Cole, B.J., Ciobanu, D., Lin, J., Yoshinaga, Y. *et al.* (2019) High-throughput single-cell transcriptome profiling of plant cell types. *Cell Reports* 27, 2241–2247.

Singh, R.K., Jaishankar, J., Muthamilarasan, M., Shweta, S., Dangi, A. *et al.* (2016) Genome-wide analysis of heat shock proteins in C4 model, foxtail millet identifies potential candidates for crop improvement under abiotic stress. *Science Reports* 6, 32641.

Singh, R.K., Prasad, A., Muthamilarasan, M., Parida, S.K. and Prasad, M. (2020) Breeding and biotechnological interventions for trait improvement: Status and prospects. *Planta* 252, 54.

Singh, R.K., Muthamilarasan, M. and Prasad, M. (2021) Biotechnological approaches to dissect climate-resilient traits in millets and their application in crop improvement. *Journal of Biotechnology* 327, 64–73.

Singh, R.K., Dhaka, A., Muthamilarasan, M. and Prasad, M. (2022a) Nutritional improvement of cereal crops to combat hidden hunger during COVID-19 pandemic: Progress and prospects. *Advances in Food Security and Sustainability* 7, 61–82.

Singh, U.M., Dixit, S., Alam, S., Yadav, S., Prasanth, V.V. *et al.* (2022b) Marker-assisted forward breeding to develop a drought-, bacterial-leaf-blight-, and blast-resistant rice cultivar. *The Plant Genome* 15, e20170.

Sood, P., Pandey, S. and Prasad, M. (2021) Biotechnological advances and advanced mutation breeding techniques for crop plants. In: Kumar, P. and Kumar Thakur, K. (eds) *Crop Improvement*. CRC Press, Boca Raton, pp. 139–165.

Sun, G., Xia, M., Li, J., W., M.A., Li, Q. *et al*. (2022) The maize single-nucleus transcriptome comprehensively describes signaling networks governing movement and development of grass stomata. *Plant Cell* 34, 1890–1911.

Tettelin, H., Masignani, V., Cieslewicz, M.J., Donati, C., Medini, D. *et al*. (2005) Genome analysis of multiple pathogenic isolates of Streptococcus agalactiae: Implications for the microbial "pan-genome". *Proceedings of the National Academy of Sciences USA* 102(39), 13950–13955.

Varshney, R.K. (2016) Exciting journey of 10 years from genomes to fields and markets: Some success stories of genomics-assisted breeding in chickpea, pigeonpea and groundnut. *Plant Science* 242, 98–107.

Varshney, R.K., Roorkiwal, M., Sun, S., Bajaj, P., Chitikineni, A. *et al*. (2021) A chickpea genetic variation map based on the sequencing of 3,366 genomes. *Nature* 599, 622–627.

Vernikos, G., Medini, D., Riley, D.R. and Tettelin, H. (2015) Ten years of pan-genome analyses. *Current Opinion in Microbiology* 23, 148–154.

Wang, W., Mauleon, R., Hu, Z., Chebotarov, D., Tai, S. *et al*. (2018) Genomic variation in 3,010 diverse accessions of Asian cultivated rice. *Nature* 557, 43–49.

Xiao, Q., Bai, X., Zhang, C. and He, Y. (2021) Advanced high-throughput plant phenotyping techniques for genome-wide association studies: A review. *Journal of Advanced Research* 35, 215–230.

Yocca, A.E. and Edger, P.P. (2022) Machine learning approaches to identify core and dispensable genes in pangenomes. *Plant Genome* 15, e20135.

Zampieri, E., Volante, A., Marè, C., Orasen, G., Desiderio, F. *et al*. (2023) Marker-assisted pyramiding of blast-resistance genes in a *Japonica* elite rice cultivar through forward and background selection. *Plants (Basel)* 12, 757.

Zhang, T.Q., Chen, Y., Liu, Y., Lin, W.H. and Wang, J.W. (2021) Single-cell transcriptome atlas and chromatin accessibility landscape reveal differentiation trajectories in the rice root. *Nature Communications* 12, 2053.

Zhao, J., Bayer, P.E., Ruperao, P., Saxena, R.K., Khan, A.W. *et al*. (2020) Trait associations in the pangenome of pigeon pea (*Cajanus cajan*). *Plant Biotechnology Journal* 18, 1946–1954.

Zhu, H., Li, C. and Gao, C. (2020) Applications of CRISPR-Cas in agriculture and plant biotechnology. *Nature Reviews Molecular Cell Biology* 21, 661–677.

3 Transcriptomics Approaches for Understanding the Dynamics of Climate-resilient Traits

Aaqif Zaffar[1], Abhinav[2], Suman Dutta[3], Vishal Balasaheb Mhetre[4], Reena Prusty[5], Ritwika Das[6], Soumya Sharma[6], Abhishek Chitranshi[7], Amarjeet Kumar[8], Mohammad Irfan[5], Alka Bharati[9] and Vishal Dinkar[10]*

[1]Division of Genetics & Plant Breeding, SKUAST – Kashmir, Faculty of Agriculture, Wadura, Sopore, India; [2]Department of Biotechnology, Motilal Nehru National Institute of Technology, Prayagraj, India; [3]IRDM Faculty Centre, Ramakrishna Mission Vivekananda Educational and Research Institute, Kolkata, India; [4]Defence Institute of High Altitude Research, Leh, India; [5]ICAR-Central Institute of Temperate Horticulture (CITH), Srinagar, India; [6]Division of Bioinformatics, ICAR-Indian Agricultural Statistics Research Institute (IASRI), New Delhi, India; [7]ICAR-Central Island Agricultural Research Institute, Sri Vijaya Puram, India; [8]Department of Genetics and Plant Breeding, College of Agriculture (CAU, Imphal), Kyrdemkulai, India; [9]ICAR-Research Complex for NEH Region, Umiam, India; [10]ICAR-Indian Institute of Pulses Research (IIPR), Kanpur, India

Abstract

Climate change poses a significant threat to global agriculture, necessitating innovative strategies to ensure food security. Transcriptomics, the study of RNA transcripts under specific conditions, emerges as a pivotal tool for unravelling gene expression patterns that govern stress tolerance. The chapter delves into the role of transcriptomic approaches in understanding the dynamics of climate-resilient traits in crops. It discussed a comprehensive overview of transcriptomic techniques such as RNA-Seq, microarrays, and single-cell RNA-Seq, highlighting their applications in identifying candidate genes associated with traits like drought and heat tolerance. The integration of transcriptomics with other 'omics' technologies, including proteomics and metabolomics, is explored, showcasing its potential to elucidate complex gene–environment interactions. Case studies illustrate how transcriptomic data have accelerated breeding programs for climate-smart crops by identifying key regulatory pathways and stress-responsive genes. Advanced bioinformatics and machine learning applications further enhance data analysis, enabling precise identification of genetic targets for crop improvement. Here, we emphasized the transformative impact of transcriptomics in breeding climate-resilient crops. By leveraging these approaches, researchers and breeders can develop sustainable agricultural systems capable of withstanding the challenges posed by a rapidly changing climate. The insights gained underscore the importance of transcriptomics as a cornerstone of modern agricultural innovation.

*Corresponding author: vishaldinkar1991@gmail.com

 From Gene Discovery to Climate-Resilient Crops: Omics in Crop Improvement (eds S. Pandey *et al.*)
DOI: 10.1079/9781800628717.0003

3.1 Introduction

The agriculture industry has been threatened by numerous factors created directly or indirectly by humans, making the most vulnerable situation for humankind. Among these, extreme climatic fluctuations stand as a major threat posing profound implications on environmental sustainability, farmer livelihoods, and food security affecting the world economy. According to the United Nations Framework Convention on Climate Change (UNFCCC), climate change is defined as, 'any change in the composition of the global atmosphere that can be directly or indirectly attributed to human activity' (Mahato, 2014). It implies different climate emergencies such as rising temperatures, erratic weather patterns, and extremity in climatic events (droughts, floods, fires, hurricanes, cyclones, etc.) adversely impacting crop health, yield and productivity from the last many decades (Pandey and Singh, 2024; Singh *et al.*, 2025). Each 1°C increase in global temperatures could reduce yields of staple crops like corn (maize), soybeans, and wheat by 16–20%, with a significant decline in farm income and resilience against economic shocks (Gashler, 2024). As per the Indian Council of Agricultural Research (ICAR), forced maturity in states like Andhra Pradesh, Punjab, and Rajasthan has resulted in increased crop duration and high evaporative demand, which have caused farming to use up to 30% more water (Sahu *et al.*, 2020). Due to a lack of adequate cold weather, even apple orchards in Himachal Pradesh have shifted to a higher elevation (Sahu *et al.*, 2020). The Intergovernmental Panel on Climate Change (IPCC) has released its sixth assessment report stated that over the next ten years, the global mean temperature will surpass the 1.5°C threshold, and in 2040–2060, it will surpass the 2°C threshold (Koll, 2021). According to a Ministry of Agriculture estimate, between 2015–16 and 2021–22, floods caused damage to over 340 lakh hectares of cultivated land, while drought caused an equivalent loss of cropped land. The farming community, particularly the smallholders who make up more than 80% of India's farmers, are in a perilous situation (Reddy *et al.*, 2025). This poses a question mark on the production of increasing demand of agricultural produce to feed the resurgent population.

The answer lies in the adoption of climate-smart/resilient agriculture approaches. A route to development and food security, climate-smart agriculture is based on three pillars – raising incomes and productivity, strengthening ecosystems and livelihoods' resilience, and lowering and eliminating greenhouse gas emissions from the environment (FAO, 2011). Adoption of climate-resilient crop varieties, precision farming, agroforestry, plant genetic resource (PGR) conservation, livestock, pastoral, grazing, forestry, soil and water management strategies are central to reducing vulnerability and improving resilience in the face of longer-term climate stressors (Safdar *et al.*, 2024). The National Mission for Sustainable Agriculture (NMSA) focuses on climate-resilient farming, sustainable land and water use, and efficient resource practices. To address local challenges, India has developed more than 2000 high-yielding, climate-resilient crop varieties resistant to various stresses that ensures food security and sustainability. Government promotes climate-smart mechanization to improve efficiency, lower costs, and support eco-friendly practices like residue management. These measures, supported by substantial government investment and policy backing, demonstrate India's commitment to building climate resilience in agriculture while addressing the economic and ecological challenges posed by climate change.

Amid these challenges, transcriptomics offers a powerful tool for advancing crop resilience and productivity. Transcriptomics, the study of RNA transcripts produced by the genome under specific conditions, provides detailed insights into gene expression patterns. By understanding how crops respond to environmental stressors such as drought, salinity, and heat, researchers can identify key genes and regulatory networks that drive resilience. Because the transcriptome varies depending on several factors (such as developmental stage, tissue type, environmental stressors, etc.), it is a promising molecular level for investigating an organism's stress responses in contrast to the genome, which is stable (Escandón *et al.*, 2021). For instance, transcriptomic analysis enables the identification of different stress-tolerant genes in staple crops like rice and maize. These research efforts have revealed genes and metabolic pathways that can be targeted in marker-assisted breeding (MAB) initiatives and for creating variations through selective breeding or genetic engineering that flourish in increasingly erratic climates. Moreover, transcriptomics facilitates high-throughput screening of germplasm, accelerating the breeding process for climate-smart crop varieties.

Emerging research has demonstrated the integration of transcriptomics with other 'omics' technologies, such as proteomics, metabolomics, ionomics and phenomics to create a holistic understanding of crop stress responses (Wang *et al.*, 2014a). This systems biology approach helps identify complex gene–environment interactions, paving the way for precision agriculture and targeted interventions. Presently, the idea of DNA-free genome editing also uses CRISPR-ribonucleoprotein complexes to create transgenic organisms instead of DNA. Therefore, knowledge of this broad field is crucial for not only determining a gene's function but also for providing answers to basic biological queries that eventually help us comprehend an organism's phenomics (Bhattacharjee, 2021). Furthermore, advancements in bioinformatics and machine learning have streamlined the analysis of large transcriptomic datasets, making it easier to translate these findings into practical agricultural innovations. This chapter will delve deeper into these themes, emphasizing the intersection of climate science and cutting-edge transcriptomic research as pivotal in reshaping the future of agriculture.

3.2 Overview of Transcriptomics

3.2.1 Definition and scope

Transcriptomics provides a snapshot of gene expression and regulation through the in-depth analysis of RNA transcripts within a cell or organism at a particular moment or circumstance (Han *et al.*, 2015). The area of study is comprehending the process of gene transcription into RNA and the degradation and processing of these transcripts. Transcriptomics shows dynamic variations in gene activity in response to internal or external stimuli, in contrast to genomics, which offers static information about DNA (Massaia *et al.*, 2018). Transcriptomics encompasses the study of transcript types, amounts, and modifications using a variety of methods, including microarrays and RNA sequencing (RNA-Seq). These technologies provide insights into cellular activities, disease causes, and developmental processes by allowing researchers to detect genes with variable expression, alternative splicing events, and RNA alterations (Dalal *et al.*, 2020). With applications in biotechnology, medicine, and agriculture, transcriptomics is an essential tool for finding biomarkers, deciphering complicated features, and creating treatment plans.

3.2.2 Historical development and technological advances

The historical development of sequencing saw first-generation platforms sequencing for clonal DNA populations (Shendure *et al.*, 2017). Second-generation platforms increased throughput by parallelizing reactions, revolutionizing large-scale genomic studies. Third-generation platforms advanced further by enabling direct sequencing of single DNA molecules, improving accuracy, read length, and enabling real-time sequencing with minimal sample preparation. When Walter Fiers and associates sequenced the whole transcriptome of the MS2 bacteriophage (3569 nucleotides) in the 1970s, they initiated the field of RNA-Seq research (Crowgey and Mahajan, 2019). Because RNA is single-stranded, it is extremely unstable and easily broken down by RNases that are linked to cell membranes. Consequently, it will take a lot of work to precisely sequence the whole transcriptome. The discovery of reverse transcriptase made it possible to convert messenger and non-coding RNA into stabilized DNA (Zhang, 2019). Complementary DNA (cDNA) is the term used to describe the reverse-synthesized DNA. DNA sequencing technique was created in 1975 by British scientist Sanger

(Sanger *et al.*, 1977; Zhang, 2019). Iscove (Hwang *et al.*, 2018) amplified cDNA exponentially with the use of RNA-Sequencing PCR. Later on, microarray (chip) technology was created as a result of these innovative efforts (Bolón-Canedo *et al.*, 2019). Molecular hybridization techniques may be used to mount hundreds of known partial DNA sequences on nylon membranes or solid support slides using arrayed technology. Hybridization allows for the quantitative detection of several genes. Microarrays are a great tool for investigating gene function since the transcriptomes of various biological samples can be quickly sequenced using reverse synthesis of cDNA (Clark *et al.*, 2002). Currently, known genes are studied using this technique more often than undiscovered ones. The microchip also has trouble identifying a large number of transcripts produced by alternative splicing.

Sanger sequencing is an expensive, time-consuming, and labor-intensive method. However, since its introduction in 2006, next-generation sequencing (NGS) has satisfied several crucial demands for molecular biologists. Fast, high-throughput, and affordable, NGS technology outperforms first-generation sequencing technology. NGS makes it feasible to sequence billions or even millions of nucleic acids at once, which makes it possible to analyze the transcriptomes and genomes of every species. Currently, NGS is employed to create organism genomes to gather the entire gene sequences of different plant species (Bräutigam and Gowik, 2010). Most of the time, NGS is used to sequence and examine transcriptome-derived messenger RNA (mRNA) and small RNA (small RNA). The Roche 454, Illumina Solexa, and ABI SOLiD are three well-known NGS systems (Zhang *et al.*, 2016). Based on the pyro-sequencing approach, Roche launched the high-throughput Roche 454 sequencing technology in 2005 (Margulies *et al.*, 2005). Compared to other NGS methods, 454 sequencing has a greater read length and a shorter run time. Nucleic acid probes are not needed for this non-fluorescent method. Nevertheless, due to the sequencing process throughout, it may add mistakes such deletions or insertions (Liang *et al.*, 2017). DNA fragments are randomly joined on a flow cell according to the 'sequencing by synthesis' idea, which is the foundation of Solexa's (Illumina) methodology. The surface of the glass produces millions of clusters made up of thousands of similar DNA fragments after extension and amplification. On the stretched DNA strand, labeled dNTPs (fluorescence) are sequenced. According to Liu *et al.* (2012), this method enables the employment of synthetic probes and reference sequences in genome-wide expression analysis. The platform has several shortcomings, such as its shorter read duration and the difficulty of piecing together reads from the start. Solid sequencing uses extremely parallel data sequencing using magnetic beads. Because of ongoing ligation and the creation of fluorescently labeled oligonucleotides, this method enables large DNA amplification without resulting in significant expenses (Hedges *et al.*, 2011). Although the main advantage of the technology is its great precision, its main drawback is it is vulnerable to chain decoding errors. Improvements in sequencing technologies have made third-generation sequencing possible. Template displacement, mismatches in nucleotides and GC percentages, and amplification needs specific to NGS platforms are typical obstacles associated with these methods. Sequencing data are less accurate and comprehensive as a result of these difficulties. Third-generation sequencing has been extensively employed in structural variation, methylation, transcriptome, and genome research, among other applications, since it provides such lengthy reads. Recently, two important technologies for third-generation sequencing have emerged: nanopore sequencing and single-molecule real-time (SMRT) sequencing (Ma *et al.*, 2021). The SMRT sequencing technique makes use of the 'synthesis-based principle.' When opposed to traditional techniques, SMRT technology offers two distinct advantages. To remove the background noise, the fluorescent group was then added and linked to the phosphoric acid (H_3PO_4) group. A further advantage is that the SMRT self-corrects, eliminating the requirement for amplification and enabling more accurate readings (Li *et al.*, 2018). However, there are several drawbacks to the SMRT sequencing technique, one of which is the incorporation of random mistakes. Compared to second-generation sequencing, the accuracy of a single read may now be increased to 99.8% by employing circular consensus sequencing (CCS) and enhancing sequence coverage (Wenger *et al.*, 2019). With Nanopore, an advanced SMRT sequencing device, variations in electrical impulses are primarily used to ascertain the base composition. Compared to other systems, nanopore sequencing is more affordable and provides longer reads. But the biggest drawback of the technology is its high mistake rate, which can vary from 5 to 20% (Huang *et al.*, 2021). Notwithstanding its benefits, there are many drawbacks, such as the increased likelihood of random and single-base mistakes.

Without a reference sequence, using the above discussed methods may be prohibitively expensive and time-consuming. An essential component of NGS technology and a key instrument for transcriptome study is RNA-Seq. It gets over the drawbacks of microarray analysis and advances our knowledge of transcriptome research.

3.2.3 Transcriptomics vis-à-vis other omics approaches

Proteomics, transcriptomics, and genomes are complementing omics techniques that delve into several biological levels (Tolani *et al.*, 2021). A static map of possible gene functions and genetic information is provided by genomics, which focuses on the complete collection of DNA sequences, or the genome (Dutta *et al.*, 2024a). It does not disclose gene activity or expression levels, but rather gene structure, variants, and evolutionary trends. On the other hand, transcriptomics uses RNA transcript analysis to investigate the dynamic nature of gene expression (Wang *et al.*, 2014b). It provides information on which genes are actively expressed in a given situation and provides context for understanding regulatory mechanisms, environmental reactions, and temporal fluctuations in gene activity. In contrast, proteomics looks at a cell, tissue, or whole protein complement of an organism (Patterson and Aebersold, 2003). It provides information on the functions, relationships, and signaling pathways of proteins by capturing the ultimate functional products of gene expression and post-translational modifications (Graves and Haystead, 2002). Transcriptomics and proteomics offer a more in-depth, current understanding of how genetic information is produced and used in living systems, while genomics lays the groundwork.

3.3 Transcriptomic Techniques

3.3.1 RNA sequencing (RNA-Seq)

RNA Sequencing (RNA-Seq) has transformed agricultural research by giving a complete and high-resolution picture of gene expression, revealing insights into plant growth, stress responses, and crop improvement (Roy *et al.*, 2024). RNA-Seq is an effective technique for comprehending intricate gene regulation networks because it enables precise measurement of transcript levels, identification of new transcripts, and detection of alternative splicing processes (Costa *et al.*, 2010). This technique has been used in crops to investigate a variety of processes, including blooming, abiotic and biotic stress tolerance, seed development, and efficient nutrient usage (Sahraeian *et al.*, 2017; Dutta *et al.*, 2021). For instance, major crops like rice, maize, and wheat have had their molecular processes of heat and drought tolerance uncovered by RNA-Seq, which has also identified important transcription factors (TFs) and signaling pathways that impart resilience (Dutta *et al.*, 2019a; Das *et al.*, 2022). Researchers have identified DEGs linked to adaptive features by comparing transcriptomes under stress and non-stress settings (Evans, 2015). This has made MAB and transgenic techniques easier to implement. RNA-Seq data combined with genomic selection models have improved the ability of crop improvement programs to predict attributes like yield and quality (Dutta *et al.*, 2019b; Das *et al.*, 2023; Maman *et al.*, 2023). Additionally, RNA-Seq has aided in the investigation of plant–pathogen interactions by exposing transcripts linked to plant defense as well as pathogen specificity, which is essential for creating cultivars that are resistant to disease (Nejat *et al.*, 2017). Furthering our understanding of post-transcriptional regulation in crops is the capacity of RNA-Seq to analyze both coding and non-coding RNAs, including long non-coding RNAs (lncRNAs) and microRNAs (miRNAs) (Wang *et al.*, 2017a, b). These tiny regulatory RNAs have a significant influence on crop development and stress responses through their critical functions in translational inhibition, mRNA degradation, and gene silencing (Li *et al.*, 2017). With the development in single-cell RNA sequencing (scRNA-Seq), scientists can now investigate the dynamics of gene expression at the cellular level, providing previously unattainable insights into the tissue-specific expression patterns in crops (Bawa *et al.*, 2022). The discovery of important metabolic pathways and the creation of comprehensive models for plant systems biology have resulted from the integration of RNA-Seq data with other omics platforms, such as proteomics and metabolomics (Wanichthanarak *et al.*, 2015). Moreover, RNA-Seq has become widely used in agronomic research due to the advancement of high-throughput sequencing technology and decreased sequencing prices. These factors have also made RNA-Seq more accessible to agricultural scientists. Building transcriptome databases particular to crops has been made possible by data produced by RNA-Seq, which has aided in the process of finding and annotating genes (Costa *et al.*, 2010). Additionally, *in silico* methods for RNA-Seq data processing have developed, allowing scientists to quickly process and understand big datasets and identify discoveries that have biological significance. Moreover, transgene validation which verifies the expression of inserted genes in genetically modified crops can benefit from the use of RNA-Seq (Pauwels *et al.*, 2015). It is now feasible to study low-abundance transcripts that are essential for certain agricultural features because of the ongoing improvement of RNA-Seq procedures, which include strand-specific sequencing and ribosomal RNA depletion (Brown *et al.*, 2017). All things considered, RNA-Seq has emerged as a crucial technique in crop genomics, speeding up our knowledge of plant biology and assisting in the creation of better crop varieties that are more resilient, productive, and sustainable in the face of international agricultural difficulties.

3.3.2 Single-cell RNA sequencing (scRNA-Seq)

Single-cell transcriptomics analysis measures gene expression at the cell level, enabling scientists to make a concentration measurement (usually mRNA) of hundreds to thousands of genes within a population (Shaw *et al.*, 2021). The method thus has opportunities for the study of heterogeneous cell populations; reconstruction of cellular developmental pathways; and modeling of transcriptional dynamics that often remain hidden in bulk RNA sequencing. Single-cell RNA sequencing (scRNA-Seq) is based on the idea that the population of RNAs is transcribed into a library of complementary DNA fragments, which can then be sequenced using high-throughput next-generation sequencing. Sequence reads are mapped to a reference genome to provide gene-specific read counts. Since cell-to-cell variation is present during the formation of cDNA libraries and sequencing itself, normalization techniques are applied. One approach involves the use of RNA spike-ins: a known RNA sequence added in equal amounts to every cell lysate that help normalize the read count with comparative spike-in mRNA reads (Lun *et al.*, 2017). Another approach makes use of unique molecular identifiers (UMIs) that are very short DNA sequences tacked onto each cDNA before amplification as barcodes (Girardot *et al.*, 2016). The technique makes counts of unique UMIs tied to each gene and considers variations in the efficiencies at which amplifications are carried out. Spike-ins along with UMIs and other methods may further be used to improve normalization of the count.

3.3.3 Microarrays (hybridization-based transcriptomics)

Microarrays remain a pivotal tool in transcriptomics for climate-resilient crop breeding by profiling gene expression under various stress conditions, including drought, salinity, and heat. They are a proven cost-effective breeder-friendly genomic resource for gene/quantitative trait loci (QTL) mapping, genome-wide association study (GWAS) and diversity assessment for various agronomic features related to biotic and abiotic stress tolerance, particularly in non-model crops (Dinkar *et al.*, 2020; Verma *et al.*, 2020; Devate *et al.*, 2022; Ahmed *et al.*, 2023; Ragini *et al.*, 2024; Dinkar *et al.*, 2024). Several crop-specific microarrays have been developed to facilitate research and breeding for climate resilience and other traits in plants. These include the very first single nucleotide polymorphism (SNP) arrays, i.e.Affymetrix Arabidopsis ATH1 GeneChip® contain about 22,750 probe sets represents about 23,750 genes (Redman *et al.*, 2004). Among cereals, the Rice 50 K chip represents 50,051 SNPs from 18,980 genes (Singh *et al.*, 2015), the most recent 580 K Axiom Rice Genotyping Chip (580 K_KNU chip) represents 581,006 high-quality SNPs by integrating data from the Korean World Rice Collection (KRICE), 3 K-RGP data, 700 K HD rice array, Affymetrix 44 K SNP array, QTARO, Reactome, and plastid and GMO information (Kim *et al.*, 2022), and the Rice Pangenome Genotyping Array (RPGA) containing 80K SNP probes and presence/absence variants covering the entire 3K rice pangenome (Daware *et al.*, 2023). Winfield *et al.* (2012) used targeted exome capture and re-sequencing protocol based on NimbleGen array technology and characterized 56.5 Mb of genomic DNA, which resulted in the development of the high-density 820K Axiom® genotyping array contains 819,571 SNPs (Winfield *et al.*, 2016) and its breeder-friendly subset Wheat Breeders' 35K SNP Array (Allen *et al.*, 2017). Likewise, the 15K array comprises the common set of 12,806 SNP markers from 90K arrays, making it amenable for cost-effective wheat improvement programs (Soleimani *et al.*, 2020). Moreover, for wheat, Wheat SNP iSelect array 90K (Wang *et al.*, 2014b), Affymetrix Wheat 660K SNP array (Cui *et al.*, 2017), and TaBW280K high-throughput genotyping array containing 280,226 SNPs (Rimbert *et al.*, 2018) are also available. Likewise, the Oat 6K BeadChip design containing 5743 SNPs (Tinker *et al.*, 2014), Maize 600K Affymetrix Axiom® Genotyping Array comprising 609,442 SNPs and 6759 indels (Unterseer *et al.*, 2014) and Rye high-density 600K SNP array containing 600,843 SNPs (Bauer *et al.*, 2017) are available for breeders.

Studies using rice (*Oryza sativa*) under salinity stress identified specific QTLs for grain yield under salt stress and gene families associated with HKT ion transport and osmotic balance, critical for tolerance (Mishra *et al.*, 2016; Tiwari *et al.*, 2016). Similarly, wheat SNP microarrays are very popular among breeders and have shown their utility in nearly all type of breeding programs. Pang *et al.* (2022) identified three marker-trait associations (MTAs) using the 55K SNP array for waterlogging tolerance in wheat. Likewise, Ding *et al.* (2023) identified three stable QTLs with pleiotropic and additive effects for post-anthesis waterlogging tolerance in wheat using a 660K SNP array. Ahmed *et al.* (2023) performed GWAS using a 90K array and identified a stable and consistent locus (*Excalibur_c20796_395* on chromosome 7A), associated grain morphology and yield-related traits under drought and heat-stress environments. Lei *et al.* (2023) used wheat line H261 and its two EMS mutants (LF2099 and XiNong 239) for biparental mapping and identified SNPs from a 660K SNP array for two novel cold-resistance genes.

Among horticultural crops, the Peach Infinium® 9K SNP array (Verde *et al.*, 2012), Grape (Vitis) 9K SNP Array containing 8898 SNPs (Myles *et al.*, 2011) and 18K chip comprising 18,775 SNPs (De Lorenzis *et al.*, 2015), Apple 20 K Illumina Infinium SNP array comprising 18,019 SNPs (Bianco *et al.*, 2014), Rose WagRhSNP 68K Axiom SNP array containing 68,893 SNPs including information from 13,390 expressed full-length genes (Koning-Boucoiran *et al.*, 2015), and Potato 20k Infinium SNP array (SolSTW array) harboring 19,592 SNPs (Vos *et al.*, 2015) are available. Among legumes, the Illumina Cowpea iSelect Consortium Array represents 51,128 SNPs (Muñoz-Amatriaín *et al.*, 2017), the Axiom®*CicerSNP* 50K Array comprises 50 590 SNPs (Roorkiwal *et al.*, 2018), the Pigeonpea 62 K SNP chip array represents 62,053 SNPs from 9629 genes (Singh *et al.*, 2020), the Groundnut high-density 58 K SNP array (Axiom_*Arachis*) has 163,782 SNPs representing 58,233 unique and informative SNPs (Pandey *et al.*, 2017), the SoySNP50K iSelect BeadChip comprises 52,041 SNPs (Song *et al.*, 2013) and the low-cost high-efficiency BARC SoySNP6K array (Song *et al.*, 2020) is also available in soybean.

The utility of microarrays is enhanced by integrating them with computational tools for pathway analysis and data interpretation. For example, combining microarray data with Gene Ontology (GO) and KEGG pathway analyses has enabled the identification of enriched pathways related to stress responses. Furthermore, hybridization-based methods have been effectively applied to pinpoint QTLs for stress tolerance, which accelerates MAB programs. Additionally, the integration of microarray-derived data with proteomics can enable a comprehensive understanding of stress adaptation mechanisms. These tools continue to be pivotal in crop genomics, bridging knowledge gaps in gene function and enabling the development of improved, climate-resilient cultivars. For further details, one should explore technology providers like Affymetrix and CapitalBio. Despite being somewhat eclipsed by RNA sequencing, microarrays offer unparalleled reliability in targeted experiments and remain indispensable in initial screenings for large populations. With innovations in bioinformatics and hybrid technologies like the Nanostring platform, microarrays continue to be a cornerstone in breeding strategies for climate-resilient agriculture. Their role in addressing food security amidst global climate challenges is both significant and indispensable.

3.3.4 Quantification and validation of candidate gene expression

Understanding gene function and its significance in certain biological processes depends on the quantification and validation of candidate gene expression (Cookson *et al.*, 2009). Gene expression levels are often measured using methods like Northern blotting, RNA-Seq, and quantitative PCR (qPCR) (Nonis *et al.*, 2014). Because of its precision and sensitivity, qPCR is very helpful for verifying gene expression. During this procedure, candidate genes found using high-throughput techniques are examined in further detail to verify that they express themselves differently in diverse settings (Vieira *et al.*, 2016). While validation aids in confirming gene function and contributes to breakthroughs in functional genomics and crop enhancement research, proper normalization using reference genes assures dependable findings.

3.3.4.1 Quantitative real-time PCR (qRT-PCR)

A potent method for measuring gene expression levels in real time during the PCR amplification process is quantitative real-time PCR, or qRT-PCR (Jozefczuk and Adjaye, 2011). By combining conventional PCR with fluorescence detection, it makes it possible to estimate the amount of DNA or RNA amplified in a sample. Generally, complementary DNA (cDNA) is created by reverse transcription of RNA (Provenzano and Mocellin, 2007). This cDNA is then amplified using certain primers. Target gene quantification is made possible by the correlation between the rise in fluorescence and the quantity of PCR product produced. Because of its exceptional sensitivity and capacity to identify transcripts with low abundance, qRT-PCR is a vital tool for a wide range of applications, such as gene expression profiling, microarray result validation, and evaluating treatment effects in experimental and clinical trials (Singh and Roy-Chowdhuri, 2016; Mani *et al.*, 2021). Reliable outcomes depend on proper normalization and experimental design.

3.3.4.2 Serial analysis of gene expression (SAGE)

A high-throughput method called serial analysis of gene expression (SAGE) provides a detailed picture of the transcript levels in a sample, allowing for the analysis of gene expression (Velculescu *et al.*, 1995). In SAGE, the 3′ end of mRNA transcripts are used to create short sequence tags, which are subsequently concatenated to produce long sequences. Thousands of genes may be concurrently identified and quantified thanks to the cloning and sequencing of these concatenated sequences (Yamamoto *et al.*, 2001). SAGE is especially useful for researching intricate biological systems, including disease progression, development, and therapeutic responses (Matsumura *et al.*, 1999). Even though RNA-Seq technologies have largely replaced it in recent years, its capacity to catch uncommon transcripts and offer a quantitative profile of gene expression has made it an invaluable tool in genomics and transcriptomics research.

3.3.4.3 Northern blotting

Using a molecular biology technique called northern blotting, one may identify and measure certain RNA molecules in a complicated mixture (Pall and Hamilton, 2008). RNA samples are separated using gel electrophoresis and then transferred to a membrane, usually composed of nitrocellulose or nylon. A labeled probe complementary to the target RNA is hybridized to the membrane after the RNA has been immobilized (Blevins, 2017). To enable visualization, the probe might be marked with radioisotopes or non-radioactive markers. Researchers can examine gene expression patterns, evaluate the size and integrity of RNA, and find alternative splicing variants by using northern blotting. Although it is a labor-intensive method, it is a fundamental tool in molecular biology and has yielded important insights into gene regulation and expression.

3.3.4.4 In situ *hybridization (ISH)*

One effective method for identifying and locating certain nucleic acid sequences inside of fixed tissues or cells is called *in situ* hybridization, or ISH (Cassidy and Jones, 2014). ISH makes it possible to see patterns of gene expression in cells and tissues by using labeled complementary RNA or DNA probes. Permeabilizing the sample to facilitate probe access is the first step in the process. To guarantee specificity, hybridization under carefully regulated circumstances is the next step. Following the removal of unbound probes, several techniques, including fluorescence and chromogenic labeling, are used to identify targets. ISH is very useful for understanding disease pathophysiology, identifying certain cell types, and researching the dynamics of gene expression during development (Cassidy and Jones, 2014). ISH is a vital tool in the fields of developmental biology, neurology, and cancer research because of its capacity to offer geographical context.

3.3.4.5 Nanostring nCounter

A state-of-the-art tool for multiplexed gene expression analysis is Nanostring nCounter (Kulkarni, 2011). This technique ensures great specificity and sensitivity by enabling the simultaneous detection and measurement of hundreds of RNA targets without the need for amplification. By using colour-coded barcodes that hybridize to target RNA molecules, the nCounter technology makes it possible to quantify up to 800 genes simultaneously in a single procedure (Molania *et al.*, 2019). Following hybridization, the samples are run through a dedicated scanner that counts each unique barcode signal, enabling accurate target quantification. Studies with limited input RNA, such as those involving valuable tissues or clinical samples, benefit greatly from this method (Brumbaugh *et al.*, 2011). Because of its dependability and user-friendliness, the nCounter system has been extensively used in scientific domains like as immunology, developmental biology, and cancer.

3.3.4.6 Digital droplet PCR (ddPCR)

A sophisticated quantitative PCR method that offers extremely sensitive and accurate nucleic acid measurement is called digital droplet PCR, or ddPCR (Hudecova, 2015). In contrast to conventional PCR, ddPCR divides a sample into thousands of droplets the size of nanoliters, each of which serves as a separate reaction chamber. With positive and negative droplets signifying the target's existence, this partitioning enables the simultaneous amplification and detection of target sequences (Long and Berkemeier, 2022). Following amplification, fluorescence analysis of the droplets is performed, eliminating the requirement for standard curves and allowing for an absolute quantification of the target DNA or RNA. When it comes to identifying uncommon mutations, determining copy number variations, and gauging gene expression levels, ddPCR is especially helpful (Yin *et al.*, 2019). Because of its great sensitivity and accuracy, ddPCR is a vital tool for clinical diagnosis, research, and personalized medicine.

3.3.4.7 Functional studies

Understanding gene function and regulation requires functional investigations like gene knockout and knockdown. Through the introduction of small interfering RNA (siRNA) or short hairpin RNA (shRNA), which targets complementary mRNA for destruction and consequently reduces gene expression, RNA interference (RNAi) is a method used to silence particular genes (Chakraborty, 2007). This method is frequently applied to the functional investigation of genes in a variety of species. CRISPR (clustered regularly interspaced short palindromic repeats) technology uses the Cas9 enzyme and a guide RNA to precisely edit genes by causing double-strand breaks at particular genomic sites (Dutta *et al.*, 2024b). Gene knockout or alteration by sequence insertion or deletion may arise from this. Functional genomics has been completely transformed by these techniques, which make it possible for scientists to efficiently study the functions of genes in cellular processes, illness, and development. Table 3.1 details recent transcriptomics studies uncovering the genes and pathways regulating abiotic stresses in horticultural crops.

3.4 Data Analysis in Transcriptomics

Transcriptome data handling and analysis involves stepwise processing and interpreting high-throughput RNA data to understand gene expression dynamics (Fig. 3.1). It begins with experimental design, ensuring proper sample collection under defined environmental conditions. Preprocessing steps such as quality control, read trimming, and alignment to reference genomes

Table 3.1. Recent transcriptomics studies uncovering the genes and pathways regulating the abiotic stresses in horticultural crops.

Crop	Key genes and transcription factors (TFs)	Pathways	References
Drought stress in fruit crops			
Almond–Peach Rootstock (Garnem)	HSPs (Heat Shock Proteins), LEA (Late Embryogenesis Abundant), dehydrins (LEA5, LEA14, LEA D34-like, ERD4, COR47, DHN2), TFs (ALFIN-LIKE 5, bZIP TRAB-1-like and DREB2B, RF023, NF-YB3), LRR receptor-like serine/threonine-kinase ERECTA	Purin metabolism, thiamine metabolism, antibiotic biosynthesis, aminobenzoate degradation, sucrose and starch metabolism, and glycogenesis	Bielsa *et al.*, 2018
	TFs: TINY-like, WRKY15, MYB3R1, HAT5, GATA2, ERF023	Binding, stimulus, and metabolism pathways	Wang *et al.*, 2022b
	MdATHB-7, MdWRKY40, MdSnRK2.6, MdYUC6, MdABF2, MdNF-YA1, MdNRT1.5., ERF1, ERD15, and GLK1	Biosynthesis of secondary metabolites, diterpenoid biosynthesis, plant hormone signal transduction, MAPK signaling pathway, and flavonoid biosynthesis	Chen *et al.*, 2022a
	O-acetylserine-(thiol)-lyase, S-adenosylmethionine synthetase, dihydroxyacid dehydratase, isopropylmalate isomerase, branched-chain aminotransferase, pyruvate kinase, 3-dehydroquinate dehydratase/shikimate 5-dehydrogenase	Amino acid biosynthesis pathways	Li *et al.*, 2024
	Phospholipase D (PLD) (MdPLD genes)	Stress response, hormone response, and growth and development-related cis-elements	Fang *et al.*, 2023
Banana	ABA 80 -hydroxylase 1, Ethylene-responsive TF ERF012, Probable WRKY- 57, Glutathione S-transferase F8, 17.5 kDa class I heat shock protein, Probable trehalose-phosphate phosphatase 6, and ATP-dependent DNA helicase Q-like 5 TFs: HSF, bHLH, bZIP, MIKC, DOF, and C2H2	Carbohydrate degradation, protein modifications, glycan metabolism, lipid metabolism, alkaloid biosynthesis, and amino acid biosynthesis	Muthusamy *et al.*, 2016
	TFs: NAC (MaNAC51, MaNAC77, MaNAC87, MaNAC95, MaNAC100, MaNAC103, MaNAC136, and MaNAC149)	ABA-dependent and ABA-independent pathways	Chaudhari *et al.*, 2019
Cherry	ERFs, brassinosteroid-regulated proteins, LRR receptors, putative receptor proteins, cytochrome P450 94C1, MYB-related proteins, galactinol–sucrose galactosyltransferase, stachyose synthase-like, dehydration-responsive protein RD22, alphatrehalose-phosphate synthase, receptor-like protein kinase	Phenylpropanoid biosynthesis and cyanoamino acid metabolism, terpenoid biosynthesis	Wan *et al.*, 2021
Citrus (*C. sunki*)	TF–AP2 family, DEGs involved in hormone biosynthesis (brassinosteroids, jasmonate, salicylic acid, auxin, and ethylene), leucine kinase receptor, wall-associated kinase (WAK), extensin motif (RLK) and lysine motif (LysM)	Ascorbate and glutathione pathways, photorespiration pathway, flavonoid and phenylpropanoid biosynthesis pathways, wax biosynthesis	Santos *et al.*, 2021

Continued

Table 3.1. Continued

Crop	Key genes and transcription factors (TFs)	Pathways	References
Carrizo and Bitters citrange	PELPK1 (*C. clementina* protein), cell division control protein homologue B (*C. sinensis* protein), G2/mitotic-specific cyclin S13-7 (*C. sinensis*), ENDOSPERM DEFECTIVE 1(*C. sinensis* protein), C. clementina microtubule-associated protein RP/EB family member 1C	Metabolic pathways, other glycan degradation, sucrose-starch metabolism, biosynthesis of secondary metabolites, phenylpropanoid biosynthesis, cutin, suberin and wax biosynthesis	Scialò *et al.*, 2024
C. sinensis	UDP-glucose, cellulose synthases, galactosyltransferases, pectinesterases, expansins, beta-fructosidase 3, sucrose synthase 3, sucrose phosphate synthase 1F, granule bound starch synthase 1, HSPs, MYB, AP2/ERF, WRKY, bZIP, NAC, ZF and bHLH	Cell wall and carbohydrate metabolism, antioxidant metabolism, ethylene, and ABA signaling pathways	Gonçalves *et al.*, 2019
C. sinensis	Very long chain aldehyde reductase, ATP-binding cassette, subfamily G members, CsFAR-like genes (fatty acyl-CoA reductases), MAP kinase kinase kinase, superoxide dismutase 1, peroxidases, TFs (NAC, WRKY, bHLH, MADS, AP2/ERF)	Cuticular wax accumulation, wax biosynthesis and export, ROS scavenging, and MAPK signaling cascade	Liang *et al.*, 2022
C. macrophylla	Potassium channel (KAT3), ABCG transporters, nitrate transporter (NRT) CDPKs, calcium/calmodulin-regulated receptor-like kinase 1, isochorismate synthase 2, 9-cis-epoxicarotenoide dioxygenase (NCED1), brassinosteroid insensitive 1, BZR1/BES1, pathogen-associated molecular patterns (PAMP)	Redox and osmotic systems, ascorbate and aldarate metabolism, starch and sucrose metabolism, amino sugar and nucleotide sugar metabolism, photosynthesis, pathogen-response system	de Silva *et al.*, 2023
Grape	HSPs, ubiquitin ligases, TFs: MYBs, HSFs, NACs, bHLH, AP2-EREBP	Photosynthesis and hormonal signaling pathways, metabolic pathways, and biosynthesis of secondary metabolites	Ju *et al.*, 2021
	Dehydration responsive proteins (RD22, RD26, DRS1, XERICO), osmotin-like genes, b-glucanase, peroxidase 5-like, TFs: DREB1A, MYB102, AP2-EREB	Oxidation–reduction process, hydrogen peroxide, amino sugar, and nucleotide sugar metabolism, cell wall organization, MAPK signaling pathway	Cardone *et al.*, 2019
	chlorophyll A–B binding protein, outer membrane protein (enzyme PagP), CHUP1 (chloroplast signal transduction pathway), OZS1, dehydrin proteins (DHN, ERD14, LEA), aquaporin (AQP)-like (NIP6;1, NIP5;1, TIP4;1), trehalose phosphatase	Abiotic stimulus, photosynthesis, antioxidant-protection system, osmotic adjustment substance, hormonal signaling pathways	Cui *et al.*, 2020
	phenylalanine ammonia lyases (VviPALs), trans-cinnamate 4-monooxygenase (VviC4H), 4-coumarate-CoA ligase (Vvi4CL), pcoumaroyl shikimate 3'-hydroxylase (VviC3H), chalcone synthases (VviCHSs), chalcone isomerases (VviCHIs), flavonoid-3′5′-hydroxylase (VviF3′5′H), dihydroflavonol reductase (VviDFR), one 1- deoxy-D-xylulose-5-phosphate synthase (VviDXS1), terpene synthases (VviTPSs)	Phenylpropanoid pathway, terpenoid pathway, flavonoid, and carotenoid pathways	Savoi *et al.*, 2016

Continued

Table 3.1. Continued

Crop	Key genes and transcription factors (TFs)	Pathways	References
	β-carotene hydroxylase (VvBCH1), ABA-responsive bZIP TF (VvbZIP25), AWPM-19-like family, ascorbate peroxidase, glyoxal oxidase, peptidyl-prolyl-trans isomerase, peroxidase, and thioredoxin	ABA pathway, carbohydrate metabolism, oxidative stress-induced protein damage repair pathway	Dal Santo *et al.*, 2016
	Anthocyanin regulatory C1, chalcone synthase, histone-lysine N-methyltransferase (SUVR3)	Flavonoid biosynthesis, plant hormone signal transduction, galactose metabolism, diterpenoid biosynthesis, starch and sucrose metabolism, MAPK-signaling pathway, methionine metabolism	Tan *et al.*, 2023
	Glutamate tRNA Ligase, dehydrogenase/ reductase SDR family member, radical S-adenosyl methionine domain-containing protein1, protochlorophyllide oxidoreductase, Glutamate tRNA reductase 1, Ferrochelatase-2, psbB, psbCs, ascorbate peroxidases, glutathione reductase, IAA synthetase (GH3), DELLA, SNF1-related protein kinases 2, jasmonate-ZIM-domain proteins, allene oxide synthase, pyroline-5-carboxylate synthetase, Prolyl hydroxylase	Metabolic pathway, biosynthesis of secondary metabolites, plant–pathogen interaction pathway, chlorophyll metabolic pathway, hormonal pathways, proline metabolism	Haider *et al.*, 2017
Mango	WRKY transcription factor 3, polyamine oxidase 4, protein MEI2-like 1	Phenylpropanoid biosynthesis, ABC transporter, plant hormones and signal transduction, biosynthesis of amino acids, starch and sugar metabolism, and MAPK signaling	Hou *et al.*, 2024
Papaya	TFs: DREB, bZIP, HSF, MYC, MYB, NAC, AP2, NFY, ERF, WRKY, C2H2, bHLH, ERD	Cellular metabolic process, catalytic activity, hydroxylase activity, and biosynthetic process	Estrella-Maldonado *et al.*, 2021
Peach	Receptor-like kinase (RLK), leucine-rich repeat receptor-like kinase (LRR-RLK), cysteine-rich receptor-like, glutathione S transferase, kinases (CRKs), serine-threonine kinases (STKs), peroxidase, thioredoxin, ascorbate peroxidase TFs: bHLH, MYB, NAC, ERT, ORG2-like, WRKY	Phenylpropanoid biosynthesis, aminobenzoate degradation, glutathione metabolism, drug metabolism associated with cytochrome P450	Ksouri *et al.*, 2016
	Group A encoded auxin transporter-like protein, auxin-responsive GH3-like protein, histidine-containing phosphotransfer proteins	Metabolic process, catalytic, oxidation reduction, oxidoreductase	Wang *et al.*, 2019
Pear	Cinnamyl-alcohol dehydrogenase, Ferulate-5-hydroxylase, caffeic acid 3-O-methyltransferase, S, aminocyclopropane-1-carboxylate synthase, cis-epoxycarotenoid dioxygenase, Lupeol synthase 2, Cinnamoyl-CoA reductase, peroxidase	Phenylpropanoid biosynthesis pathway, flavonoid biosynthesis pathway, carotenoid biosynthesis pathway, cysteine and methionine metabolism pathway, brassinosteroid biosynthesis pathway	Yang *et al.*, 2021

Continued

Table 3.1. Continued

Crop	Key genes and transcription factors (TFs)	Pathways	References
	Probable WRKY TF40, WRKY TF 18, ferritin-4, 4-hydroxycoumarin synthase 1, histidine-containing phosphotransfer protein 4, NIM1- Interacting 1	Plant monoterpenoid biosynthesis, diterpenoid biosynthesis, phenylpropanoid biosynthesis, flavonoid biosynthesis, cysteine and methionine metabolism	Yang *et al.*, 2022
Strawberry	9-cisepoxycarotenoid dioxygenase, ABA 8'-hydroxylase, Pyrabactin Resistance/ PYR-like/Regulatory Components of ABA Receptors (PYR/PYL/RCAR)	Glycerophospholipid metabolic pathway, ABA signalling pathway, MAPK signalling and ribosome biosynthesis	Cao *et al.*, 2022b
Drought stress in vegetable crops			
Capsicum (*Capsicum annum*)	Late Embryogenesis Abundant protein, ABA metabolism-related ASR1, TAS14 and ATHB-7, plant cell wall-related genes, i.e. EXLB1, AAA1 KATANIN, KETOACYL-COA SYNTHASE 3, serine protease, MYO-INOSITOL OXYGENASE 1, 3-KETOACYLCOA SYNTHASE 3 TFs: NAC, bHLH, MYB, ERF, HSF	Hydrolase activity, response to abiotic stimulus, response to water	Borràs *et al.*, 2021
Chinese cabbage (*Brassica napa* ssp. *pekinensis*)	Non-specific lipid-transfer protein 3, dehydrin Rab18-like, sucrose synthase 3, aquaporin TIP2-1	ABA biosynthesis and degradation pathway, stress-induced signal transduction pathways, multiple metabolic pathways, glucosinolate biosynthetic pathway	Eom *et al.*, 2018
Common bean (*Phaseolus vulgaris*)	EIX receptor 1/2, calmodulin, S-(hydroxymethyl) glutathione dehydrogenase / alcohol dehydrogenase, calcium-binding protein CML, SAUR family protein, peroxygenase, auxin responsive GH3 gene family	plant hormone signal transduction, MAPK signaling pathway-plant, Photosynthesis-antenna proteins, Cutin, suberin and wax biosynthesis, glyoxylate and dicarboxylate metabolism, carotenoid biosynthesis, photosynthesis	Zheng *et al.*, 2023a
	AHK1 (Histidine kinase 1), ABCG40 TFs: C2H2 family protein, MYB, WOX	Plant hormone signal transduction pathway, phenylpropanoid biosynthesis, MAPK signaling pathway, monoterpenoid biosynthesis, starch and sucrose metabolism	Subramani *et al.*, 2023
Cucumber	Sucrose-phosphate synthase, sucrose phosphatase, glucose-1-phosphate adenylyltransferase, sucrose synthase, 4-α-glucanotransferase, Calcium-dependent protein kinase, 3-Ketoacyl-CoA thiolase 1, NAC domain-containing protein, Calcium-dependent protein kinase-like protein, Circadian clock coupling factor	Biosynthesis of secondary metabolites, ABA signaling pathway, MAPK signaling pathway, metabolic pathways, plant hormone signal transduction, biosynthesis of secondary metabolites	Wang *et al.*, 2018a

Continued

Table 3.1. Continued

Crop	Key genes and transcription factors (TFs)	Pathways	References
Faba bean	Abscisic acid (ABA) receptor PYL (Pyrabactin-resistance like), ABA signaling protein phosphatase 2c (PP2C), SnRK2 (SNF1_SCHPO SNF1-like protein kinase ssp2), beta-amylase, glucose-6-phosphate 1- dehydrogenase, raffinose synthase	Cysteine and methionine metabolism, N-glycan biosynthesis, sulfur metabolism, ABC transporters, MAPK pathway, biosynthesis of secondary metabolites, pentose phosphate pathway, galactose metabolism pathway	Wen *et al.*, 2024
	ATP-dependent zinc metalloprotease, AP2-like ethylene-responsive TF, RING-type E3 ubiquitin ligase, superoxide dismutase, 1-aminocyclopropane-1- carboxylate oxidase, ABA receptor PYL9-like protein, glutamate decarboxylase, glutathione S-transferase, dehydrin 2, beta-amylase TFs: MYB, AP2-EREBP, bHLH, C3H, WRKY, NAC	Carbohydrate metabolism, amino acid metabolism, lipid metabolism, biosynthesis of other secondary metabolites	Khan *et al.*, 2019
	calcium-dependent protein kinase, LRR receptor-like kinase, histidine kinase, protein phosphatase 2C, ABA aldehyde oxidase, zeaxanthin epoxidase, 1-aminocyclopropane1-carboxylate oxidase, Ent-kaurenoic acid oxidase, lipoxygenase, indole-3-pyruvate monooxygenase, TFs: BHLH, NAC, MYP, and WRKY	Metabolism, biosynthesis of other secondary metabolites, genetic information processing, environmental information processing	Alghamdi *et al.*, 2018
Garlic	Beta-carotene hydroxylase, ABA 8'-hydroxylase, hsps, E3 ubiquitin ligases	Secondary metabolites pathway, phenylpropanoid, glycerolipid, starch and sucrose metabolism, various signal transduction pathways including plant hormone and MAPK, biosynthesis of amino acids, glyoxylate and dicarboxylate metabolism	Zhou *et al.*, 2021a
	9-cis-epoxycarotenoid dioxygenase, violaxanthin de-epoxidase, xanthoxin dehydrogenase, beta-carotene hydroxylase, ABA 8' -hydroxylase, ABA responsive element binding factor, ABA receptor PYL4-like, protein phosphatase 2C, calcium-dependent protein kinase 26, calcium-binding protein CML35, CBL-interacting serine/threonine-protein kinase 4 (CIPK4), Calcineurin B-like protein 3 (CBL3)	Plant hormone signal transduction, biosynthesis of secondary metabolites	Wang *et al.*, 2022a
Onion	Cytochrome P450 (CYP81, CYP71A, CYP85A), aquaporin NIP1-1-like and TIP3-2, 9-cis-epoxycarotenoid dioxygenase, ABC subfamily G, UDP-glycosyltransferase, anthocyanidin 5,3-O-glucosyltransferase, flavonoid glucosyltransferase, flavonol synthase, α-galactosidase, β-galactosidase, galactinol synthase, galactinol–sucrose galactosyltransferase TFs: WRKY, MYB, NAC	Glycan metabolism, lipid metabolism, pectin degradation, carbohydrate degradation	Ghodke *et al.*, 2020

Continued

Table 3.1. Continued

Crop	Key genes and transcription factors (TFs)	Pathways	References
Potato	Protein LE25, Dehydrin DH2a, Thaumatin protein, Cytochrome P450- dependent fatty acid hydroxylase, Polygalacturonase noncatalytic subunit AroGP3, Xyloglucan endotransglycosylase	Metabolism, environmental information processing, organismal systems, plant hormone signal transduction	Zhang *et al.*, 2014
	LEA domain-containing protein, CAP160 protein, Expansin-like B1, Laccase 14, Carboxyesterase 13, WRKY DNA-binding protein 33, HSP20-like chaperones superfamily protein, UDP-Glycosyltransferase superfamily protein/ Anthocyanidin 3-O-glucosyltransferase 5-like	Photosynthesis-antenna protein, protein processing in endoplasmic reticulum, biosynthesis of secondary metabolites, metabolic pathways	Moon *et al.*, 2018
	PYP/PYL receptor proteins, MKK1–MPK6 (MAPK cascade), ethylene-responsive factors (ERF1), WRKY22/29, NDPK2, CaM4, GH3, AUX1, JAZ, TCH4, TGA TFs	Mitogen-activated protein kinase (MAPK) and plant hormone transduction pathways	Qin *et al.*, 2022
Sweet potato	NITRATE TRANSPORTER 1.2, DEHYDRATION-RESPONSIVE ELEMENT-BINDING PROTEIN 2A (DREB2A), LATE EMBRYOGENESIS ABUNDANT (LEA), leucine-rich receptor-like kinases (LRR-RLK), 9-cis- epoxycarotenoid dioxygenase, SLOW ANION CHANNEL-ASSOCIATED 1 (SLAC1), LIGHT-HARVESTING CHLOROPHYLL A/B-BINDING6 (LHCB6)	ABA signaling pathway, pentose phosphate pathway	Lau *et al.*, 2018
	ABA insensitive-5 like protein (bZIP-8), 9-cis-epoxy carotenoid dehydrogenase, ATP-dependent CLP protease, homeobox-leucine zipper protein, LanC-like protein (GCL2), EID1-like F-box protein-3, GID1-like gibberellin receptor, cinnamoyl-Co-A reductase-1, tyrosine phosphatase, carboxylesterase-8	Ascorbate and aldarate metabolism, metabolic pathways, biosynthesis of secondary metabolites, plant hormone signal transduction	Arisha *et al.*, 2020
	Leucine-rich repeat (LRR) receptor kinases, lectin (G-lectin and L-lectin) RLK (receptor like kinase), wall-associated kinase (WAK), cysteine-rich RLK, leaf rust kinase-like RLK, HSPs, LEA protein TFs: AP2-EREBP, MYB, bHLH, WRKY	ABA biosynthesis and signaling pathway, osmotic homeostasis, damage control and repair, starch and sucrose metabolism, amino sugar and nucleotide sugar metabolism	Cheng *et al.*, 2024
Tomato	NCED2 (Retinal pigment epithelial membrane protein), AAO (Aldehyde oxidase and xanthine dehydrogenase), CYP707A (Cytochrome P), PYR/PYL (Polyketide cyclase / dehydrase and lipid transport), PP2C (Protein phosphatase 2C), ABF (bZIP TF)	ABA biosynthesis and signal transduction	Zhang *et al.*, 2019b
	histone encoding genes, cell wall modifying enzymes (Pectinesterase), HSPs, Chlorophyll a/b binding protein, dehydrin, 2 NAC domain encoding IPR003441, Peptidyl-prolyl cis-trans isomerase	Photosynthetic light harvesting, cell wall metabolism, ABA signal transduction pathway,	Iovieno *et al.*, 2016
	chlorophyll a-b binding protein, stress enhanced protein 2, aspartyl protease family protein 2, trehalose-phosphate phosphatase A, trehalose-6-phosphate synthase TFs: NAC, GNAT, JUNGBRUNNEN 1-like isoform X1, HB-HD-ZIP, C2C2-CO-like bZIP, C2C2-Dof, C3H, MADS-MIKC, MYB, GARPG2-like, LOB, RWP-RK, HSF, Tify	plant–pathogen interaction, MAPK signaling pathway–plant, plant hormone signal transduction	Shu *et al.*, 2024

Continued

Table 3.1. Continued

Crop	Key genes and transcription factors (TFs)	Pathways	References
Watermelon	Cellulose SynthaseInteractive Protein 1 (CSI1), Zinc-Finger Homeodomain Protein 2 (ZHD2), bZIP TF 53-like (bZIP53- like), Late Embryogenesis Abundant Protein (LEA), L-Ascorbate Oxidase (AO), Pericycle Factor Type-A 5 (PFA5)	Wnt signaling, oxidation-reduction pathway, starch and sucrose metabolism, fatty acid degradation	Mahmoud *et al.*, 2023
Heat stress in fruit crops			
Avocado	HSPs, geraniol 8-hydroxylase-like protein, putative aquaporin PIP2-8 TFs: bHLH, ERF, NAC, WRKY, MYB	Plant hormone signal transduction, protein processing in the endoplasmic reticulum, pentose and glucuronate interconversions, starch and sucrose metabolism	Zheng *et al.*, 2024
Blueberry	HSPs, protein processing and repair (GRP94, NEF, CRT, BiP, CNX, PDIs), ubiquitination genes (Doa10, Derlin, Ubx, Ufd1, SAR1, Sec 61)	Purine and thiamine metabolism pathways, biosynthesis of antibiotics, starch and sucrose metabolism, biosynthesis of secondary metabolites, porphyrin and chlorophyll metabolism, aminobenzoate degradation, pentose and glucuronate interconversions	Callwood *et al.*, 2021
Grape	Polyadenylate-binding protein, Phox and Bem1 (PB1) protein, Increased sodium tolerance protein 1 (IST1), Coiled-Coil Domain-Containing (CCDC) protein, Valine-glutamine (VQ) protein, Glycerol-3-phosphate permease (G3Pp) protein	Plant hormone signal transduction, plant–pathogen interaction, starch and sucrose metabolism, flavone and flavonol biosynthesis, plant–pathogen interaction,	Wu *et al.*, 2023
	Three Small HSPs, Hsp70, Aluminium activated malate transporter 9, Phenylalanine-ammonia-lyase (PAL), TFs: Trehlix TF, TCP family (TCP 1), HSF A3	Amino acid synthesis, phenylpropanoid related pathways, malate pathway	Rienth *et al.*, 2016
	PsbA, PsbP, PetE, F-type ATPase, PsaA, Phytoene synthase 2 (PSY2), 9-cis-epoxycarotenoid dioxygenase 2 (NCED2), ABA glucosyltransferase (ABA-GT)], four HSPs	Protein processing in endoplasmic reticulum, zeatin biosynthesis, plant–pathogen interaction, plant hormone signal transduction, MAPK signaling pathway-plant	Guo *et al.*, 2022
Jujube (*Ziziphus jujuba*)	HSP90s, HSP70s, sHSPs, ubiquitin-protein ligases TFs: C2H2, ERF, HSF, WRKY, DERB, MYB, bHLH	Carbon metabolism, protein processing in endoplasmic reticulum, biosynthesis of amino acids, plant hormone signal transduction, glycolysis / gluconeogenesis, glycine, serine and threonine metabolism	Jin *et al.*, 2020
Litchi	ZHD4 (Zinc-finger homeodomain protein), AMP1 (Probable Glutamate Carboxypeptidase 2), ABH1 (Nuclear Cap-binding Protein Subunit), LHY (Late Elongated Hypocotyls), CRY2 (Cryptochrome-2), AGL8 (Agamous-like MADS-box protein AGL8 encoding gene), ELF4 (Protein EARLY FLOWERING four encoding gene), Indole-3-acetic acid-amido synthetase (GH3.6)	salicylic acid mediated signaling pathway, metabolic pathways	Liu *et al.*, 2019a

Continued

Table 3.1. Continued

Crop	Key genes and transcription factors (TFs)	Pathways	References
Loquat	HSP70, HSP17, HSP90s, Heat Shock Cognate Protein 70 (HSC70), SMALL AUXIN UPREGULATED RNA (SAUR), GRETCHEN HAGEN (GH), Auxin Response Factors (ARF17, ARF18, ARF3), ABA-INSENSITIVE 5 (ABF5), Protein Phosphatase 2C (PP2C), (Arabidopsis Response Regulators (ARR12, ARR19)	Plant hormone signal transduction, photosynthesis, oxidative phosphorylation, protein processing in endoplasmic reticulum	Chen *et al.*, 2022b
Passion fruit	Chlorophyllide a Oxygenases (CAOs) (PesCAO1/2/3/6), auxin-responsive protein (PesSAUR1), protein phosphatase (PesPP1), xyloglucan endotransglucosylase (PesXTH1), glutathione peroxidase (GSH), glutathione S-transferases (GST)	Starch and sucrose metabolism, ribosome, fatty acid metabolism, amino sugar and nucleotide sugar metabolism, plant hormone signal transduction	Wang *et al.*, 2023a
Pineapple	HSC70, HSP83-like, probable glutathione S-transferase, protochlorophyllide reductase-like	Argine biosynthesis, monoterpenoid biosynthesis, hepatocellular carcinoma	Zhao *et al.*, 2021
Raspberry	HSP83A, HSP18.1, HSP18.5-C, HSP90s, HSP18s, HSP70s, ascorbate peroxidase (APX3), superoxide dismutase (SODA.4, SODCC.1), peroxidase (POD), catalase (CAT1/3), CYTB6F, Photosystem light harvesting complex genes (LHCA1, LHCB2, LHCB3, LHCB7)	Biosynthesis of polyketone units, biosynthesis of glucosinolates, photosynthesisantenna protein, ndole alkaloid biosynthesis	Guo *et al.*, 2023
Red pitaya	GRXC11 (glutathione-disulfide oxidoreductase activity), FER2 (chloroplastic), PFE (chloroplastic), PGM1 (2,3-bisphosphoglycerate-independent phosphoglycerate mutase), PER72 (peroxidase 72) TFs: MYB, AP2/ERF, HSP, HSF, bZIP	Metabolic pathways, biosynthesis of antibiotics, biosynthesis of secondary metabolites, photosynthesis, plant hormone signal transduction	Jiao *et al.*, 2021
Heat stress in vegetable crops			
Bottle gourd	bHLH (HG10006168, HG10004527, HG10015989), ERF, MAPK (on of HG10000206, HG10006437, HG10008395, HG1000)	MAPK signaling pathway, photosynthesis, phenylpropanoid biosynthesis	Wang *et al.*, 2024
Brinjal (aubergine)	Chalcone synthase B, Leucoanthocyanidin dioxygenase, Probable chalcone--flavonone isomerase 3, Flavonoid 3′,5′ -hydroxylase, Chalcone--flavonone isomerase, Agmatine coumaroyltransferase-2, Anthocyanidin 3-Oglucosyltransferase, Phenylalanine ammonia-lyase 2, HSPs, HSFs	Protein processing in endoplasmic reticulum, pyruvate metabolism, amino sugar and nucleotide sugar metabolism, glutathione metabolism, glycolysis/ gluconeogenesis	Zhang *et al.*, 2019a
Broccoli	Apyrases protein (AYP5), N-acetylglutamate kinase (NAGK), epithiospecifier protein (ESP), branched-chain aminotransferase 4 (BCAT4), FLOWERING LOCUS C (BoFLC1)	Nucleotide metabolism, secondary metabolism sulfur-containing glucosinolates	Lin *et al.*, 2019
Cabbage	Calcium-dependent protein kinase (CPK), pathogenesis-related gene transcriptional activator (PT16), disease resistance protein (RPS2), cathepsin F (CTSF), mitogenactivated protein kinase 1 (MAP2K1, MEK1), DELLA, protein phosphatase 2C (PP2C), ABA responsive element binding factor (ABF), gibberellin receptor (GID1), brassinosteroid insensitive one gene (BRI1)	Plant–pathogen interaction, ribosome, plant hormone signal transduction, carbon metabolism, and biosynthesis of amino acid pathways	Li *et al.*, 2023

Continued

Table 3.1. Continued

Crop	Key genes and transcription factors (TFs)	Pathways	References
Cassava	sHSPs, HSP40, HSP70 luminal-binding protein, light harvesting complex (Lhcb1, Lhcb2, Lhcb3), chloroplast sensor kinase (CSK), plastocyanin (PC), serine acetyltransferases (SERATs), aspartate aminotransferase (ASP), cysteine synthase (CYS),	Protein processing in endoplasmic reticulum, cutin, suberine, and wax biosynthesis, pyruvate synthesis pathway, flavonoid biosynthesis	Wang *et al.*, 2023b
Celery	Pyruvate kinase, phosphofructokinase-1, phosphoglycerate kinase, pyruvate dehydrogenase, isocitrate dehydrogenase ($NADP^+$), β-glucosidase, trans-cinnamate 4-monooxygenase, cinnamyl-coenzyme A reductase, cinnamyl alcohol dehydrogenase, peroxidase	Pyruvate metabolic pathway, phenylpropanoid biosynthesis pathway,	Li *et al.*, 2022a
Chinese cabbage	Cytochrome P450 (CYP450), photosystem II (PSBQ3, PSBQ2, Psb27), PSII oxygen-evolving enhancer protein 2 (PSBQ2)	MAPK signaling pathway, biosynthesis of secondary metabolites, plant hormone signal transduction	Ikram *et al.*, 2022
	Hexokinase-Like 1 (HKL1), light-harvesting chlorophyll a/bbinding (LHC) (LHCB3, LHCB5, LHCB1.4, LHCB1.3, LHCB2.4), Peroxidases (Prx50/52/54), Phospholipid hydroperoxide glutathione peroxidase 1 (GPX1), Superoxide dismutases (SOD1/2), Brassica_rapa_newGene_7058, Brassica_rapa_newGene_7942	Starch and sucrose metabolism, glycolysis/gluconeogenesis, galactose metabolism	Zhang *et al.*, 2022
	Shoot Meristemless (STM), Light Sensitive Hypocotyls 3 (LSH3), Knotted 1-Like Homeoboxs (KNAT1/2/6), Syntaxin of Plants 111 (SYP111), Sucrose-Proton Symporter 2 (SUC2), Fasciclin-Like Arabinogalactan-Protein 12 (FLA12), Protodermal Factor 1 (PDF1), RUBISCO Small Subunit 3B (RBCS3B)	Photosynthesis, fatty acid metabolism pathway, heat stress response pathway,	Sun *et al.*, 2022a
Chinese Kale (Brassica alboglabra)	Methionine aminotransferase (BCAT4), Methylthioalkylmalate synthase 1/2 (MAM1/2), Dihomomethionine N-hydroxylase (CYP79F1), S-alkyl-thiohydroximate lyase (SUR1), thioglucoside glucohydrolase (TGG), glucosinolate GS-Alk (AOP2), MYB28/29/76, HSFs, HSPs	Glucosinolate biosynthesis, brassinosteroid biosynthesis, oxidative phosphorylation, lipid metabolism, plant hormone signal transduction	Guo *et al.*, 2019
Cucumber	chorismate mutase, threonine aldolase, pectate lyase, hexosyltransferase, xyloglucan endotransglucosylase, syntaxin, Reactive intermediate/imine deaminase, mandelonitrile lyase, wound-responsive family protein, glycine-rich protein-2, sucrose transporters, invertase, sucrose synthase	Anther development, carbohydrate metabolism	Chen *et al.*, 2021
	Auxin transporter-like protein 2, Cyclin-D3–3, ABA 8'-hydroxylase 1, NAC domain-containing protein 2-like, Ethylene-responsive TF ANT, Receptor-like protein 12, LRR receptor-like serine/threonine-protein kinase, Phenylalanine ammonia-lyase, Lipase-like isoform X2, Branched-chain-amino-acid aminotransferase 2	Plant hormones signal transduction, MAPK signaling pathway, biosynthesis of secondary metabolites	Wang *et al.*, 2020

Continued

Table 3.1. Continued

Crop	Key genes and transcription factors (TFs)	Pathways	References
Garlic	HSP40, HSP70, HSP90, ABA, PYL, AUX, IAA, SAUR, ARF, PP2C	Protein processing in endoplasmic reticulum, plant hormone signal transduction	Yang *et al.*, 2024
Lettuce	Hydroxyl Cinnamoyl transferase, Flavonol synthase, caffeoyl coenzyme A methyltransferase, Dihydro-flavonol reductase, chalcone synthase, Colourless anthocyanin dioxygenase	Flavonoid related metabolic pathways	Sun *et al.*, 2022b
Pakchoi (*Brassica rapa* L. *chinensis*)	Heat-shock 70 kDa protein1/8, HSP70-interacting protein, Ethylene-responsive TF 1, phospholipase D	Metabolism, amino-acid metabolism, biosynthesis of other secondary metabolites, xenobiotics biodegradation and metabolism	Xu *et al.*, 2016
Pepper (*Capsicum annuum*)	Photosystem I P700 chlorophyll a apoprotein A1, photosystem II protein D1, acetyl-CoA carboxylase beta subunit, phosphate dikinase, psbQ-like protein2, ATP synthase, sucrose and glucose synthase genes (UDP-glucuronate, trehalose-6-phophate synthase), HSPs	Protein processing in endoplasmic reticulum, starch and sucrose metabolism, photosynthesis	Li *et al.*, 2015
	Mitogen-activated protein kinase kinase kinase a-like, eid1-like F-box protein 3, L-ascorbate oxidase homologue	Biosynthesis of amino acids, starch and sucrose metabolism, carbon metabolism, DNA replication and plant hormone signal transduction pathways	Liu *et al.*, 2021a
	Photosystem II CP43 reaction center protein, Photosystem II protein D1, Photosystem II D2 protein, potassium (K^+) channel genes (KAT2, AKT2), PPIases (CYP21–1, CYP59, CYP26–2), ER lumen protein retaining receptor (ERD2)	Photosynthesis, ion transmembrane transport and ATP synthesis coupled proton transport, protein glycosylation and sulfur compound biosynthetic process	Liu *et al.*, 2022a
Potato	SuSy degradation, beta-amylase, beta-1,4-glucanases, pectate lyases, polygalacturonases, glutamate decarboxylase, dUTP diphosphatase, aspartate transcarbamoylase, cellulose synthase, pectate lyases, HSPs	Biosynthesis of secondary metabolites, flavone and flavonol biosynthesis, biosynthesis of amino acids, fatty acid biosynthesis	Liu *et al.*, 2021b
Radish	MAPK17/19, Calcium-dependent protein kinases (CDPKs), TFs: WRKY, APETALA2, basic helix–loop–helix, basic leucine zipper, heat shock factor, ethylene responsive factor, auxin response factor, myeloblastosis protein, HSPs and HSFs	Protein processing in endoplasmic reticulum, stilbenoid, diarylheptanoid and gingerol biosynthesis, metabolic pathways, biosynthesis of secondary metabolites	Wang *et al.*, 2018b
Spinach	HSPs, sHSPs, Dihydroflavonol 4-reductase, protein-S-isoprenylcysteine O-methyltransferase B, 4-coumarate CoA ligase (4CL), 3-hydroxy-3-methy lglutaryl-coenzyme A reductase, geranylgeranyl hydrogenase, Cu/Zn superoxide dismutase (Cu/Zn-SOD), Dehydroascorbate reductase (DHAR)	Terpenoid biosynthesis, heat stress resistance pathways, biosynthesis of secondary metabolites, metabolic pathways	Yan *et al.*, 2016
Sweet potato	4-hydroxyphenylpyruvate dioxygenase, HSPs, isocitrate lyase (ICL), malate synthase (MS), ADP-glucose pyrophosphorylase beta subunit IbAGPb1A, starch branching enzyme I, starch phosphorylase	Biosynthesis of secondary metabolites, protein processing in the endoplasmic reticulum, glyoxylate pathway	Tang *et al.*, 2023

Continued

Table 3.1. Continued

Crop	Key genes and transcription factors (TFs)	Pathways	References
Tomato	Invertase, sucrose synthase, hexokinase (HK), Glucose 6-phosphate dehydrogenase (G6PD), 6-phosphogluconate dehydrogenase (6PGD), phosphofructokinase (PFK), pyruvate kinase (PK), phosphoenolpyruvate carboxylase (PEPC), isocitrate dehydrogenase (IDH)	Sugar metabolism, organic acid metabolism	Zheng *et al.*, 2023b
Watermelon	serine acetyltransferase (SATase), tyrosine aminotransferase (TAT), spermidine synthase, 1-aminocyclopropane1-carboxylate synthase, hydroxypyruvate reductase (HPR), aspartate kinase, bifunctional aspartokinase/ homoserine dehydrogenase 1, threonine dehydratase/deaminase (TD)	Cysteine and methionine metabolism pathway, biosynthesis of amino acids, glycine, serine and threonine metabolism signal pathways	Zhu *et al.*, 2020
Salinity stress in fruit crops			
Apple	Salt Overly Sensitive (SOS1, SOS2, SOS3)	Cellular processes, organismal systems, environmental information processing, phosphatidylinositol signaling system	Li *et al.*, 2019
	Pyrabactin resistance 1-like (PYL), ABA signaling protein phosphatase 2c (PP2C), mitogen-activated protein kinase 3 (MPK3), SNF1_SCHPO SNF1-like protein kinase ssp2 (SnRK2)	Signaling pathways of plant hormones, carbohydrate metabolism, environmental information processing	Huang *et al.*, 2023
	Chalcone synthase (CHS), flavanone 3-hydroxylase (F3H), anthocyanidin synthase (ANS), flavonoid-3-O-glucosyltransferase (UFGT), copper-zinc superoxide dismutases (Cu/Zn-SODs), manganese superoxide dismutases (MnSODs), Cation/H^+ Exchanger 20 (CHX20), K^+ efflux antiporter 5 (KEA5)	Phenylpropanoid, flavonoid, and anthocyanin biosynthesis pathway	Jiang *et al.*, 2024
	Auxin response factor (ARF5), Indole-3-acetic acid-amido synthetase (GH3.6), SMALL AUXIN UPREGULATED RNAs (SAUR32/36), ISOPENTENYLTRANSFERASE (IPT5), CBL1/9-CIPK1 calcium sensor	Plant hormonal pathways	Xuan *et al.*, 2019
	Auxin resistant 1 (AUX1), SAURs, Auxin early response genes (Aux/IAAs), B-type ARR TFs (B-ARRs, cytokinin receptor (CRE1), histidine-containing phosphotransfer proteins (AHPs), Probable indole-3-pyruvate monooxygenases (YUCCA8/10), adenylate isopentenyltransferase, Indole-3-pyruvate monooxygenase YUCCA2-like, Cytokinin riboside 5' -monophosphate phosphoribohydrolase LOG8-like	Plant hormone signaling transduction pathways, indoleacetic acid and cytokinin biosynthesis	Wang *et al.*, 2023c
Banana	9-cis-epoxycarotenoid dioxygenase, Violaxanthin de-epoxidase, SODs, PODs, omega-hydroxyPALmitate O-feruloyl transferase-like, phenylalanine ammonia-lyaselike, fatty acid desaturase 4, 4-coumarate–CoA ligase-like 3, chalcone synthase 2-like	Plant metabolism, phenylpropane biosynthesis pathway	Wei *et al.*, 2022

Continued

Table 3.1. Continued

Crop	Key genes and transcription factors (TFs)	Pathways	References
	Heat shock cognate 70 kDa protein, 2-oxoglutarate 3-dioxygenase, Dolichyldiphosphatase, Glutamate dehydrogenase, Vacuolar amino acid transporter, Ribulose bisphosphate carboxylase, Photosystem II reaction center W protein, ethylene-responsive TFs (ERFs), WRKY	ABA-activated signaling, ROS signal transduction pathway, calcium and phospholipid signalling pathways	Hu *et al.*, 2017
Citrus junos (Citrus)	Nine-cis-epoxycarotenoid dioxygenase 3 (NCED3), PYR1-like 4, Highly ABA-induced PP2C gene 2, Lipoxygenase 2/3, Jasmonate-zim-domain protein 1/8/10, ACC synthase1/6, ACC oxidate1/4, Auxin-responsive GH3 family protein, Auxin efflux carrier family protein (PIN), SAUR-like auxin-responsive protein, Auxin response factor 4, Xyloglucan endotransglucosylase/hydrolase 5/6/9/16/26/32/33, Expansins, Calcium-binding EF-hand protein (SOS3), Glutathione peroxidase 6, Peroxidases	Metabolic pathways, biosynthesis of secondary metabolites, plant hormone signal transduction, biosynthesis of antibiotics, Phenylpropanoid biosynthesis, carbon metabolism, plant–pathogen interaction	Xie *et al.*, 2018
	S-locus lectin protein kinase, Leucine-rich repeat protein kinase, Calcineurin-like phosphoesterase, Transducin/WD40 repeat-like, P-loop containing nucleoside triphosphate hydrolase, amino acid transporter 1, sucrose synthase, Calcineurin-like phosphoesterase	Plant hormone signal transduction, tyrosine and tryptophan biosynthesis, starch and sucrose metabolism, arginine and proline metabolism	Xie *et al.*, 2017
	Orange1.1 (CHIV) TFs: bHLH, MYB, ERF, WRKY, NAC	Chitin and secondary metabolism processes	Song *et al.*, 2023
Date palm	ABA-Insensitive 5 (ABI5), sulfate bicarbonate oxalate exchanger and transporter sat-1, dead-box RNA helicase, Chitinase 1, Chitinase 1, xyloglucan endotransglycosylase hydrolase, cellulose synthase-like protein, cinnamoyl-reductase 1-like, U-box–containing E3 ubiquitin ligase gene family TFs: WRKY, MYB	ABA-independent mechanisms, ABA signaling, cell wall and cytoskeleton biosynthesis, secondary metabolism pathways	Radwan *et al.*, 2015
Fig	Tubby-like protein 8, Nodulation signalling pathway (RBK2), terminal-fower1-like, delta-1-pyrroline-5-carboxylate synthetase, Expansin-like protein, Expansin-like protein, copper/zinc superoxide dismutase copper chaperone, 2-oxoglutarate-dependent dioxygenase, HXXXD-type acyl-transferase family, cytochrome P450 98A2/71D10, UDP-glycosyltransferase superfamily protein 2-oxoglutarate, ECERIFERUM 1, late embryogenesis abundant, Clp-N motif-containing P-loop nucleoside triphosphate hydrolase, early-responsive to dehydration stress protein	ROS signaling, amino acids metabolism, biosynthesis of secondary metabolites, glutamate and proline pathways, glycolysis pathway	Vangelisti *et al.*, 2019
Grape	phenylalanine ammonia lyase, strictosidine synthase, terpene synthases, chalcone synthase, carotenoid oxygenase, oxoglutarate dioxygenases, SAM dependent carboxyl methyltransferase, allene oxide cyclase, amidohydrolase, NADPH dehydrogenase, ribulose-1, 5- bisphosphate carboxylase/oxygenase activator, HSPs, TFs: WRKY, bZIP, NAC, AP2/ERF	Biosynthesis of secondary metabolites, phytohormone synthesis and/or hormone signaling, photosynthesis	Upadhyay *et al.*, 2018

Continued

Table 3.1. Continued

Crop	Key genes and transcription factors (TFs)	Pathways	References
	Receptor-like kinases (RLKs), hyperosmotically inducible periplasm protein, Cl- loading to xylm/SLAH, protein kinases, protein phosphatases 2C (PP2C), NCED (9-cis-epoxycarotenoid dioxygenase, dehydrin, LEA, Osmotin *Osmy*	Osmotic stress and osmoregulation pathways, phosphoinositide signalling pathway, Ca^{2+} signaling pathway, ABA biosynthesis/ transport	Daldoul *et al.*, 2022
Jujube	Glutamyl-tRNA reductase, pheophorbide a oxygenase, ferrochelatase, photosystem I chlorophyll a-b binding protein 6 (LHCA2), chlorophyll a-b binding proteins, oxygen-evolving enhancer proteins, photosynthetic NDH subunit of lumenal locations, ferredoxin C 2 (PetF), pyruvate, phosphate dikinase (ppdK), malate dehydrogenases, ribulose bisphosphate carboxylase small chain, glyceraldehyde-3-phosphate dehydrogenase GAPCP2 (GAPDH), fructose-1,6-bisphosphatase 1 (FBP)	Porphyrin and chlorophyll synthesis pathways, photosynthesis-related pathways, photosynthesis-antenna proteins pathways, carbon fixation pathways	Chang *et al.*, 2020
Mango	Fructose-bisphosphate aldolase 2, NADH dehydrogenase (ubiquinone) one beta subcomplex subunit 3-B, Photosystem I reaction center subunit IV, NADP-dependent D-sorbitol-6-phosphate dehydrogenase, Pyruvate dehydrogenase E1 component subunit beta-1, NADP-dependent glyceraldehyde-3- phosphate dehydrogenase, Photosystem I reaction center subunit IV, Auxin-repressed 12.5 kDa protein, Auxin response factor 2B, Ethylene receptor 2, Protein phosphatase 2C 51, Serine/threonine-protein phosphatase PP2A-4 catalytic subunit, Nodulin-related protein 1, Superoxide dismutase (Cu-Zn)	Oxidative phosphorylation, photosynthesis, amyotrophic lateral sclerosis, galactose metabolism	Perveen *et al.*, 2023
Pistachio (*Pistacia vera*)	Pentatricopeptide repeat, Protein kinase domain, Leucine-rich repeat, RNA recognition motif domain, Serine-threonine/ tyrosine-protein kinase catalytic domain, Zinc finger, RING-type, Helicase, ABC transporter-like, Cytochrome P450, Oxoglutarate/iron-dependent dioxygenase, ABC transporter type	Biosynthesis of secondary metabolites, biosynthesis of antibiotics, genetic information processing, cellular processes and environmental information processing	Moazzzam Jazi *et al.*, 2017
Pomegranate	ABA receptors (PYR/PYL/RCAR), sucrose non-fermenting1-related protein kinase 2 (SnRK2s), type 2C protein phosphatase (PP2Cs), Ca^{2+}-ATPases, cation/H^+ antiporters, cation/calcium exchangers, CBL-interacting protein kinases	ABA signaling and Ca^{2+}-related signaling pathways	Liu *et al.*, 2019b
Salinity stress in vegetable crops			
Common bean	Phvul.011G202000 (AT5G44640; response to salt stress), Phvul.011G105900 (AT1G05260; hyperosmotic salinity response), Phvul.011G056000 (AT5G42260; response to salt stress), Phvul.005G151800, Phvul.005G151700, Phvul.002G300800	Glucosinolate catabolic process, phenylpropanoid biosynthesis, biosynthesis of secondary metabolites	Zhang *et al.*, 2021a

Continued

Table 3.1. Continued

Crop	Key genes and transcription factors (TFs)	Pathways	References
Cucumber	9-cis-epoxycarotenoid dioxygenases (NCEDs), 1-Aminocyclopropane-1-carboxylate oxidase (ACO), S-Adenosyl-methionine (SAM), EBF1/2, EIL1, EIN3	ABA biosynthesis and degradation pathways, ethylene and polyamine biosynthesis, hormonal signal transduction pathways	Du *et al.*, 2021
Garlic	Potassium channel like, Potassium transporter, Two-pore potassium channel like, Potassium transporter, phytoene synthase (PSY), Lycopene beta-cyclase (LCYB), ABA 8'-hydroxylase, Auxin/Indole-3-Acetic Acid (Aux/IAA), SAUR	Ribosome biogenesis in eukaryotes, phenylpropanoid biosynthesis, alpha-linolenic acid metabolism, carotenoid biosynthesis, galactose metabolism, starch and sucrose metabolism	Zhou *et al.*, 2021b
Potato	Delta-1-pyrroline-5-carboxylate synthetase, pyrroline-5-carboxylate reductase, proline dehydrogenase, ARG, AST AP2/ERF, WRKY, MYB, NAC	Arginine and proline metabolism, phenylpropanoid biosynthesis, flavonoid biosynthesis	Jing *et al.*, 2022
Sugarbeet	Fatty acyl-CoA reductase (FAR), squalene monooxygenase, b-amyrin synthase, chalcone synthase, flavonol synthase, linalool synthase, pectinesterase, minichromosome maintenance (MCM), DNA replication ATP-dependent helicase, linoleate 9S-lipoxygenase 5	Cutin, suberine and wax biosynthesis, biosynthesis of secondary metabolites, sesquiterpenoid and triterpenoid biosynthesis, starch and sucrose metabolism	Geng *et al.*, 2020
Sweet potato	Protein Phosphatase 2C (PP2Cs), serine/threonine kinases (SnRKs), EIN2, EIN3, Receptor-like Kinases (RLKs), glutathione S-transferase (GST), respiratory burst oxidase homologs (RBOHs), xyloglucan endotransglucosylase/hydrolase, expansin	Salt-related signaling, ABA-signalling and ethylene signaling, ROS scavenging, cell wall organization or biogenesis pathways	Luo *et al.*, 2017
Tomato (*Solanum chilense*; wild tomato)	Polyamine oxidase one isoform X1, Polyamine oxidase, Arginine decarboxylase, Trehalose 6-phosphate phosphatase, Pyrroline-5-carboxylate reductase, Cytochrome P450, Cytochrome P450-type monooxygenase 97C11, L-ascorbate oxidase, 12-oxophytodienoate reductase 1, 1-aminocyclopropane-1-carboxylate oxidase homologue, Auxin response factor, Cytokinin hydroxylase-like, Cytokinin riboside 5′-monophosphate phosphoribohydrolase , Gibberellin 20-oxidase-1/3, allene oxide synthase 2-like	Amine and polyamine biosynthesis and metabolism, oxidation–reduction pathways, hormone related pathways	Kashyap *et al.*, 2020
Watermelon	MYC2, β-glucosidases, KUP system potassium uptake protein, cellulose synthase A, cytochrome P450	Jasmonic acid signaling pathway, MAPK signaling pathway, plant hormone signal transduction, and phenylpropanoid biosynthesis	Yuan *et al.*, 2022
Waterlogging/ flood stress in fruit crops			
Apple	Chalcone synthase, leucoanthocyanidin reductase, flavanone 4-reductase, flavanone 3 0 -hydroxylase, flavonoid three beta-hydroxylase, dihydroflavonol 4-reductase, anthocyanidin reductase, probable ethylene-response sensor, ethylene receptor, auxin-transport-like proteins, auxin-induced proteins, jasmonic acid-amido synthetase	Flavonoid biosynthesis, plant hormone signal transduction, carbon metabolism, glycolysis/gluconeogenesis, biosynthesis of amino acids	Zhang *et al.*, 2023

Continued

Table 3.1. Continued

Crop	Key genes and transcription factors (TFs)	Pathways	References
Avocado	Isoflavone reductase, Glutathione S-transferase, Cinnamyl alcohol dehydrogenase, Cinnamoyl-CoA reductase, Cysteine synthase, Elicitor-responsive protein 3, Chalcone synthase, Superoxide dismutase, Phosphoglucose isomerase, Alanine aminotransferase, Pyruvate decarboxylase, Phosphofructokinase, Sucrose synthase, Alcohol dehydrogenase 1	Carbohydrate metabolism, amino acid metabolism, iosynthesis of other secondary metabolites, glycan biosynthesis and metabolism, energy metabolism	Reeksting *et al.*, 2014
Citrus (*Citrus junos*)	ABA-Insensitive 5-like protein, ABA-responsive element binding factor, Auxin/ Indole3-Acetic Acid, histidinecontaining phosphotransfer protein, epoxide hydrolase	Cutin, suberine, and wax biosynthesis, glycerophospholipid metabolism, plant hormone signal transduction	He *et al.*, 2023
Grape	Glutamyl-tRNA synthetase, glutamyl-tRNA reductase, 5-aminolevulinic acid dehydratase, uroporphyrinogen-III synthase, coproporphyrinogen III oxidase, magnesium-chelatase subunit D, magnesium-protoporphyrin O-methyltransferase, psbOs, psbPs, psbS, psbW, psbY, psb27, cytochrome b6-f complex, α-amylase, β-amylase, hexokinase, trehalose 6-phosphate synthase, glyceraldehyde 3-phosphate dehydrogenase, lactate dehydrogenase	Chlorophyll metabolic pathway, photosynthesis, starch and sucrose metabolism, glycolysis	Zhu *et al.*, 2018
Kiwi	Alcohol dehydrogenase, metallothionein-like protein, 1-aminocyclopropane-1-carboxylate oxidase 1, defensin-like protein, superoxide dismutase, dehydration-responsive protein RD22, pyruvate decarboxylase, stem-specific protein tsjt1, wound-induced protein	Carbohydrate metabolism, amino acids pathways, lipid metabolism, other secondary metabolite pathways, plant hormone signal transduction, starch and sucrose metabolism, unsaturated fatty acid pathways	Zhang *et al.*, 2015
	Trehalose-6- phosphate synthase (TPS), alcohol dehydrogenase (ADH), peroxidase, catalase TFs: AP2/ERF-ERF, C3H, Trihelix, LATERAL ORGAN BOUNDARIES (LOB)	Carbohydrate metabolism, free amino acid metabolism, reactive oxygen species scavenging pathways, nicotinamide adenine dinucleotidedependent glutamate synthase/alanine aminotransferase (NADH-GOGAT/AlaAT) cycle	Li *et al.*, 2022b
	UDP-glycosyltransferase, oxidoreductase, alcohol dehydrogenase (AdADH1, AdADH2) TFs: NAC, MYB, ERF, WRKY, bHLH	Metabolic pathways, stress response pathways, molecular transport, mitotic organization	Xing *et al.*, 2023
	Pyruvate decarboxylases (AcPDC1, AcPDC2), alcohol dehydrogenases (AcADH1, AcADH2) TFs: AcERF4/6/74/75, AcNAC22/84/143	Glycolysis/ gluconeogenesis pathway	Liu *et al.*, 2022b

Continued

Table 3.1. Continued

Crop	Key genes and transcription factors (TFs)	Pathways	References
Mulberry	Respiratory burst oxidase (RBOH), glutathione peroxidase (GPX), ascorbate peroxidase, glutathione S-transferase, mitogen-activated protein kinase kinase 17, MAP kinase 4 substrate 1, Mulatexin (MLX), epidermal patterning factor 1, ethylene-responsive TF, 1-aminocyclopropane-1-carboxylic oxidase, auxin-responsive proteins, beta-glucosidase, glucose-1-phosphate adenylyltransferase	ROS, MAPK and plant hormone signaling, starch and sucrose metabolism, glycolysis and gluconeogenesis	Hu *et al*., 2023
Prunus sp. (Rootstocks)	Alcohol dehydrogenase-like, pyruvate decarboxylase (PDC), RRTF-1 (Redox responsive TF 1), leucine-rich repeat (LRR), hexose-phosphorylating enzymes hexokinases, fructokinases, ascorbate peroxidase 2, ACC oxidase 1	Oxide reduction pathway, glycolytic pathway, ROS scavenging, fermentative pathway, anaerobic pathways	Arismendi *et al*., 2015
Waterlogging/ flood stress in vegetable crops			
Cassava	Cytochrome b6/f, F-type ATPase, glyceraldehyde 3-phosphate dehydrogenase (GAPDH), phosphofructokinase, alcohol dehydrogenase (ADH), pyruvate kinase (PK), L-lactate dehydrogenase (LDH), glucuronosyltransferase, Galactoside-sucrose galactosyltransferase, β-galactosidase	Photosynthesis pathway, glycolysis pathway, galactose metabolic pathway	Cao *et al*., 2022a
Cucumber	Glucose-6-phosphate 1- epimerase, pyruvate kinase, 6-phosphogluconate dehydrogenase, triose phosphate isomerase, phosphofructokinase, alcohol dehydrogenase, aspartate aminotransferase, alanine aminotransferase, nitrate transporters, non-symbiotic hemoglobins, Indole-3-acetic acid-amido synthetase, Auxin efflux carrier components	Carbohydrate mobilization, plant hormone biosynthesis and signalling, nitrate assimilation, cell division	Xu *et al*., 2017
	Alpha-amylase/subtilisin inhibitor, beta-amyrin 11-oxidase, probable isoaspartyl peptidase/L-asparaginase 2, sugar transporter ERD6-like 16, calmodulin-like protein 8, SMAX1-LIKE 4, cinnamoyl-CoA reductase 1, calcium-binding protein PBP1,	Plant hormone signal transduction pathway, environmental information processing, glycerophospholipid metabolism, amino sugar and nucleotide sugar metabolism	Kęska *et al*., 2021
Hot pepper (*Capsicum annuum*)	Beta glucosidase, shikimate O-hydroxycinnamoyl transferase 1, RBOH1/2, WRKY, calmodulins, vetispiradiene, premnaspirodiene oxygenase,	Phenylpropanoid biosynthesis pathways, stress-related pathway	Gong *et al*., 2022

Continued

Table 3.1. Continued

Crop	Key genes and transcription factors (TFs)	Pathways	References
Melon	α-amylase, α-glucan phosphorylase, phosphoglucan phosphatase, sucrose synthase, invertase, hexokinase, glucose6-phosphate isomerase, phosphofructokinase, glyceraldehyde-3-phosphate dehydrogenase, enolase, alanine aminotransferase, lactate dehydrogenase, pyruvate decarboxylase	Energy pathways, glycolysis pathway, anaerobic respiration, ROS scavenging	Zhang *et al.*, 2021b
Sweet potato	Constitutive Triple Response 1, MAPK kinase 4/5 (SlMKK), EIN3, ERFs	Pant hormone signaling, carbohydrate transport, MAPK signaling	Park *et al.*, 2020

are followed by quantification using tools like HTSeq or Kallisto. Differential expression analysis identifies key genes under stress, employing statistical methods such as DESeq2 or EdgeR. Functional annotation and pathway analysis, aided by databases like KEGG and GO, provide biological context. Finally, visualization techniques such as heatmaps and principal component analysis (PCA) summarize data, offering insights into mechanisms for climate resilience in crops. All the steps are discussed in the forthcoming sections.

3.4.1 Experimental design

3.4.1.1 Research objectives and hypotheses

Before starting any transcriptomic study, well-defined research objectives must be established. For example, these could be understanding the molecular responses of particular crops to different abiotic stresses like cold stress, salinity stress, heat stress and drought stress; identification of differentially expressed genes (DEGs) that are associated with enhanced the climate resilience; identification of different biochemical pathways and genes associated with regulatory networks involved in the stress-tolerance mechanism; and identification of different genetic markers and regulatory elements that could be preferential targets for genetic improvement of crop through conventional and advanced breeding programs (genetic engineering). These research efforts are based on significant changes in gene expression profiles due to exposure to different abiotic stresses. The mechanisms behind crop resilience can be better understood because of these transcriptome changes, which also make it feasible to forecast.

3.4.1.2 Choice of crops/specialty and their varieties

The selection of an appropriate crop species and its different varieties for climate resiliency studies is influenced by several factors such as its agricultural significance, its tolerance or susceptibility to particular abiotic and biotic stresses and availability of genomic resource. Plant species that have a well annotated genome paired with a representative transcriptome and wide-spread reference gene set make the analytical part of any dataset more manageable. In general, contrast varieties with respect to a particular trait are used to understand the change in transcriptome data. This kind of comparative analysis is very useful in narrowing down the candidate genes exerting resilience.

3.4.1.3 Sample collection and preparation

For transcriptomics study, sample collections play a pivotal role and the quality as well as consistency of the samples are very important for obtaining reliable results. Robust biological replication is prerequisite, and usually it involves 3–6 replicates per condition to optimize the biological variability. Sample size played crucial role in understanding the molecular behaviour under different biotic and abiotic stresses in order to obtain a statistically significant result. One of the

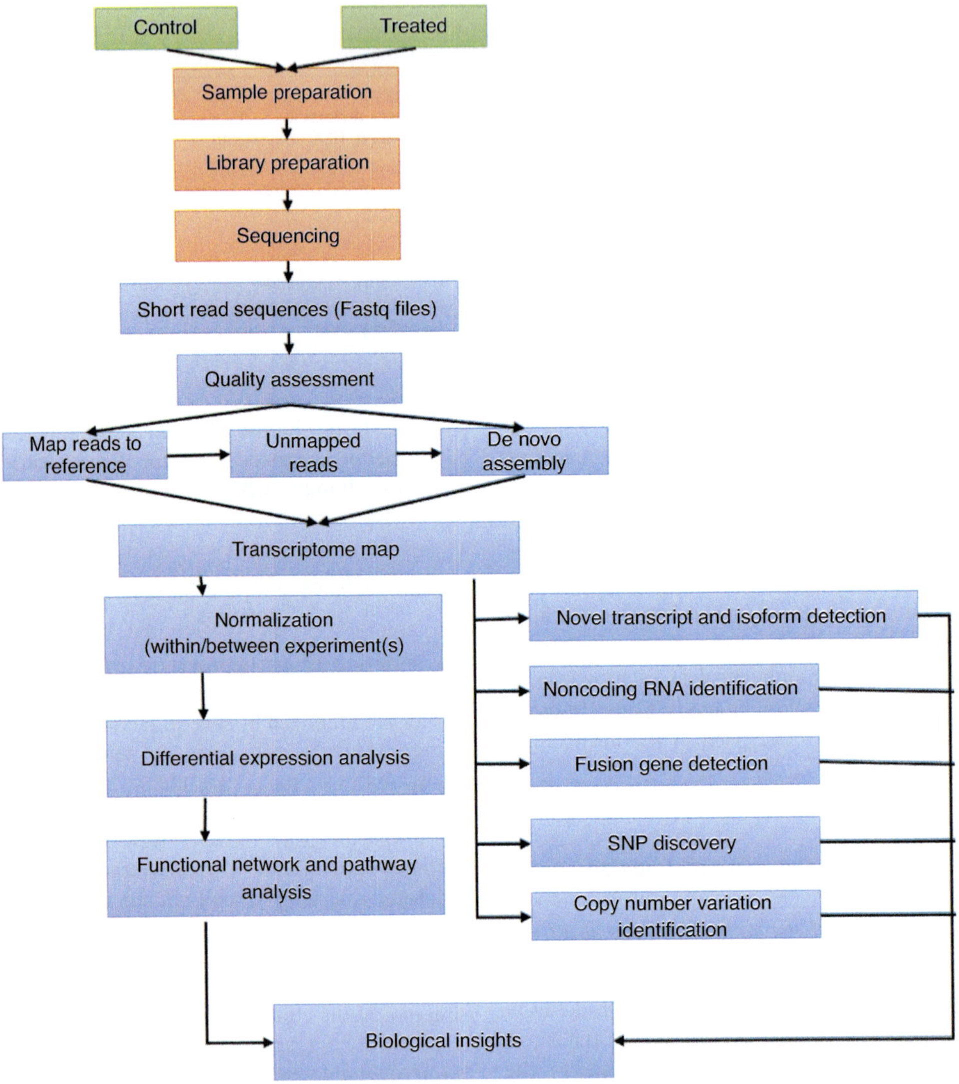

Fig. 3.1. Pipeline of comparative transcriptomic data analysis. Author's own image.

main mechanisms for controlling variables is through randomly assigning sample collection. Methods of sample collection for RNA isolation is very crucial. Tissue samples should be snap-frozen in liquid nitrogen directly after harvesting to maintain the RNA integrity and can then be stored at −80°C to maintain the integrity and intactness of sample for RNA isolation, or processed for RNA extraction right away.

The transcriptomic responses show spatio-temporal variability, i.e. different responses with respect to tissue, and timing of sample collection, making careful selection crucial for obtaining meaningful gene expression changes. The selection of tissue types is critical because different plant parts respond differently to different biotic and abiotic stresses. For example, the first organs to perceive and respond in drought are roots; as a result, roots frequently are the first organs to perceive and respond to conditions of deficit water; in leaves, changes in photosynthesis, water-use efficiency or stomatal regulation would be an observable expression of transcriptomic reorientation. The seeds or stems respond differently to the type and duration of stress. Another factor that could be crucial in understanding the potential impact of transcriptomics is the developmental stage of the plant. Sampling at different growth stages, such as seedling, flowering, or grain filling, may give comprehensive insights into critical periods in which the plant becomes more susceptible to environmental stresses or is more capable of adaptive responses. For instance, the flowering stage might be very responsive to drought stress, wherein a change in the patterns of gene expression could be correlated with reproductive development and yield formation. Similarly, anthesis and grain filling stages are very crucial in wheat during heat stress as pollen and enzymes related to starch biosynthesis, i.e. AGPase, are heat susceptible.

3.4.1.4 RNA extraction and quality control

The construction of an RNA-Seq library requires the acquisition of a pure, high-quality RNA sample. Since RNA is more unstable than DNA, tissues from which RNA is to be extracted must be stored with extreme caution (Farrell Jr, R.E, 2009). Generally, RNA integrity is best preserved at temperatures below −80°C to reduce degradation. RNA stabilization reagents can be used to maintain the RNA integrity during freezing and thawing. For standard RNA-Seq library preparation, it is advisable to begin with 100 ng to 1 μg of purified RNA. Different methods for RNA extraction include organic extraction using reagents such as TRIzol, filter-based spin column methods involving lysis buffers, or magnetic particle-based techniques. RNA samples treated with DNase to remove genomic DNA contamination which can interfere downstream applications.

Once RNA is extracted, it must be stored in an RNase free environment or the product will degrade rapidly. Several reagents are commercially available to facilitate this practice, including Invitrogen THE RNA Storage Solution, DEPC-treated ultrapure H_2O containing 0.1 mM EDTA, TE Buffer, and Invitrogen RNAsecure Resuspension Solution. To improve the number of relevant RNA-Seq reads, it is often necessary to enrich the sample for specific RNA regions or targets. This step decreases the non-transcriptome RNA content by removing RNAs like residual ribosomal RNA (rRNA), and enhances the discovery of low copy number sequences. The RiboMinus technology enriches RNA reads along the entire length of transcripts independent of polyadenylation and this method can remove up to 99.9% of rRNA, allowing for a more thorough investigation of less abundant transcripts.

The quality of mRNA sample is checked by running mRNA on formaldehyde gels to check dark 18S and 28S rRNA bands and high molecular weight mRNA smear. If these bands are faint or there is a lot of low molecular weight smearing, then the sample is likely to be degraded. The Agilent Bioanalyzer also gives an indication of the quality and an RNA Integrity Number (RIN) total score of above eight is considered a high-quality input. Chloroplast rRNA can interrupt on RIN accuracy, so the interpretation of the electropherogram needs to be performed cautiously. mRNA or cDNA fragmentation is done with nebulization to provide sequence uniformity. The Nanodrop Spectrophotometer also measures RNA purity by checking absorbance ratios (A260/A280 and A260/A230) and ideal ratios are approximately 2.0 for both, while the Qubit Flurometer measures RNA concentration more accurately

Transcriptome analysis using NGS follows two main approaches: align-then-assemble (alignment-first) or assemble-then-align (assemble-first). Alignment-first methods rely on an existing reference genome, while assemble-first methods are de novo assembly algorithms that may use a reference genome only for identifying exon–exon junctions. Alignment-first algorithms, i.e. Scripture and Cufflinks, and several R packages for differential expression analysis, including DEGseq, and the two-stage Poisson model rely on tools like Bowtie (Langmead *et al.*, 2009) and TopHat (Trapnell *et al.*, 2009) to map sequences to a reference genome (Guttman *et al.*, 2010; Wang *et al.*, 2010). Assemble-first methods involve de novo assembly, with alignment to a close relative aiding in result validation and exon–exon boundary identification, crucial for detecting alternative splicing. Knowing these boundaries also helps design primers for RT-PCR to study specific gene expression across tissues, time points, or environments.

Although RNA can be sequenced directly, most RNA-Seq experiments employ DNA sequencing machines because DNA sequencing technology is more advance and sophisticated. Conversion of RNA into complementary DNA (cDNA) by using reverse transcriptase enzyme and building a cDNA library is crucial step in RNA-Seq. Adapter sequences are appended to both ends of each cDNA molecule in the library, giving it the appropriate length for amplification and sequencing on a particular platform. The method for creating a cDNA library can change based on the kind of RNA being investigated because RNA molecules differ in terms of size, quantity, sequence, and structure. Capturing the target RNA, converting it into the appropriately sized double-stranded cDNA, and adding the required adaptor sequences for amplification and sequencing are crucial steps.

Since the majority of mRNAs that code for proteins and a large number of long noncoding RNAs (lncRNAs) in eukaryotes contain poly(A) tails, polyadenylated RNA sequencing (RNA-Seq) is often applied. The tail helps to separate poly(A)+ RNAsfrom total cellular RNA, which accounts for 1–5% of the total RNA pool. Enrichment can be obtained by reverse transcription (RT) utilizing oligo-dT priming or by coating magnetic or cellulose beads with oligo-dT molecules. rRNA is eliminated by using biotinylated DNA or LNA (locked nucleic acid) probes followed by depleting with streptavidin probes. Also, rRNA can be removed by using a specific NSR (not so random) primer commercialized by NuGEN.

After poly(A) selection and rRNA depletion, desired RNA samples are fragmented to fit within the size limitations of most sequencing platforms, such as < 600 bp for Illumina sequencers. RNA fragmentation can be performed by using alkaline solutions, divalent cations, or enzymes like RNase III, often at elevated temperatures to reduce structural interference. Also, intact RNA can be reverse transcribed and cDNA can be fragmented using acoustic shearing, DNAse treatment and the tagmentation method. In the tagmentation method, Tn5 transposase mediated DNA fragmentation is followed by adaptor ligation.

3.4.2 Preprocessing and quality control

Preprocessing and quality control of transcriptomic data are crucial steps to ensure the reliability of downstream analyses.

3.4.2.1 Raw data quality assessment

Raw data quality assessment in transcriptomic studies involves several key parameters, each with specific threshold values to ensure data integrity. The Phred quality score, which indicates the accuracy of base calling, should generally be above 30 for the majority of the bases, indicating a base call accuracy of 99%. Reads with quality scores below this threshold are often trimmed or discarded. Adapter contamination should be minimal, typically less than 5% of the total reads. Sequence duplication levels, which indicate the redundancy of reads, should ideally be below 10%, although higher levels might be acceptable in low-complexity libraries. GC content should fall within a normal distribution range specific to the organism, typically between 40 and 60% for most species. Additionally, the percentage of reads mapped to the reference genome should be high, usually over 80%, to ensure comprehensive coverage. The percentage of reads aligned to rRNA should be low, generally less than 5%, as high levels indicate contamination. Overall read length distribution should match the expected library size, and any significant deviation might indicate technical issues. These threshold parameters are crucial for filtering out low-quality data and ensuring the reliability of subsequent transcriptomic analyses. The FastQC (Andrews *et al.*, 2010) tool has been used for this quality assessment.

3.4.2.2 Read trimming and filtering

Low-quality reads with a Phred score < 30 are trimmed off. Additionally, adapter sequences, which are remnants from the sequencing process, are also removed to prevent mapping errors. After trimming, reads are filtered based on length and overall quality. A common threshold is to retain reads longer than 50–100 base pairs, as shorter reads may map ambiguously. Both trimming and filtering are vital for minimizing technical artifacts and ensuring that the data used in downstream analyses accurately represents the biological samples. Properly processed reads lead to more reliable and reproducible results in transcriptomic studies.

3.4.3 Alignment and quantification

The process begins with aligning sequencing reads to a reference genome or transcriptome, followed by quantifying the abundance of genes or transcripts.

3.4.3.1 Mapping reads to reference genomes or transcriptomes

Reads are mapped to a reference genome or transcriptome using aligners such as:

- BWA (Burrows–Wheeler Aligner) (Li and Durbin, 2009): efficient for aligning short reads to a large genome.
- Bowtie (Langmead *et al.*, 2009): known for its speed and accuracy, suitable for short-read alignment.
- STAR (Spliced Transcripts Alignment to a Reference) (Dobin *et al.*, 2013): excellent for RNA-Seq data due to its handling of spliced reads.

Aligners typically produce sequence alignment/map (SAM) and binary alignment/map (BAM) files, which store alignment information. SAM files are converted to BAM, then sorted and indexed using Samtools.

3.4.3.2 Quantification tools and methods

Quantification of mapped reads is a crucial step in RNA-Seq data analysis, allowing researchers to measure gene expression levels. Tools like HTSeq (Anders *et al.*, 2015) and Kallisto (Bray *et al.*, 2016) are widely used for this purpose, each offering unique features and methodologies.

HTSeq is a Python-based tool that counts the number of reads mapping to each gene in an aligned RNA-Seq dataset. It requires an alignment file (e.g. in BAM format) and a corresponding gene annotation file (e.g. in GTF format). HTSeq works in two main modes: union, intersection-strict, and intersection-nonempty. The union mode, which is the default, assigns reads to genes if they overlap any exon of a gene, ensuring no reads are counted multiple times. HTSeq is particularly useful for its flexibility and detailed reporting of ambiguous reads and various count modes.

Kallisto, on the other hand, employs a different approach called pseudo-alignment. Instead of mapping reads to the genome, Kallisto maps reads directly to a transcriptome, allowing for faster and more memory-efficient processing. This method uses k-mer hashing to quickly determine the compatibility of reads with transcripts, resulting in a probabilistic assignment of reads to transcripts. Kallisto provides both raw read counts and normalized expression values, such as transcripts per million (TPM) and estimated counts, making it suitable for transcript-level expression analysis. While HTSeq relies on precise genome annotations and can handle complex genomic regions well, Kallisto's speed and efficiency make it ideal for large-scale transcriptomic studies.

3.4.4 Differential expression analysis

Differential expression analysis is a crucial technique in transcriptomics, used to identify changes in gene expression levels across different experimental conditions. R packages such as DESeq2 (Love *et al.*, 2014), EdgeR (Robinson *et al.*, 2010), and Limma (Ritchie *et al.*, 2015) are widely employed for this purpose, each offering unique features and advantages. DESeq2 is based on the negative binomial distribution and is specifically designed for RNA-Seq data. DESeq2 models the count data using a generalized linear model and employs shrinkage estimators to handle the variability in dispersion estimates. This results in more accurate identification of DEGs, especially in datasets with small sample sizes. EdgeR also uses the negative

binomial distribution to model count data. It is particularly well-suited for experiments with biological replicates. EdgeR includes robust normalization procedures to account for differences in library sizes and uses empirical Bayes methods to estimate dispersions, enhancing its power to detect DEGs in noisy data. Originally developed for microarray data, Limma has been adapted for RNA-Seq through the voom transformation, which converts count data to log-counts per million (log-CPM) and estimates the mean-variance relationship.
Here, a demonstration of R package DESeq2 for differential gene expression analysis is given using a simple dataset present in 'airway' R package.

1. Download the sample dataset from the 'airway' package. The package contains two data files. One file contains read counts of 64,102 genes in eight samples obtained from the RNA-Seq experiment on four primary human airway smooth muscle cell lines treated with one micromolar dexamethasone for 18 hours. Another file contains sample-wise metadata information, namely treated or untreated (Box 3.1; Fig. 3.2).
2. Import the count matrix and metadata file into RStudio. After that, we have to check whether the number and order of column names in the count matrix are the same as the number and order of row names in the metadata file. If it is so, then we have to load the package '**DESeq2**' to perform the subsequent differential gene expression analysis. We have to create a DESeqDataSet object and then run the 'DESeq()' function to perform the said analysis (Box 3.2).

Here, we are trying to find the genes that are differentially expressed in dexamethasone-treated conditions as compared to untreated conditions. Hence, the reference level is set as 'untreated'. After the analysis, the result contains base means, log2FoldChange values, p-values, adjusted p-values, etc. for each gene. An adjusted p-value greater than 0.01 indicates that the result is likely to have occurred by chance and is considered non-significant. Otherwise, that gene is differentially expressed if the adjusted p-value is < 0.01. In the latter case, if the log2FoldChange value is > 0, the gene is upregulated and if it is < 0, then that gene is downregulated. Thus, we can find out DEGs using R. After identifying DEGs, results can be visualized in terms of various plots such as MA plot, volcano plot, heatmap, etc.

3.4.5 Functional annotation and pathway analysis

Functional annotation and pathway analysis are essential steps in understanding how organisms respond to environmental stressors and identifying the molecular basis of climate-resilient traits. These methods help to interpret transcriptomics data by assigning biological meaning to differentially expressed genes and mapping them to known biological pathways. This process is crucial for understanding the dynamics of how organisms adapt to or withstand climate-related stresses such as drought, heat, or cold.

3.4.5.1 Gene ontology (GO) enrichment

Gene ontology (GO) is a widely used framework for the standardized representation of gene and gene product attributes across species. It consists of three main categories:

- Biological process (**BP**): The biological objectives or processes in which a gene or gene product is involved (e.g. apoptosis, cell division).
- Molecular function (MF): The biochemical activity of a gene product (e.g. enzyme activity, binding).
- Cellular component (CC): The location within the cell where a gene product is active (e.g. nucleus, mitochondrion).

GO enrichment analysis aims to identify GO terms that are over-represented among a list of genes compared to what would be expected by chance. This provides insights into the biological processes, molecular functions, or cellular components that may be involved in the phenotype of interest. Several tools and software packages are available to perform functional annotation and enrichment analysis:

Box 3.1. R code to collect sample dataset from "airway" package

```
> BiocManager::install("airway")
> library(airway)
> data(airway)
> airway
class: RangedSummarizedExperiment
dim: 64102 8
metadata(1): ''
assays(1): counts
rownames(64102): ENSG00000000003 ENSG00000000005 ... LRG_98 LRG_99
rowData names(0):
colnames(8): SRR1039508 SRR1039509 ... SRR1039520 SRR1039521
colData names(9): SampleName cell ... Sample BioSample
> sample_info <- as.data.frame(colData(airway))
> sample_info <- sample_info[,c(2,3)]
> sample_info$dex <- gsub('trt', 'treated', sample_info$dex)
> sample_info$dex <- gsub('untrt', 'untreated', sample_info$dex)
> names(sample_info) <- c('cellLine', 'dexamethasone')
> # Get the samplewise metadata file
> write.table(sample_info, file = "/sample_info.csv", sep = ',', col.names = T, row.names = T, quote = F)
> # Get the matrix of read counts for each gene in every sample
> countsData <- assay(airway)
> write.table(countsData, file = "/counts_data.csv", sep = ',', col.names = T, row.names = T, quote = F)
```

1. **DAVID (Database for Annotation, Visualization, and Integrated Discovery).** DAVID (Dennis *et al.*, 2003) is a comprehensive tool for functional annotation and enrichment analysis. It integrates biological data to help understand the biological meaning behind a large list of genes. It can be used for GO enrichment, pathway analysis (KEGG), protein–protein interactions, and functional annotation clustering. It provides a ranked list of GO terms, pathways, and other annotations that are enriched in the gene list.
2. **GSEA (Gene Set Enrichment Analysis).** GSEA (Subramanian *et al.*, 2005) is used to determine whether predefined gene sets (e. g.those associated with specific biological pathways) show statistically significant differences between two biological states (e.g.diseased versus healthy).It does not require a predefined threshold for significance; instead, it analyzes all genes in the context of gene sets. It provides enrichment scores and p-values indicating whether a gene set is significantly enriched in the data.
3. **Enrichr**. Enrichr (Chen *et al.*, 2013) is a powerful and intuitive tool for GO enrichment analysis and gene set enrichment analysis. It is integrated with multiple gene set libraries, including GO, KEGG, Reactome, and more. Output: Interactive results, with visualization options such as bar plots and network diagrams.
4. **PANTHER (Protein Analysis Through Evolutionary Relationships)**. PANTHER (Mi *et al.*, 2016) is a classification system that combines gene function, pathways, and protein classes to perform enrichment analysis. It performs GO enrichment, pathway analysis, and classifies genes/proteins into families and subfamilies. It provides enrichment results with a focus on biological pathways and protein classes.

3.4.5.2 Databases for pathway analysis

Pathway analysis involves mapping the identified genes to known biological pathways to understand the broader biological context. Several databases provide curated pathway information:

1. **KEGG (Kyoto Encyclopedia of Genes and Genomes)**. KEGG (Kanehisa and Goto, 2000) is a comprehensive resource

(a)

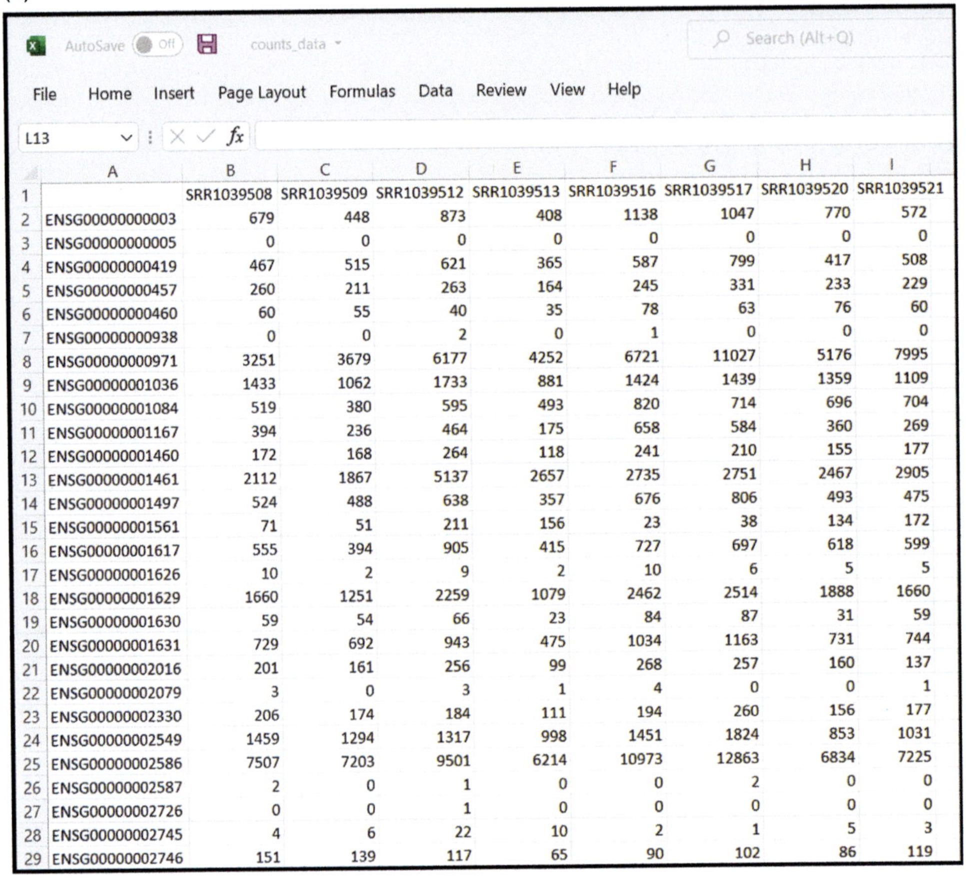

	SRR1039508	SRR1039509	SRR1039512	SRR1039513	SRR1039516	SRR1039517	SRR1039520	SRR1039521
ENSG00000000003	679	448	873	408	1138	1047	770	572
ENSG00000000005	0	0	0	0	0	0	0	0
ENSG00000000419	467	515	621	365	587	799	417	508
ENSG00000000457	260	211	263	164	245	331	233	229
ENSG00000000460	60	55	40	35	78	63	76	60
ENSG00000000938	0	0	2	0	1	0	0	0
ENSG00000000971	3251	3679	6177	4252	6721	11027	5176	7995
ENSG00000001036	1433	1062	1733	881	1424	1439	1359	1109
ENSG00000001084	519	380	595	493	820	714	696	704
ENSG00000001167	394	236	464	175	658	584	360	269
ENSG00000001460	172	168	264	118	241	210	155	177
ENSG00000001461	2112	1867	5137	2657	2735	2751	2467	2905
ENSG00000001497	524	488	638	357	676	806	493	475
ENSG00000001561	71	51	211	156	23	38	134	172
ENSG00000001617	555	394	905	415	727	697	618	599
ENSG00000001626	10	2	9	2	10	6	5	5
ENSG00000001629	1660	1251	2259	1079	2462	2514	1888	1660
ENSG00000001630	59	54	66	23	84	87	31	59
ENSG00000001631	729	692	943	475	1034	1163	731	744
ENSG00000002016	201	161	256	99	268	257	160	137
ENSG00000002079	3	0	3	1	4	0	0	1
ENSG00000002330	206	174	184	111	194	260	156	177
ENSG00000002549	1459	1294	1317	998	1451	1824	853	1031
ENSG00000002586	7507	7203	9501	6214	10973	12863	6834	7225
ENSG00000002587	2	0	1	0	0	2	0	0
ENSG00000002726	0	0	1	0	0	0	0	0
ENSG00000002745	4	6	22	10	2	1	5	3
ENSG00000002746	151	139	117	65	90	102	86	119

(b)

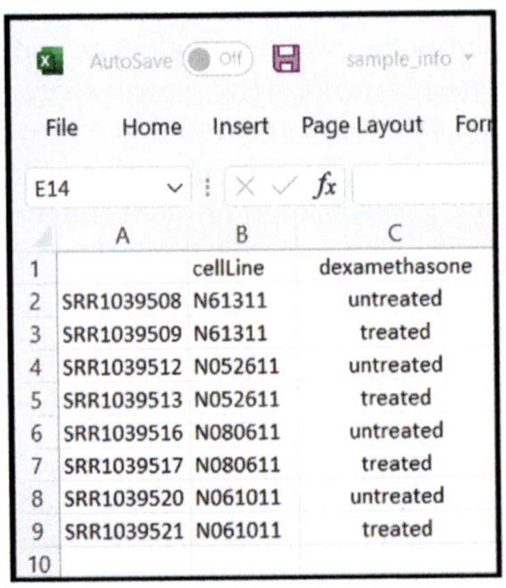

	cellLine	dexamethasone
SRR1039508	N61311	untreated
SRR1039509	N61311	treated
SRR1039512	N052611	untreated
SRR1039513	N052611	treated
SRR1039516	N080611	untreated
SRR1039517	N080611	treated
SRR1039520	N061011	untreated
SRR1039521	N061011	treated

Fig. 3.2. (a) Screenshot of the count matrix. (b) Screenshot of the metadata file. Author's own image.

Box 3.2. Differential gene expression analysis using the "DESeq2" package in R

```
> BiocManager::install("DESeq2")
> library(DESeq2)
> # read in counts data
> counts_data <- read.csv('/counts_data.csv')
> # read in sample info
> colData <- read.csv('/sample_info.csv')
> # making sure the row names in colData matches to column names in counts_data
> all(colnames(counts_data) %in% rownames(colData))
[1] TRUE
> # are they in the same order?
> all(colnames(counts_data) == rownames(colData))
[1] TRUE
> dds <- DESeqDataSetFromMatrix(countData = counts_data,
+                         colData = colData,
+                         design = ~ dexamethasone)
> dds
class: DESeqDataSet
dim: 64102 8
metadata(1): version
assays(1): counts
rownames(64102): ENSG00000000003 ENSG00000000005 ... LRG_98 LRG_99
rowData names(0):
colnames(8): SRR1039508 SRR1039509 ... SRR1039520 SRR1039521
colData names(2): cellLine dexamethasone
> # pre-filtering: removing rows with low gene counts
> # keeping rows that have at least 10 reads total
> keep <- rowSums(counts(dds)) >= 10
> dds <- dds[keep,]
> # set the factor level
> dds$dexamethasone <- relevel(dds$dexamethasone, ref = "untreated")
> # --------Run DESeq ---------------------
> dds <- DESeq(dds)
estimating size factors
estimating dispersions
gene-wise dispersion estimates
mean-dispersion relationship
final dispersion estimates
fitting model and testing
> res <- results(dds)
> res
log2 fold change (MLE): dexamethasone treated vs untreated
Wald test p-value: dexamethasone treated vs untreated
DataFrame with 22369 rows and 6 columns
          baseMean log2FoldChange lfcSE stat pvalue padj
          <numeric> <numeric> <numeric> <numeric> <numeric>
<numeric>
ENSG00000000003 708.5979   -0.3788229 0.173155 -2.187769 0.0286865 0.138470
ENSG00000000419 520.2963   0.2037893 0.100742 2.022878 0.0430857 0.182998
ENSG00000000457 237.1621   0.0340631 0.126476 0.269325 0.7876795 0.929805
ENSG00000000460 57.9324   -0.1171564 0.301583 -0.388472 0.6976669 0.894231
ENSG00000000971 5817.3108   0.4409793 0.258776 1.704099 0.0883626 0.297042
...          ...     ...     ...     ...     ...     ...
ENSG00000273483 2.68955   0.600441 1.084447 0.553684 0.5797949 NA
```

Continued

Box 3.2. *(Continued)*

```
ENSG00000273485 1.28646  0.194074 1.346550 0.144127 0.8854003 NA
ENSG00000273486 15.45244  -0.113321 0.426034 -0.265991 0.7902460 0.930697
ENSG00000273487 8.16327  1.017800 0.575797 1.767637 0.0771216 0.271627
ENSG00000273488 8.58437  0.218105 0.570714 0.382161 0.7023421 0.896550
> summary(res)

out of 22369 with nonzero total read count
adjusted p-value < 0.1
LFC > 0 (up) : 1884, 8.4%
LFC < 0 (down) : 1502, 6.7%
outliers [1] : 51, 0.23%
low counts [2] : 3903, 17%
(mean count < 4)
[1] see 'cooksCutoff' argument of ?results
[2] see 'independentFiltering' argument of ?results

> res0.01 <- results(dds, alpha = 0.01) # When padj = 0.01
> summary(res0.01)

out of 22369 with nonzero total read count
adjusted p-value < 0.01
LFC > 0 (up)      : 1030, 4.6%
LFC < 0 (down)      : 708, 3.2%
outliers [1]      : 51, 0.23%
low counts [2]      : 5200, 23% (mean count < 6)
[1] see 'cooksCutoff' argument of ?results
[2] see 'independentFiltering' argument of ?results
```

for understanding high-level functions and utilities of the biological system, such as cells, organisms, and ecosystems. It contains pathway maps for metabolic, genetic, environmental information processing, cellular processes, and human diseases. Mapping genes to KEGG pathways can reveal the biochemical pathways that may be disrupted in a disease or influenced by genetic variation.

2. **Reactome**. Reactome (Fabregat *et al.*, 2016) is an open-source, curated database of pathways and reactions in human biology. It covers a wide range of biological processes, including signal transduction, metabolism, and DNA repair. Pathway enrichment analysis using Reactome can help to identify the biological pathways that are significantly associated with a set of genes.
3. **BioCyc**. BioCyc (Paley and Karp, 2021) is a collection of pathway/genome databases (PGDBs) that provide information on metabolic pathways and the genomes that encode them. It includes detailed information on enzymes, metabolic pathways, and gene regulation. It is useful for studying metabolism and microbial genomics.

3.4.5.3 *Identifying and interpreting biological pathways*

The identification and interpretation of biological pathways involve several steps:

1. **Gene Mapping.** The first step is to map the list of genes or genetic variants to their corresponding pathways using databases like KEGG or Reactome.
2. **Enrichment Analysis.** Next, enrichment analysis is performed to identify which pathways are over-represented in the gene list compared to what would be expected by chance. Tools like DAVID, GSEA, or Enrichr can be used for this purpose.

3. **Interpretation**. The enriched pathways are then interpreted in the context of the biological question at hand. For instance:
   - Disease mechanisms: Pathways enriched in disease-associated genes may point to underlying biological mechanisms or potential therapeutic targets.
   - Drug response: Pathways involved in drug metabolism or signaling may be enriched in genes associated with drug response phenotypes.
   - Biological functions: Enriched pathways may provide insights into the biological processes that are active in certain cell types or conditions.
4. **Visualization**. Visualization tools, such as Cytoscape (Kohl *et al.*, 2011) for network analysis or PathVisio (Kutmon *et al.*, 2015) for pathway diagrams, can be used to create visual representations of the enriched pathways, making it easier to interpret complex biological data.

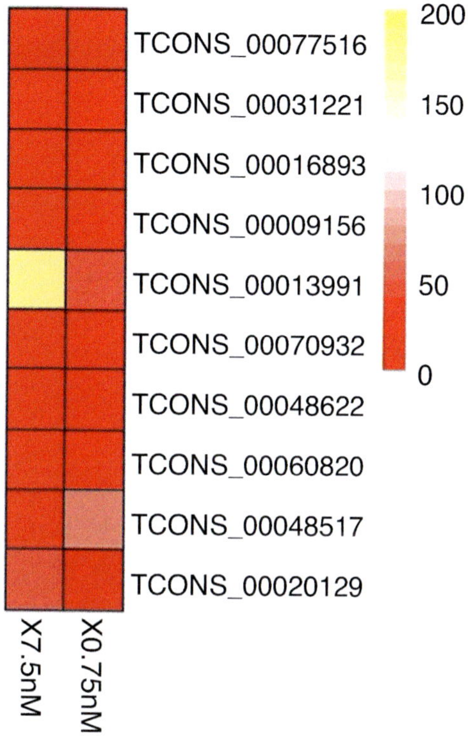

Fig. 3.3. Illustrative heatmap. Heatmaps consists of rows representing genes or transcripts and columns representing samples or conditions, with colors indicating gene expression levels, where each color corresponds to a specific range of values (yellow for high expression and red for low expression). Author's own image.

3.4.6 Visualization of transcriptomic data

Visualization plays a crucial role in the analysis and interpretation of transcriptomics data. It helps in identifying patterns, trends, and relationships within large datasets, making complex biological insights more accessible. Common visualization techniques include heatmaps, volcano plots, principal component analysis (PCA), and network diagrams, each serving different purposes in transcriptomic studies.

3.4.6.1 Heatmaps

Heatmaps are widely used to visualize the expression levels of genes across different samples. They provide an intuitive way to represent data by using color gradients to indicate the magnitude of expression (Fig. 3.3). Heatmaps are primarily used to identify patterns of gene expression, such as clusters of co-regulated genes or differences between experimental conditions. The visual patterns in a heatmap can indicate biologically relevant gene sets, such as those involved in specific pathways or cellular processes.

Applications:

- Cluster analysis: Heatmaps can be used in conjunction with hierarchical clustering to group genes and samples based on similar expression patterns.
- Differential expression: Highlighting DEGs between conditions, such as treatment versus control, can reveal biologically significant changes.

3.4.6.2. Volcano plots

Volcano plots are scatter plots that visualize the results of differential expression analysis by plotting statistical significance against fold change (Fig. 3.4.). Volcano plots help in identifying

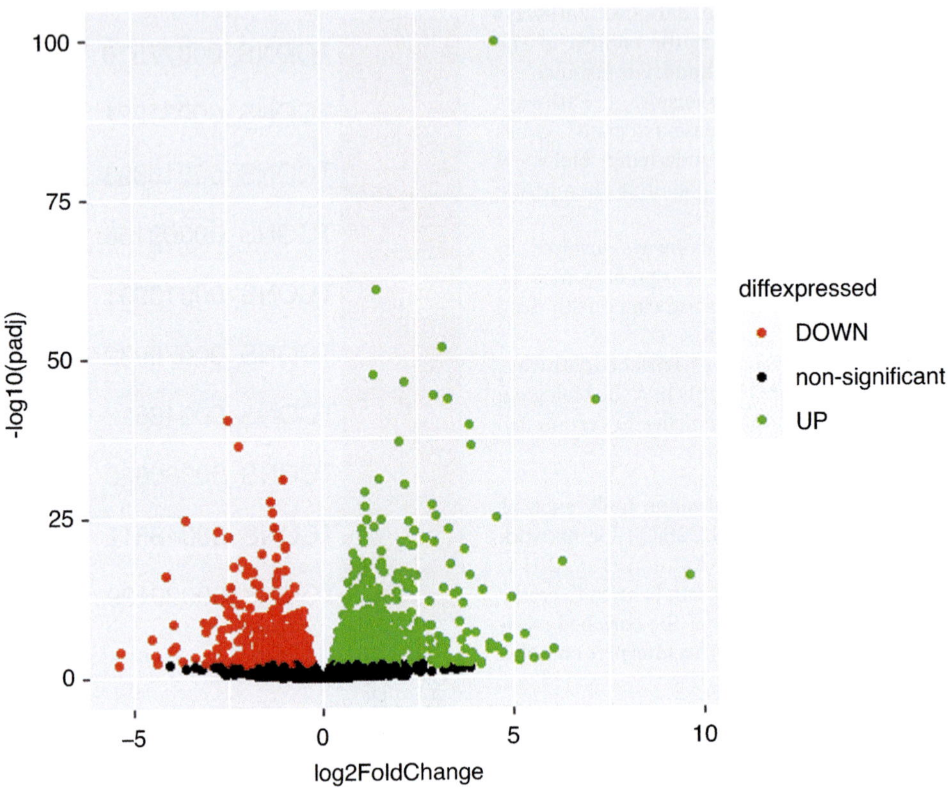

Fig. 3.4. Volcano plot. The X-axis represents the log2 fold change, indicating the magnitude of gene expression change, and the Y-axis represents the –log10 p-value, indicating the statistical significance, where genes in the upper left quadrant are significantly downregulated, those in the upper right quadrant are significantly upregulated, and the center represents genes with non-significant changes. Author's own image.

genes that are significantly differentially expressed between two conditions. Users can set thresholds for fold change and p-value to focus on the most relevant genes. The plot highlights the genes that exhibit both large fold changes and strong statistical significance, making it easier to pinpoint key drivers of the biological response.

3.4.6.3 Principal component analysis (PCA)

Principal component analysis (PCA) is a dimensionality reduction technique that transforms high-dimensional data into a smaller set of principal components while preserving as much variance as possible. PCA is used to reduce the complexity of transcriptomic data and to visualize patterns, such as similarities or differences between samples. PCA plot includes principal components PC1, PC2, etc., which are new axes that capture the most variance in the data and samples are plotted in the space of the first two or three PCs as scatter plot with each point representing a sample. The spread of samples along the principal components reflects the underlying variability in the dataset, with close clustering indicating similarity and separation suggesting differences.

Applications:

- Sample clustering: PCA can reveal natural groupings of samples, such as separating treated versus untreated samples or distinguishing different cell types.

- Outlier detection: Samples that do not cluster with others may be outliers, indicating potential issues or unique biology.

The spread of samples along the principal components reflects the underlying variability in the dataset, with close clustering indicating similarity and separation suggesting differences.

3.4.6.4 Network diagrams

Network diagrams are used to visualize the relationships and interactions between genes, proteins, or other molecular entities. Network diagrams illustrate complex interactions, such as gene regulatory networks, protein–protein interactions, or pathway connections. They consist of nodes representing genes, proteins, or other entities, and edges representing interactions or relationships between the nodes (e.g. co-expression, physical interaction), with attributes like color-coding or sizing of nodes and edges based on properties such as expression levels or interaction strength. Network diagrams provide a holistic view of molecular interactions, helping researchers understand the interconnectedness of biological processes and identify critical hubs or bottlenecks in the network.

Applications:

- Pathway analysis: Networks can highlight key pathways or modules that are active in the data.
- Interaction mapping: Visualize how different genes or proteins interact, potentially revealing novel regulatory mechanisms.

3.5 Multiomics Approaches

High-throughput technologies have given rise to a range of approaches characterized by the 'omics' suffix, such as genomics, transcriptomics, proteomics, metabolomics, etc., to measure various biological entities, including DNA, RNA, proteins, and metabolites, within a species. The big data from omics experiments are analyzed with advanced software programs and analytical methods to understand the complexity of biological systems. This new area of research has come to be known as either integrative biology or systems biology. Systems biology aims to integrate various omics data to construct models that provide a holistic picture of biological systems. It is essential for truly capturing the complexity of biological traits, particularly those that are multifaceted and intricate. By examining the roles of various molecular components (genes, transcripts, proteins, and metabolites) and how they respond to environmental stimuli, researchers can piece together a more complete picture of the regulatory processes and mechanism of complex traits like drought resistance. In recent years, multiomics approaches have significantly advanced our understanding of drought tolerance, particularly in crops like rice. Genomic studies, such as those conducted by Vikram *et al.* (2012) and Fleury *et al.* (2010), have identified numerous QTLs that are critical for drought resistance. Also, comparative genomic analyses showed the role of OsCPK9 and AtCPK10 genes in augmenting the abscisic acid (ABA)-dependent drought responsive mechanism (Zou *et al.*, 2010; Wei *et al.*, 2014), shedding light on the genetic basis of stress responses. Beyond genomics, transcriptome-wide studies revealed that a large number of TFs have helped to confer drought tolerance in rice. These factors are involved in both ABA-dependent and independent pathways, which regulate complex networks of gene expression in response to drought (Hu *et al.*, 2006; Xu *et al.*, 2008; Xiao *et al.*, 2009; Wu *et al.*, 2009; Yang *et al.*, 2010). Ma *et al.* (2009) analyzed the transcriptomic and metabolic responses of two rice cultivars with differing levels of drought tolerance under prolonged drought conditions. Their study revealed that a significant number of DEGs associated with photosynthesis were upregulated in the drought-tolerant variety. Proteomic analyses, such as those by Wang *et al.* (2017b), have provided further insights by detailing the physiological disruptions that occur under drought conditions and highlighting the activation of antioxidant defenses that help mitigate these effects. Additionally, metabolomic studies have offered a deeper understanding of the biochemical adjustments plants make when faced with drought stress. These studies emphasize the critical roles of primary and secondary metabolites in maintaining essential processes like photosynthesis and overall plant health during periods of water scarcity (Ma *et al.*, 2009; Khakimov *et al.*, 2014). When combined, these multiomics approaches provide a comprehensive framework for understanding the molecular and physiological mechanisms that underpin drought tolerance. This holistic understanding is not just academic; it holds the key to develop more resilient crop varieties that can thrive in increasingly unpredictable environmental conditions. By unraveling the complex story of how plants respond to drought or any other environmental stress conditions, scientists are paving the way for innovations in agriculture that will help secure food supplies in the face of global climate change.

3.6 Conclusion

Transcriptomics has been instrumental in unraveling and understanding the dynamics of gene expression and interacting pathways under various environmental stresses, and helps in the dissection of mechanisms driving climate resilience. Techniques like RNA-Seq, scRNA-Seq, and microarrays facilitate the identification of key players and pathways involved in the regulation of plant responses under drought, heat, and salinity. Advances in computational tools and statistical methods, such as differential expression analysis and pathway enrichment, streamline the discovery of actionable targets for crop breeding programs. Integrating transcriptomics with genomics, proteomics, and metabolomics enriches the systems biology perspective, provides a holistic overview of plant responses to climatic challenges. While transcriptomics faces technical challenges like scalability, data interpretation and cost-effectiveness, new emerging technologies such as single-cell and long-read sequencing hold promise to overcome these hurdles. Leveraging single-cell transcriptomics can offer unprecedented insights into cell-specific gene expression under stress, enabling more targeted interventions. Long-read technologies, such as those offered by PacBio or Oxford Nanopore, can enhance transcriptome assembly accuracy and facilitate the detection of alternative splicing events. Developing experimental protocols to simulate diverse climate stress conditions, including combinations like drought and heat, will refine the understanding of stress tolerance mechanisms. Expanding efforts in multiomics integration with machine learning and artificial intelligence tools can provide predictive models for trait performance and accelerate breeding programs. Application of transcriptomics to neglected and underutilized crop species can unlock their potential as resilient food sources under changing climates. Collaborations between molecular biologists and field agronomists can ensure the translation of transcriptomic findings into practical breeding outcomes. Publicly available databases for transcriptomic data specific to climate resilience will empower global research collaborations. These directions position transcriptomics as a cornerstone in achieving sustainable agriculture and food security amidst the challenges posed by climate change.

References

Ahmed, H.G.M.D., Zeng, Y., Khan, M.A., Rashid, M.A.R., Ameen, M. *et al.* (2023) Genome-wide association mapping of bread wheat genotypes using yield and grain morphology-related traits under different environments. *Frontiers in Genetics* 13, 1008024. DOI: 10.3389/fgene.2022.1008024.

Alghamdi, S.S., Khan, M.A., Ammar, M.H., Sun, Q., Huang, L. *et al.* (2018) Characterization of drought stress-responsive root transcriptome of faba bean (*Vicia faba* L.) using RNA sequencing 1–19.

Allen, A.M., Winfield, M.O., Burridge, A.J., Downie, R.C. and Benbow, H.R. (2017) Characterization of a Wheat Breeders' Array suitable for high-throughput SNP genotyping of global accessions of hexaploid bread wheat (*Triticum aestivum*). *Plant Biotechnology Journal* 15(3), 390–401. DOI: 10.1111/pbi.12635.

Anders, S., Pyl, P.T. and Huber, W. (2015) HTSeq—a python framework to work with high-throughput sequencing data. *Bioinformatics* 31(2), 166–169. DOI: 10.1093/bioinformatics/btu638.

Andrews, S., Krueger, F., Segonds-Pichon, A., Biggins, L., Krueger, C. *et al.* (2010) FastQC. A quality control tool for high throughput sequence data. Available at: http://www.bioinformatics.babraham.ac.uk/projects/fastqc (accessed 12 June 2025).

Arisha, M.H., Ahmad, M.Q., Tang, W., Liu, Y., Yan, H. *et al.* (2020) RNA-sequencing analysis revealed genes associated drought stress responses of different durations in hexaploid sweet potato. *Scientific Reports* 10(1), 12573. DOI: 10.1038/s41598-020-69232-3.

Arismendi, M.J., Almada, R., Pimentel, P., Bastias, A., Salvatierra, A. *et al.* (2015) Transcriptome sequencing of *Prunus* sp. rootstocks roots to identify candidate genes involved in the response to root hypoxia. *Tree Genetics & Genomes* 11, 1–16.

Bauer, E., Schmutzer, T., Barilar, I., Mascher, M., Gundlach, H. *et al.* (2017) Towards a whole-genome sequence for rye (*Secale cereale* L.). *The Plant Journal* 89(5), 853–869. DOI: 10.1111/tpj.13436.

Bawa, G., Liu, Z., Yu, X., Qin, A. and Sun, X. (2022) Single-cell RNA sequencing for plant research: Insights and possible benefits. *International Journal of Molecular Sciences* 23(9), 4497.

Bhattacharjee, S. (2021) Advances of transcriptomics in crop improvement: A review. *Journal of Emerging Technologies and Innovative Research* 8(6), 22–37.

Bianco, L., Cestaro, A., Sargent, D.J., Banchi, E., Derdak, S. *et al.* (2014) Development and validation of a 20K single nucleotide polymorphism (SNP) whole genome genotyping array for apple (*Malus* × *domestica* Borkh). *PloS One* 9(10), e110377. DOI: 10.1371/journal.pone.0110377.

Bielsa, B., Hewitt, S., Reyes-Chin-Wo, S., Dhingra, A. and Rubio-Cabetas, M.J. (2018) Identification of water use efficiency related genes in 'Garnem' almond-peach rootstock using time-course transcriptome analysis. *PLoS One* 13(10), e0205493.

Blevins, T. (2017) Northern blotting techniques for small RNAs. *Plant Epigenetics: Methods and Protocols* 631, 141–162.

Bolón-Canedo, V., Alonso-Betanzos, A., López-de-Ullibarri, I. and Cao, R. (2019) Challenges and future trends for microarray analysis. *Microarray Bioinformatics* 283–293.

Borràs, D., Barchi, L., Schulz, K., Moglia, A., Acquadro, A. *et al.* (2021) Transcriptome-based identification and functional characterization of NAC transcription factors responsive to drought stress in *Capsicum annuum* L. *Frontiers in Genetics* 12, 743902. DOI: 10.3389/fgene.2021.743902.

Bräutigam, A. and Gowik, U. (2010) What can next generation sequencing do for you? Next generation sequencing as a valuable tool in plant research. *Plant Biology* 12(6), 831–841.

Bray, N.L., Pimentel, H., Melsted, P. and Pachter, L. (2016) Near-optimal probabilistic RNA-seq quantification. *Nature Biotechnology* 34(5), 525–527. DOI: 10.1038/nbt.3519.

Brown, J.W., Calixto, C.P. and Zhang, R. (2017) High-quality reference transcript datasets hold the key to transcript-specific RNA-sequencing analysis in plants. *New Phytologist* 213(2), 525–530.

Brumbaugh, C.D., Kim, H.J., Giovacchini, M. and Pourmand, N. (2011) NanoStriDE: Normalization and differential expression analysis of NanoString nCounter data. *BMC Bioinformatics* 12, 1–4.

Callwood, J., Melmaiee, K., Kulkarni, K.P., Vennapusa, A.R., Aicha, D. *et al.* (2021) Differential morpho-physiological and transcriptomic responses to heat stress in two blueberry species. *International Journal of Molecular Sciences* 22(5), 2481.

Cao, M., Zheng, L., Li, J., Mao, Y., Zhang, R. *et al.* (2022a) Transcriptomic profiling suggests candidate molecular responses to waterlogging in Cassava. *PloS One* 17(1), e0261086. DOI: 10.1371/journal.pone.0261086.

Cao, Q., Huang, L., Li, J., Qu, P., Tao, P. *et al.* (2022b) Integrated transcriptome and methylome analyses reveal the molecular regulation of drought stress in wild strawberry (*Fragaria nilgerrensis*). *BMC Plant Biology* 22(1), 613.

Cardone, M.F., Perniola, R., Catacchio, C.R., Alagna, F., Rotunno, S. *et al.* (2019) Grapevine adaptation to drought: New candidate genes for the genotype-dependent response. In: *BIO Web of Conferences*, Vol. 15. EDP Sciences, p. 01016.

Cassidy, A. and Jones, J. (2014) Developments in *in situ* hybridisation. *Methods* 70(1), 39–45.

Chakraborty, C. (2007) Potentiality of small interfering RNAs (siRNA) as recent therapeutic targets for gene-silencing. *Current Drug Targets* 8(3), 469–482. DOI: 10.2174/138945007780058988.

Chang, X., Sun, J., Liu, L., He, W. and Zhao, B. (2020) Transcriptome analysis of differentially expressed genes in wild jujube seedlings under salt stress. *Journal of the American Society for Horticultural Science* 145(3), 174–185.

Chaudhari, R.S., Jangale, B.L., Azeez, A., Krishna, B., Sane, P.V. *et al.* (2019) Differential regulation of the banana stress NAC family by individual and combined stresses of drought and heat in susceptible and resistant genotypes. *Plant Physiology and Biochemistry* 145, 184–194.

Chen, E.Y., Tan, C.M., Kou, Y., Duan, Q., Wang, Z. *et al.* (2013) Enrichr: Interactive and collaborative *HTML5* gene list enrichment analysis tool. *BMC Bioinformatics* 14, 1–14.

Chen, L., Yun, M., Cao, Z., Liang, Z., Liu, W. *et al.* (2021) Phenotypic characteristics and transcriptome of cucumber male flower development under heat stress. *Frontiers in Plant Science* 12, 758976. DOI: 10.3389/fpls.2021.758976.

Chen, P., Zhi, F., Li, X., Shen, W., Yan, M. *et al.* (2022a) Zinc-finger protein *MdBBX7/MdCOL9*, a target of *MdMIEL1 E3* ligase, confers drought tolerance in apple. *Plant Physiology* 188(1), 540–559.

Chen, Y., Deng, C., Xu, Q., Chen, X., Jiang, F. *et al.* (2022b) Integrated analysis of the metabolome, transcriptome and miRNome reveals crucial roles of auxin and heat shock proteins in the heat stress response of loquat fruit. *Scientia Horticulturae* 294, 110764.

Cheng, P., Kong, F., Han, Y., Liu, X. and Xia, J. (2024) Transcriptome analysis reveals genes and pathways associated with drought tolerance of early stages in sweet potato (*Ipomoea batatas* (L.) Lam. *Genes* 15(7), 948.

Clark, T.A., Sugnet, C.W. and Ares Jr, M. (2002) Genomewide analysis of mRNA processing in yeast using splicing-specific microarrays. *Science* 296(5569), 907–910.

Cookson, W., Liang, L., Abecasis, G., Moffatt, M. and Lathrop, M. (2009) Mapping complex disease traits with global gene expression. *Nature Reviews Genetics* 10(3), 184–194.

Costa, V., Angelini, C., De Feis, I. and Ciccodicola, A. (2010) Uncovering the complexity of transcriptomes with RNA-seq. *BioMed Research International* 2010(1), 853916.

Crowgey, E.L. and Mahajan, N. (2019) Advancements in next-generation sequencing for detecting minimal residual disease. In: Druley, T.E. (ed.) *Minimal Residual Disease Testing: Current Innovations and Future Directions*. Springer, Cham, pp. 159–192.

Cui, F., Zhang, N., Fan, X.L., Zhang, W., Zhao, C.H. *et al.* (2017) Utilization of a Wheat660k SNP array-derived high-density genetic map for high-resolution mapping of a major QTL for kernel number. *Scientific Reports* 7(1), 3788. DOI: 10.1038/s41598-017-04028-6.

Cui, X., Zhang, B., Chen, C., Tang, Y., Zhang, P. *et al.* (2020) Physiological change and screening of differentially expressed genes of wild Chinese *Vitis yeshanensis* and American *Vitis riparia* in response to drought stress. *Scientia Horticulturae* 266, 109140.

Dal Santo, S., Palliotti, A., Zenoni, S., Tornielli, G.B., Fasoli, M. *et al.* (2016) Distinct transcriptome responses to water limitation in isohydric and anisohydric grapevine cultivars. *BMC Genomics* 17, 1–19.

Dalal, N., Jalandra, R., Sharma, M., Prakash, H., Makharia, G.K. *et al.* (2020) Omics technologies for improved diagnosis and treatment of colorectal cancer: Technical advancement and major perspectives. *Biomedicine & Pharmacotherapy* 131, 110648. DOI: 10.1016/j.biopha.2020.110648.

Daldoul, S., Hanzouli, F., Hamdi, Z., Chenenaoui, S., Wetzel, T. *et al.* (2022) The root transcriptome dynamics reveals new valuable insights in the salt-resilience mechanism of wild grapevine (*Vitis vinifera subsp. sylvestris*). *Frontiers in Plant Science* 13, 1077710.

Das, A., Singh, S., Islam, Z., Munshi, A.D., Behera, T.K. *et al.* (2022) Current progress in genetic and genomics-aided breeding for stress resistance in cucumber (*Cucumis sativus* L.). *Scientia Horticulturae* 300, 111059.

Das, A., Rout, B.M., Datta, S., Singh, S., Munshi, A.D. *et al.* (2023) Spinach (*Spinacia oleracea* L.) breeding: From classical to genomics-centric approach. In: *Smart Plant Breeding for Vegetable Crops in Post-Genomics Era*. Springer Nature, Singapore, pp. 117–142.

Daware, A., Malik, A., Srivastava, R., Das, D., Ellur, R.K. *et al.* (2023) Rice pangenome genotyping array: An efficient genotyping solution for pangenome-based accelerated genetic improvement in rice. *The Plant Journal* 113(1), 26–46. DOI: 10.1111/tpj.16028.

De Lorenzis, G., Chipashvili, R., Failla, O. and Maghradze, D. (2015) Study of genetic variability in *Vitis vinifera* L. germplasm by high-throughput Vitis18kSNP array: The case of Georgian genetic resources. *BMC Plant Biology* 15, 1–14. DOI: 10.1186/s12870-015-0510-9.

de Silva, M.C., Pinto, P.I., Guerra, R., Duarte, A., Power, D.M. *et al.* (2023) Gene transcripts responsive to drought stress identified in *Citrus macrophylla* bark tissue transcriptome have a modified response in plants infected by Citrus tristeza virus. *Scientia Horticulturae* 307, 111526.

Dennis, G., Sherman, B.T., Hosack, D.A., Yang, J., Gao, W. *et al.* (2003) DAVID: Database for annotation, visualization, and integrated discovery. *Genome Biology* 4, 1–11.

Devate, N.B., Krishna, H., Parmeshwarappa, S.K.V., Manjunath, K.K., Chauhan, D. *et al.* (2022) Genome-wide association mapping for component traits of drought and heat tolerance in wheat. *Frontiers in Plant Science* 13, 943033. DOI: 10.3389/fpls.2022.943033.

Ding, F., Tong, J., Xu, R., Chen, J., Xu, X. *et al.* (2023) Identification of stable quantitative trait loci underlying waterlogging tolerance post-anthesis in common wheat (*Triticum aestivum*). *The Crop Journal* 11(4), 1163–1170. DOI: 10.1016/j.cj.2023.06.004.

Dinkar, V., Jha, S.K., Mallick, N., Niranjana, M., Agarwal, P. *et al.* (2020) Molecular mapping of a new recessive wheat leaf rust resistance gene originating from *Triticum spelta*. *Scientific Reports* 10(1), 22113. DOI: 10.1038/s41598-020-78679-3.

Dinkar, V., Jha, S.K., Niranjana, M., Mallick, N., Agarwal, P. *et al.* (2024) Molecular mapping of *Triticum monococcum* derived leaf rust resistance gene in durum wheat. *Physiological and Molecular Plant Pathology* 134, 102467. DOI: 10.1016/j.pmpp.2024.102467.

Dobin, A., Davis, C.A., Schlesinger, F., Drenkow, J., Zaleski, C. *et al.* (2013) STAR: Ultrafast universal RNA-seq aligner. *Bioinformatics* 29(1), 15–21.

Du, C., Li, H., Liu, C. and Fan, H. (2021) Understanding of the post-germinative development response to salinity and drought stresses in cucumber seeds by integrated proteomics and transcriptomics analysis. *Journal of Proteomics* 232, 104062.

Dutta, S., Muthusamy, V., Zunjare, R.U. and Hossain, F. (2019a) Analysis of paralogous genes of Carotenoid dioxygenase affecting carotenoid biosynthesis pathway in maize (*Zea mays* L.). *Journal of Pharmacognosy and Phytochemistry* 8(2), 524–530.

Dutta, S., Muthusamy, V., Zunjare, R.U., Bhowmick, R. and Hossain, F. (2019b) Genome wide study of fatty acid hydroxylase (FAH) superfamily containing β-carotene hydroxylase (*crtRB1*) in maize (*Zea mays* L.). *Pharma Innovation International Journal* 8(1), 422–428.

Dutta, S., Muthusamy, V., Chhabra, R., Baveja, A., Zunjare, R.U. *et al.* (2021) Low expression of carotenoids cleavage dioxygenase 1 (*ccd1*) gene improves the retention of provitamin-A in maize grains during storage. *Molecular Genetics and Genomics* 296, 141–153.

Dutta, S., Chhabra, R., Muthusamy, V., Gain, N., Subramani, R. *et al.* (2024a) Allelic variation and haplotype diversity of Matrilineal (*MTL*) gene governing *in vivo* maternal haploid induction in maize. *Physiology and Molecular Biology of Plants* 30(5), 823–838.

Dutta, S., Paul, D., Das, A., Kumar, S., Roy, A. *et al.* (2024b) Directed mutagenesis in plants through genome editing using guide RNA library. *The Nucleus* 67, 61–73.

Eom, S.H., Baek, S.A., Kim, J.K. and Hyun, T.K. (2018) Transcriptome analysis in Chinese cabbage (*Brassica rapa ssp. pekinensis*) provides the role of glucosinolate metabolism in response to drought stress. *Molecules* 23(5), 1186.

Escandón, M., Castillejo, M.Á., Jorrín-Novo, J.V. and Rey, M.D. (2021) Molecular research on stress responses in *Quercus* spp.: From classical biochemistry to systems biology through omics analysis. *Forests* 12(3), 364.

Estrella-Maldonado, H., Ramírez, A.G., Ortiz, G.F., Peraza-Echeverria, S., Vega, O. *et al.* (2021) Transcriptomic analysis reveals key transcription factors associated to drought tolerance in a wild papaya (*Carica papaya*) genotype. *PloS One* 16(1), e0245855.

Evans, T.G. (2015) Considerations for the use of transcriptomics in identifying the 'genes that matter' for environmental adaptation. *The Journal of Experimental Biology* 218(12), 1925–1935.

Fabregat, A., Sidiropoulos, K., Garapati, P., Gillespie, M., Hausmann, K. *et al.* (2016) The reactome pathway knowledgebase. *Nucleic Acids Research* 44(D1), D481–D487.

Fang, S., Han, X., Yuan, P., Song, C., Song, S. *et al.* (2023) Genome-wide analysis of the apple PLD gene family and a functional characterization of *MdPLD17* in drought tolerance. *Scientia Horticulturae* 321, 112311.

FAO (2011) FAO statistical yearbook 2009, agricultural production. Available at: http://www.fao.org/docrep/014/am079m/am079m00.htm (accessed 18 July 2025).

Farrell Jr, R.E. (2009) *RNA Methodologies: Laboratory Guide for Isolation and Characterization*. Academic Press, San Diego, California.

Fleury, D., Jefferies, S., Kuchel, H. and Langridge, P. (2010) Genetic and genomic tools to improve drought tolerance in wheat. *Journal of Experimental Botany* 61(12), 3211–3222.

Gashler, K. (2024) Report: Warmer planet will trigger increased farm losses. Cornell Chronicle. Available at: https://news.cornell.edu/stories/2024/01/report-warmer-planet-will-trigger-increased-farm-losses (accessed 12 June 2025).

Geng, G., Li, R., Stevanato, P., Lv, C., Lu, Z. *et al*. (2020) Physiological and transcriptome analysis of sugar beet reveals different mechanisms of response to neutral salt and alkaline salt stresses. *Frontiers in Plant Science* 11, 71864.

Ghodke, P., Khandagale, K., Thangasamy, A., Kulkarni, A., Narwade, N. *et al*. (2020) Comparative transcriptome analyses in contrasting onion (*Allium cepa* L.) genotypes for drought stress. *PLoS One* 15(8), e0237457.

Girardot, C., Scholtalbers, J., Sauer, S., Su, S.Y. and Furlong, E.E. (2016) Je, a versatile suite to handle multiplexed NGS libraries with unique molecular identifiers. *BMC Bioinformatics* 17, 1–6.

Gonçalves, L.P., Boscariol Camargo, R.L., Takita, M.A., Machado, M.A., dos Soares Filho, W.S. *et al*. (2019) Rootstock-induced molecular responses associated with drought tolerance in sweet orange as revealed by RNA-seq. *BMC Genomics* 20, 1–14.

Gong, X., Xu, Y., Li, H., Chen, X. and Song, Z. (2022) Antioxidant activation, cell wall reinforcement, and reactive oxygen species regulation promote resistance to waterlogging stress in hot pepper (*Capsicum annuum* L.). *BMC Plant Biology* 22(1), 425.

Graves, P.R. and Haystead, T.A. (2002) Molecular biologist's guide to proteomics. *Microbiology and Molecular Biology Reviews* 66(1), 39–63.

Guo, J., Zhang, R., Cheng, S., Fu, Z., Jia, P. *et al*. (2023) Physiological and transcriptomic analysis reveal the crucial factors in heat stress response of red raspberry 'Polka' seedlings. *Frontiers in Plant Science* 14, 1233448.

Guo, R., Wang, X., Han, X., Li, W., Liu, T. *et al*. (2019) Comparative transcriptome analyses revealed different heat stress responses in high-and low-GS *Brassica alboglabra* sprouts. *BMC Genomics* 20, 1–17.

Guo, R., Lin, L., Huang, G., Shi, X., Wei, R. *et al*. (2022) Transcriptome analysis of 'Kyoho' grapevine leaves identifies heat response genes involved in the transcriptional regulation of photosynthesis and abscisic acid. *Agronomy* 12(10), 2591.

Guttman, M., Garber, M., Levin, J.Z., Donaghey, J., Robinson, J. *et al*. (2010) Ab initio reconstruction of cell type–specific transcriptomes in mouse reveals the conserved multi-exonic structure of lincRNAs. *Nature Biotechnology* 28(5), 503–510.

Haider, M.S., Zhang, C., Kurjogi, M.M., Pervaiz, T., Zheng, T. *et al*. (2017) Insights into grapevine defense response against drought as revealed by biochemical, physiological and RNA-Seq analysis. *Scientific Reports* 7(1), 13134.

Han, Y., Gao, S., Muegge, K., Zhang, W. and Zhou, B. (2015) Advanced applications of RNA sequencing and challenges. *Bioinformatics and Biology Insights* 9, BBI–S28991.

He, W., Luo, L., Xie, R., Chai, J., Wang, H. *et al*. (2023) Transcriptome sequencing analyses uncover mechanisms of citrus rootstock seedlings under waterlogging stress. *Frontiers in Plant Science* 14, 1198930.

Hedges, D.J., Guettouche, T., Yang, S., Bademci, G., Diaz, A. *et al*. (2011) Comparison of three targeted enrichment strategies on the SOLiD sequencing platform. *PloS One* 6(4), e18595.

Hou, X., Kong, Y., Teng, Z., Yang, C., Li, Y. *et al*. (2024) Integrating genes and metabolites: Unraveling mango's drought resilience mechanisms. *BMC Plant Biology* 24(1), 208.

Hu, H., Dai, M., Yao, J., Xiao, B., Li, X. *et al*. (2006) Overexpressing a *NAM, ATAF, and CUC (NAC)* transcription factor enhances drought resistance and salt tolerance in rice. *Proceedings of the National Academy of Sciences* 103(35), 12987–12992.

Hu, J., Duan, Y., Yang, J., Gan, L., Chen, W. *et al*. (2023) Transcriptome analysis reveals genes associated with flooding tolerance in mulberry plants. *Life* 13(5), 1087.

Hu, W., Ding, Z., Tie, W., Yan, Y., Liu, Y. *et al*. (2017) Comparative physiological and transcriptomic analyses provide integrated insight into osmotic, cold, and salt stress tolerance mechanisms in banana. *Scientific Reports* 7(1), 43007.

Huang, N., Nie, F., Ni, P., Luo, F., Gao, X. *et al*. (2021) NeuralPolish: A novel Nanopore polishing method based on alignment matrix construction and orthogonal Bi-GRU Networks. *Bioinformatics* 37(19), 3120–3127.

Huang, X., Zhang, M., Wang, L., Zhang, X., Wu, R. *et al*. (2023) Time-course RNA-sequencing and co-expression modules revealed a critical salt response regulatory network in apple. *Journal of the American Society for Horticultural Science* 148(2), 53–63.

Hudecova, I. (2015) Digital PCR analysis of circulating nucleic acids. *Clinical Biochemistry* 48(15), 948–956.

Hwang, B., Lee, J.H. and Bang, D. (2018) Single-cell RNA sequencing technologies and bioinformatics pipelines. *Experimental & Molecular Medicine* 50(8), 1–14.

Ikram, M., Chen, J., Xia, Y., Li, R., Siddique, K.H. *et al.* (2022) Comprehensive transcriptome analysis reveals heat-responsive genes in flowering Chinese cabbage (*Brassica campestris* L. ssp. *chinensis*) using RNA sequencing. *Frontiers in Plant Science* 13, 1077920.

Iovieno, P., Punzo, P., Guida, G., Mistretta, C., Oosten, M.J. *et al.* (2016) Transcriptomic changes drive physiological responses to progressive drought stress and rehydration in tomato. *Frontiers in Plant Science* 7, 371.

Jiang, L., Yang, Y., Zhou, Z. and Chen, X. (2024) Physiological studies and transcriptomic analysis reveal the mechanism of saline-alkali stress resistance of *Malus sieversii f. niedzwetzkyan*. *Horticulturae* 10(5), 510.

Jiao, Z., Xu, W., Nong, Q., Zhang, M., Jian, S. *et al.* (2021) An integrative transcriptomic and metabolomic analysis of red pitaya (*Hylocereus polyrhizus*) seedlings in response to heat stress. *Genes* 12(11), 1714.

Jin, J., Yang, L., Fan, D., Liu, X. and Hao, Q. (2020) Comparative transcriptome analysis uncovers different heat stress responses in heat-resistant and heat-sensitive jujube cultivars. *PLoS One* 15(9), e0235763.

Jing, Q., Hou, H., Meng, X., Chen, A., Wang, L. *et al.* (2022) Transcriptome analysis reveals the proline metabolic pathway and its potential regulation *TF-hub* genes in salt-stressed potato. *Frontiers in Plant Science* 13, 1030138.

Jozefczuk, J. and Adjaye, J. (2011) Quantitative real-time PCR-based analysis of gene expression. In: Jameson, D., Verma, M. and Westerhoff, H.V. (eds) *Methods in Enzymology*. Academic Press, pp. 99–109.

Ju, Y.L., Min, Z., Zhang, Y., Zhang, K.K., Liu, M. *et al.* (2021) Transcriptome profiling provide new insights into the molecular mechanism of grapevine response to heat, drought, and combined stress. *Scientia Horticulturae* 286, 110076.

Kanehisa, M. and Goto, S. (2000) KEGG: Kyoto encyclopedia of genes and genomes. *Nucleic Acids Research* 28(1), 27–30.

Kashyap, S.P., Prasanna, H.C., Kumari, N., Mishra, P. and Singh, B. (2020) Understanding salt tolerance mechanism using transcriptome profiling and de novo assembly of wild tomato *Solanum chilense*. *Scientific Reports* 10(1), 15835.

Kęska, K., Szcześniak, M.W., Makałowska, I. and Czernicka, M. (2021) Long-term waterlogging as factor contributing to hypoxia stress tolerance enhancement in cucumber: Comparative transcriptome analysis of waterlogging sensitive and tolerant accessions. *Genes* 12(2), 189.

Khakimov, B., Møller Jespersen, B. and Balling Engelsen, S. (2014) Comprehensive and comparative metabolomic profiling of wheat, barley, oat and rye using gas chromatography-mass spectrometry and advanced chemometrics. *Foods* 3(4), 569–585.

Khan, M.A., Alghamdi, S.S., Ammar, M.H., Sun, Q., Teng, F. *et al.* (2019) Transcriptome profiling of faba bean (*Vicia faba* L.) drought-tolerant variety hassawi-2 under drought stress using RNA sequencing. *Electronic Journal of Biotechnology* 39, 15–29.

Kim, K.W., Nawade, B., Nam, J., Chu, S.H., Ha, J. *et al.* (2022) Development of an inclusive 580K SNP array and its application for genomic selection and genome-wide association studies in rice. *Frontiers in Plant Science* 13, 1036177. DOI: 10.3389/fpls.2022.1036177.

Kohl, M., Wiese, S. and Warscheid, B. (2011) Cytoscape: Software for visualization and analysis of biological networks. In: Hamacher, M., Eisenacher, M. and Stephan, C. (eds) *Data Mining in Proteomics*. Humana Press, New York and Heidelberg, pp. 291–303.

Koll, R.M. (2021). Available at: https://www.climate.rocksea.org/images/climate/roxy_business_standard_news_13aug2021-scaled.jpeg (accessed 28 November 2024).

Koning-Boucoiran, C.F., Esselink, G.D., Vukosavljev, M., van't Westende, W.P. and Gitonga, V.W. (2015) Using RNA-seq to assemble a rose transcriptome with more than 13,000 full-length expressed genes and to develop the WagRhSNP 68k Axiom SNP array for rose (Rosa L.). *Frontiers in Plant Science* 6, 249. DOI: 10.3389/fpls.2015.00249.

Ksouri, N., Jiménez, S., Wells, C.E., Contreras-Moreira, B. and Gogorcena, Y. (2016) Transcriptional responses in root and leaf of *Prunus persica* under drought stress using RNA sequencing. Frontiers in Plant Science 7, 1715.

Kulkarni, M.M. (2011) Digital multiplexed gene expression analysis using the NanoString nCounter system. Current Protocols in Molecular Biology 94(1), 25B-10.

Kutmon, M., van Iersel, M.P., Bohler, A., Kelder, T., Nunes, N. *et al.* (2015) PathVisio 3: An extendable pathway analysis toolbox. *PLoS Computational Biology* 11(2), e1004085.

Langmead, B., Trapnell, C., Pop, M. and Salzberg, S.L. (2009) Ultrafast and memory-efficient alignment of short DNA sequences to the human genome. *Genome Biology* 10, 1–10.

Lau, K.H., Rosario Herrera, M., Crisovan, E., Wu, S., Fei, Z. *et al.* (2018) Transcriptomic analysis of sweet potato under dehydration stress identifies candidate genes for drought tolerance. *Plant Direct* 2(10), e00092.

Lei, C., Li, M., Cheng, Z., He, W., Liu, B. *et al.* (2023) Molecular mapping of two novel frost resistance genes in common wheat by 660k SNP array. *Molecular Breeding* 43, 83. DOI: 10.1007/s11032-023-01425-w.

Li, H. and Durbin, R. (2009) Fast and accurate short read alignment with Burrows–Wheeler transform. *Bioinformatics* 25(14), 1754–1760.

Li, T., Xu, X., Li, Y., Wang, H., Li, Z. *et al.* (2015) Comparative transcriptome analysis reveals differential transcription in heat-susceptible and heat-tolerant pepper (*Capsicum annum* L.) cultivars under heat stress. *Journal of Plant Biology* 58, 411–424.

Li, S., Castillo-González, C., Yu, B. and Zhang, X. (2017) The functions of plant small RNAs in development and in stress responses. *The Plant Journal* 90(4), 654–670.

Li, Y.M., Li, S.X., Li, X.S. and Li, C.Y. (2018) Transcriptome studies with the third-generation sequencing technology. *Life Science Instruments* 16, 114–121.

Li, X., Li, M., Zhou, B., Yang, Y., Wei, Q. *et al.* (2019) Transcriptome analysis provides insights into the stress response crosstalk in apple (*Malus* × *domestica*) subjected to drought, cold and high salinity. *Scientific Reports* 9(1), 9071.

Li, M., Zhou, J., Du, J., Li, X., Sun, Y. *et al.* (2022a) Comparative physiological and transcriptomic analyses of improved heat stress tolerance in celery (*Apium graveolens* L.) caused by exogenous melatonin. *International Journal of Molecular Sciences* 23(19), 11382.

Li, Z., Bai, D., Zhong, Y., Lin, M., Sun, L. *et al.* (2022b) Full-length transcriptome and RNA-seq analyses reveal the mechanisms underlying waterlogging tolerance in kiwifruit (*Actinidia valvata*). *International Journal of Molecular Sciences* 23(6), 3237.

Li, Q., Zhang, G., Zhao, Y., Gu, L., Wang, Y. *et al.* (2023) Hub gene identification and heat-stress-related transcriptional regulation mechanism in cabbage (*Brassica oleracea* L.). *Horticulturae* 9(9), 977.

Li, X., Liu, Y., Hu, W., Yin, B., Liang, B. *et al.* (2024) Integrative physiological, metabolomic, and transcriptomic analysis reveals the drought responses of two apple rootstock cultivars. *BMC Plant Biology* 24(1), 219.

Liang, S.B., Liu, J.Y., Liu, J., Yang, J.T., Li, J.L. *et al.* (2017) Next-generation sequencing applications for crop genomes. *China Biotechnology* 37(2), 111–120.

Liang, B., Wan, S., Ma, Q., Yang, L., Hu, W. *et al.* (2022) Transcriptome and physiological analyses of a navel orange mutant with improved drought tolerance and water use efficiency caused by increases of cuticular wax accumulation and ROS scavenging capacity. *International Journal of Molecular Sciences* 23(10), 5660.

Lin, C.W., Fu, S.F., Liu, Y.J., Chen, C.C., Chang, C.H. *et al.* (2019) Analysis of ambient temperature-responsive transcriptome in shoot apical meristem of heat-tolerant and heat-sensitive broccoli inbred lines during floral head formation. *BMC Plant Biology* 19, 1–16.

Liu, L., Li, Y., Li, S., Hu, N., He, Y. *et al.* (2012) Comparison of next-generation sequencing systems. *BioMed Research International* 2012(1), 251364.

Liu, H., Wang, C., Chen, H. and Zhou, B. (2019a) Genome-wide transcriptome analysis reveals the molecular mechanism of high temperature-induced floral abortion in litchi chinensis. *BMC Genomics* 20, 1–15.

Liu, C., Zhao, Y., Zhao, X., Wang, J., Gu, M. *et al.* (2019b) Transcriptomic profiling of pomegranate provides insights into salt tolerance. *Agronomy* 10(1), 44.

Liu, R., Yang, B., Mao, L., Huang, Y., Dai, X. *et al.* (2021a) Evaluation of the effect of high temperature on the full-length transcriptome and biochemical parameters of chili pepper fruit (*Capsicum annuum* L.) at different developmental stages. *European Journal of Horticultural Science* 86(6), 651–662.

Liu, B., Kong, L., Zhang, Y. and Liao, Y. (2021b) Gene and metabolite integration analysis through transcriptome and metabolome brings new insight into heat stress tolerance in potato (*Solanum tuberosum* L.). *Plants* 10(1), 103.

Liu, C., Luo, S., Zhao, Y., Miao, Y., Wang, Q. *et al.* (2022a) Multiomics analyses reveal high temperature-induced molecular regulation of ascorbic acid and capsaicin biosynthesis in pepper fruits. *Environmental and Experimental Botany* 201, 104941.

Liu, J., Chen, Y., Wang, W.Q., Liu, J.H., Zhu, C.Q. *et al.* (2022b) Transcription factors *AcERF74/75* respond to waterlogging stress and trigger alcoholic fermentation-related genes in kiwifruit. *Plant Science* 314, 111115.

Long, S. and Berkemeier, B. (2022) Ultrasensitive detection and quantification of viral nucleic acids with Raindance droplet digital PCR (ddPCR). *Methods* 201, 49–64.

Love, M.I., Huber, W. and Anders, S. (2014) Moderated estimation of fold change and dispersion for RNA-seq data with DESeq2. *Genome Biology* 15, 1–21.

Lun, A.T., Calero-Nieto, F.J., Haim-Vilmovsky, L., Göttgens, B. and Marioni, J.C. (2017) Assessing the reliability of spike-in normalization for analyses of single-cell RNA sequencing data. *Genome Research* 27(11), 1795–1806.

Luo, Y., Reid, R., Freese, D., Li, C., Watkins, J. *et al.* (2017) Salt tolerance response revealed by RNA-seq in a diploid halophytic wild relative of sweet potato. *Scientific Reports* 7(1), 9624.

Ma, L., Dong, C., Song, C., Wang, X., Zheng, X. *et al.* (2021) De novo genome assembly of the potent medicinal plant *Rehmannia glutinosa* using nanopore technology. *Computational and Structural Biotechnology Journal* 19, 3954–3963.

Ma, Y., Szostkiewicz, I., Korte, A., Moes, D., Yang, Y. *et al.* (2009) Regulators of PP2C phosphatase activity function as abscisic acid sensors. *Science* 324(5930), 1064–1068.

Mahato, A. (2014) Climate change and its impact on agriculture. *International Journal of Scientific and Research Publications* 4(4), 1–6.

Mahmoud, A., Qi, R., Chi, X., Liao, N., Malangisha, G.K. *et al.* (2023) Integrated bulk segregant analysis, fine mapping, and transcriptome revealed QTLs and candidate genes associated with drought adaptation in wild watermelon. *International Journal of Molecular Sciences* 25(1), 65.

Maman, S., Muthusamy, V., Katral, A., Chhabra, R., Gain, N. *et al.* (2023) Low expression of lipoxygenase 3 (LOX3) enhances the retention of kernel tocopherols in maize during storage. *Molecular Biology Reports* 50(11), 9283–9294.

Mani, S.D., Pandey, S., Govindan, M., Muthamilarasan, M. and Nagarathnam, R. (2021) Transcriptome dynamics underlying elicitor-induced defense responses against septoria leaf spot disease of tomato (*Solanum lycopersicum* L.). *Physiology and Molecular Biology of Plants* 27, 873–888.

Margulies, M., Egholm, M., Altman, W.E., Attiya, S., Bader, J.S. *et al.* (2005) Genome sequencing in microfabricated high-density picolitre reactors. *Nature* 437(7057), 376–380.

Massaia, A., Chaves, P., Samari, S., Miragaia, R.J., Meyer, K. *et al.* (2018) Single cell gene expression to understand the dynamic architecture of the heart. *Frontiers in Cardiovascular Medicine* 5, 167.

Matsumura, H., Nirasawa, S. and Terauchi, R. (1999) Transcript profiling in rice (*Oryza sativa* L.) seedlings using serial analysis of gene expression (SAGE). *The Plant Journal* 20(6), 719–726.

Mi, H., Poudel, S., Muruganujan, A., Casagrande, J.T. and Thomas, P.D. (2016) PANTHER version 10: Expanded protein families and functions, and analysis tools. *Nucleic Acids Research* 44(D1), D336–D342.

Mishra, S., Singh, B., Panda, K., Singh, B.P., Singh, N. *et al.* (2016) Association of SNP haplotypes of HKT family genes with salt tolerance in Indian wild rice germplasm. *Rice* 9, 1–13. DOI: 10.1186/s12284-016-0083-8.

Moazzzam Jazi, M., Seyedi, S.M., Ebrahimie, E., Ebrahimi, M., Moro, G. *et al.* (2017) A genome-wide transcriptome map of pistachio (*Pistacia vera* L.) provides novel insights into salinity-related genes and marker discovery. *BMC Genomics* 18, 1–21.

Molania, R., Gagnon-Bartsch, J.A., Dobrovic, A. and Speed, T.P. (2019) A new normalization for nanostring ncounter gene expression data. *Nucleic Acids Research* 47(12), 6073–6083.

Moon, K.B., Ahn, D.J., Park, J.S., Jung, W.Y., Cho, H.S. *et al.* (2018) Transcriptome profiling and characterization of drought-tolerant potato plant (*Solanum tuberosum* L.). *Molecules and Cells* 41(11), 979–992.

Muthusamy, M., Uma, S., Backiyarani, S., Saraswathi, M.S. and Chandrasekar, A. (2016) Transcriptomic changes of drought-tolerant and sensitive banana cultivars exposed to drought stress. *Frontiers in Plant Science* 7, 1609.

Muñoz-Amatriaín, M., Mirebrahim, H., Xu, P., Wanamaker, S.I., Luo, M. *et al*. (2017) Genome resources for climate-resilient cowpea, an essential crop for food security. *The Plant Journal* 89(5), 1042–1054. DOI: 10.1111/tpj.13404.

Myles, S., Boyko, A.R., Owens, C.L., Brown, P.J., Grassi, F. *et al*. (2011) Genetic structure and domestication history of the grape. *Proceedings of the National Academy of Sciences* 108(9), 3530–3535. DOI: 10.1073/pnas.1009363108.

Nejat, N., Rookes, J., Mantri, N.L. and Cahill, D.M. (2017) Plant–pathogen interactions: toward development of next-generation disease-resistant plants. *Critical Reviews in Biotechnology* 37(2), 229–237.

Nonis, A., Nardi, B. and Nonis, A. (2014) Choosing between RT-qPCR and RNA-seq: A back-of-the-envelope estimate towards the definition of the break-even-point. *Analytical and Bioanalytical Chemistry* 406, 3533–3536.

Paley, S. and Karp, P.D. (2021) The BioCyc metabolic network explorer. *Bmc Bioinformatics* 22, 1–6.

Pall, G.S. and Hamilton, A.J. (2008) Improved northern blot method for enhanced detection of small RNA. *Nature Protocols* 3(6), 1077–1084.

Pandey, S. and Singh, A. (2024) Climate change: Effects and need of climate-resilient agriculture. In: *Climate-Resilient Agriculture*. Apple Academic Press, Florida, pp. 3–8.

Pandey, M.K., Agarwal, G., Kale, S.M., Clevenger, J., Nayak, S.N. *et al*. (2017) Development and evaluation of a high density genotyping 'Axiom_ Arachis' array with 58 K SNPs for accelerating genetics and breeding in groundnut. *Scientific Reports* 7(1), 40577. DOI: 10.1038/srep40577.

Pang, Y., Wang, X., Zhao, M., Lu, Y., Yan, Q. *et al*. (2022) Identification and validation of the genomic regions for waterlogging tolerance at germination stage in wheat. *Agronomy* 12(8), 1848. DOI: 10.3390/agronomy12081848.

Park, S.U., Kim, Y.H., Lee, C.J., Kim, S.E., Lim, Y.H. *et al*. (2020) Comparative transcriptome profiling of two sweet potato cultivars with contrasting flooding stress tolerance levels. Plant Biotechnology Reports 14, 743–756.

Patterson, S.D. and Aebersold, R.H. (2003) Proteomics: The first decade and beyond. Nature Genetics 33(3), 311–323.

Pauwels, K., De Keersmaecker, S.C., De Schrijver, A., du Jardin, P., Roosens, N.H. *et al*. (2015) Next-generation sequencing as a tool for the molecular characterisation and risk assessment of genetically modified plants: Added value or not? *Trends in Food Science & Technology* 45(2), 319–326.

Perveen, N., Dinesh, M.R., Sankaran, M., Ravishankar, K.V., Krishnajee, H.G. *et al*. (2023) Comparative transcriptome analysis provides novel insights into molecular response of salt-tolerant and sensitive polyembryonic mango genotypes to salinity stress at seedling stage. Frontiers in Plant Science 14, 1152485.

Provenzano, M. and Mocellin, S. (2007) Complementary techniques: Validation of gene expression data by quantitative real time PCR. In: Mocellin, S. (ed.) Microarray Technology and Cancer Gene Profiling. Springer, New York, pp. 66–73.

Qin, T., Sun, C., Kazim, A., Cui, S., Wang, Y. *et al*. (2022) Comparative transcriptome analysis of deep-rooting and shallow-rooting potato (*Solanum tuberosum* L.) genotypes under drought stress. Plants 11(15), 2024.

Radwan, O., Arro, J., Keller, C. and Korban, S.S. (2015) RNA-Seq transcriptome analysis in date palm suggests multi-dimensional responses to salinity stress. Tropical Plant Biology 8, 74–86.

Ragini, R., Murukan, N., Sekhon, N.K., Chugh, C., Agarwal, P. *et al*. (2024) Breaking the association between gametocidal gene (s) and leaf rust resistance gene (*LrS2427*) in *Triticum aestivum-Aegilops speltoides* derivative by gamma irradiation. Molecular Breeding 44(8), 54. DOI: 10.1007/s11032-024-01491-8.

Reddy, R.V., Rahut, D. and Sonobe, T. (2025) The Future of Smallholder Farming in Asia. Asian Development Bank.

Redman, J.C., Haas, B.J., Tanimoto, G. and Town, C.D. (2004) Development and evaluation of an Arabidopsis whole genome Affymetrix probe array. The Plant Journal 38(3), 545–561. DOI: 10.1111/j.1365-313X.2004.02061.x.

Reeksting, B.J., Coetzer, N., Mahomed, W., Engelbrecht, J. and Berg, N. (2014) De novo sequencing, assembly, and analysis of the root transcriptome of *Persea americana* (Mill.) in response to *Phytophthora cinnamomi* and flooding. *PLoS One* 9(2), e86399.

Rienth, M., Torregrosa, L., Sarah, G., Ardisson, M., Brillouet, J.M. *et al.* (2016) Temperature desynchronizes sugar and organic acid metabolism in ripening grapevine fruits and remodels their transcriptome. *BMC Plant Biology* 16, 1–23.

Rimbert, H., Darrier, B., Navarro, J., Kitt, J., Choulet, F. *et al.* (2018) High throughput SNP discovery and genotyping in hexaploid wheat. *PloS One* 13(1), e0186329. DOI: 10.1371/journal.pone.0186329.

Ritchie, M.E., Phipson, B., Wu, D.I., Hu, Y., Law, C.W. *et al.* (2015) Limma powers differential expression analyses for RNA-sequencing and microarray studies. *Nucleic Acids Research* 43(7), e47–e47.

Robinson, M.D., McCarthy, D.J. and Smyth, G.K. (2010) EdgeR: A bioconductor package for differential expression analysis of digital gene expression data. *Bioinformatics* 26(1), 139–140.

Roorkiwal, M., Jain, A., Kale, S.M., Doddamani, D., Chitikineni, A. *et al.* (2018) Development and evaluation of high-density Axiom® *Cicer* SNP Array for high-resolution genetic mapping and breeding applications in chickpea. *Plant Biotechnology Journal* 16(4), 890–901. DOI: 10.1111/pbi.12836.

Roy, A., Dutta, S., Das, S. and Choudhury, M.R. (2024) Next-generation sequencing in the development of climate-resilient and stress-responsive crops–A review. *The Open Biotechnology Journal* 18(1).

Safdar, M., Shahid, M.A., Yang, C., Rasul, F., Tahir, M. *et al.* (2024) *Emerging Technologies and Marketing Strategies for Sustainable Agriculture*. IGI Global Scientific Publishing, Pennsylvania. DOI: 10.4018/979-8-3693-4864-2.ch002.

Sahraeian, S.M.E., Mohiyuddin, M., Sebra, R., Tilgner, H., Afshar, P.T. *et al.* (2017) Gaining comprehensive biological insight into the transcriptome by performing a broad-spectrum RNA-seq analysis. *Nature Communications* 8(1), 59.

Sahu, N., Saini, A., Behera, S.K., Sayama, T., Sahu, L. *et al.* (2020) Why apple orchards are shifting to the higher altitudes of the himalayas? *PLoS One* 15(7), e0235041.

Sanger, F., Nicklen, S. and Coulson, A.R. (1977) DNA sequencing with chain-terminating inhibitors. *Proceedings of the National Academy of Sciences* 74(12), 5463–5467.

Santos, A.S., de Andrade, E.M., de Oliveira Sousa, A.R., do Carmo Costa, M.M., Togawa, R. *et al.* (2021) Transcriptome profile of drought responsive candidate genes in varieties of citrus rootstocks with different tolerance strategies. *Scientia Horticulturae* 277, 109838.

Savoi, S., Wong, D.C., Arapitsas, P., Miculan, M., Bucchetti, B. *et al.* (2016) Transcriptome and metabolite profiling reveals that prolonged drought modulates the phenylpropanoid and terpenoid pathway in white grapes (*Vitis vinifera* L.). *BMC Plant Biology* 16, 1–17.

Scialò, E., Sicilia, A., Continella, A., Gentile, A. and Lo Piero, A.R. (2024) Transcriptome profiling and weighted gene correlation network analysis reveal hub genes and pathways involved in the response to polyethylene-glycol-induced drought stress of two citrus rootstocks. *Biology* 13(8), 595.

Shaw, R., Tian, X. and Xu, J. (2021) Single-cell transcriptome analysis in plants: Advances and challenges. *Molecular Plant* 14(1), 115–126.

Shendure, J., Balasubramanian, S., Church, G.M., Gilbert, W., Rogers, J. *et al.* (2017) DNA sequencing at 40: Past, present and future. *Nature* 550(7676), 345–353.

Shu, J., Zhang, L., Liu, G., Wang, X., Liu, F. *et al.* (2024) Transcriptome analysis and metabolic profiling reveal the key regulatory pathways in drought stress responses and recovery in tomatoes. *International Journal of Molecular Sciences* 25(4), 2187.

Singh, C. and Roy-Chowdhuri, S. (2016) Quantitative real-time PCR: Recent advances. Humana Press, pp. 161–176.

Singh, S., Mahato, A.K., Jayaswal, P.K., Singh, N., Dheer, M. *et al.* (2020) A 62K genic-SNP chip array for genetic studies and breeding applications in pigeonpea (*Cajanus cajan* L. Millsp.). *Scientific Reports* 10(1), 4960. DOI: 10.1038/s41598-020-61889-0.

Singh, N., Jayaswal, P.K., Panda, K., Mandal, P., Kumar, V. *et al.* (2015) Single-copy gene based 50 k SNP chip for genetic studies and molecular breeding in rice. *Scientific Reports* 5(1), 11600. DOI: 10.1038/srep11600.

Singh, A., Pandey, S. and Kumar, A. (eds) (2025) *Climate Change and Biotic Factors: A Molecular Approach*. CRC Press, Florida.

Soleimani, B., Lehnert, H., Keilwagen, J., Plieske, J., Ordon, F. *et al.* (2020) Comparison between core set selection methods using different Illumina marker platforms: A case study of assessment of diversity in wheat. *Frontiers in Plant Science* 11, 1040. DOI: 10.3389/fpls.2020.01040.

Song, Q., Hyten, D.L., Jia, G., Quigley, C.V., Fickus, E.W. *et al.* (2013) Development and evaluation of SoySNP50K, a high-density genotyping array for soybean. *PloS One* 8(1), e54985. DOI: 10.1371/journal.pone.0054985.

Song, Q., Yan, L., Quigley, C., Fickus, E., Wei, H. *et al.* (2020) Soybean BARCSoySNP6K: An assay for soybean genetics and breeding research. *The Plant Journal* 104(3), 800–811. DOI: 10.1111/tpj.14960.

Song, X., Duan, Y.Y., Tan, F.Q., Ren, J., Cao, H.X. *et al.* (2023) Comparative transcriptome analysis of salt tolerance of roots in diploid and autotetraploid citrus rootstock (*C. junos cv. Ziyang xiangcheng*) and identification of salt tolerance-related genes. *Scientia Horticulturae* 317, 112083.

Subramani, M., Urrea, C.A., Habib, R., Bhide, K., Thimmapuram, J. *et al.* (2023) Comparative transcriptome analysis of tolerant and sensitive genotypes of common bean (*Phaseolus vulgaris* L.) in response to terminal drought stress. *Plants* 12(1), 210.

Subramanian, A., Tamayo, P., Mootha, V.K., Mukherjee, S., Ebert, B.L. *et al.* (2005) Gene set enrichment analysis: A knowledge-based approach for interpreting genome-wide expression profiles. *Proceedings of the National Academy of Sciences* 102(43), 15545–15550.

Sun, X., Feng, D., Liu, M., Qin, R., Li, Y. *et al.* (2022a) Single-cell transcriptome reveals dominant subgenome expression and transcriptional response to heat stress in Chinese cabbage. *Genome Biology* 23(1), 262.

Sun, W., Hao, J., Fan, S., Liu, C. and Han, Y. (2022b) Transcriptome and metabolome analysis revealed that exogenous spermidine-modulated flavone enhances the heat tolerance of lettuce. *Antioxidants* 11(12), 2332.

Tan, J.W., Shinde, H., Tesfamicael, K., Hu, Y., Fruzangohar, M. *et al.* (2023) Global transcriptome and gene co-expression network analyses reveal regulatory and non-additive effects of drought and heat stress in grapevine. *Frontiers in Plant Science* 14, 1096225.

Tang, W., Arisha, M.H., Zhang, Z., Yan, H., Kou, M. *et al.* (2023) Comparative transcriptomic and proteomic analysis reveals common molecular factors responsive to heat and drought stresses in sweet potato (*Ipomoea batatas*). *Frontiers in Plant Science* 13, 1081948.

Tinker, N.A., Chao, S., Lazo, G.R., Oliver, R.E., Huang, Y.F. *et al.* (2014) A SNP genotyping array for hexaploid oat. *The Plant Genome* 7(3), 2014–03. DOI: 10.3835/plantgenome2014.03.0010.

Tiwari, S., SI, K., Kumar, V., Singh, B., Rao, A.R. *et al.* (2016) Mapping QTLs for salt tolerance in rice (*Oryza sativa* L.) by bulked segregant analysis of recombinant inbred lines using 50K SNP chip. *PloS One* 11(4), e0153610. DOI: 10.1371/journal.pone.0153610.

Tolani, P., Gupta, S., Yadav, K., Aggarwal, S. and Yadav, A.K. (2021) Big data, integrative omics and network biology. *Advances in Protein Chemistry and Structural Biology* 127, 127–160.

Trapnell, C., Pachter, L. and Salzberg, S.L. (2009) TopHat: Discovering splice junctions with RNA-seq. *Bioinformatics* 25(9), 1105–1111.

Unterseer, S., Bauer, E., Haberer, G., Seidel, M., Knaak, C. *et al.* (2014) A powerful tool for genome analysis in maize: Development and evaluation of the high density 600 k SNP genotyping array. *BMC Genomics* 15, 1–15. DOI: 10.1186/1471-2164-15-823.

Upadhyay, A., Gaonkar, T., Upadhyay, A.K., Jogaiah, S., Shinde, M.P. *et al.* (2018) Global transcriptome analysis of grapevine (*Vitis vinifera* L.) leaves under salt stress reveals differential response at early and late stages of stress in table grape cv. *Thompson Seedless. Plant Physiology and Biochemistry* 129, 168–179.

Vangelisti, A., Zambrano, L.S., Caruso, G., Macheda, D., Bernardi, R. *et al.* (2019) How an ancient, salt-tolerant fruit crop, *Ficus carica* L., copes with salinity: A transcriptome analysis. *Scientific Reports* 9(1), 2561.

Velculescu, V.E., Zhang, L., Vogelstein, B. and Kinzler, K.W. (1995) Serial analysis of gene expression. *Science* 270(5235), 484–487.

Verde, I., Bassil, N., Scalabrin, S., Gilmore, B., Lawley, C.T. *et al.* (2012) Development and evaluation of a 9K SNP array for peach by internationally coordinated SNP detection and validation in breeding germplasm. *PloS One* 7(4), e35668. DOI: 10.1371/journal.pone.0035668.

Verma, A., Niranjana, M., Jha, S.K., Mallick, N., Agarwal, P. *et al.* (2020) QTL detection and putative candidate gene prediction for leaf rolling under moisture stress condition in wheat. *Scientific Reports* 10(1), 18696. DOI: 10.1038/s41598-020-75703-4.

Vieira, A., Cabral, A., Fino, J., Azinheira, H.G., Loureiro, A. *et al.* (2016) Comparative validation of conventional and RNA-seq data-derived reference genes for qPCR expression studies of *Colletotrichum kahawae*. *PLoS One* 11(3), e0150651.

Vikram, P., Kumar, A., Singh, A. and Singh, N.K. (2012) Rice: Genomics-assisted breeding for drought tolerance. In: Tuteja, N., Gill, S.S., Tiburcio, A.F. and Tuteja, R. (eds) *Improving Crop Resistance to Abiotic Stress*. Wiley, Weinheim, Germany, pp. 715–731.

Vos, P.G., Uitdewilligen, J.G., Voorrips, R.E., Visser, R.G. and Eck, H.J. (2015) Development and analysis of a 20K SNP array for potato (*Solanum tuberosum*): An insight into the breeding history. *Theoretical and Applied Genetics* 128, 2387–2401. DOI: 10.1007/s00122-015-2593-y.

Wan, T., Feng, Y., Liang, C., Pan, L., He, L. *et al.* (2021) Metabolomics and transcriptomics analyses of two contrasting cherry rootstocks in response to drought stress. *Biology* 10(3), 201.

Wang, L., Feng, Z., Wang, X., Wang, X. and Zhang, X. (2010) DEGseq: An R package for identifying differentially expressed genes from RNA-seq data. *Bioinformatics* 26(1), 136–138.

Wang, K., Huang, C. and Nice, E.C. (2014a) Proteomics, genomics and transcriptomics: Their emerging roles in the discovery and validation of colorectal cancer biomarkers. *Expert Review of Proteomics* 11(2), 179–205.

Wang, S., Wong, D., Forrest, K., Allen, A., Chao, S. *et al.* (2014b) Characterization of polyploid wheat genomic diversity using a high-density 90 000 single nucleotide polymorphism array. *Plant Biotechnology Journal* 12(6), 787–796. DOI: 10.1111/pbi.12183.

Wang, J., Meng, X., Dobrovolskaya, O.B., Orlov, Y.L. and Chen, M. (2017a) Non-coding RNAs and their roles in stress response in plants. *Genomics, Proteomics and Bioinformatics* 15(5), 301–312.

Wang, Y., Xu, C., Zhang, B., Wu, M. and Chen, G. (2017b) Physiological and proteomic analysis of rice (*Oryza sativa* L.) in flag leaf during flowering stage and milk stage under drought stress. *Plant Growth Regulation* 82, 201–218.

Wang, M., Jiang, B., Peng, Q., Liu, W., He, X. *et al.* (2018a) Transcriptome analyses in different cucumber cultivars provide novel insights into drought stress responses. *International Journal of Molecular Sciences* 19(7).

Wang, R., Mei, Y., Xu, L., Zhu, X., Wang, Y. *et al.* (2018b) Genome-wide characterization of differentially expressed genes provides insights into regulatory network of heat stress response in radish (*Raphanus sativus* L.). *Functional & Integrative Genomics* 18, 225–239.

Wang, Q., Cao, K., Zhu, G., Fang, W., Chen, C. *et al.* (2019) Comparative transcriptome analysis of genes involved in the response of resistant and susceptible peach cultivars to water stress. *Scientia Horticulturae* 245, 29–38.

Wang, M., He, X., Peng, Q., Liang, Z., Peng, Q. *et al.* (2020) Understanding the heat resistance of cucumber through leaf transcriptomics. *Functional Plant Biology* 47(8), 704–715.

Wang, G.L., Liu, L.Y., Wang, Q.Z., Ren, X.Q., Xiong, A.S. *et al.* (2022a) Transcriptome characterization of the roles of abscisic acid and calcium signaling during water deficit in garlic. *Applied Sciences* 12(5), 2440.

Wang, S., He, J., Deng, M., Wang, C., Wang, R. *et al.* (2022b) Integrating ATAC-seq and RNA-seq reveals the dynamics of chromatin accessibility and gene expression in apple response to drought. *International Journal of Molecular Sciences* 23(19), 11191.

Wang, H., Zhao, J., Lai, M., Zhang, Y., Qiu, W. *et al.* (2023a) Differential gene expression analysis and physiological response characteristics of passion fruit (*Passiflora edulis*) buds under high-temperature stress. *PeerJ* 11, e14839.

Wang, S., Zhou, X., Pan, K., Zhang, H., Shen, X. *et al.* (2023b) Distinct heat response molecular mechanisms emerge in cassava vasculature compared to leaf mesophyll tissue under high temperature stress. *Frontiers in Plant Science* 14, 1281436.

Wang, X.L., Peng, L., Wang, J., Liu, J.L., Jia, J.J. *et al.* (2023c) Transcriptome analyses reveal the effects of mixed saline–alkali stress on indoleacetic acid and cytokinins in *Malus hupehensis* Rehd. leaves. *Physiology and Molecular Biology of Plants* 29(1), 11–22.

Wang, M., Liu, W., Peng, Q., Shi, S., Wang, Y. *et al.* (2024) Excavation of genes response to heat resistance by transcriptome analysis in bottle gourd (*Lagenaria siceraria* (Mol.) Standl). *Agronomy* 14(2), 299.

Wanichthanarak, K., Fahrmann, J.F. and Grapov, D. (2015) Genomic, proteomic, and metabolomic data integration strategies. *Biomarker Insights* 10, BMI–S29511.

Wei, S., Hu, W., Deng, X., Zhang, Y., Liu, X. *et al.* (2014) A rice calcium-dependent protein kinase *OsCPK9* positively regulates drought stress tolerance and spikelet fertility. *BMC Plant Biology* 14, 1–13.

Wei, J., Liu, D., Liu, Y. and Wei, S. (2022) Physiological analysis and transcriptome sequencing reveal the effects of salt stress on banana (*Musa acuminata cv. BD*) leaf. *Frontiers in Plant Science* 13, 822838.

Wen, X., Liu, C., Yang, F., Wei, Z., Li, L. *et al.* (2024) Accurate long-read RNA sequencing analysis reveals the key pathways and candidate genes under drought stress in the seed germination stage in faba bean. *International Journal of Molecular Sciences* 25(16), 8875.

Wenger, A.M., Peluso, P., Rowell, W.J., Chang, P.C., Hall, R.J. *et al.* (2019) Accurate circular consensus long-read sequencing improves variant detection and assembly of a human genome. *Nature Biotechnology* 37(10), 1155–1162.

Winfield, M.O., Wilkinson, P.A., Allen, A.M., Barker, G.L., Coghill, J.A. *et al.* (2012) Targeted re-sequencing of the allohexaploid wheat exome. *Plant Biotechnology Journal* 10(6), 733–742. DOI: 10.1111/j.1467-7652.2012.00713.x.

Winfield, M.O., Allen, A.M., Burridge, A.J., Barker, G.L., Benbow, H.R. *et al.* (2016) High-density SNP genotyping array for hexaploid wheat and its secondary and tertiary gene pool. *Plant Biotechnology Journal* 14(5), 1195–1206. DOI: 10.1111/pbi.12485.

Wu, J., Zhang, F., Liu, G., Abudureheman, R., Bai, S. *et al.* (2023) Transcriptome and coexpression network analysis reveals properties and candidate genes associated with grape (*Vitis vinifera* L.) heat tolerance. *Frontiers in Plant Science* 14, 1270933.

Wu, X., Shiroto, Y., Kishitani, S., Ito, Y. and Toriyama, K. (2009) Enhanced heat and drought tolerance in transgenic rice seedlings overexpressing *OsWRKY11* under the control of *HSP101* promoter. *Plant Cell Reports* 28, 21–30.

Xiao, B.Z., Chen, X., Xiang, C.B., Tang, N., Zhang, Q.F. *et al.* (2009) Evaluation of seven function-known candidate genes for their effects on improving drought resistance of transgenic rice under field conditions. *Molecular Plant* 2(1), 73–83.

Xie, R., Zhang, J., Ma, Y., Pan, X., Dong, C. *et al.* (2017) Combined analysis of mRNA and miRNA identifies dehydration and salinity responsive key molecular players in citrus roots. *Scientific Reports* 7(1), 42094.

Xie, R., Pan, X., Zhang, J., Ma, Y., He, S. *et al.* (2018) Effect of salt-stress on gene expression in citrus roots revealed by RNA-seq. *Functional & Integrative Genomics* 18, 155–173.

Xing, M., Huang, K., Zhang, C., Xi, D., Luo, H. *et al.* (2023) Transcriptome analysis reveals the molecular mechanism and responsive genes of waterlogging stress in *Actinidia deliciosa* planch kiwifruit plants. *International Journal of Molecular Sciences* 24(21), 15887.

Xu, D.Q., Huang, J., Guo, S.Q., Yang, X., Bao, Y.M. *et al.* (2008) Overexpression of a *TFIIIA*-type zinc finger protein gene *ZFP252* enhances drought and salt tolerance in rice (*Oryza sativa* L.). *FEBS Letters* 582(7), 1037–1043.

Xu, H., Chen, L., Song, B., Fan, X., Yuan, X. *et al.* (2016) De novo transcriptome sequencing of pakchoi (*Brassica rapa* L. *chinensis*) reveals the key genes related to the response of heat stress. *Acta Physiologiae Plantarum* 38, 1–12.

Xu, X., Chen, M., Ji, J., Xu, Q., Qi, X. *et al.* (2017) Comparative RNA-seq based transcriptome profiling of waterlogging response in cucumber hypocotyls reveals novel insights into the de novo adventitious root primordia initiation. *BMC Plant Biology* 17, 1–13.

Xuan, L.I.U., Liang, W., Li, Y.X., Li, M.J., Liu, C.H. *et al.* (2019) Transcriptome analysis reveals the effects of alkali stress on root system architecture and endogenous hormones in apple rootstocks. *Journal of Integrative Agriculture* 18(10), 2264–2271.

Yamamoto, M., Wakatsuki, T., Hada, A. and Ryo, A. (2001) Use of serial analysis of gene expression (SAGE) technology. *Journal of Immunological Methods* 250(1–2), 45–66.

Yan, J., Yu, L., Xuan, J., Lu, Y., Lu, S. *et al.* (2016) De novo transcriptome sequencing and gene expression profiling of spinach (*Spinacia oleracea* L.) leaves under heat stress. *Scientific Reports* 6(1), 19473.

Yang, S., Vanderbeld, B., Wan, J. and Huang, Y. (2010) Narrowing down the targets: Towards successful genetic engineering of drought-tolerant crops. *Molecular Plant* 3(3), 469–490.

Yang, S., Bai, M., Hao, G., Zhang, X., Guo, H. *et al.* (2021) Transcriptome survey and expression analysis reveals the adaptive mechanism of 'Yulu Xiang' pear in response to long-term drought stress. *PloS One* 16(2), e0246070.

Yang, S., Bai, M., Hao, G., Guo, H. and Fu, B. (2022) Transcriptomics analysis of field-droughted pear (*Pyrus spp.*) reveals potential drought stress genes and metabolic pathways. *PeerJ* 10, e12921.

Yang, Q.Q., Yang, F., Liu, C.Y., Zhao, Y.Q., Li, M.Y. *et al.* (2024) Transcriptome sequencing of garlic reveals key genes related to the heat stress response. *Scientific Reports* 14(1), 15956.

Yin, J., Hu, J., Sun, J., Wang, B. and Mu, Y. (2019) A fast nucleic acid extraction system for point-of-care and integration of digital PCR. *Analyst* 144(23), 7032–7040.

Yuan, G., Sun, D., An, G., Li, W., Si, W. *et al.* (2022) Transcriptomic and metabolomic analysis of the effects of exogenous trehalose on salt tolerance in watermelon (*Citrullus lanatus*). *Cells* 11(15), 2338.

Zhang, H. (2019) The review of transcriptome sequencing: Principles, history and advances. In: *IOP Conference Series: Earth and Environmental Science 332*, Vol. 4. IOP Publishing, Bristol, UK, p. 042003.

Zhang, N., Liu, B., Ma, C., Zhang, G., Chang, J. *et al.* (2014) Transcriptome characterization and sequencing-based identification of drought-responsive genes in potato. *Molecular Biology Reports* 41, 505–517.

Zhang, J.Y., Huang, S.N., Mo, Z.H., Xuan, J.P., Jia, X.D. *et al.* (2015) De novo transcriptome sequencing and comparative analysis of differentially expressed genes in kiwifruit under waterlogging stress. *Molecular Breeding* 35, 1–12.

Zhang, D.Y., Zhang, T.X. and Wang, G.X. (2016) Development and application of second-generation sequencing technology. *Environmental Science and Technology* 39, 96–102.

Zhang, S., Zhang, A., Wu, X., Zhu, Z., Yang, Z. *et al.* (2019a) Transcriptome analysis revealed expression of genes related to anthocyanin biosynthesis in eggplant (*Solanum melongena* L.) under high-temperature stress. *BMC Plant Biology* 19, 1–13.

Zhang, Z., Cao, B., Li, N., Chen, Z. and Xu, K. (2019b) Comparative transcriptome analysis of the regulation of ABA signaling genes in different rootstock grafted tomato seedlings under drought stress. *Environmental and Experimental Botany* 166, 103814.

Zhang, Q., Li, M., Xia, C.Y., Zhang, W.J., Yin, Z.G. *et al.* (2021a) Transcriptome-based analysis of salt-related genes during the sprout stage of common bean (*Phaseolus vulgaris*) under salt stress conditions. *Biotechnology & Biotechnological Equipment* 35(1), 1086–1098.

Zhang, H., Li, G., Yan, C., Cao, N., Yang, H. *et al.* (2021b) Depicting the molecular responses of adventitious rooting to waterlogging in melon hypocotyls by transcriptome profiling. *Biotech* 11(7), 351.

Zhang, L., Dai, Y., Yue, L., Chen, G., Yuan, L. *et al.* (2022) Heat stress response in Chinese cabbage (*Brassica rapa* L.) revealed by transcriptome and physiological analysis. *PeerJ* 10, e13427.

Zhang, K., Chen, X., Yuan, P., Song, C., Song, S. *et al.* (2023) Comparative physiological and transcriptome analysis reveals potential pathways and specific genes involved in waterlogging tolerance in apple rootstocks. *International Journal of Molecular Sciences* 24(11), 9298.

Zhao, W., Ma, Z., Liu, S., Yang, W. and Ma, J. (2021) Transcriptome profiling reveals potential genes and pathways supporting *Ananas comosus* L. Merr's high temperature stress tolerance. *Tropical Plant Biology* 14(2), 132–142.

Zheng, S., Zhou, D., Liu, T., Wang, T., Zhai, X. *et al.* (2023a) Comparative transcriptome analysis of two varieties of common bean (*Phaseolus vulgaris* L.) to identify candidate drought resistance genes. *Biotechnology & Biotechnological Equipment* 37(1), 2238847.

Zheng, X., Zhu, Q., Liu, Y., Chen, J., Wang, L. *et al.* (2024) Combined analysis of transcriptome and metabolome provides insights in response mechanism under heat stress in avocado (*Persea americana* Mill.). *International Journal of Molecular Sciences* 25(19), 10312.

Zheng, Y., Yang, Z., Luo, J., Zhang, Y., Jiang, N. *et al.* (2023b) Transcriptome analysis of sugar and acid metabolism in young tomato fruits under high temperature and nitrogen fertilizer influence. *Frontiers in Plant Science* 14, 1197553.

Zhou, X., Condori-Apfata, J.A., Liu, X., Condori-Pacsi, S.J., Valencia, M.V. *et al.* (2021a) Transcriptomic changes induced by drought stress in hardneck garlic during the bolting/bulbing stage. *Agronomy* 11(2), 246.

Zhou, X., Dou, Y., Huang, X., Li, G., Zhang, H. *et al.* (2021b) Using principal component analysis and RNA-seq to identify candidate genes that control salt tolerance in garlic (*Allium sativum* L.). *Agronomy* 11(4), 691.

Zhu, X., Li, X., Jiu, S., Zhang, K., Wang, C. *et al.* (2018) Analysis of the regulation networks in grapevine reveals response to waterlogging stress and candidate gene-marker selection for damage severity. *Royal Society Open Science* 5(6), 172253.

Zhu, Y.C., Sun, D.X., Yun, D.E.N.G., An, G.L., Li, W.H. *et al.* (2020) Comparative transcriptome analysis of the effect of different heat shock periods on the unfertilized ovule in watermelon (*Citrullus lanatus*). *Journal of Integrative Agriculture* 19(2), 528–540.

Zou, J.J., Wei, F.J., Wang, C., Wu, J.J., Ratnasekera, D. *et al.* (2010) *Arabidopsis* calcium-dependent protein kinase *CPK10* functions in abscisic acid-and Ca^{2+}-mediated stomatal regulation in response to drought stress. *Plant Physiology* 154, 1232–1243.

4 Introduction to Omics and Its Application in Crop Improvement

Anuradha Pandey, Neeraj Unnikrishnan, Aparna Ramakrishnan P.N. and Dipak Gayen*

Department of Biochemistry, Central University of Rajasthan, Rajasthan, India

Abstract

The escalating effects of climate change constitute a huge threat to world agriculture, demanding the development of crops that can grow in increasingly harsh climatic conditions. Traditional breeding procedures, while effective, are sometimes time-consuming and have limited genetic diversity. The development of omics technologies—genomics, transcriptomics, proteomics, metabolomics, and phenomics—has transformed crop improvement by offering a comprehensive approach to understanding and improving plant resilience. This book chapter focuses on the critical function of omics in finding and modifying important genetic, molecular, and biochemical processes that underpin crop stress tolerance. Genomics has allowed for the quick identification of genes associated with desirable qualities, whereas transcriptomics has provided insight into gene expression patterns during stress. Proteomics and metabolomics have provided vital insights into protein function and metabolic alterations, respectively, allowing for a more comprehensive understanding of plant responses to environmental stresses. Phenomics, or high-throughput phenotyping, combines phenotypic data with other omics information to aid in the identification of superior genotypes. Despite these advances, issues such as data integration, scalability, and ethical considerations persist. This book chapter emphasizes the revolutionary potential of omics technologies in the development of climate-resilient crops, and it advocates for continuous multidisciplinary research and collaboration to transfer scientific discoveries into sustainable agricultural solutions.

4.1 Introduction

Climate change is one of the most significant threats to global food security and economic stability. Its impact on agriculture—the backbone of many economies worldwide—will potentially cause major disturbances in food production and delivery. The World Bank stated that agriculture contributes about 4% of global GDP; however, in certain developing countries like sub-Saharan Africa and South Asia, agriculture constitutes up to 25% of these economies' GDP. Crop failures due to climatic change have wide-ranging consequences both on food security and economic performance, especially for low-income, agrarian countries that rely heavily on crop exports due to the important role of agriculture in the global economy. In areas where the agricultural industries are serious contributors to their economies, extreme climatic events can cause massive losses in terms of revenue. For instance, extreme heatwaves, droughts, and floods in a country like India where agriculture contributes approximately 17% of its GDP have lost billions of dollars in terms of revenue and depressed crop yields (Lobell *et al.*, 2011). It is further claimed that the world GDP

*Corresponding author: dipak.gayen@curaj.ac.in,gayen.dipak@gmail.com

DOI: 10.1079/9781800628717.0004

may decline by as much as 10–25% by 2050 if climate change goes unchecked—impacting the global agricultural output. While the impacts must be mitigated, climate-resilient crops form the essential foundation to ensure economic stability, particularly for countries highly dependent on agriculture. While these crops will secure food production for communities, they will also protect the economies of countries if and when agriculture remains dominant. In this sense, climate-resilient crops can be seen as vital not only for food security but to maintain global economic stability in a warmer world (FAO, 2017). Traditional breeding methods have been in use for generations to enhance crop yield and durability through the selection of characteristics such as resistance of these crops to pests and tolerance to drought. Though those methods had worked well long ago, they are too slow and labor intensive. Traditional breeding in most cases may take a long time of 40 years or so before it produces crops with the required effect and this is a significant restriction when considering climate change effects on agriculture in the rest of the world (Tester and Langridge, 2010). In addition, methods that have been applied for generations by breeders are often constrained by the genomic variation that appears in the gene pool, meaning that breeders are somewhat limited to exploiting natural diversity in the crop species of focus. This gradual improvement has economic repercussions. Farming is one of the largest contributors to the world's GDP, and slow crop improvement would continue to leave farmers vulnerable to climate-induced crop failure. For instance, annual losses caused by extreme weather have increased over the past few decades in the United States, which is still the world's largest producer of maize and soybean. This translates into reduced agricultural productivity and billions of dollars annually lost in the US economy. This will affect the farmers and also the global economy, which will lead to inflation.

The use of omics technologies thus represents a powerful and efficient approach to crop improvement in the face of limitations in traditional breeding. In this regard, technologies such as genomics, transcriptomics, proteomics, and metabolomics enable scientists to question the underlying mechanisms of control in plant responses to environmental stressors. Omics technologies would enable crops to be developed with enhanced features of resistance and tolerance to stresses, thereby making them more resilient to climate change compared with the usual period of breeding as dictated by conventional methods (Hirayama and Shinozaki, 2010). For instance, genomics information about genetic crop composition provides knowledge for the identification of stress-resistance genes that are used in breeding programs. Genomic studies in rice, a staple food for more than half of the world's population, have even identified drought-resistance genes that may confer large yield improvements in regions with water scarcity. This could save not just the world's food but also the economies of countries where rice happens to be one of its principal agricultural products, such as India, China, and many South-east Asian countries (Mochida and Shinozaki, 2011). Going beyond genomics, transcriptomics, and proteomics opens further avenues in improving crop resilience through understanding how environmental stress affects gene expression in some scenarios. The examination of how genes are expressed under different environmental conditions will allow scientists to pinpoint key regulatory networks driving the plant's response to stress. This new understanding will help breed crops more resilient to heat, drought, and other climatic challenges, and stabilize global agricultural output, thereby stabilizing global GDP (Zhu *et al.*, 2017). However, the economic benefits of omics technologies do not stop at national borders. Climate-resilient crops have the potential to stabilize global trade in agricultural products, which remains one of the heavy lifters in world GDP. International trade in food and agricultural products was valued at over US$1.5 trillion by the United Nations in 2020. The importance of stabilizing agriculture for the global economy is self-evident. By developing more resilient crops with the power of omics technologies, countries can now stand to protect not only their economies but also the very sanity of global markets (Canton, 2021). With agriculture continuing as an important part of the global GDP, the omics technologies could be used to develop climate-resilient crops with immense economic value in ensuring that the resilience of the world's food systems and economies continues into a plausible climate future.

4.2 Omics Technologies

4.2.1 Genomics

Genomics is a branch of science focusing on the complicated relationships between genes, their cumulative impact on biological functions, the effect of external influences on gene activity, and the functioning of non-coding portions of the genome. In contrast to classical genetics, genomics and transcriptomics have marked a dramatic shift by giving detailed insights into gene structure and epigenetic variation alterations. These advancements have enabled the examination of a broad range of genetic variations and the provision of the precise quantification of the activity levels of individual genes (Singh *et al.*, 2015). Next-generation sequencing (NGS) is an innovative technology that has dramatically enhanced the understanding of genomics, allowing the collection and evaluation of data on genome structure, performance, interactions, and many more features across numerous omics areas of research. NGS is a low-cost technology that rapidly generates enormous quantities of sequence data, allowing for the precise analysis of nucleotide variants and the identification of marker genes on a broad scale (Schuster, 2008). These are used in the indirect selection of economically desirable features centered on gene mapping and quantitative trait locus (QTL) mapping, and genome-wide association studies (GWAS). The most advanced technique in crop polymorphism research entails sequencing or resequencing the whole genome or particular sections of it over an extensive range of accessions. This was not conceivable prior to the introduction of NGS systems, which have transformed genomic research by significantly boosting the efficiency of the sequencing process while lowering prices. Additionally, notable NGS technologies such as S-FLX and Illumina HiSeq (Illumina High-Throughput Sequencing), are employed to sequence various crop varieties using the whole genome shotgun (WGS) technique to facilitate the rapid generation of large datasets, and some other firms have even recently launched products in third-generation sequencing technology. NGS technology has allowed molecular marker mining to implement a novel technique known as coverage-based consensus calling (CbCC), which has assisted in the discovery of single nucleotide polymorphisms (SNPs) in chickpeas (Azam *et al.*, 2012). NGS has also streamlined access to organellar genomes, such as mitochondria and chloroplasts, as illustrated by the entire sequencing of the chloroplast and mitochondrial genomes of *Boea hygrometrica*, which demonstrates that the smaller chloroplast genome (150 kb) incorporates a greater number of coding genes than the larger genome of the mitochondria (510.5 kb) (Roychoudhury, 2020).

Molecular genomics is a major area within plant biotechnology, providing critical tools and procedures that considerably advance plant breeding operations. Marker-assisted selection (MAS) and genomic selection are examples of techniques that are used in several kinds of applications, which promote and advance plant breeding processes (Varshney and Tuberosa, 2007). However, MAS is crucial in plant breeding as it handles traits that cannot be studied by using traditional phenotyping methods, where those traits include complex inheritance patterns, climate-dependent characteristics, or those that would cost considerable money to investigate (Xu and Crouch, 2008). MAS additionally allows the ultimate integration of desired traits within a single genotype, reducing unnecessary losses and requiring fewer selection cycles. Recent research suggests that genomics-assisted methodologies have been used in marker-assisted sorghum breeding to aid in the selection of parent lines, maintain the genetic integrity of F1 hybrids, and determine traits for insertion. Furthermore, molecular markers have been frequently utilized to assess genetic variation within sorghum, resulting in the generation of hybrids with higher grain yield and resistance (Baloch *et al.*, 2023). MAS promotes breeders to indirectly locate genetic variables correlated with progeny qualities by crossing suitable parents (De Mori and Cipriani, 2023). The effectiveness of MAS is governed by a variety of parameters, including trait complexity and marker proximity to the targeted trait. In addition, MAS offers tremendous prospects for enhancing plant breeding since it allows for the precise transfer of specific genomic regions and accelerates the restoration of genome segments encoding prominent characteristics.

Alternatively, by integrating substantial genome-wide data, genomic selection facilitates breeding evaluation of the genetic potential of complex features in organisms. (Verma *et al.*, 2024) highlight the noteworthy overlap between MAS and genomic selection (GS), as they open new avenues for developing desired plant traits. With the decrease in genotyping costs, GS has become an effective approach in numerous plant and animal breeding programs, particularly for minimizing the period needed to complete breeding cycles. In addition, Vivek *et al.* (2017) conducted a comparative study of GS and conventional phenotypic selection in corn (maize) breeding, concentrating on performance under drought and optimum conditions. The findings revealed that GS consistently obtained better genetic gains every cycle than phenotypic selection, which resulted in more significant annual improvements in breeding performance. This demonstrates the efficacy of GS in accelerating breeding efforts while also boosting crop endurance and productivity, making it a crucial approach to fostering crop growth under changing environmental circumstances.

4.2.2 Transcriptomics

Transcriptomics elucidates the highly complex processes underpinning gene expression and transcriptional level monitoring via analyzing RNA transcripts in specific cells or tissues under specified conditions. According to Horgan and Kenny (2011), the transcriptome consists of all the mRNA in a living thing or cell that functions as a template for protein translation and subsequently indicates the genes that are expressed differently at any particular time. Transcriptomics has been effectively implemented in the field of environmental physiology to explore how different species adapt or become accustomed to their peculiar abiotic situation. Transcriptome sequencing technology facilitates the rapid generation of a wide range of expressed sequence tags (ESTs) at minimal expense. The utility of these ESTs lies in their ability to detect molecular markers such as SNPs and simple sequence repeats (SSRs) in vegetables such as corn (Mukesh Kumar *et al.*, 2018).

Furthermore, cap analysis of gene expression (CAGE) is a technique that analyzes the 5′ end of transcripts to identify transcription initiation sites and promoter domains. Unlike other transcriptome checks, which focus on the 3′ end, CAGE catches the 5′ end of mRNA, converts it to cDNA by RT-PCR, and then utilizes biotinylated linkers with restriction enzyme sites for additional analysis (Bhattacharjee, 2021). Microarray technology has proven to be extremely effective in executing systematic analyses of large-scale gene expression. According to Lee *et al.* (2005), this technique has revealed multiple Arabidopsis genes that respond to drought, cold, and high salinity stress at the transcriptional phase. In keeping with this perspective, Salvi and Tuberosa (2005) emphasize that microarrays can be used to enhance QTL cloning operations by precisely selecting relevant genetic resources and revealing the functional bases of cloned plant QTLs (Blanca *et al.*, 2012). However, RNA-Seq has gained prominence over microarray approaches because it produces read counts as non-negative integers with a discrete range. On the other hand, microarray data provide consistent intensity values with a log-normal dispersion, highlighting RNA-sequencing superior dependability for transcriptome analysis. RNA-Seq is a method for detecting and quantifying RNA transcripts in a sample that uses rapid sequencing and powerful computational algorithms for producing sequences of nucleotide as short as 30 bp and as long as 10,000 bp (Ozsolak and Milos, 2011). RNA-Seq is essential for analyzing formerly unidentified genes and newly recognized transcript isoforms that immediately display sequence identities. RNA-Seq technology has less uncertainty than microarray technology, notably its ability to identify novel transcripts and transcript isoforms, discern closely related paralogous patterns, and measure gene expression digitally, as opposed to the analog method. Nevertheless, there are concerns about the preciseness of RNA-Seq for evaluating absolute levels of transcripts, as studies demonstrate that sequencing reads can vary over the transcript length (Fu *et al.*, 2009).

(Li, 2020) investigated how *Salvia miltiorrhiza* responds to moderate drought stress at the transcriptome level. Among the 58,085 unigenes discovered, there was a notable rise in genes associated with metabolic pathways, particularly those coding for enzymes that govern phenylpropanoid and terpenoid biosynthesis, which were more highly expressed during stress. This study showcases how *Salvia miltiorrhiza* controls medicinal chemical production dependence on drought levels. Moreover, Feng ChuHang *et al.* (2019) performed third-generation sequencing to assess the complete transcriptome of *Angelica sinensis*, and discovered 25,463 transcripts with substantial variance in expression levels by performing differential expression evaluation on wild-type samples. Subsequent observations found significant variations in transcripts correlated with plant–pathogen interactions and plant hormone signal transduction pathways, unveiling important molecular processes driving plant defensive responses and hormonal control in *Angelica sinensis*.

4.2.3 Metabolomics

Metabolomics represents a detailed investigation of small molecules in cells, tissues, or organisms. These small metabolites have been reported to be sugars, lipids, amino acids, and secondary metabolites that work as the basic input for the biochemical mechanisms associated with development, growth, and responses to environmental cues. Studying the whole set of metabolites present in a biological sample allows researchers to gain insights into the metabolic state of an organism and how it is modified by genetic or environmental perturbations. (Fiehn, 2006). Metabolomics has emerged as a potent technology with huge potential for advancing improvements in crop species within the field of plant biology. By relating metabolic profiles to phenotypic variables, breeders and researchers can understand better the plant physiology and genetic determination of crucial agricultural traits, such as yield, stress tolerance, and nutritional quality (Arbona *et al.*, 2013). Metabolomics is a high-throughput biochemical trait screening platform that uses cutting-edge analytical technologies, including mass spectrometry (MS) and nuclear magnetic resonance (NMR), to target beneficial traits (Singh *et al.*, 2023). In crop improvement, metabolomics can be applied in several ways such as trait discovery and improvement. Metabolomics can identify metabolic markers of traits for desirable properties such as disease resistance, drought tolerance, or enhanced nutrition value. Metabolomics benefits crop improvement in many ways. From an understanding of the metabolic pathways contributing to these traits, scientists can design genetic manipulations and targeted breeding for better crop performance (Razzaq *et al.*, 2019). Plants are often exposed to numerous biotic and abiotic stresses, which exert great effects on crop quality and yield. The metabolomics approach allows detailed descriptions of metabolite changes induced by stresses, thus facilitating an understanding of the mechanisms underlying stress tolerance. This information can help in the development of crops that are more resistant to environmental stresses (Arbona *et al.*, 2013). Metabolic fingerprinting can be applied to quality control of production in agriculture and assessment of the nutritional composition of crops. Breeders could select cultivars possessing enhanced nutritional content—higher levels of certain vitamins, minerals, or beneficial phytochemicals—by investigating metabolic composition (Singh *et al.*, 2020). Thanks to metabolomic insights, metabolic engineering has reached a level where modification of pathways for the generation of new compounds or the enhancement of useful metabolite production can be offered. Thereby, crops with enhanced fragrances, tastes, or medicinal characteristics could be bred (Bahl *et al.*, 2018). In all, metabolomics is a comprehensive and integrated approach to the understanding of plant metabolism and its effects on crop traits. Metabolomics combined with other omics technologies, like transcriptomics and genomics, will be able to provide full insights into plant biology. This will open up possibilities for new and inventive methods of crop improvement that can meet the needs of a growing population worldwide (Shiferaw *et al.*, 2011). Metabolic profiling has turned out to be an exceptionally strong tool in crop science, enabling the analysis of complicated biochemical networks in plants and the identification of those metabolites related to important agricultural traits. State-of-the-art analytical techniques integrate plant physiology, stress responses, and the entire potential for crop improvement (Yang *et al.*, 2021).

MS is one of the dominant techniques in metabolic profiling due to its high degree of sensitivity and ability to detect a wide range of metabolites. In MS, chemical compounds get ionized to generate charged molecules, that are then measured based on the mass-to-charge ratio. Gas chromatography (GC)–MS, liquid chromatography (LC)–MS, and MS/MS are some examples of MS techniques applied in measuring plant metabolites (Vinaixa *et al.*, 2016). GC–MS will, hence, be very useful in the analysis of volatile and semi-volatile compounds. In the determination of primary metabolites such as sugars and organic acids, this technique offers very high resolution and sensitivity. It LC–MS is used for non-volatile and thermally labile molecules, with the ability to detect a huge spectrum of secondary metabolites, including flavonoids and alkaloids. MS/MS enables the identification and quantification of metabolites based on fragmentation patterns, as well as detailed information on their structure (Xiao *et al.*, 2012).

One of the other powerful techniques within metabolic profiling is NMR spectroscopy, which offers non-destructive metabolite analysis with quantitative information on the molecular structure of compounds. The power of NMR thus lies in the identification of unknown metabolites and in good reproducibility across different samples (Emwas *et al.*, 2019). The most used techniques include ^{1}H and ^{13}C NMR, which provide information on hydrogen and carbon backbones of metabolites, respectively. Frequently, NMR has been coupled to other analysis techniques for the validation and complementation of observations from MS, so allowing for a more detailed view on plant metabolism (Salem *et al.*, 2020). Chromatography forms an integral part of metabolic profiling and is a separation technique used to resolve complex mixtures of metabolites for MS or NMR analysis. The most common methods are those of GC and LC, but LC is more versatile in that it can manage a broader range of compound polarities and sizes. High-performance LC (HPLC) and, more recently, ultra-performance LC (UPLC) offer very wide resolution and speed and thus are very useful in handling and analyzing complex plant extracts (Steinmann and Ganzera, 2011).

Large datasets that are generated in metabolic profiling require sophisticated data analysis tools. The interpretation of these complex data thus involves the application of multivariate statistical analysis, machine learning, and bioinformatics techniques to identify key metabolic markers and finally metabolomic data integration with genomic and transcriptomic information (Ogunjobi *et al.*, 2024). Plants are constantly bombarded with various stresses coming from drought, salinity, extreme temperatures, and biotic factors. These stresses can have a very negative effect on plant growth and productivity. Plants have evolved complex physiological and biochemical mechanisms to resist such stresses, including specific metabolite accumulation to enhance their tolerance to stress. Understanding the metabolites conferring such properties can lead to crop breeding for better adaptation to environmental stress (Marone *et al.*, 2022).

Proline is an amino acid that is essential for the tolerance of plants to a wide variety of environmental stresses. Mass accumulation has been shown to occur under different abiotic stresses in plants, such as drought, salinity, and extreme temperatures. This protection by proline makes it an osmoprotectant through direct or indirect functions associated with the stabilization of proteins and membranes, free radical-scavenging properties, or rebalancing cellular redox status (Yancey, 2005). Its accumulation is often correlated with enhanced stress resistance, and manipulation of the proline biosynthesis pathways has been one of the targets for improving crop resilience. Trehalose is a non-reducing sugar that serves as a protectant in plants. It tends to stabilize proteins and cellular structures against various stresses, such as drought and high salinity. In crop production, this synthesis has been related to enhanced stress tolerance while it takes part in the regulation of carbon allocation and energy balance (Newman *et al.*, 1993). Glycine betaine is a quaternary ammonium compound that accumulates under abiotic stresses, mainly salinity and drought. It functions as osmoprotectant in the osmotic balance maintenance of cells, protection of photosynthetic machinery, and protein stability enhancement. Small organic cations known to be involved in plant growth and stress responses are the polyamines putrescine, spermidine, and spermine (Ali *et al.*, 2020). They interact with membrane stabilization, reactive oxygen species scavenging, and ion channel regulation during stress. Polyamines also modulate the expression of stress-responsive genes. Vitamin C (ascorbate) and glutathione are natural antioxidants that have crucial functions in the detoxification of reactive oxygen species (ROS) produced in plants under stress conditions. They become major constituents of the ascorbate–glutathione cycle, which provides protection to plants against oxidative damage (Hasanuzzaman *et al.*, 2019). Plants can withstand more stress by developing higher concentrations of these antioxidants through breeding or genetic engineering, which strengthens the plant defences against oxidative stress. Abscisic acid (ABA) is a phytohormone that regulates responses against abiotic stresses focused on drought and salinity, controlling closure of stomata to reduce water loss due to transpiration, modulating gene expression for stress tolerance, and cross-talking with other hormones and signaling pathways in the orchestration of complex stress responses (Cabot *et al.*, 2014). The alteration of ABA signaling pathways has been discussed to improve the stress tolerance of crops (Vishwakarma *et al.*, 2017). Metabolic engineering of metabolic pathways in organisms allows for overproduction of specific compounds of interest or improvements in certain traits. In crop improvement for stress tolerance, metabolic engineering is applied to all those pathways that produce protective metabolites and stress-responsive compounds. Already, there have been quite a number of promising solutions provided toward developing crops with resilience to various environmental stresses such as drought, salinity, and temperature extremes through this approach.

For example, under stressful circumstances, osmoprotectants such as proline, trehalose, and glycine betaine are essential for preserving the osmotic balance of cells. By increasing the biosynthesis of these compounds in plants, metabolic engineering can improve their ability to withstand stress. For instance, it has been demonstrated that increasing the expression of proline biosynthesis-related genes enhances a crop's ability to withstand salt and drought (Per *et al.*, 2017). Under stressed conditions, ROS accumulate in plants, causing cellular damage. Engineering antioxidant pathways to overproduce compounds such as ascorbate and glutathione can improve the detoxification ability of plants against ROS. Improved detoxification of ROS protects cellular structures from oxidative damage, thus improving stress tolerance (Hossain *et al.*, 2015).

Plant responses to environmental stressors are regulated by stress hormones like jasmonic acid (JA), salicylic acid (SA), and ABA. The biosynthesis and signaling pathways of these hormones can be changed through metabolic engineering, improving the capacity of the plant to withstand stress. For instance, by controlling stomatal closure and water-use efficiency, ABA level changes can enhance drought resistance (Negin and Moshelion, 2016). Secondary metabolites, among which are flavonoids and alkaloids, participate in the plant stress responses and defense mechanisms. Manipulation of pathways increasing the production of these metabolites can make crops more resistant to stresses. For example, enhanced flavonoid synthesis may generally enhance UV tolerance and pathogen resistance (Ferreyra *et al.*, 2021). One can engineer transcription factors targeting stress-responsive genes to improve tolerance against these stresses. For example, overexpression of DREB, a transcription factor from the Dehydration-Responsive Element Binding family, led to the induction of downstream genes regulating stress responses, such as against drought and cold stresses (Zhou *et al.*, 2010). Integration of genomics, transcriptomics, proteomics, and metabolomics with metabolic engineering has enabled a better understanding of the stress responses and identification of key targets for manipulation. Only in this way, via such a holistic approach, can crops be made tolerant to stresses by designing manipulations across different pathways and processes.

4.2.4 Phenomics

Phenomics assesses immense quantities of phenotypic data to determine how the genome expresses itself as a phenotype in certain environmental circumstances. Environment plays an essential role in plant phenomics since many significant plant features are quantifiable and significantly affected by environmental influences (Masuka *et al.*, 2012). Forward phenomics uses innovative phenotyping technology to recognize germplasm with appealing characteristics among an expansive collection, hence assisting in the choice of high-performing plant types. This method is refined using high levels of throughput, and computerized screening tools, which permit the prompt identification of favorable traits at the phase of seedling development, limiting the necessity for nurturing plants to complete maturation (Kumar *et al.*, 2015b). Reverse phenomics, however, is an analysis of numerous genotypes and genome products known to have desired phenotypic features, which aims to determine the associated biochemical and genomic footing of the associated traits. By discovering and leveraging these methods, this technique can improve breeding procedures, such as uncovering and adding genes with drought-resistant features in plants.

Plant high-throughput phenotyping is an important agricultural technique that enables quick and successful evaluation of a wide range of plant properties directly in the field, where it combines modern equipment to collect ample data on plant features such as structure, chemical constitution, and physiological responses, allowing further insights into how genes and environmental variables interact (Haghighattalab *et al.*, 2016). The utilization of automated vision, imaging spectroscopy, and thermal infrared imaging help in assessing plant attributes such as yield, disease-resistant qualities, endurance to drought, and salt tolerance in high-throughput phenotyping (Li *et al.*, 2014). Similarly, Andrade-Sanchez *et al.* (2013) emphasize the importance of collecting high-resolution data for precisely assessing and choosing optimal cultivars in crop improvement and phenotype analysis. High-throughput phenotyping platforms (HTPPs) allow the non-intrusive monitoring of plants along their developmental phases in regulated environments. Nonetheless, the QTLs and potential genes found in these controlled situations have scarcely contributed to increased yields of grains in the field.

Plant phenotyping leans extensively on imaging technology for recognizing and quantifying an array of morphological features, including the color, shape, size, and texture of seeds, leaves, canopies, fruits, and roots. Furthermore, the incorporation of two- and three-dimensional methods of imaging facilitates more precise estimates of canopy volume and fruit distribution.

4.2.4.1 Fluorescence imaging

Fluorescence imaging on a larger scale enables the rapid and distant acquisition of fluorescence data with fine spatial resolution, ranging from leaf to canopy-level observations. Unlike reflectance signals, fluorescence imaging provides plant-specific spectral categories that function as distinctive markers. Fluorescence occurs when a substance absorbs radiation at wavelengths that are shorter, while chlorophyll in plants is the primary source of this emission. When exposed to blue or actinic light, chlorophyll re-emits the absorbed light, leading to luminescence, which reveals the plant's light metabolism and photo assimilation capacity (Gorbe and Calatayud, 2012). Fluorescence imaging, which uses charge-coupled device (CCD) cameras, collects these emissions and measures critical physiological factors including non-photochemical quenching. Additionally, this imaging method is facilitated by exposing specimens to visible or UV light from sources such as pulsed lasers or light-emitting diodes (LEDs). Fluorescence imaging has been used in genotype screening as well as detecting stress-induced biomarkers in plants. Bürling *et al.* (2010) demonstrated that it was useful in measuring the resistance of wheat varieties to *Puccinia triticina*, with the quantum yield of non-regulated dissipated energy in PSII serving as a reliable metric for separating across varieties. Susceptible wheat cultivars, in particular, revealed significant alterations in fluorescence characteristics between control and infected leaves, which were increasingly noticeable over time.

4.2.4.2 Thermal imaging

Thermal imaging requires infrared radiation to recognize and map temperature variations between surfaces. The spectral range of 3–14 μm allows for precise temperature surveillance, making it an invaluable resource for plant phenotyping. In contrast to traditional measurement approaches, thermal imaging allows for the non-invasive examination of plant water reserves and stress by spotting changes in stomatal activities. Thermal imaging technology has proven beneficial in measuring fruit tree health, examining three-dimensional structures, and locating fruits. For example, Narváez *et al.* (2016) used ground-based thermal imaging to analyze heat-related morphological properties in fruit plants.

4.2.4.3 Tomographic imaging

Tomographic imaging tools, such as magnetic resonance imaging (MRI), X-ray computed tomography (X-ray CT), and positron emission tomography (PET), are capable of creating high-resolution 3D imagery of specific plants or plant portions. MRI scanners use magnetic fields, electric gradients, and radiofrequency waves to generate complete interior imagery of plants, enabling the non-invasive analysis of properties such as development, moisture dispersion, and response to stress. Furthermore, positron emission tomography (PET) creates 3D images by capturing gamma rays generated by a positron-emitting tracer administered into the plant. In contrast, X-ray computed tomography (CT) uses a collimated X-ray beam to acquire and recreate 3D pictures from many 2D estimations, allowing for clear visibility of both interior and exterior plant attributes (Crain *et al.*, 2018). Jones *et al.* (2014) outline Förster resonance energy transfer (FRET) sensing technology as an advanced, non-invasive method that is capable of accurately measuring minuscule molecules inside living tissue. Utilizing genetically encoded ratiometric fluorescence sensors that bind preferentially to specific molecules, FRET allows molecular phenotyping by presenting a detailed analysis of molecular fluctuations and permitting real-time monitoring of biochemical interactions.

4.2.4.4 Phenotype–genotype associations

Phenotype–genotype association in plant breeding refers to the connection between genetic makeup and observable traits. There are multiple benefits such as disease resistance and stress tolerance. Shakoor *et al.* (2017) mentioned that the disease monitoring and resistance in such breeding is done with the help of HTPPs, allowing the most accurate and repeatable measures and methods for boosting the quality of the plant. These are also used in monitoring the physiological traits of the plant, which helps in the improved identification of changes. Stress tolerance can be improved for various conditions, including drought, abiotic stress, and other types. According to Bansal *et al.* (2014), for drought stress, numerous polymorphisms of candidate genes were identified, which helps in improving the tolerance. Through several studies, it has been found that the level of crop production needs to be at an increasing percentage of between 59 and 98% by 2050, compared to levels in 2005 to meet the requirements of the global population. The studies establish an estimated 2.4% crop increase is needed every year but the recent status is less than 1% (Crain *et al.*, 2018). Hence, a high yield using the phenotypic breeding method is a must. The other advantage of genotype–phenotype breeding is the precision breeding method, in which plant breeding is done in a precise way to make required changes in the DNA (Hackauf *et al.*, 2021). According to Sinha *et al.* (2021), recent advancements in next-generation DNA sequencing have brought significant progress in understanding genomes and hence shifting from genomics to phenomics. More precise phenotypic data can be captured with noninvasive high-throughput phenotypic technologies, but the advancements of the technology in real-life applications have to be improved. However, Rauf *et al.* (2010) have mentioned that genetic diversity will be affected by constant plant breeding and the consequences should be constantly monitored.

4.2.5 Proteomics

Proteomics refers to large-scale research on proteins, which are important constituents of living organisms and play a significant role in various biological processes. It may be described as comprehensive analytical research into the protein content of crops and other plants, focusing on their structure, function, and interactions. This became an important field in recent years because of its potential for giving valuable insights into plant physiology, stress responses, and genetic diversity—information crucial for crop improvement (Aslam *et al.*, 2016). By using proteomics, researchers are able to detect and quantify proteins in various tissues, developmental stages, and under environmental conditions. The inferences regarding the molecular mechanisms underlying plant growth, development, and response to biotic and abiotic stresses can be obtained from protein expression profiles. This information is crucial to understanding how plants adapt to changing environments and identifying key proteins involved in stress tolerance and resistance (Raza *et al.*, 2020). The identification of stress-responsive proteins is one of the major applications of proteomics in crop improvement. Plants are always subjected to different environmental stresses such as drought, salinity, and extreme temperatures, which adversely affect plant growth and productivity. Proteomics helps in identifying proteins that get either upregulated or downregulated in plants due to such stresses and provides important targets for genetic engineering and breeding programs. For example, some commonly identified factors contributing to stress tolerance include heat shock proteins, antioxidants, and enzymes involved in osmoprotection (Priya *et al.*, 2019). Proteomics can be applied to improve crop resistance to various diseases and pests through characterization of proteins associated with the plant's defense mechanisms. With proper research in this direction, the protein networks developed to enhance crop resilience can be understood. Moreover, studies on proteins related to nutritional value, flavor, and shelf life can help in enhancing crop quality. This is quite important for developing crops with enhanced nutritional value and better marketability (Wei *et al.*, 2009). It provides molecular markers so that scientists can speed up marker-assisted breeding programs: in this case, protein markers can be used to select plants with improved characteristics for the traits of interest. This accelerates the process of developing new crop varieties with increased yield, improved stress tolerance, and enhanced quality (Singh *et al.*, 2010). The integration of proteomics with other omics technologies, like genomics, transcriptomics, and metabolomics, has provided all-around knowledge on plant biology. This approach—multiomics—further allows one to study intricate interactions between genes, proteins, and metabolites for a holistic view of the plant responses against environmental change. Such integrative studies could identify key regulatory networks and pathways to become targets for crop improvement (Kumar *et al.*, 2015a).

4.2.5.1 Two-dimensional gel electrophoresis (2-DE)

Two-dimensional gel electrophoresis is a classical technique for the separation of proteins into their components, using the variables that are the isoelectric points and molecular weights. In this process, there are two steps: isoelectric focusing, which separates the proteins in accordance with charge, and SDS–PAGE, which separates according to size.

Application: 2-DE is a significant tool that has been applied to the analysis of changes in protein expression under stress conditions like drought, salinity, and heat stress. This technique compares protein profiles of stressed and non-stressed plants and helps in the identification of stress-responsive proteins.

Advantages: 2-DE allows the researcher to visualize thousands of proteins simultaneously, thus enabling an overview of the proteome. It is therefore very useful in the detection of post-translational modifications. This method has limited power in the resolution of highly hydrophobic proteins and low-abundance proteins (Penque, 2009).

4.2.5.2 Mass spectrometry (MS)

MS is an analytical method of high power for the identification and quantification of proteins and peptides. It can offer data related to protein structure, function, and interactions when linked to techniques such as liquid chromatography.

Application: MS-based proteomics identifies and quantifies stress-responsive proteins in crops. Using this technique allows the identification of novel proteins involved in stress responses and enables high-throughput analysis. This technique offers high sensitivity, specificity, and accuracy, hence it is very appropriate for the analysis of complex proteomes. It can also identify proteins of low abundance and analyze post-translational modifications.

Disadvantages: The equipment used in MS is very expensive, and specialized expertise is needed (Jayathirtha *et al.*, 2021).

4.2.5.3 ICAT: Isotope-coded affinity tag labeling

ICAT is another approach to quantitative proteomics whereby the proteins in two comparative samples are differentially labeled with isotope-labeled tags. This enables relative quantification between the two samples according to mass differences.

Application: ICAT is used in the identification of proteins for which expression levels change under stress, helping to identify proteins up- or down-regulated under stress.

Advantages: ICAT is quantitative and efficient for the analysis of complex samples. It can be combined with MS for comprehensive proteome analysis.

Limitations: The technique is limited only to the cysteine-containing proteins, which may miss proteins without accessible cysteine residues (Deracinois *et al.*, 2013).

4.2.5.4 Stable isotope labelling by amino acids in cell culture (SILAC)

SILAC is a metabolic labeling method used in quantitative proteomics. Cells are grown in media containing heavy or light isotopes, allowing the incorporation of labeled amino acids into proteins.

Application: SILAC has been developed for application in plants for investigating dynamics of proteins upon stress responses. It allows the identification of stress-responsive proteins by comparisons between labeled and unlabeled samples.

Advantages: SILAC includes absolute quantification and is amenable to dynamic studies. It is MS compatible for high-throughput analyses.

Limitations: Metabolic labeling is required, and this is not possible in all plant species or tissues for SILAC (Harsha *et al.*, 2008).

4.2.5.5 Protein microarrays

Protein microarrays are applied for high-throughput analysis of protein–protein interactions, antibody–antigen interactions, and enzyme–substrate interactions. Proteins are immobilized on a solid surface and probed with specific ligands or antibodies.

Application: Protein microarrays can be used in screening stress-responsive proteins through investigating changes in protein interactions and modifications against stress.

Advantages: Protein microarrays allow the processing of thousands of proteins in parallel. This makes it possible to gain an overview of a proteome.

Limitations: Protein microarrays may have issues with protein stability and activity on a solid surface (Cahill, 2001).

4.2.5.6 Post-translational modifications (PTMs)

Chemical changes known as post-translational modifications (PTMs) occur following protein synthesis and impact the stability, interactions, structure, and function of proteins. PTMs are important targets for crop improvement strategies because they regulate how plants respond to environmental stressors. Gaining knowledge about PTMs can help us cultivate crops that are more resilient to stress and more adaptive (Pandey and Gayen, 2024).

Kinases mediate phosphorylation, which is the addition of a phosphate group to serine, threonine, or tyrosine residues. For signal transduction pathways that control stress responses, this alteration is essential. Phosphorylation controls the activity of enzymes involved in stress signaling pathways and transcription factors that respond to stress. For instance, phosphorylation triggers the activation of mitogen-activated protein kinases (MAPKs) in response to drought stress, thereby controlling the expression of genes associated with stress tolerance (Li *et al.*, 2017). Researchers can engineer plants with increased tolerance to abiotic stresses like salinity and drought by understanding the phosphorylation networks involved in stress responses. This is in turn achieved by attaching ubiquitin molecules to lysine residues, a process called ubiquitination, a relevant mechanism involved in quality control and protein turnover. Ubiquitination mediates proteolysis of potentially damaged or misfolded proteins under stressful conditions. It also modulates the abundance of some principal regulatory proteins implicated in stress responses.

A ubiquitin proteasome system can improve resilience in plants by ensuring the timely degradation of stress-induced proteins, thus maintaining cellular homeostasis (Xu *et al.*, 2024). Acetylation is the addition of an acetyl group to lysine residues, which can modulate protein stability and interactions. However, histone acetylation is one of the well-known PTMs regulating gene expression through an alteration in chromatin structure. Histone and non-histone protein acetylation modulates stress-responsive gene expression. It is involved in the regulation of heat shock and other stress pathways. Modifying gene expression profiles by targeting the mechanisms of acetylation can impart stress tolerance to crops (Yadav *et al.*, 2024).

Glycosylation is the covalent attachment of sugar moieties to proteins, modulating their folding, stability, and activity. Glycosylation plays a role in stability and activity for stress-responsive proteins concerned with defense against pathogens and adaptation to environmental stresses. Knowing the patterns of glycosylation can be helpful in developing crops that are more resistant to biotic and abiotic stresses through the stabilization of important proteins involved in their defense mechanisms (Qazi *et al.*, 2019). Sumoylation is the process whereby small ubiquitin-like modifier proteins are attached to the V-amino group of lysine residues in target proteins, which then affects protein localization, stability, and interaction. Sumoylation modulates the action of transcriptional factors and signaling proteins involved in stress responses. It is also involved in the modulation of plant responses to drought, salinity, and temperature stresses. By modifying sumoylation pathways, the stability and activity of stress-responsive proteins will be enhanced, resulting in improved stress tolerance of crops.

4.3 Integration of Omics Technology for Crop Improvement

The integration of omics techniques, such as genomics, proteomics, metabolomics, and phenomics, provides a powerful approach for improving crop health by offering comprehensive insights into plant biology. Genomics, for instance, forms the foundation by providing a detailed understanding of a plant's genetic makeup. Advances in NGS have enabled the rapid identification of genes associated with desirable traits such as disease resistance, drought tolerance, and nutrient efficiency. For example, GWAS have uncovered specific loci that control pathogen resistance in crops like rice and wheat. Genomic information is also invaluable for precision breeding, including genome editing techniques such as CRISPR-Cas9, which can be used to modify specific genes, leading to enhanced crop health and stress resilience (Bhattacharya *et al.*, 2021). Proteomics complements genomics through its emphasis on the proteins that conduct the functions encoded by genes. Given that proteins perform most biological activities, proteomic analyses help researchers relate genotypes to phenotypes. For example, using MS, proteomic studies have identified networks of proteins interacting to survive drought and respond to pathogens in plants, thus providing targets for breeding programs. Proteins involved in detoxification processes, especially regulatory proteins for ROS, have been highly induced by drought stress conditions and therefore appear to serve as promising markers of drought tolerance. Understanding protein expression and post-translational modifications directly feeds into the objectives of breeding crop varieties with enhanced tolerance to adverse environmental conditions (Zenda *et al.*, 2018).

Metabolomics, a significant technique within the omics field, pertains to the analysis of metabolites—small molecules resulting from cellular activities. These metabolites play a vital role in elucidating plant responses to an array of stresses, encompassing biotic influences such as pathogens and abiotic conditions like drought or elevated temperatures. The application of metabolomic profiling has facilitated the discovery of critical biomarkers that signify a plant's resilience or vulnerability to stressors. For example, flavonoids and alkaloids have been linked to resistance to pathogens in most crops. Such metabolites can be found, enabling plant breeders to select plants with higher levels of natural defense, thus healthier crops with resilience (Kang *et al.*, 2019). Phenomics is the high-throughput and elaborate measurement of plant characteristics, linking genetic and metabolic information to observable traits. The devices that are equipped with sensors and imaging technologies allow scientists to determine traits like growth, photosynthetic efficiency, and water-use efficiency in a variety of environmental contexts. The high level of data aggregation can be combined with other genomic and proteomic data to gain a better understanding of how specific genes and proteins impact observable traits. For example, phenomics has been used to screen the performance of rice varieties under drought conditions, which helps to identify genes linked to stability in grain yield when challenged environmentally (Yang *et al.*, 2020). Omics technologies are highly effective if synthesized. Approaches based on systems biology integrate information on various types of omics, including genomics, proteomics, metabolomics, and phenomics, to develop an integrated model that explains plant functionality under different conditions. The integrative approach helps in identifying molecular pathways that may regulate complicated traits, such as stress resilience, nutrient utilization efficiency, and resistance to diseases. For example, combining metabolomic and proteomic data has helped elucidate metabolic pathways relevant for drought tolerance in crops like corn. Genomic and phenomic data integration also identified stress-resilience-determining specific genomic regions of appreciable yield stability. The outcomes of this study may, therefore, inform specialized breeding efforts directed toward the improvement of specific traits, which would eventually support crop varieties that are better adapted to endure environmental stress factors that characterize climate change (Varshney *et al.*, 2005).

4.4 Challenges in GMO Crops

Editing genes is difficult, especially when using CRISPR-Cas9 because of the off-targeting of the cells that leads to some mutation reactions that could have occurred in any other part of the genes' structure, causing other unintended traits in these modified plants. Off-target mutations are a type of spontaneous genetic alteration that occurs inadvertently out of the desired area within a given genetic sequence without any constraints. Off targets that are within the amino acid domain constitute these alterations taking place other than where intended. This may cause unexpected characters or health complications in genetically modified plants (Guo *et al.*, 2023). Delivering gene-editing components into plant cells is a major barrier, and most of the techniques presently used, such as Agrobacterium-mediated transformation, biolistics, and direct DNA uptake, have variable efficiencies with respect to plant species and tissue types. It has remained a technical challenge to edit multiple genes with high precision simultaneously in the case of polygenic traits like drought resistance and yield improvement, where many different genes have intricate interactions (Mitra, 2001). One cannot turn a blind eye to the challenges of transition from success in a laboratory to field success, where edited crops are sure to undergo rigorous field trials to prove their worth in different environmental conditions and for generation after generation. The step of introducing gene-edited crops into existing breeding programs is associated with considerable effort to ensure the stable inheritance of the edited traits, and their compatibility with other desirable traits, which creates a significant practical challenge (Park *et al.*, 2019).

There is considerable public doubt and ethical concern relating to GMOs and gene-edited crops, covering issues of the naturalness of crops, the morality of manipulating nature, and long-term ecological effects (Shinwari *et al.*, 2018). One critical ethical dimension related to gene editing of crops is ensuring the safety of the crop for consumption. Some of the concerns that come into play are related to possible allergens, toxins, or other unintended effects that might be created by gene edits. There are concerns over the naturalness of GMOs and gene-edited crops and possible long-term ecological effects, which affect public perception and acceptance (Nawaz, 2021). The development of gene-edited crops is too expensive in the initial stages, entailing heavy investments in research, development, and compliance with regulatory requirements; thus, this forms a barrier to smaller companies and public research institutions (Marone *et al.*, 2023). To achieve commercial viability of production of gene-edited seeds and to be able to sell them globally, there is a need for strong supply chains and distribution networks, which are very costly to establish and maintain. The high development cost of gene-edited crops—in terms of the actual investment in research and development and regulatory compliance—puts them beyond the reach of most small companies and public research institutions (Brookes and Smyth, 2024). Countries across the world have a divergent array of regulatory frameworks on gene-edited crops, making navigation through the regulations cumbersome and time-consuming. For instance, European Union regulations against gene editing are much tighter than those in the United States (Voigt and Münichsdorfer, 2019). The regulatory process of gene-edited crops is complicated and lengthy due to varying regulatory environments in different countries. Coupled with this are legal challenges related to the ownership and patenting of gene-editing technologies and, subsequently, crops produced, which raise intellectual property concerns for sharing beneficial traits among researchers and farmers. Gene-edited crops could have a serious impact, either positive or negative, on local biodiversity (Turnbull *et al.*, 2021).

To be more specific, a huge concern of public and regulatory concern is the possible hybridization with wild relatives, resulting in the transfer of engineered traits into wild populations. This gene flow might disrupt local ecosystems through the introgression of hybrid plants with superior competitive abilities, replacing native species and losing biodiversity (Verhoeven *et al.*, 2011). The constant use of these transgenic crops for pest and herbicide resistance can ultimately lead to the development of resistant pest and weed populations. What is at work here is the same principle of how bacteria become resistant to antibiotics. As pests and weeds adapt, one can see farmers resorting to a higher dosage or finding new strategies; at worst, this starts a cycle of escalating interventions with potential harm to the environment. Socio-economic issues include the adoption of gene-edited seeds, often patented, that can increase farmers' dependence on biotech companies. For example, farmers may be forced to buy new seeds every seasonal change, rather than using some from their harvest, which traditional practice allows. This makes them dependent and may mean increased costs and economic disadvantages, especially for small-scale farmers who cannot afford this yearly. It can also eventually shift the control of agricultural practices and plant breeding from farmers to huge corporations and reduce farmers' autonomy and traditional knowledge (Altieri *et al.*, 2012). Market access may be substantially impeded for gene-edited crops. In countries where there is considerable resistance to GMOs, like in parts of the European Union, the regulatory hurdles and issues linked to perception by the public basically limit commercialization regarding gene-edited crops. This kind of resistance can result from ethical misgivings, health risks, or environmental impacts (Lindberg *et al.*, 2023). International collaboration in gene editing and GMO research is key to combating food insecurity. However, these are greatly hampered due to the lack of harmonization in regulatory and ethical standards among different countries. Harmonization in this area and international partnerships will help fast-track the development and adoption of beneficial technologies. There are special challenges that developing countries face in their efforts to adopt gene-editing technologies. These technologies are pretty hard to execute with resource-constrained infrastructures and regulatory frameworks. Moreover, access to training and education in advanced agricultural practices is also limited, hence hindering adoption. Ensuring the traceability and proper labeling of gene-edited products along the supply chain is key to both consumer trust and regulatory compliance, and it empowers consumers to make informed choices about the products they buy, thereby addressing the concerns related to transparency and safety. Gene editing and GMOs possess the transformative power of crop improvement; they hold potential solutions for food security, enhancing crop resilience, and improving nutritional profiles (Ndudzo *et al.*, 2024). However, a host of challenges and limitations emanate from these new technologies across technical, practical, ethical, socio-economic, and regulatory dimensions. In other words, although gene editing and GMOs do hold a wonderful future for agriculture, multifaceted challenges require a balanced approach. For this to happen, collaborative efforts from scientists, policy makers, farmers, and the public are necessary for unlocking the full potential of these technologies while adhering to ethical standards, environmental sustainability, and socio-economic equity (Hefferon and Herring, 2017).

4.5 Conclusions and Future Perspectives

During the past couple of decades, omics techniques involving genomics, transcriptomics, proteomics, and metabolomics have emerged as crucial techniques for examining the genetic and molecular mechanisms underlying crop growth and resilience to various environmental and biological stressors (Yang *et al.*, 2021). The development of high-throughput omics technologies took a major step ahead with the introduction of DNA microarray technology. In accordance with the central dogma, these techniques have been applied to study static changes in the genome, temporary variations in transcriptomics, splicing mechanisms, and the complex spatiotemporal aspects of proteomics, with a focus on post-translational modifications. The potential for overcoming the shortcomings of current methods is considerable as sequencing technology progresses, particularly with the development of third-generation sequencing. Extended read lengths offered by third-generation sequencing improve the capacity to identify repetitive patterns and structural variations. With further advancements, these technologies should enable greater detail in genome assembling as well as offer a greater grasp of genomic organization (Jain *et al.*, 2015). Considerable possibilities exist for improving crop development through the combination of several omics methods, particularly through the analysis of transcriptome and metabolomic profiles. Implementing these integrated approaches in the field of crop sciences has become feasible with recent improvements in computational techniques and omics technologies. As an illustration, the regulatory mechanisms governing lipid production all throughout the growing process of cotton fibers have been elucidated through the merging of multiomics and epigenetic information, offering vital details that could guide breeding endeavors (Wang *et al.*, 2016). These technologies will hopefully enhance understanding of the intricate relationships that shape the characteristics of plants as they evolve, which will further help to generate crops with higher yields and more tolerance to environmental stress.

In addition, the coupling of panomics with methods of genome editing including TALENs and CRISPR-Cas9 has been identified as a successful platform for fostering breakthroughs in precision breeding initiatives (Weckwerth *et al.*, 2020). Research on genomic selection and marker-assisted selection has identified significant genes that account for about 40% of heritable phenotypic variation. In response, it is postulated that the remaining 60% of phenotypic variance may be revealed through combining genome editing technology with panomics, which would simplify the process of identifying pertinent agronomic genes and increase the effectiveness of breeding initiatives. Single-cell multiomics, or the integration of multiomics data at the single-cell level, holds immense potential for future progress in crop development. This technique makes it less difficult to find non-anatomical markers for a range of cell populations and assists in the tracking of specific cell phases in crop plants when they differentiate. As single-cell multiomics technologies keep making progress, it will eventually be possible to do the concurrent analysis of the genome, transcriptome, metabolome, and epigenome from individual cells (Shaw *et al.*, 2021). These thorough analyses offer important new information on how genotype affects phenotype at the single-cell level. Furthermore, the ongoing development of omics technologies will lead to accurate modifications in plant genomes, which will allow complete biochemical pathways associated with beneficial traits to be transferred between different species. With global populations growing ever faster, these techniques are necessary for creating transgenic crops that are capable of mitigating the issues of future food shortages.

The integration of high-throughput technologies and multiomics approaches is poised to revolutionize crop improvement, encouraging multifaceted innovation and the advancement of sustainable farming practices.

References

Ali, S., Abbas, Z., Seleiman, M.F., Rizwan, M., YavaŞ, İ. *et al.* (2020) Glycine betaine accumulation, significance and interests for heavy metal tolerance in plants. *Plants* 9(7), 896.

Altieri, M.A., Funes-Monzote, F.R. and Petersen, P. (2012) Agroecologically efficient agricultural systems for smallholder farmers: Contributions to food sovereignty. *Agronomy for Sustainable Development* 32(1), 1–13.

Andrade-Sanchez, P., Gore, M.A., Heun, J.T., Thorp, K.R., Carmo-Silva, A.E. *et al.* (2013) Development and evaluation of a field-based high-throughput phenotyping platform. *Functional Plant Biology* 41(1), 68–79.

Arbona, V., Manzi, M., Ollas, C. and Gómez-Cadenas, A. (2013) Metabolomics as a tool to investigate abiotic stress tolerance in plants. *International Journal of Molecular Sciences* 14(3), 4885–4911.

Aslam, B., Basit, M., Nisar, M.A., Khurshid, M. and Rasool, M.H. (2016) Proteomics: Technologies and their applications. *Journal of Chromatographic Science* 1–15.

Azam, S., Thakur, V., Ruperao, P., Shah, T., Balaji, J. *et al.* (2012) Coverage-based consensus calling (CbCC) of short sequence reads and comparison of CbCC results to identify SNPs in chickpea (*Cicer arietinum*; *Fabaceae*), a crop species without a reference genome. *American Journal of Botany* 99(2), 186–192.

Bahl, J.R., Singh, A.K., Lal, R.K. and Gupta, A.K. (2018) High-yielding improved varieties of medicinal and aromatic crops for enhanced income New Age Herbals.

Baloch, F.S., Altaf, M.T., Liaqat, W., Bedir, M., Nadeem, M.A. *et al.* (2023) Recent advancements in the breeding of sorghum crop: Current status and future strategies for marker-assisted breeding. *Frontiers in Genetics* 14, 1150616.

Bansal, K.C., Lenka, S.K. and Mondal, T.K. (2014) Genomic resources for breeding crops with enhanced abiotic stress tolerance. *Plant Breeding* 133(1), 1–11.

Bhattacharjee, S. (2021) Advances of transcriptomics in crop improvement: A review. *Journal of Emerging Technologies and Innovative Research* 8(6), 22–37.

Bhattacharya, A., Parkhi, V. and Char, B. (2021) Genome editing for crop improvement: A perspective from India. *Vitro Cellular & Developmental Biology-Plant* 57(4), 565–573.

Blanca, J., Esteras, C., Ziarsolo, P., Pérez, D., Fernãndez-Pedrosa, V. *et al.* (2012) Transcriptome sequencing for SNP discovery across *Cucumis melo*. *BMC Genomics* 13, 1–18.

Brookes, G. and Smyth, S.J. (2024) Risk-appropriate regulations for gene-editing technologies. *GM Crops & Food* 15(1), 1–14.

Bürling, K., Hunsche, M. and Noga, G. (2010) Quantum yield of non-regulated energy dissipation in PSII (Y (NO)) for early detection of leaf rust (*Puccinia triticina*) infection in susceptible and resistant wheat (*Triticum aestivum* L.) cultivars. *Precision Agriculture* 11, 703–716.

Cabot, C., Sibole, J.V., Barceló, J. and Poschenrieder, C. (2014) Lessons from crop plants struggling with salinity. *Plant Science* 226, 2–13.

Cahill, D.J. (2001) Protein and antibody arrays and their medical applications. *Journal of Immunological Methods* 250(1–2), 81–91.

Canton, H. (2021) United nations conference on trade and development—Unctad. In: *The Europa Directory of International Organizations 2021*. Routledge, London, pp. 172–176.

Crain, J., Mondal, S., Rutkoski, J., Singh, R.P. and Poland, J. (2018) Combining high-throughput phenotyping and genomic information to increase prediction and selection accuracy in wheat breeding. *The Plant Genome* 11(1), 170043.

De Mori, G. and Cipriani, G. (2023) Marker-assisted selection in breeding for fruit trait improvement: A review. *International Journal of Molecular Sciences* 24(10), 8984.

Deracinois, B., Flahaut, C., Duban-Deweer, S. and Karamanos, Y. (2013) Comparative and quantitative global proteomics approaches: An overview. *Proteomes* 1(3), 180–218.

Emwas, A.-H., Roy, R., McKay, R.T., Tenori, L., Saccenti, E. *et al.* (2019) NMR spectroscopy for metabolomics research. *Metabolites* 9(7), 123.

FAO (2017) In: *The State of Food and Agriculture 2016: Climate Change, Agriculture and Food Security*. Food and Agriculture Organization of the UN (FAO), Rome.

Feng ChuHang, F.C., He CaiYun, H.C., Wang Ying, W.Y., Zeng YanFei, Z.Y. and Zhang JianGuo, Z.J. (2019) Phylogenetic position of *Chosenia arbutifolia* in the Salicaceae inferred from whole chloroplast genome.

Ferreyra, M.L.F., Serra, P. and Casati, P. (2021) Recent advances on the roles of flavonoids as plant protective molecules after UV and high light exposure. *Physiologia Plantarum* 173(3), 736–749.

Fiehn, O. (2006) Study of metabolic control in plants by metabolomics. *Annual Plant Reviews* Volume 22, 60–84.

Fu, X., Fu, N., Guo, S., Yan, Z., Xu, Y. *et al.* (2009) Estimating accuracy of RNA-seq and microarrays with proteomics. *BMC Genomics* 10, 1–9.

Gorbe, E. and Calatayud, A. (2012) Applications of chlorophyll fluorescence imaging technique in horticultural research: A review. *Scientia Horticulturae* 138, 24–35.

Guo, C., Ma, X., Gao, F. and Guo, Y. (2023) Off-target effects in CRISPR/Cas9 gene editing. *Frontiers in Bioengineering and Biotechnology* 11, 1143157.

Hackauf, B., Rabanus-Wallace, M.T. and Korzun, V. (2021) Bridging the genotype–phenotype gap for precision breeding in rye. In: Rabanus-Wallace, M.T. and Stein, N. (eds) *The Rye Genome*. Springer, Cham, pp. 135–180.

Haghighattalab, A., González Pérez, L., Mondal, S., Singh, D., Schinstock, D. *et al.* (2016) Application of unmanned aerial systems for high throughput phenotyping of large wheat breeding nurseries. *Plant Methods* 12, 1–15.

Harsha, H.C., Molina, H. and Pandey, A. (2008) Quantitative proteomics using stable isotope labeling with amino acids in cell culture. *Nature Protocols* 3(3), 505–516.

Hasanuzzaman, M., Bhuyan, M.H.M.B., Anee, T.I., Parvin, K., Nahar, K. *et al.* (2019) Regulation of ascorbate-glutathione pathway in mitigating oxidative damage in plants under abiotic stress. *Antioxidants* 8(9), 384.

Hefferon, K.L. and Herring, R.J. (2017) The end of the GMO? Genome editing, gene drives and new frontiers of plant technology. *Review of Agrarian Studies* 7(1).

Hirayama, T. and Shinozaki, K. (2010) Research on plant abiotic stress responses in the post-genome era: Past, present and future. *The Plant Journal* 61(6), 1041–1052. DOI: 10.1111/j.1365-313X.2010.04124.x.

Horgan, R.P. and Kenny, L.C. (2011) 'Omic'technologies: Genomics, transcriptomics, proteomics and metabolomics. *Obstetrician & Gynaecologist* 13(3), 189–195.

Hossain, M.A., Bhattacharjee, S., Armin, S.-M., Qian, P., Xin, W. *et al.* (2015) Hydrogen peroxide priming modulates abiotic oxidative stress tolerance: Insights from ROS detoxification and scavenging. *Frontiers in Plant Science* 6, 420.

Jain, M., Fiddes, I.T., Miga, K.H., Olsen, H.E., Paten, B. *et al.* (2015) Improved data analysis for the MinION nanopore sequencer. *Nature Methods* 12(4), 351–356.

Jayathirtha, M., Dupree, E.J., Manzoor, Z., Larose, B., Sechrist, Z. *et al.* (2021) Mass spectrometric (MS) analysis of proteins and peptides. *Current Protein and Peptide Science* 22(2), 92–120.

Jones, A.M., Danielson, J.Å.H., ManojKumar, S.N., Lanquar, V., Grossmann, G. *et al.* (2014) Abscisic acid dynamics in roots detected with genetically encoded FRET sensors. *Elife* 3, e01741.

Kang, K., Yue, L., Xia, X., Liu, K. and Zhang, W. (2019) Comparative metabolomics analysis of different resistant rice varieties in response to the brown planthopper *Nilaparvata lugens* Hemiptera: Delphacidae. *Metabolomics* 15, 1–13.

Kumar, A., Pathak, R.K., Gupta, S.M., Gaur, V.S. and Pandey, D. (2015a) Systems biology for smart crops and agricultural innovation: Filling the gaps between genotype and phenotype for complex traits linked with robust agricultural productivity and sustainability. *Omics: A Journal of Integrative Biology* 19(10), 581–601.

Kumar, J., Pratap, A. and Kumar, S. (2015b) Plant phenomics: An overview. In: *Phenomics in Crop Plants: Trends, Options and Limitations*. pp. 1–10.

Lee, B., Henderson, D.A. and Zhu, J.-K. (2005) The *Arabidopsis* cold-responsive transcriptome and its regulation by ICE1. *The Plant Cell* 17(11), 3155–3175.

Li, X.-Y. (2020) Sequencing and analysis of transcriptome to reveal regulation of gene expression in *Salvia miltiorrhiza* under moderate drought stress. *Chinese Traditional and Herbal Drugs*.

Li, L., Zhang, Q. and Huang, D. (2014) A review of imaging techniques for plant phenotyping. *Sensors* 14(11), 20078–20111.

Li, F., Li, M., Wang, P., Cox Jr, K.L., Duan, L. *et al.* (2017) Regulation of cotton (*Gossypium hirsutum*) drought responses by mitogen-activated protein (MAP) kinase cascade-mediated phosphorylation of Gh WRKY 59. *New Phytologist* 215(4), 1462–1475.

Lindberg, S., Bain, C. and Selfa, T. (2023) Regulating gene editing in agriculture and food in the European Union: Disentangling expectations and path dependencies. *Sociologia Ruralis* 63(2), 348–369.

Lobell, D.B., Schlenker, W. and Costa-Roberts, J. (2011) Climate trends and global crop production since 1980. *Science* 333(6042), 616–620. DOI: 10.1126/science.1204531.

Marone, D., Mastrangelo, A.M., Borrelli, G.M., Mores, A., Laidò, G. *et al.* (2022) Specialized metabolites: Physiological and biochemical role in stress resistance, strategies to improve their accumulation, and new applications in crop breeding and management. *Plant Physiology and Biochemistry* 172, 48–55.

Marone, D., Mastrangelo, A.M. and Borrelli, G.M. (2023) From transgenesis to genome editing in crop improvement: Applications, marketing, and legal issues. *International Journal of Molecular Sciences* 24(8), 7122.

Masuka, B., Araus, J.L., Das, B., Sonder, K. and Cairns, J.E. (2012) Phenotyping for abiotic stress tolerance in maize F. *Journal of Integrative Plant Biology* 54(4), 238–249.

Mitra, J. (2001) Genetics and genetic improvement of drought resistance in crop plants. *Current Science* 758–763.

Mochida, K. and Shinozaki, K. (2011) Advances in omics and bioinformatics tools for systems analyses of plant functions. *Plant and Cell Physiology* 52(12), 2017–2038. DOI: 10.1093/pcp/pcr153.

Mukesh Kumar, M.K., Veena Chaudhary, V.C., Vijai Kumar, V.K., Ujjwal Sirohi, U.S. and Sharma, V.R. (2018) Genetic studies in common vegetable crops by using biochemical and DNA markers: A review. *HortFlora Research Spectrum* 7, 177–191.

Narváez, F.J.Y., Pedregal, J.S., Prieto, P.A., Torres-Torriti, M. and Cheein, F.A.A. (2016) LiDAR and thermal images fusion for ground-based 3D characterisation of fruit trees. *Biosystems Engineering* 151, 479–494.

Nawaz, S. (2021) *Beyond naturalness?: Social dimensions of gene editing in agriculture*. University of British Columbia, Canada.

Ndudzo, A., Makuvise, A.S., Moyo, S. and Bobo, E.D. (2024) CRISPR-Cas9 genome editing in crop breeding for climate change resilience: Implications for smallholder farmers in Africa. *Journal of Agriculture and Food Research* 101132.

Negin, B. and Moshelion, M. (2016) The evolution of the role of ABA in the regulation of water-use efficiency: From biochemical mechanisms to stomatal conductance. *Plant Science* 251, 82–89.

Newman, Y.M., Ring, S.G. and Colaco, C. (1993) The role of trehalose and other carbohydrates in biopreservation. *Biotechnology and Genetic Engineering Reviews* 11(1), 263–294.

Ogunjobi, T.T., Ohaeri, P.N., Akintola, O.T., Atanda, D.O., Orji, F.P. *et al.* (2024) Bioinformatics applications in chronic diseases: A comprehensive review of genomic, transcriptomics, proteomic, metabolomics, and machine learning approaches. *Medinformatics*. DOI: 10.47852/bonviewMEDIN42022335.

Ozsolak, F. and Milos, P.M. (2011) RNA sequencing: Advances, challenges and opportunities. *Nature Reviews Genetics* 12(2), 87–98.

Pandey, A. and Gayen, D. (2024) Decoding post-translational modifications for understanding stress tolerance in plant. *Crop Design* 100077.

Park, S.-C., Joung, Y.-H., Kim, K.-M., Kim, J.-K. and Koh, H.-J. (2019) Gene-edited crops: Present status and their future. *Korean Society of Breeding Science* 51(3), 175–183.

Penque, D. (2009) Two-dimensional gel electrophoresis and mass spectrometry for biomarker discovery. *Proteomics–Clinical Applications* 3(2), 155–172.

Per, T.S., Khan, N.A., Reddy, P.S., Masood, A., Hasanuzzaman, M. *et al.* (2017) Approaches in modulating proline metabolism in plants for salt and drought stress tolerance: Phytohormones, mineral nutrients and transgenics. *Plant Physiology and Biochemistry* 115, 126–140.

Priya, M., Dhanker, O.P., Siddique, K.H.M., HanumanthaRao, B., Nair, R.M. *et al.* (2019) Drought and heat stress-related proteins: An update about their functional relevance in imparting stress tolerance in agricultural crops. *Theoretical and Applied Genetics* 132, 1607–1638.

Qazi, H.A., Jan, N., Ramazan, S. and John, R. (2019) Protein modification in plants in response to abiotic stress. In: *Protein Modificomics*. Academic Press, pp. 171–201.

Rauf, S., Silva, J.T., Khan, A.A. and Naveed, A. (2010) Consequences of plant breeding on genetic diversity. *International Journal of Plant Breeding* 4(1), 1–21.

Raza, A., Ashraf, F., Zou, X., Zhang, X. and Tosif, H. (2020) Plant adaptation and tolerance to environmental stresses: Mechanisms and perspectives. In: Hasanuzzaman, M. (ed.) *Plant Ecophysiology and Adaptation under Climate Change: Mechanisms and Perspectives I: General Consequences and Plant Responses*. Springer, Singapore, pp. 117–145.

Razzaq, A., Sadia, B., Raza, A., Khalid Hameed, M. and Saleem, F. (2019) Metabolomics: A way forward for crop improvement. *Metabolites* 9(12), 303.

Roychoudhury, A. (2020) Next generation sequencing: Prospects in plant breeding and crop improvement. *SF Journal of Agriculture and Crop Management* 1(1), 1004.

Salem, M.A., Souza, L., Serag, A., Fernie, A.R., Farag, M.A. *et al.* (2020) Metabolomics in the context of plant natural products research: From sample preparation to metabolite analysis. *Metabolites* 10(1), 37.

Salvi, S. and Tuberosa, R. (2005) To clone or not to clone plant QTLs: Present and future challenges. *Trends in Plant Science* 10(6), 297–304.

Schuster, S.C. (2008) Next-generation sequencing transforms today's biology. *Nature Methods* 5(1), 16–18. DOI: 10.1038/nmeth1156.

Shakoor, N., Lee, S. and Mockler, T.C. (2017) High throughput phenotyping to accelerate crop breeding and monitoring of diseases in the field. *Current Opinion in Plant Biology* 38, 184–192.

Shaw, R., Tian, X. and Xu, J. (2021) Single-cell transcriptome analysis in plants: Advances and challenges. *Molecular Plant* 14(1), 115–126.

Shiferaw, B., Prasanna, B.M., Hellin, J. and Bänziger, M. (2011) Crops that feed the world 6. Past successes and future challenges to the role played by maize in global food security. *Food Security* 3, 307–327.

Shinwari, Z.K., Tanveer, F. and Khalil, A.T. (2018) Ethical issues regarding CRISPR mediated genome editing. *Current Issues in Molecular Biology* 26(1), 103–110.

Singh, R.K., Redoña, E. and Refuerzo, L. (2010) Varietal improvement for abiotic stress tolerance in crop plants: Special reference to salinity in rice. *Abiotic Stress Adaptation in Plants: Physiological, Molecular and Genomic Foundation* 387–415.

Singh, V.K., Singh, A.K., Singh, S. and Singh, B.D. (2015) Next-generation sequencing (NGS) tools and impact in plant breeding. In: Al-Khayri, J., Jain, S. and Johnson, D. (eds) *Advances in Plant Breeding Strategies: Breeding, Biotechnology and Molecular Tools*. Springer, Cham, pp. 563–612.

Singh, S., Selvakumar, R., Mangal, M. and Kalia, P. (2020) Breeding and genomic investigations for quality and nutraceutical traits in vegetable crops-A review. *Indian Journal of Horticulture* 77(1), 1–40.

Sinha, P., Singh, V.K., Bohra, A., Kumar, A., Reif, J.C. *et al.* (2021) Genomics and breeding innovations for enhancing genetic gain for climate resilience and nutrition traits. *Theoretical and Applied Genetics* 134(6), 1829–1843.

Singh, S., Sarma, D.K., Verma, V., Nagpal, R. and Kumar, M. (2023) Unveiling the future of metabolic medicine: Omics technologies driving personalized solutions for precision treatment of metabolic disorders. *Biochemical and Biophysical Research Communications* 682, 1–20.

Steinmann, D. and Ganzera, M. (2011) Recent advances on HPLC/MS in medicinal plant analysis. *Journal of Pharmaceutical and Biomedical Analysis* 55(4), 744–757.

Tester, M. and Langridge, P. (2010) Breeding technologies to increase crop production in a changing world. *Science* 327(5967), 818–822. DOI: 10.1126/science.1183700.

Turnbull, C., Lillemo, M. and Hvoslef-Eide, T.A.K. (2021) Global regulation of genetically modified crops amid the gene edited crop boom–A review. *Frontiers in Plant Science* 12, 630396.

Varshney, R.K., Graner, A. and Sorrells, M.E. (2005) Genomics-assisted breeding for crop improvement. *Trends in Plant Science* 10(12), 621–630.

Varshney, R.K. and Tuberosa, R. (2007) Genomics-assisted crop improvement: An overview. In: Varshney, R.K. and Tuberosa, R. (eds) *Genomics-Assisted Crop Improvement Vol. 1: Genomics Approaches and Platforms*. Springer, Dordrecht, pp. 1–12.

Verhoeven, K.J.F., Macel, M., Wolfe, L.M. and Biere, A. (2011) Population admixture, biological invasions and the balance between local adaptation and inbreeding depression. *Proceedings of the Royal Society B: Biological Sciences* 278(1702), 2–8.

Verma, S., Gupta, A., Yalla, S., Shreya Patel, P.J., Sharma, R. *et al.* (2024) Integrating marker-assisted (MAS) and genomic selection (GS) for plant functional trait improvement. In: *Plant Functional Traits for Improving Productivity*. Springer, Cham, pp. 203–215.

Vinaixa, M., Schymanski, E.L., Neumann, S., Navarro, M., Salek, R.M. *et al.* (2016) Mass spectral databases for LC/MS-and GC/MS-based metabolomics: State of the field and future prospects. *TrAC Trends in Analytical Chemistry* 78, 23–35.

Vishwakarma, K., Upadhyay, N., Kumar, N., Yadav, G., Singh, J. *et al.* (2017) Abscisic acid signaling and abiotic stress tolerance in plants: A review on current knowledge and future prospects. *Frontiers in Plant Science* 8, 161.

Vivek, B.S., Krishna, G.K., Vengadessan, V., Babu, R., Zaidi, P.H. *et al.* (2017) Use of genomic estimated breeding values results in rapid genetic gains for drought tolerance in maize. *The Plant Genome* 10(1), lantgenome2016–07.

Voigt, B. and Münichsdorfer, A. (2019) Regulation of genome editing in plant biotechnology: European Union. In: Dederer, H.-G. and Hamburger, D. (eds) *Regulation of Genome Editing in Plant Biotechnology: A Comparative Analysis of Regulatory Frameworks of Selected Countries and the EU*. Springer, Cham, pp. 137–238.

Wang, M., Wang, P., Tu, L., Zhu, S., Zhang, L. *et al.* (2016) Multi-omics maps of cotton fibre reveal epigenetic basis for staged single-cell differentiation. *Nucleic Acids Research* 44(9), 4067–4079.

Weckwerth, W., Ghatak, A., Bellaire, A., Chaturvedi, P. and Varshney, R.K. (2020) PANOMICS meets germplasm. *Plant Biotechnology Journal* 18(7), 1507–1525.

Wei, Z., Hu, W., Lin, Q., Cheng, X., Tong, M. *et al.* (2009) Understanding rice plant resistance to the Brown Planthopper (*Nilaparvata lugens*): A proteomic approach. *Proteomics* 9(10), 2798–2808.

Xiao, J.F., Zhou, B. and Ressom, H.W. (2012) Metabolite identification and quantitation in LC-MS/MS-based metabolomics. *TrAC Trends in Analytical Chemistry* 32, 1–14.

Xu, Y. and Crouch, J.H. (2008) Marker-assisted selection in plant breeding: From publications to practice. *Crop Science* 48(2), 391–407.

Xu, J., Liu, H., Zhou, C., Wang, J., Wang, J. *et al.* (2024) The ubiquitin-proteasome system in the plant response to abiotic stress: Potential role in crop resilience improvement. *Plant Science* 112035.

Yadav, S., Meena, S., Kalwan, G. and Jain, P.K. (2024) DNA methylation: An emerging paradigm of gene regulation under drought stress in plants. *Molecular Biology Reports* 51(1), 311.

Yancey, P.H. (2005) Organic osmolytes as compatible, metabolic and counteracting cytoprotectants in high osmolarity and other stresses. *Journal of Experimental Biology* 208(15), 2819–2830.

Yang, W., Feng, H., Zhang, X., Zhang, J., Doonan, J.H. *et al.* (2020) Crop phenomics and high-throughput phenotyping: Past decades, current challenges, and future perspectives. *Molecular Plant* 13(2), 187–214.

Yang, Y., Saand, M.A., Huang, L., Abdelaal, W.B., Zhang, J. *et al.* (2021) Applications of multi-omics technologies for crop improvement. *Frontiers in Plant Science* 12, 563953.

Zenda, T., Liu, S., Wang, X., Jin, H., Liu, G. *et al.* (2018) Comparative proteomic and physiological analyses of two divergent maize inbred lines provide more insights into drought-stress tolerance mechanisms. *International Journal of Molecular Sciences* 19(10), 3225.

Zhou, M.-L., Ma, J.-T., Pang, J.-F., Zhang, Z.-L., Tang, Y.-X. *et al.* (2010) Regulation of plant stress response by dehydration responsive element binding (DREB) transcription factors. *African Journal of Biotechnology* 9(54), 9255–9269.

Zhu, L., Ni, W., Liu, S., Cai, B., Xing, H. *et al.* (2017) Transcriptomics analysis of apple leaves in response to *Alternaria alternata* apple pathotype infection. *Frontiers in Plant Science* 8, 22. DOI: 10.3389/fpls.2017.00022.

5 Advances in Plant Phenomics with Integration of Artificial Intelligence for Precise Phenotyping and Climate-smart Breeding

Tabassum[1*], Anand Singh Jeena[2], Gurdeep Singh[1] and Rohit[2]
[1]Baba Farid College, Baba Farid Group of Institutions, Bathinda, Punjab, India;
[2]G. B. Pant University of Agriculture & Technology, Pantnagar, Uttarakhand, India

Abstract

As an emerging technology, artificial intelligence (AI) uplevels the agriculture sector by improving crop production through enhanced prompted monitoring with latest computerized data and structures obtained by drones or remote sensing. AI supports the development of farming applications through numerous sophisticated computer-supported structures, generated for the determination of various factors such as plant yield, disease incidence, and growth stage detection. AI brings immense computational ability with novel tools and techniques for smart breeding especially for high-throughput genotyping and phenotyping. It has been developed into a transformative field for shaping recent crop-breeding activities. Climate-smart breeding with enhanced efficiency, sustainability and adaptability to the adverse climatic conditions has become a challenge due to a continuously changing environment. The omics approaches, namely crop genomics, phenomics with enviromics, are playing key role for illuminating the complicated biological procedures and channels that drive crop functions in response to varying environmental conditions. The three omics approaches provide precise tools for the evaluation of various plant characters of large populations in a relatively short duration. Omics-empowered AI can pave the way forward for rapid gene identification, association of observable phenotypes with genotypes and eventually accelerate breeding programs. As compared to conventional approaches, AI is quick and less labor intensive, thereby complementing speed breeding. Speed breeding is being utilized to proliferate the reproductive cycles of crops rapidly by growing the crop off-season under controlled conditions. Together with AI, molecular marker supported selection and advanced genome editing techniques, speed breeding playing important role for crop improvement in early selection, genetic gain and targeted genetic manipulations.

5.1 Introduction

Agriculture mainly depends on agro-climatic conditions for production. Predicting various traits like disease and insect-pest resistance, and crop yield in advance is a daunting task in subsistence and precision agriculture because it involves complex steps requiring the use of many datasets from various sources such as agro-climatic conditions, soil, rainfall, the variety used for sowing, and fertilizer use, etc. (Xu *et al.*, 2019; Klompenburg *et al.*, 2020). There are many factors like shortage of available labor, uncertainty in annual rainfall and

*Corresponding author: tabassum12081992@gmail.com

 From Gene Discovery to Climate-Resilient Crops: Omics in Crop Improvement (eds S. Pandey *et al.*)
DOI: 10.1079/9781800628717.0005

temperature fluctuations during crop seasons, all these factors in combination impose a new challenge (Ennouri *et al.*, 2021). Various methods, such as manual surveys, models of agro-meteorology, remote sensing techniques and satellite imaging have been used so far to predict the crop yield. As farming operations are vast, manual field surveys are neither practical nor efficient, and modeling-based methods are also on check because of difficulties in determining model parameters, scaling to huge areas and using advanced computational resources (Khaki *et al.*, 2021). Data mining is used to accumulate previously forecast particulars, information from huge databases, and utilize the information for analyzing future patterns and traits. Data analysis is modeling the available data in such a way to gain useful knowledge and draw conclusions. Data analysis and mining are both being utilized for trait prediction in recent times. Also, prediction models are being used to estimate the genuine yield, but it is still expedient to find more accurate performing tools and techniques for trait prediction (Filippi *et al.*, 2019; Klompenburg *et al.*, 2020). Precision agriculture is a whole-farm management approach that uses information from multiple sources like remote sensing and satellite enabled data for cost-effective crop management in the field to improve crop yield with optimized returns on inputs with reduced environmental impacts (Cavalli *et al.*, 2021).

Artificial intelligence (AI) is described as a set of techniques and learnings aimed at duplicating human intelligence by creating intelligent computers and mathematical features of statistical modeling to develop technological systems to solve complications in various fields and to execute functions normally not performed by the human brain (Yang and Siau, 2018; Kumar *et al.*, 2022). AI has the ability to analyze big datasets quickly, allowing researchers in agriculture and plant breeding to systematize disarticulated information on various factors available in the public domain and transform it into a breeding decision, thus enabling precise decision making in plant breeding (Khan *et al.*, 2022). AI technologies can play a major role in overcoming obstacles related to high-efficiency phenotyping and gene functional studies in present-day breeding operations and can bring new progress in constructing envirotyping data, which can be utilized in future plant breeding programs. The phenotypic, genotypic and environmental data are collected (Fig. 5.1), knowledge of phenomics, genomics along with enviromics are consolidated with

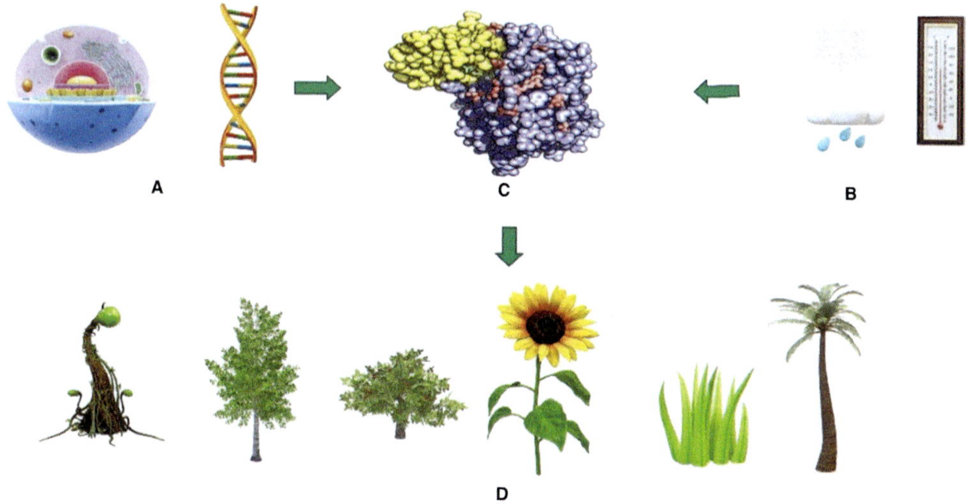

Fig. 5.1. Representation of multiomics. A, Gene and genotype (genomics); B, environmental factors influencing the genotypic expression in the form of proteins (enviromics); C, altered protein development under variable environmental conditions (proteomics); D, resultant phenotype of the plant (phenomics). Author's own image.

AI technologies (such as a high efficiency/throughput phenotyping (HTP) server with the Internet), and AI-assisted breeding systems are developed that verify through simulation whether the chosen varieties/cultivars are appropriate for cultivation under all the major environments or in environments with certain limited factors (Khan *et al.*, 2022). Khan *et al.* (2022) took multiple datasets of phenotypic and genotypic data along with environmental data, where genotype ×environment (G × E) interactions were quantified in several environments using AI technologies such as machine learning (ML) or deep learning (DL) for the selection of cultivars to be grown under specific as well as multiple environments. The end results showed that this approach was more accurate in making decisions on whether the selected cultivar is perfectly fitted to cultivate in a specific environment or in multiple environments.

5.2 What Is Plant Phenomics?

Wilhelm Johannsen introduced the term 'phenotype' in 1911, which came from two Greek words phainein (to show) and typos (type). The term 'phenomics' refers to the high-throughput (HT) analysis of phenotypes, while 'phenotyping' encompasses the methods used to measure plant traits. Classic plant trait phenotyping approaches tend to be destructive to plants, labor-intensive, time-consuming, and subjective (Bohra *et al.*, 2021). Recent advancements have caused the progression and integration of automated phenotyping techniques that provide precise, high-resolution, and HT data for various traits. Today, the term phenotyping often brings to mind images of drones or pheno-carts quickly traversing research plots to gather high-resolution data on a wide range of traits. This capability has been made possible by recent developments in sensing technologies and data processing. Increased investment in phenotyping is gaining acceptance as researchers seek to leverage the latest developments in crop genomics and predictive models, which can be constructed from HT and highly detailed phenotypic data. Phenotyping can be applied in various contexts, including breeding, exploring genetic resources, developing parent lines, and conducting translational research to create new breeding resources.

Phenotyping continues to be a fundamental aspect of plant breeding. Even with progress in genetics and the use of molecular technologies in crop research, crop breeding still primarily depends on the assessment of grain yield and a select few agronomically significant traits to guide selections and define commercial products (Reynolds *et al.*, 2020). In field crops, grain yield is the most critical phenotypic trait for breeders, as it encapsulates all significant traits and genes into a biologically and economically relevant metric. This remains true despite the considerable variation in how cultivars respond to management practices, fluctuations in weather and soil conditions, and seasonal influences, including both biotic and abiotic components (Kang, 2004). Moreover, there is often a substantial discrepancy between yields observed in experimental plots and those achieved by farmers. To date, no 'genetic index' has succeeded in supplanting grain yield as the primary driver of economic impact in breeding, despite the importance of major-effect genes and their markers, although advancements are being made (Reynolds *et al.*, 2020).

Global warming and unpredictable climatic fluctuations, driven by various biotic and abiotic stresses, have significantly impacted crop growth and productivity worldwide, resulting in losses amounting to several billion dollars. Unlike animals, plants are unable to flee from unfavorable climatic/environmental factors like drought, heat, salinity, and cold stress. With ongoing global climate changes, these stresses are anticipated to intensify, significantly impacting G × E interactions and modifying the plant's basic metabolism, as seen in its phenotype (Gupta *et al.*, 2015). The response of any genotype to stress can be analyzed by observing its phenotypic changing in controlled environments, which helps evaluate the regulatory mechanisms involved in the plant's stress response. In the postgenomic era, effectively utilizing crop genetic resources depends on accurately analyzing phenotypic response databases for specific stresses, which are becoming increasingly complex due to the surge in multiomics datasets (Bohra *et al.*, 2021).

5.3 AI in Agriculture

Artificial intelligence or machine intelligence is a branch of computer science aimed at developing technologies, software or hardware components able to make choices and fulfill the functions to replace active human efforts (Dwivedi *et al.*, 2021). Although worldwide in agriculture AI is applied to solve a number of problems, it is still sparsely known to solve various problems related to crops. AI can aid at various levels in agriculture and plant breeding in effective selection and yield improvement by assisting in overcoming conventional difficulties. It has potential applications in the field including use of cameras and sensor-based equipment for data collection and analysis. These AI-based techniques help in analyzing the data from different sources such as agroclimatic/weather conditions, crop production and productivity, soil health and soil temperature, etc. to draw results that can serve the farming community to produce healthy crops with more yield (Ennouri *et al.*, 2021).

In agriculture, there are day-to-day activities like field preparation, sowing, fertilizer application, irrigation, crop harvesting and many more that need to be completed in every crop season and some of them are a great challenge for growers. Changes in various climatic factors, such as soil/atmospheric temperature and moisture, and precipitation, have a variable impact on agricultural activities. Increased pollution with environmental degradation also contributes to changing the climate, which is a great issue for raising crops. Each plant requires precise irrigation as well as fertigation; nitrogen, phosphorus and potassium (NPK) are considered as basic nutrients needed to crop. Any soil lacking these basic NPK elements may result in lower production and productivity of the crop. The changing climatic conditions also play an important role in the emergence of new agricultural landscapes. Each growing season introduces cutting-edge tools for maximizing yields; AI is one of these, playing an important part in improving agricultural processes (irrigation, and fertilizer, insecticide, pesticide, and weedicide application) by giving farmers real-time crop information (Mishra *et al.*, 2024). It has an important role in the analysis of produced data to increase the yield prediction proficiency and decrease the undesirable impact of adverse environmental conditions (Bannerjee *et al.*, 2018; Jha *et al.*, 2019). The predictions about these abiotic stresses are beneficial for growers to avoid hazards and yield losses (Ennouri *et al.*, 2021).

The incorporation of remote sensing and AI allows us to measure phenotypic data precisely at the field level and can integrate big data into predictive and prescriptive management tools (Jung *et al.*, 2021).

5.4 Traditional Plant Breeding Before the Artificial Intelligence Era

Owing to crop genetics and the interaction with external environmental elements, breeding to increase yield is a more complex and unpredictable task (Collins *et al.*, 2008). In plants, a primary character like seed yield is dependent on various quantitative characters and is highly influenced by the environment. Dealing with huge breeding populations of thousands of genotypes is also a time-consuming and labor-intensive assignment (Cai *et al.*, 2016; Xiong *et al.*, 2018). Therefore, breeding procedures utilizing secondary characteristics (grain yield contributing characteristics, such as number of grains per plant, grain weight, harvest index, etc.), which are significantly associated with primary character (grain yield), empowered plant breeders to recognize propitious genotypes even at an early growth stage (Jin *et al.*, 2010; Montesinos-Lopez *et al.*, 2018). The generation of superior varieties with respect to several traits is always the main objective of plant breeding and this process continues with the development of traditional as well as modern plant breeding approaches. Systematic plant breeding started with the domestication of cereals and pulses; after that, many advances were made in the conduction of trials in field as well as in labs, crossing designs, breeding designs for statistical analysis, etc. Various achievements were documented in plant breeding; some of the most important include the green revolution (the development of semi-dwarf, nutrient-responsive varieties in cereals), the development of hybrid cultivars in various crops, transgenics development, the development of synthetic and composite varieties, etc. (Bhat *et al.*, 2021).

In plant breeding, the prediction of performance of different lines, and yield prediction for the progeny plants, is a central activity. Advancements in phenomics and genomics have aided positively in early generation selection and empowered plant breeders to increase economic yield. In addition to phenotypic data, the use of molecular markers and genomics then began, in which genomic selection utilized molecular markers to identify the breeding value of individual plant genotypes and assisted traditional plant breeding methods for selection, popularly known as marker assisted selection (MAS). Now, with vast quantity of omics data and the highly advanced information technologies, a new phase of plant breeding has begun (Zargar *et al.*, 2015; Khan *et al.*, 2022).

5.5 Plant Phenotyping and Climate-Smart Plant Breeding with AI

The world population is increasing and arable land is decreasing with time. Continuously changing climatic conditions required accurate, error-free high-efficiency approaches for assisting in the development of crop varieties more rapidly with changing requirements. In this context, AI technologies have remarkable potential to help in the development of climate-resilient crop varieties to aid in smart crop breeding under an adverse climate. Such varieties hold promise to perform well, giving appreciable crop yields, and providing tolerance towards numerous biotic/abiotic stresses (Zohry and Ouda, 2022).

Many economically important plant characteristics like grain yield follow complex quantitative inheritance and the variation observed among individuals can be genotypic, environmental or due to the G × E interaction. This means performance of a genotype (line/variety/cultivar) in one environment may be as predicted but could be poor in other environments. From comparative and mapping studies, it is now understood that although much phenotypic variation is heritable, only a small quantity can be associated with recognized genetic elements such as simple sequence repeats (SSRs), inter simple sequence repeats (ISSRs), single nucleotide polymorphism (SNPs)/point mutation and small insertion/deletion (indel) or macro mutation, etc. (Harfouche *et al.*, 2019). This could be partly because most quantitative traits are controlled by polygenes that are affected by the environment and contribute more to the G × E interaction, thereby generating variation.

It has been observed that there is not always a straight relationship between the genotype of an individual and its associated phenotype; even a tiny alteration at the genotypic level may lead to huge effect on the phenotypic level. Within a biological system, genes encounter a rednexus regulatory network, and the synthesized proteins undergo several modifications by interacting with other molecules or proteins that are not easy to identify in studies (Harfouche *et al.*, 2019). Solution to this problem appeared with the availability of thorough and precise data in different individuals, leading to the capture of elementary information on biological and environmental systems. Breeders now have access to high-efficiency sensors and high-resolution image capturing methods for a variety of characteristics as well as environmental factors in field conditions and can consistently gather multiomics big-data information such as phenomics, genomics, epigenetics, transcriptomics, metabolomics, proteomics, enviromics and others, in relation to the plant conditions, its performance with reference to micro-environment and open-field conditions. However, the potential of precise prediction and selection of the top-performing genotypes (principally for a specific environment) mainly relies on the plant breeder's ability to model the information collected from extremely complex genomic and phenomic data generation systems (Fig. 5.2).

In the area of plant breeding, AI techniques with multiomics data can expedite varietal development for combating various biotic and abiotic stresses (disease/insect-pest resistance and tolerance, cold stress, heat stress, drought stress, etc.), resulting in higher yield and stability in performance across environments followed by increased production/productivity under field conditions (Harfouche *et al.*, 2019). The amalgamation of omics data with AI can help in accelerating varietal development when

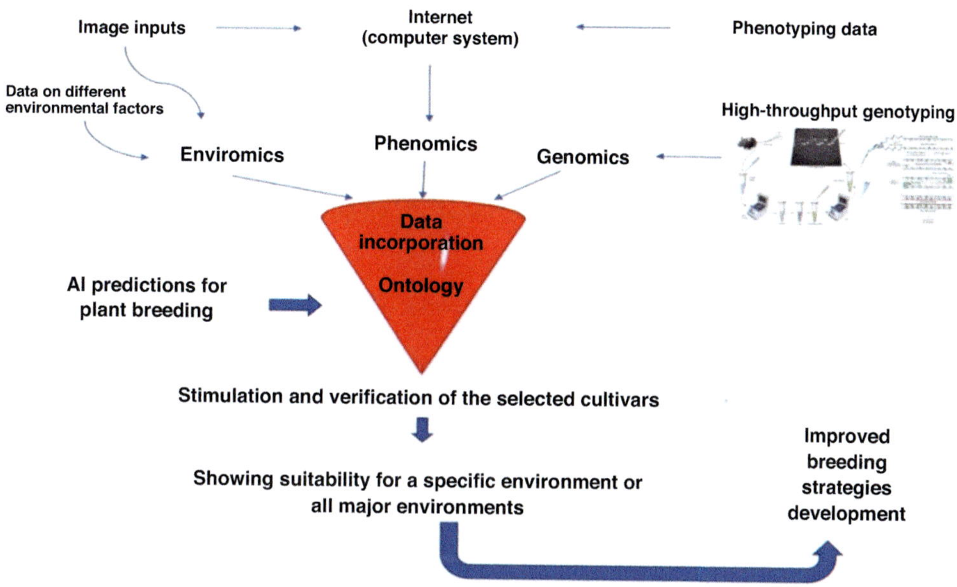

Fig. 5.2. Integration and utilization of phenomics, genomics and enviromics data by AI technologies for the creation of optimized plant breeding strategies. Author's own image.

large amounts of heterogeneous and complex omics data are analyzed using AI techniques in breeding programs.

5.6 Picture Identification Approach in Phenotyping

AI has enabled applications like the picture identification approach (PIA) to be effective for determining potential irregularities in healthy/diseased plants and soil health. With PIA, pictures caught by the camera are used to recognize various imperfect conditions in the plants. Deep learning tools have been developed for exploring plants and herbage models of agriculture developed to understand soil deficiencies, plant problems and sickness (Perea *et al.*, 2018). For example, one can assemble information recorded by means of AI approaches to verify weeds in the field for the application of weedicides precisely on the pinpointed location of weeds, which can ultimately limit the application of these chemicals in a particular area and thereby reduce the cost of cultivation.

Earlier work on image-based yield forecasting was based on vegetative indices of agriculture, such as the normalized difference vegetation index (NDVI) at the regional level, which was grounded on low-resolution data collected by remote sensing or by using robotics (Baghdasaryan *et al.*, 2022). NDVI is an established scale of the green biome because it shows a strong correlation with green canopy cover and vegetation greening. NDVI data furnish images in visible red and near-infrared bands often obtained from satellites such as Landsat, Pleiades, Sentinel-2, SPOT, etc. (Cavalli *et al.*, 2021). NDVI forecasting can be applied as a huge-area forecast, which uses a model for making predictions over a huge and homogeneous region (examples include an entire irrigated district or a complete forest cover), or in field-level areas of one or more hectares (Ahmad *et al.*, 2020).

A multi-variable, multi-phase NDVI prediction approach was put forward by Cavalli *et al.* (2021) based on long short-term memory (LSTM) networks. In this publicly available approach, relevant input data were used and the model was competent in generating high-accuracy predictions of NDVI values for next 1–3 days. To conduct field predictions, Baghdasaryan *et al.* (2022) enforced an encoder–decoder framework that not only provided in-field outstanding performance but also identified low-performing regions of the field, which can be improved from further inspection and treatments. During their experiment conducted in 602 corn (maize) fields, they collected high-resolution (10 cm/pixel) and multi-spectral red–green–blue +near-infra-red (RGB + NIR) images over two seasons (2020–2021) and used these data to predict the final yield density at the pixel level, thus demonstrated the performance at the field level. In recent times, technology like remote sensing has been improved with the rapid advancement of ML approaches. Data collection through satellites, unmanned aerial vehicles (UAVs) and manned aircraft increased sensor resolution and channels beyond RGB have become more common.

Using data from IT, satellite, and remote sensing, one can increase crop yield with cost-effective crop management inputs like fertigation and irrigation, thus achieving the precision agriculture goal of optimizing returns on inputs. For the implementation of data-centric PIA, suitable metrics of crop vigor, health, or development are needed. As mentioned, NDVI is an indicator of the greenness of the biomes that has been widely used for ecosystem monitoring and specifically as a proxy for vegetation vigor, especially for crop monitoring in agriculture, and can be calculated as:

$$NDVI = \frac{\rho NIR - \rho R}{\rho NIR + \rho R}$$

where ρNIR and ρR are the spectral reflectance measured in the near-infrared and visible red wavebands, respectively.

5.7 High-Efficiency Plant Phenotyping with AI

In modern plant breeding, AI serves as useful tool for high-efficiency plant phenotyping and prediction of gene functional analysis where the phenotypic and genotypic data (such as SNPs, QTLs, GWAS, genomics selection and genome sequenced data) are collected from huge collections of germplasm or advanced breeding populations integrated with AI technology (Khan *et al.*, 2022). This has shown impressive results in field and has a prime focus on accuracy in prediction. Next-generation AI equips us for fine data integration, analysis and interpretation, which ultimately improves the dynamics of experiments with effective and rapid outcomes (Khaki *et al.*, 2021). Next-generation AI refers to advanced AI systems that leverage cutting-edge technologies and methodologies to enhance their capabilities beyond current models. It is currently utilized for the prediction of breeding value for complex polygenic characters highly influenced by different environments and time scales. This needs smart and systematic data mining related to both plant biology and environmental systems and their involvement with the human population. It is aiding better trait prediction related to various factors, along with which plant variety would perform better in specified environment and soil conditions.

Numerous research studies have been conducted to analyze the potential of AI and next-generation AI in crop phenotyping and yield predictions. For example, Selvaraj *et al.* (2020) reported that three ML algorithms, namely, kNN, RF, and SVM, recorded the best predictions of root yield in cassava (*Manihot esculenta* Crantz), with the highest accuracy of R2 = 0.67, 0.66, and 0.64, respectively. Recording phenotypic traits such as growth stage is an important aspect of precision agriculture, crop breeding and phenotyping. In the past decade, computer vision (CV) that extracts valuable information from RGB images has become an essential technique for recognizing phenotypic changes in plants. The application of CV to digital images provides a HT and gentle non-disruptive alternative to manual recordings that is highly utilized in agriculture these days. Sadeghi-Tehran *et al.* (2017) introduced an automated technique for identifying the heading and flowering stages of wheat by applying CV to digital images. They used the Field Scanalyzer phenotyping platform to acquire RGB images at high resolution. The results indicated that this method surpassed existing algorithms, achieving accuracies of 95.24%, 97.79%, and 99.59% at the early, medium, and late stages of heading, respectively, along with an accuracy of 85.45% for flowering detection.

Silk growth is a key characteristic for drought tolerance in maize, but assessing it phenotypically poses a great challenge. Brichet *et al.* (2017) combined CV, ML and robotics to establish a non-invasive, consistent, and automated framework for identifying maize ears and silks and monitoring silk growth dynamics in a HT phenotyping platform. The framework resulted in a robust tool for extensive genetic studies on the regulation of reproductive growth to changes in environmental conditions in a non-invasive and automatized way. By utilizing an ear detection algorithm to operate a robotic camera, this framework captured high-resolution images to monitor ear and silk growth daily across hundreds of plants. Images were fragmented, with stem pixels labeled, and the ear position was determined by observing variations in width along the stem. A mobile camera is automatically positioned in real time at a distance of 30 cm from the ear, capturing detailed images in which silks are identified by their texture and color. This enabled the analysis of the growth timeline for ears and silks in thousands of plants. The pipeline was evaluated using a panel of more than 360 plants from 60 maize hybrids, successfully estimating ear position in 86% of cases before visual evaluation could take place. The estimated silk growth rate for all plants decreased over time, aligning with existing literature. The pipeline enabled a clear assessment of how genotypes and water deficit influenced both the rate and duration of silk growth. The accurate image segmentation is a critical and indispensable step in image-based plant phenotyping (Zhu *et al.*, 2015). Numerous studies have applied ML and CV to separate plants from weeds (Abdalla *et al.*, 2019). A common image analysis pipeline to separate crops amongst weeds is first to segment green vegetation regions from images and then classify the segmented regions into targeted crops and weeds. Features like color, shape, and texture, etc. have been used on an individual basis or as a combination of multiple features. Zhang *et al.* (2017) proposed a method to separate maize from weeds using high-dimensional color index features; the highest accuracy achieved was 93.87% using the support vector machine. Abdalla *et al.* (2019) conducted pixel-wise segmentation of oilseed rape images in a field densely populated with weeds and minimized both the amount of training data required and the training duration using transfer learning. They evaluated three different transfer learning methods using the visual geometry group (VGG) 16-based encoder model and assessed their performance differences with the VGG19-based encoder model. Relying on transfer learning and intensive use of data augmentation, they proved that the model can be trained end-to-end using a small number of image datasets and can separate value crops among weeds in an automated fashion. The highest accuracy of 96% was obtained by the VGG16-based encoder.

Identifying and characterizing the resistance responses of crops to pathogens is crucial for selecting resistant genotypes. During resistance reactions and pathogenesis, plants activate various structural and biochemical defense mechanisms that partially influence the optical properties of their organs. Innovative digital phenotyping methods have been developed to access compatible and incompatible plant–pathogen interactions. Classical phytopathology has been combined with digital phenotyping methods such as sensors, robotics, ML, and artificial intelligence. Together, they can non-invasively and objectively detect pathogen-induced changes in the physiology of both susceptible and resistant plants (Mahlein *et al.*, 2019).

Table 5.1 includes the various high-throughput phenotyping (HTP) platforms utilized by different researchers. Phenotypic data are crucial for crop breeding, yet they have considerable obstacles that impede their complete utilization. Numerous large-scale, HT, and automated phenotyping platforms have been developed so far to enhance plant analysis. Following are the notable examples. Traitmill is a bioinformatics tool designed for HT phenotyping for morphometric yield-related traits (Reuzeau *et al.*, 2010). Automated weighing and RGB imaging systems and PHENOPSIS have been used to study water deficit responses in Arabidopsis (Granier *et al.*, 2006) without any data integration features. The PHENOPSIS DB was introduced in 2011 with a facility to store hundreds of gigabytes of metadata and images for diverse analysis modules (Fabre *et al.*, 2011). GROWSCREEN has been used for color imaging to evaluate seedling growth dynamics (Walter *et al.*, 2007) and GROWSCREEN FLUORO for chlorophyll fluorescence imaging (Jansen *et al.*, 2009). The phenoscope rotating platform was designed to facilitate large-scale phenotyping of shoot growth and water content (Tisné *et al.*, 2013), while the phenovator (a high-speed rail system) was designed to collect data on photosynthesis and plant growth parameters (Flood *et al.*, 2016). The Bellwether Phenotyping

Table 5.1 Important HTP platforms utilized by various researchers in plant phenotyping. Author's own table.

S.no	HTP platform	Description	Source
1	PHENOPSIS	Designed an automated system to analyze how Arabidopsis plants respond to soil water stress	Sheikh *et al.*, 2024
2	GROWSCREEN FLUORO	Engineered to assess leaf growth and chlorophyll fluorescence, facilitating the identification of abiotic stress tolerance in *Arabidopsis*	Sheikh *et al.*, 2024
3	LemnaTec 3D Scanalyzer System	Utilized for non-invasive evaluation of salinity tolerance traits in rice, enabling thorough assessments	Sheikh *et al.*, 2024
4	HyperART	Allowed for non-destructive measurement of leaf characteristics, such as chlorophyll content and disease severity, across different crops	Sheikh *et al.*, 2024
5	PhenoBox	Identified diseases such as head smut in *Brachypodium* and corn smut in maize (corn), along with the assessment of salt stress responses in Tobacco	Sheikh *et al.*, 2024
6	PHENOVISION	Created to identify drought stress and recovery patterns in corn plants, aiding in the analysis of stress responses	Sheikh *et al.*, 2024
7	PhenoField	Focused on characterizing different abiotic stresses in wheat, enhanced understanding of stress-related plant traits	Sheikh *et al.*, 2024
8	PlantScreen™ Robotic XYZ System	Employed to examine various traits linked to drought tolerance in rice, improved understanding of resilience mechanisms in plants	Sheikh *et al.*, 2024
9	PocketPlant3D Tool	Utilized to assess the overall canopy structure	Confalonieri *et al.*, 2017
10	Phenoscope rotating platform	Designed to facilitate large-scale phenotyping of shoot growth and water content	Tisné *et al.*, 2013
11	Phenovator	A high-speed rail system to collect data on photosynthesis and plant growth parameters	Yang *et al.*, 2020
12	Traitmill	HT Phenotyping tool for morphometric yield-related traits	Reuzeau *et al.*, 2005
13	Integrated Analysis Platform	A Java-based application for analyzing drought responses and genetic variation	Chen *et al.*, 2014
14	Scanalyzer3D system and PlantCV	For processing RGB, fluorescent, and NIR images	Fahlgren *et al.*, 2015
15	PlaRoM	To analyze root hair development and growth dynamics	Yazdanbakhsh and Fisahn, 2009
16	RADIX	A rhizoslide platform for screening both roots and shoots	Le Marié *et al.*, 2016
17	GROWSCREEN-Rhizo	For simultaneous imaging of roots and shoots in transparent soil-filled rhizotrons	Nagel *et al.*, 2012
18	RhizoTubes	An automated plant-sensing platform	Jeudy *et al.*, 2016

Continued

Table 5.1 Continued

S.no	HTP platform	Description	Source
19	RootNav, SmartRoot, RootTrace,	Semi-automatic software for 2D image analysis of root architecture	French *et al*., 2009; Lobet *et al*., 2011; Pound *et al*., 2013
20	EZ-Root-VIS	Fully automatic software for 2D image analysis of root architecture	Shahzad *et al*., 2018
21	GiARoots, RootReader3D, RooTrak	For 3D image reconstruction and analysis of roots	Clark *et al*., 2011; Galkovskyi *et al*., 2012; Mairhofer *et al*., 2015; Yang *et al*., 2020
22	Field Scanalyzer phenotyping platform	A fully automated, HT, fixed-field phenotyping platform, carrying multiple sensors for non-invasive monitoring of plant growth, morphology, physiology and health	Sadeghi-Tehran *et al*., 2017

Platform developed the Scanalyzer3D system and PlantCV for processing RGB, fluorescent, and NIR images (Fahlgren *et al.*, 2015). The Scanalyzer3D platform has been utilized in phenomics data analysis, and the Plant Accelerator provided a computer-controlled conveyor system for assessing nitrogen, water deficiencies, and salinity stress through automated weighing, watering, and imaging (Neilson *et al.*, 2015; Atieno *et al.*, 2017).

The HT Rice Phenotyping Facility (HRPF) was developed to analyze plant architecture using X-ray CT and color imaging (Guo *et al.*, 2018; Duan *et al.*, 2018). 3D canopy modeling provided high-resolution insights into aboveground biomass and plant height (Maimaitijiang *et al.*, 2019), while Drones equipped with computer vision and photogrammetric techniques facilitate rapid data capture for remote sensing applications, including drought stress analysis and QTL identification (Bohra *et al.*, 2021). Advanced deep learning methods further enhanced the analysis of remote sensing images (Li *et al.*, 2020), Overall, UAV-based remote sensing showing significant promise for advancing crop functional genomics and breeding initiatives (Yang *et al.*, 2020).

5.8 AI-Empowered Pocket Phenotyping Approach

Pocket phenotyping is a rapidly evolving field that uses pocket or portable devices and technologies to accelerate and change phenotyping by assessing and analyzing plant traits in real time, facilitating field-based phenotypic measurements (Fahlgren *et al.*, 2015). These tools use the high-resolution cameras and computing power of smartphones to capture data and perform analysis. This approach allows researchers to gather data on various characteristics such as growth, health, and yield potential without the need for extensive equipment. By integrating tools like smartphones, drones, and sensors, pocket phenotyping enhances precision agriculture, enabling more efficient breeding programs and better crop management. It supports quick decision making and contributes to sustainable agricultural practices by optimizing resource use and improving crop performance in diverse environments. For many years, researchers have employed specialized handheld portable tools for phenotyping and standardized data collection. The integration of multiple sensors, data analysis software, AI, and fifth-generation mobile networks allowed for comprehensive analysis in field settings (Yang *et al.*, 2020). Additionally, a wide variety of optical sensors have significantly improved the portability and connectivity of phenotyping instruments. These advancements not only streamline the phenotyping process but also foster the development of resilient crop varieties that can adapt to changing environmental conditions, ultimately contributing to sustainable agricultural practices (Fahlgren *et al.*, 2015).

This concept has been used in various crops for effective phenotyping of different traits. Confalonieri *et al.* (2017) utilized a PocketPlant3D tool to assess the overall canopy structure. In wheat breeding, handheld devices with multi-spectral sensors have been deployed to assess plant height, biomass, and disease resistance in real time, facilitating informed selections based on growth dynamics. In case of barley also, portable devices were used to monitor water status and leaf temperature for selection of drought-resistant varieties (Thomas *et al.*, 2018). Likewise, Wungrampha et al. (2019) used a handy plant efficiency analyzer (PEA) to evaluate chlorophyll fluorescence kinetics in both controlled and stress conditions in greenhouses and fields. Likewise, in rice breeding programs, portable sensors have been used to monitor chlorophyll content and stress responses, enabling breeders to select varieties that sustain productivity under challenging conditions (Zhou *et al.*, 2019). In maize, hand-held devices were used to evaluate canopy structure and leaf area index, which helped in the identification of superior genotypes with enhanced photosynthetic capacity. In soybean, portable chlorophyll meters have been used to measure leaf chlorophyll concentration, offering valuable insights into plant health and photosynthetic efficiency. This technology helps breeders in identifying and selecting high-yielding genotypes more effectively. Overall, these innovations in pocket phenotyping not only improving breeding efficiency but also aid in developing crops that can thrive in various environmental conditions.

5.9 AI-Aided Belowground Phenotyping

Belowground phenotyping involves assessing plant traits and characteristics occurring beneath the soil surface, such as root structure, depth, and biomass. This approach is crucial for understanding how root systems influence nutrient uptake, water efficiency, and overall plant performance, especially under stress conditions like drought or poor soil quality. Advanced techniques, including soil imaging, minirhizotrons, and root scanning, enable researchers to analyze root traits in real time (York and Lynch, 2015). By integrating belowground data with aboveground traits, breeders can develop crop varieties better adapted to their environments, ultimately enhancing agricultural productivity and sustainability in the face of climate challenges (Lynch, 2013).

Despite its critical role in water and nutrient absorption, *in situ* phenotyping of the underground root system is limited, which directly impacts yield and biomass (Xiong *et al.*, 2021). A variety of growth media, including transparent options like gel-based, aeroponic, and hydroponic systems, along with PlaRoM equipped with RGB cameras, have been employed to analyze root hair development and growth dynamics (Yazdanbakhsh and Fisahn, 2009). Initially, root system architecture at the field level was analyzed destructively by excavating, washing, and examining the roots, a method known as 'shovelomics' (Trachsel *et al.*, 2011). Subsequently, non-destructive methods were developed, including *RADIX*, a rhizoslide platform for screening both roots and shoots (Le Marié *et al.*, 2016), GROWSCREEN-Rhizo, for simultaneous imaging of roots and shoots in transparent soil-filled rhizotrons (Nagel *et al.*, 2012), and RhizoTubes, an automated plant-sensing platform (Jeudy *et al.*, 2016). Furthermore, semi-automatic software like RootNav (Pound *et al.*, 2013), SmartRoot (Lobet *et al.*, 2011), and RootTrace (French *et al.*, 2009), as well as fully automatic software such as EZ-Root-VIS (Shahzad *et al.*, 2018), have been created for 2D image analysis of root architecture. For 3D image reconstruction and analysis, tools like GiARoots (Galkovskyi *et al.*, 2012), RootReader3D (Clark *et al.*, 2011), an open-source software, and RooTrak have also been introduced (Mairhofer *et al.*, 2015) (Table 5.1). Alternative techniques for reconstructing 3D root morphology include magnetic resonance imaging (MRI) and non-destructive scanning using positron emission tomography (PET) for assessing roots and carbon transportation (Atkinson *et al.*, 2019). Despite these advancements, non-destructive root phenotyping remains challenging, and there is still a need for breakthrough technologies in belowground imaging (Yang *et al.*, 2020).

5.10 Linking Genomics with Phenomics and Use of ML Algorithms

High-speed genotyping techniques have empowered breeders for more precise selection at early growth stages with a reduction in the time and the cost of assessment in the experiments (Rutkoski *et al.*, 2016). In recent times, we have made a significant progress with current plant breeding problems through HT genotyping coupled with highly efficient phenotyping techniques like remote sensing, used at different plant growth and development stages in huge plant breeding populations (Araus and Cairns, 2014; Rutkoski *et al.*, 2016; Tardieu *et al.*, 2017). These techniques are being used to calculate spectral reflectance properties that are genotype specific and affected by plant morphological, anatomical, and physiological characteristics of the plant, so could be efficiently utilized to screen genotypes with potential agronomic features (Kycko *et al.*, 2018; Schweiger *et al.*, 2018) (Fig. 5.2).

With the aim of improving the efficiency of the existing crop improvement methods, researchers used ML algorithms for the development of model-based breeding procedures (Hesami *et al.*, 2020a). Non-linear computational methods like artificial neural networks (ANNs), including multilayer perceptron (MLP), evolved by Pal and Mitra (1992) have been widely utilized for modeling and anticipating polygenic complex quantitative characters like yield (Geetha, 2020), determining the association among the input and targeted variables and for the quantification of the intrinsic cognizance existing in the database by parallel use of some highly interconnected processing neurons to solve specific problems (Ghorbani *et al.*, 2016; Geetha, 2020; Hesami *et al.*, 2020b). Support vector machines (SVMs) developed by Vapnik (2000) are another type of ML algorithm able to recognize patterns and behavior of non-linear relationships (Auria and Moro, 2008; Su *et al.*, 2017). SVMs use structural risk minimization inductive principles and mass formulations of learning problems in order to solve a quadratic optimization problem (Feng *et al.*, 2020; Hesami *et al.*, 2020b). In addition to tSVMs and MLP, another data modeling method used is Randomforest (RF) by Breiman (2001), equipped with a computational efficient training phase, more generalization accuracy and widely applied for tasks such as object recognition, skin detection, plant phenomics, and genomics (Mokry *et al.*, 2013). The past successful use of these ML algorithms in the performance assessment and prediction of various agronomic characters including plant yield are reported in *Senecio* species (Carvalho *et al.*, 2013), grassland (Rocha *et al.*, 2018), chrysanthemum (Hesami *et al.*, 2019) and Alfalfa (Feng *et al.*, 2020).

ML discovers relationships from massive training datasets. For environment and weather applications, first, big data involves the collection of meteorological or earth system-related measurements, as well as high spatial and temporal resolution earth system model (ESM) outputs for analysis; second, ML is the refining or discovery of new linkages between locations, times, and quantities in the datasets (e.g.where sea surface temperature features aid the weather prediction for months over land regions); third, AI is a means of providing automatic warnings and guidance to society in the event of oncoming weather extremes, based on the links discovered by ML. The use of computer graphics processing units (GPUs) in place of ordinary central processing units, has improved the speed and application of ML methods (Khan *et al.*, 2022).

5.11 AI-based Genomic Selection for Linking Genotype with Phenotype

Genomic selection models use genetic data to estimate relatedness between individuals. The genetic (heritable) constituent of the characters can be identified by the joint effect of all genome-wide variation on the phenotype along with environmental factors. Mapping between genotype and phenotype is a highly challenging statistical task, particularly for complex characters like grain yield, generally governed by polygenes, regulatory interactions with alternating environmental components (Heslot *et al.*, 2012). A statistical model was developed by Hayes *et al.* (2009) for the purpose of mapping one omic data layer (genotyping, e.g. SNPs) to another (phenotyping, e.g. drought tolerance) in a training population, which then used in an already genotyped independent breeding population in order to predict the potential phenotype years before the actual phenotype becomes available to measure (Hayes *et al.*, 2009). More accurate predictions can aid into increased yield along with reduced breeding cycles.

Remote sensing mechanisms and satellites have been used to draw out some illustrative indicators, namely leaf area index, normalized difference vegetation index (NDVI), enhanced vegetation index, temperature condition index, surface temperature and precipitation (from satellite images), which played crucial and helpful roles in monitoring the crop growth during the cropping season (Liou and Kar, 2014; Song *et al.*, 2015). The information collected from all these AI tools can be used to monitor various abiotic factors such as drought, excessive soil moisture detection, weather impacts on vegetation in an area, vegetation health, production and productivity of the crops and many others (Singh *et al.*, 2003; Khaki *et al.*, 2021).

Field-level precise and salubrious yield prediction at an early growth stage plays an indispensable part in site-specific crop management continuously in a growing period enabling farmers, breeders, and government organizations for decision making to maximize yield output and provide further economic benefit (Khaki *et al.*, 2021). Different vegetation indices have been used in previous studies for yield estimation in various crops (maize, rice, wheat, grapes, soybean, etc.) with the help of random forest and crop models, neural networks, convolutional neural networks (CNNs, both 2D and 3D), multiple linear regression and partial least squares regression, etc. Recently, researchers have combined CNN with satellite images for the prediction of crop yield (Khaki *et al.*, 2021). The connectivity pattern of CNN is influenced by biological processes like the organization of animal visual cortex where each cortical neuron responds to stimuli in a restricted region only of the visual field known as the receptive field. The receptive fields of different neurons slightly overlap in order to cover the complete visual field. Also called shift invariant or space invariant artificial neural networks (SIANNs), CNN is a form of artificial neural network generally used in computer vision, employed for the purpose of analyzing a grid-like matrix dataset, for example, imagery from sources like satellites. CV is an area of AI that empowers a computer to understand the input image or visual data and give interpretations. In place of general matrix multiplication, CNN applied convolution (a mathematical operation), which is particularly designed for image/video recognition, segmentation, classification processing, and medical image analysis with pixel data processing. Rather than analyzing the whole input/image in one go, 3D CNN utilizes 3D filters to perform complex convolutions by breaking the input/image into simple and small fragments, followed by progress through the CNN architecture (input layer, convolutional layer, pooling layer, fully connected dense layers and output layer), allowing the network to efficiently learn increasingly abstract representations of the input, complex patterns in data with minimum risk of overfitting.

5.12 Speed Breeding Empowered with AI

Speed breeding replicates natural day and night cycles by exposing crop plants to artificial light with various combinations and wavelengths, as well as temperature conditions, which significantly boosts crop photosynthetic, physiological and metabolic functions, promoting earlier flowering, fruiting, and seed development (Ahmar *et al.*, 2020). Speed breeding has become an innovative method for hastening the breeding cycles of crop plants by cultivating them under controlled artificial light and temperature conditions. Together with marker-assisted selection and advanced gene-editing technologies, it can enable early selection and modification of important crops. Additionally, AI can combine, process, and interpret complex omics datasets, which is a crucial requirement for effectively implementing speed breeding protocols to develop crop plants with enhanced yield and adaptability (Rai, 2022). For the utilization of next-generation AI in plant breeding, there is a need for smart and effective analysis of breeding datasets through the use of relevant models and precise algorithms. AI tools such as neural networks and DL have been used to improve the efficiency and accuracy of multiomics data. Researchers are continuously striving to enhance AI capabilities, particularly for high-definition image recognition and to analyze complex datasets more effectively. Consequently, this has become a key focus for speeding up the crop improvement process in recent years (Harfouche *et al.*, 2019). This system provides plant scientists with a convenient way to phenotype even the smallest variations. Researchers are now combining genotypic and environmental data with observed phenotypes to enhance agronomic programs aimed at breeding for abiotic stresses and to identify the optimal genotype with essential agronomic traits. Given that these traits are influenced by multiple genes, a sophisticated monitoring system is necessary to track minute changes or alterations in the plants. To achieve this, an AI-driven physiological gravimetric system has been developed which detects even the smallest changes in plants related to soil and atmospheric conditions, known as the soil–plant–atmosphere continuum. This system monitors complex traits throughout various stages of plant growth and development.

5.13 Role of Precise Phenotyping in Climate-smart Breeding

Abiotic stresses resulting from climate change are major contributors to crop biomass reduction and global yield losses (Verma and Deepti, 2016). These stresses are complex and dynamic, involving osmotic and metabolic interactions. To fully understand the manifold responses of crops to abiotic stresses, an integrated approach combining phenomics and genomics is essential. The phenotypic effects of abiotic stress depend on the specific organ and tissue along with duration and intensity of that stress. Visualizing abiotic stress-responsive traits and their correlation with related phenotypes is crucial for developing stress-tolerant plant varieties; the simplest examples include examining green vs yellow leaf areas under salinity stress and variations in canopy temperature during drought (Munns, 2010).

Recent advancements in genomics have significantly enhanced crop breeding with a focus on trait enrichment for improved productivity and climate resilience. A deeper understanding of plant mechanisms and the creation of innovative strategies to engineer stress tolerance are vital for crop improvement, utilizing genetic diversity, genome-scale breeding and precise agronomic practices (Bailey-Serres *et al.*, 2019). HT phenotyping is crucial in developing robust breeding models across diverse environments due to its cost-effectiveness and ease of use (Crossa *et al.*, 2017). However, the slower progress in phenomics compared to genomics has hindered the detection of potential in various crop genotypes. For the development of high yielding, climate-smart varieties there is an urgent need for HT, non-destructive phenomics approaches that measure subtle phenotypic changes and integrate technologies for consistent analysis in both controlled and field environments (Singh *et al.*, 2018).

The expression of traits is influenced by growth stage, which is evident in phenology and growth analysis. However, this influence can be subtler for traits where morphological or anatomical changes are less visible. For instance, canopy temperature is affected by ground cover, the presence of spikes that retain heat, and chlorosis due to senescence, which influences stomatal behavior. One effective approach is to create sub-blocks that isolate these factors. Alternatively, measuring traits across different genotypes at similar growth stages can help, although this may introduce variability due to weather and irrigation (Reynolds *et al.*, 2020). Measuring a large number of genotypes quickly often leads to lower spatial resolution and precision. Terrestrial vehicles and UAVs offer a good balance for breeding applications; While terrestrial vehicles provide high resolution and data quality for hundreds of plots, UAVs can assess thousands of plots more efficiently, reducing environmental variability impacts and avoiding soil compaction. Currently, UAVs face challenges with resolution and payload, but advancements in sensor technology may address these issues. Compared to manned aircraft and satellite imagery, both UAVs and terrestrial vehicles are preferred for small-plot genotypic screening, aided by increasingly user-friendly data analysis software.

5.14 AI Technologies Benefiting Agriculture and Crop Breeding

Numerous abiotic stresses, like drought, heat stress, heavy/untimely rain, and cold/chilling

Table 5.2 Summary of research conducted for phenotyping and climate resilient breeding using AI. Author's own table.

Study	Data source	Data type	Method/model/AI tool used	Key findings
Akintuyi (2024)	IoT (Internet of Things) sensors (soil moisture sensors, weather stations), satellite imagery, real-time farm data	Quantitative	Machine learning algorithms, IoT, cloud computing	Significant improvement in decision-making processes related to planting, watering, and harvesting; increased production rates and reduced operational costs.
Mishra *et al.* (2024)	Sensor, drone, satellite data	Quantitative	Machine learning, computer vision, remote sensing	AI-driven precision agriculture has improved irrigation, reduced input costs, and increased yields.
Rai (2022)	Various AI systems, field phenomics from different previous studies	Quantitative	AI-based physiological gravimetric system, UAV and ground-based equipment, MAS, GWAS	Identification of stress-responsive QTLs Improved phenotyping, linking phenotypic data with genotypic data Speed breeding with MAS and AI tools were found effective for early selection and manipulation of essential crops with superior agronomic traits.
Cavalli *et al.* (2021)	Multi-spectral remote sensing data related to agronomic indices	Quantitative	Regression models: Lasso Regression, Random Forest, Light GBM Agronomic indices: NDVI, NDWI, SAVI, EVI, GRNDVI	Incorporated agronomic indices along with stress and growth features which significantly improved model performance for crop yield forecasting Enhanced model accuracy and robustness in various environmental conditions
Ennouri *et al.* (2021)	Experiment data on morphological parameters of crop, Remote sensing	Quantitative	Multiple Linear Regression, Artificial Neural Networks	Improved crop production and enhanced monitoring, processing, and collection using AI and remote sensing tools.
Pasala and Pandey (2020)	High throughput phenotyping, Remote sensing and meteorological data	Quantitative	3D imaging, Infrared Imaging, Fluorescence Imaging, Visible Light Scanning, MRI	Remote sensing enabled efficient and rapid screening of plants in field conditions for traits such as drought and disease resistance
Saad *et al.* (2022)	Genomic data	Quantitative	CRISPR-Cas9, TALENs, ZFNs	Gene editing using CRISPR-Cas9 allows precise trait stacking in crops, enhancing resilience and yield

stress, are majorly caused by climate change occurring globally, having a huge effect on ecosystems. Climate change is very closely related to water supply and food security, occurrences of which are now frequent, often leading to increases in crop prices resulting in an unstable global food supply (Lesk *et al.*, 2016). Sudden change in climate, such as drought and heatwaves, especially at the grain-filling stage of crops, have a severe impact on plant growth and crop production (Kim *et al.*, 2020). This adverse effect has been mentioned by various scientists, and as Lobell *et al.* (2011) reported, because of climate change, the yield of maize and soybean could drop by approximately 5.50 and 3.80%, respectively. In addition, Lesk *et al.* (2016) approximated losses in cereal crop production and reported that, due to drought and heatwave, annual national cereal production was reduced by 9–10% in Canada and the UK. They further concluded that due to drought both plant yield and harvested area were reduced, while heavy heat primarily reduced yield of cereals. Similarly, Qian *et al.* (2019) scrutinized the impact of climate crisis in Canada on the yield of spring wheat, corn and canola, applying multiple crop models, and concluded that different types of yield changes occurred with respect to different crops. All these research efforts show the impact of climate change/crisis on crop yield, but were not able to show the prediction capacity of the model under stress conditions like drought and heat (Kim *et al.*, 2020).

As a rapidly emerging technology, AI is leading to super-intellective techniques, facilitating high-tech systems that provide great possibilities for agricultural applications, thus creating fresh frontiers for cybernated breeding (Harfouche *et al.*, 2019). This new era of AI has enabled us to make more precise, faster decisions than humans based on the data available (Xu *et al.*, 2021). AI is so useful for plant breeders in various tasks such as continuous farm monitoring, automation of farms and generalization of data, enabling plant breeders to devote more time to higher-value jobs such as data interpretation, analysis and high-value product development (Talaviya *et al.*, 2020). Table 5.2 summarizes some of the recent research conducted for phenotyping and climate-resilient breeding using AI. AI technology such as HT genomics and phenomics have proven useful in accelerating the development of new cultivars (Esposito *et al.*, 2020). AI tools such as ML, deep learning, etc. used in the prediction of yield aided us to think, learn, plan and take actions in order to develop systems for a better understanding of plant behavior under various climatic conditions. These technologies have served useful in genome analysis and genomic selection, as well as marker-aided selection for various traits (Crossa *et al.*, 2017; Reinoso-Pelaez *et al.*, 2022). Different organizations are investing hundreds of millions of dollars in techniques that could exploit large-scale data on types of soil, different varieties sown in an area, and their interaction with weather conditions to reduce costs and increase yield (Khan *et al.*, 2022). Due to the complicated dynamic data of agricultural experiments, sometimes even the advanced frameworks and architectonic algorithms and analytics deal with limitations in extricating the beneficial facts and information from the data that are otherwise hidden (Priya and Ramesh, 2020).

5.15 Conclusions

Crop improvement is a very challenging task that not only depends on the crop genotype but is also greatly affected by agriculture input parameters. There is much less awareness of the advantages of crop cultivation at the most suitable time and place. But many progressive growers have started using advance technologies and machinery to boost cultivation. Using data from numerous factors influencing crop production from multiple sources, one can assess crop yield as well as adaptability, which ultimately results in the improvement of crop quality, and increases economic growth and profit for growers. AI technologies are capable of decisiveness in making accurate recommendations using the huge data from the environment and plant system biology to assist with and contribute to plant breeding. Next-generation AI, on the other hand, efficiently utilizes highly complex heterogeneous data and enables smooth breeding even in adverse conditions. There is a need to develop new technologies and simplify the existing ones, which will lead to a new agricultural revolution, creating a better environment and fulfilling human needs. There is an urgent need to adapt the digital revolution to advance plant breeding programmes by researchers and plant breeders to achieve goals in climate-resilient breeding with yield enhancement.

References

Abdalla, A., Cen, H., Wan, L., Rashid, R., Weng, H. *et al.* (2019) Fine-tuning convolutional neural network with transfer learning for semantic segmentation of ground-level oilseed rape images in a field with high weed pressure. *Computers and Electronics in Agriculture* 167, 105091.

Ahmad, R., Yang, B., Ettlin, G., Berger, A. and Rodriguez-Bocca, P. (2020) A machine-learning based ConvLSTM architecture for NDVI forecasting. *International Transactions in Operational Research* 30, 2025–2048. DOI: 10.1111/itor.12887.

Ahmar, S., Gill, R.A., Jung, K.H., Faheem, A., Qasim, M.U. *et al.* (2020) Conventional and molecular techniques from simple breeding to speed breeding in crop plants: Recent advances and future outlook. *International Journal of Molecular Sciences* 21(7), 2590. DOI: 10.3390/ijms21072590.

Akintuyi, O.B. (2024) Adaptive AI in precision agriculture: A review: Investigating the use of self-learning algorithms in optimizing farm operations based on real-time data. *Open Access Research Journal of Multidisciplinary Studies* 07(2), 016–030.

Araus, J.L. and Cairns, J.E. (2014) Field high-throughput phenotyping: The new crop breeding frontier. *Trends in Plant Science* 19, 52–61. DOI: 10.1016/j.tplants.2013.09.008.

Atieno, J., Li, Y., Langridge, P., Dowling, K., Brien, C. *et al.* (2017) Exploring genetic variation for salinity tolerance in chickpea using image-based phenotyping. *Scientific Reports* 7(1), 1300.

Atkinson, J.A., Pound, M.P., Bennett, M.J. and Wells, D.M. (2019) Uncovering the hidden half of plants using new advances in root phenotyping. *Current Opinion in Biotechnology* 55, 1–8.

Auria, L. and Moro, R.A. (2008) Support vector machines (SVM) as a technique for solvency analysis. DIW Discussion Papers. Deutsches Institut Für Wirtschaftsforschung (DIW). DOI: 10.2139/ssrn.1424949.

Baghdasaryan, L., Melikbekyan, R., Dolmajain, A. and Hobbs, J. (2022) Deep density estimation based on multi-spectral remote sensing data for in-field crop yield forecasting. In: *IEEE/CVF Conference on Computer Vision and Pattern Recognition Workshops (CVPRW)*, pp. 2013–2022. DOI: 10.1109/CVPRW56347.2022.00219.

Bailey-Serres, J., Parker, J.E., Ainsworth, E.A., Oldroyd, G.E. and Schroeder, J.I. (2019) Genetic strategies for improving crop yields. *Nature* 575(7781), 109–118.

Bannerjee, G., Sarkar, U., Das, S. and Ghosh, I. (2018) Artificial intelligence in agriculture: A literature survey. *International Journal of Scientific Research in Computer Science Applications and Management Studies* 7, 1–6.

Bhat, J.A., Yu, D., Bohra, A., Ganie, S.A. and Varshney, R.K. (2021) Features and applications of haplotypes in crop breeding. *Communications Biology* 4, 121266.

Bohra, A., Satheesh Naik, S.J., Kumari, A., Tiwari, A. and Joshi, R. (2021) Integrating phenomics with breeding for climate-smart agriculture. In: *Omics Technologies for Sustainable Agriculture and Global Food Security*, Vol. II. Springer, Singapore, pp. 1–24.

Breiman, L. (2001) Random forests. *Machine Learning* 45, 5–32. DOI: 10.1023/A.

Brichet, N., Fournier, C., Turc, O., Strauss, O., Artzet, S. *et al.* (2017) A robot-assisted imaging pipeline for tracking the growths of maize ear and silks in a high-throughput phenotyping platform. *Plant Methods* 13, 96.

Cai, G., Yang, Q., Chen, H., Yang, Q., Zhang, C. *et al.* (2016) Genetic dissection of plant architecture and yield-related traits in *Brassica napus*. *Science Reports* 6, 21625. DOI: 10.1038/srep21625.

Carvalho, S., Macel, M., Schlerf, M., Moghaddam, F.E., Mulder, P.P. *et al.* (2013) Changes in plant defense chemistry (pyrrolizidine alkaloids) revealed through high-resolution spectroscopy. *ISPRS Journal of Photogrammetry and Remote Sensing* 80, 51–60. DOI: 10.1016/j.isprsjprs.2013.03.004.

Cavalli, S., Penzotti, G., Amoretti, M. and Caselli, S.A. (2021) Machine learning approach for NDVI forecasting based on sentinel-2 data. In: *Proceedings of the 16th International Conference on Software Technologies (ICSOFT 2021)*, pp. 473–480. DOI: 10.5220/0010544504730480.

Chen, D., Neumann, K., Friedel, S., Kilian, B., Chen, M, *et al.* (2014) Dissecting the phenotypic components of crop plant growth and drought responses based on high-throughput image analysis. *The Plant Cell* 26(12), 4636–4655.

Clark, R.T., MacCurdy, R.B., Jung, J.K., Shaff, J.E., McCouch, S.R. *et al.* (2011) Three-dimensional root phenotyping with a novel imaging and software platform. *Plant Physiology* 156(2), 455–465.

Collins, N.C., Tardieu, F. and Tuberosa, R. (2008) Quantitative trait loci and crop performance under abiotic stress: Where do we stand? *Plant Physiology* 147, 469–486. DOI: 10.1104/pp.108.118117.

Confalonieri, R., Paleari, L., Foi, M., Movedi, E., Vesely, F.M. *et al.* (2017) PocketPlant3D: Analysing canopy structure using a smartphone. *Biosystems Engineering* 164, 1–12.

Crossa, J., Perez-Rodriguez, P., Cuevas, J., Montesinos-Lopez, O., Jarquin, D. *et al.* (2017) Genomic selection in plant breeding: Methods, models, and perspectives. *Trends in Plant Science* 22, 961–975.

Duan, L., Han, J., Guo, Z., Tu, H., Yang, P. *et al.* (2018) Novel digital features discriminate between drought resistant and drought sensitive rice under controlled and field conditions. *Frontiers in Plant Science* 9, 492.

Dwivedi, Y.K., Hughes, L., Ismagilova, E., Aarts, G., Coombs, C. *et al.* (2021) Artificial intelligence (AI): multidisciplinary perspectives on emerging challenges, opportunities, and agenda for research, practice and policy. *International Journal of Information Management* 57, 101994.

Ennouri, K., Smaoui, S., Gharbi, Y., Cheffi, M., Braiek, O.B. *et al.* (2021) Usage of artificial intelligence and remote sensing as efficient devices to increase agricultural system yields. *Journal of Food Quality* 6242288. DOI: 10.1155/2021/6242288.

Esposito, S., Carputo, D., Cardi, T. and Tripodi, P. (2020) Applications and trends of machine learning in genomics and phenomics for next-generation breeding. *Plants* 9, 34.

Fabre, J., Dauzat, M., Nègre, V., Wuyts, N., Tireau, A. *et al.* (2011) PHENOPSIS DB: An information system for *Arabidopsis thaliana* phenotypic data in an environmental context. *BMC Plant Biology* 11(1), 77.

Fahlgren, N., Feldman, M., Gehan, M.A., Wilson, M.S., Shyu, C. *et al.* (2015) A versatile phenotyping system and analytics platform reveals diverse temporal responses to water availability in Setaria. *Molecular Plant* 8(10), 1520–1535.

Feng, L., Zhang, Z., Ma, Y., Du, Q., Williams, P. *et al.* (2020) Alfalfa yield prediction using UAV-based hyperspectral imagery and ensemble learning. *Remote Sensing* 12. DOI: 10.3390/rs12122028.

Filippi, P., Jones, E.J., Wimalathunge, N.S., Somarathna, P.D.S.N., Pozza, L.E. *et al.* (2019) An approach to forecast grain crop yield using multilayered, multi-farm data sets and machine learning. *Precision Agriculture* 20, 1015–1029. Available at: https://doi.org/10.1007/s11119-018-09628-4

Flood, P.J., Kruijer, W., Schnabel, S.K., Schoor, R., Jalink, H. *et al.* (2016) Phenomics for photosynthesis, growth and reflectance in *Arabidopsis thaliana* reveals circadian and long-term fluctuations in heritability. *Plant Methods* 12(1), 14.

French, A., Ubeda-Tomás, S., Holman, T.J., Bennett, M.J. and Pridmore, T. (2009) High-throughput quantification of root growth using a novel image-analysis tool. *Plant Physiology* 150(4), 1784–1795.

Galkovskyi, T., Mileyko, Y., Bucksch, A., Moore, B., Symonova, O. *et al.* (2012) GiA roots: Software for the high throughput analysis of plant root system architecture. *BMC Plant Biology* 12(1), 116.

Geetha, M. (2020) Forecasting the crop yield production in Trichy district using fuzzy C-means algorithm and multilayer perceptron (MLP). *International Journal of Knowledge Systems Science (IJKSS)* 11, 83–98. DOI: 10.4018/IJKSS.2020070105.

Ghorbani, M.A., Zadeh, H.A., Isazadeh, M. and Terzi, O. (2016) A comparative study of artificial neural network (MLP, RBF) and support vector machine models for river flow prediction. *Environmental Earth Sciences* 75, 476. DOI: 10.1007/s12665-015-5096-x.

Granier, C., Aguirrezabal, L., Chenu, K., Cookson, S.J., Dauzat, M. *et al.* (2006) PHENOPSIS, an automated platform for reproducible phenotyping of plant responses to soil water deficit in *Arabidopsis thaliana* permitted the identification of an accession with low sensitivity to soil water deficit. *New Phytologist* 169(3), 623–635.

Guo, Z., Yang, W., Chang, Y., Ma, X., Tu, H. *et al.* (2018) Genome-wide association studies of image traits reveal genetic architecture of drought resistance in rice. *Molecular Plant* 11(6), 789–805.

Gupta, A., Singh, M. and Laxmi, A. (2015) Interaction between glucose and brassinosteroid during the regulation of lateral root development in *Arabidopsis*. *Plant Physiology* 168(1), 307–320.

Harfouche, A.L., Jacobson, D.A., Kainer, D., Romero, J.C., Harfouche, A.H. *et al.* (2019) Accelerating climate resilient plant breeding by applying next-generation artificial intelligence. *Trends in Biotechnology* 37, 1217–1235.

Hayes, B.J., Bowman, P.J., Chamberlain, A.J. and Goddard, M.E. (2009) Invited review: Genomic selection in dairy cattle: Progress and challenges. *Journal of Dairy Science* 92, 433–443. DOI: 10.3168/jds.2008-1646.

Hesami, M., Naderi, R., Tohidfar, M. and Yoosefzadeh-Najafabadi, M. (2019) Application of adaptive neuro-fuzzy inference system-non-dominated sorting genetic algorithm-II (ANFIS-NSGAII) for modeling and optimizing somatic embryogenesis of chrysanthemum. *Frontiers in Plant Science* 10, 869. DOI: 10.3389/fpls.2019.00869.

Hesami, M., Condori-Apfata, J.A., Valderrama Valencia, M. and Mohammadi, M. (2020a) Application of artificial neural network for modeling and studying *in vitro* genotype-independent shoot regeneration in wheat. *Applied Sciences* 10, 5370. DOI: 10.3390/app10155370.

Hesami, M., Naderi, R., Tohidfar, M. and Yoosefzadeh-Najafabadi, M. (2020b) Development of support vector machine-based model and comparative analysis with artificial neural network for modeling the plant tissue culture procedures: Effect of plant growth regulators on somatic embryogenesis of chrysanthemum, as a case study. *Plant Methods* 16, 1–15. DOI: 10.1186/s13007-020-00655-9.

Heslot, N., Yang, H.P., Sorrells, M.E. and Jannink, J.L. (2012) Genomic selection in plant breeding: A comparison of models. *Crop Science* 52, 146–160. DOI: 10.2135/cropsci2011.06.0297.

Jansen, M., Gilmer, F., Biskup, B., Nagel, K.A., Rascher, U. *et al.* (2009) Simultaneous phenotyping of leaf growth and chlorophyll fluorescence via GROWSCREEN FLUORO allows detection of stress tolerance in *Abidopsis thaliana* and other rosette plants. *Functional Plant Biology* 36(11), 902–914.

Jeudy, C., Adrian, M., Baussard, C., Bernard, C., Bernaud, E. *et al.* (2016) RhizoTubes as a new tool for high throughput imaging of plant root development and architecture: Test, comparison with pot grown plants and validation. *Plant Methods* 12(1), 31.

Jha, K., Doshi, A., Patel, P. and Shah, M. (2019) A comprehensive review on automation in agriculture using artificial intelligence. *Artificial Intelligence in Agriculture* 2, 1–12.

Jin, J., Liu, X., Wang, G., Mi, L., Shen, Z. *et al.* (2010) Agronomic and physiological contributions to the yield improvement of soybean cultivars released from 1950 to 2006 in Northeast China. *Field Crops Research* 115, 116–123. DOI: 10.1016/j.fcr.2009.10.016.

Jung, J., Maeda, M., Chang, A., Bhandari, M., Ashapure, A. *et al.* (2021) The potential of remote sensing and artificial intelligence as tools to improve the resilience of agriculture production systems. *Current Opinion in Biotechnology* 70, 15–22.

Kang, M.S. (2004) Breeding: Genotype-by-environment interaction. In: *Encyclopedia of Plant and Crop Science*. CRC Press, Boca Raton, Florida, pp. 218–221.

Khaki, S., Pham, H. and Wang, L. (2021) Simultaneous corn and soybean yield prediction from remote sensing data using deep transfer learning. *Scientific Reports* 11, 11132. DOI: 10.1038/s41598-021-89779-z.

Khan, M.H.U., Wang, S., Wang, J., Ahmar, S., Saeed, S. *et al.* (2022) Applications of artificial intelligence in climate-resilient smart-crop breeding. *International Journal of Molecular Sciences* 23, 11156. DOI: 10.3390/ijms231911156.

Kim, N., Na, S., Park, C.W., Huh, M., Oh, J. *et al.* (2020) An artificial intelligence approach to prediction of corn yields under extreme weather conditions using satellite and meteorological data. *Applied Sciences* 10, 3785. DOI: 10.3390/app10113785.

Klompenburg, T.V., Kassahun, A. and Catal, C. (2020) Crop yield prediction using machine learning: A systematic literature review. *Computers and Electronics in Agriculture* 177, 105709. DOI: 10.1016/j.compag.2020.105709.

Kumar, R.S., Keerthana, G., Murali, L., Chidambaranathan, S., Premkumar, C. *et al.* (2022) Enabling artificial intelligence and cyber security in smart manufacturing. In: *Artificial Intelligent Techniques for Wireless Communication and Networking*. John Wiley & Sons, Ltd, Hoboken, New Jersey, pp. 269–286. DOI: 10.1002/9781119821809.ch17.

Kycko, M., Zagajewski, B., Lavender, S., Romanowska, E. and Zwijacz-Kozica, M. (2018) The impact of tourist traffic on the condition and cell structures of alpine swards. *Remote Sensing* 10, 220. DOI: 10.3390/rs10020220.

Le Marié, C., Kirchgessner, N., Flütsch, P., Pfeifer, J., Walter, A. *et al.* (2016) RADIX: Rhizoslide platform allowing high throughput digital image analysis of root system expansion. *Plant Methods* 12(1), 40.

Lesk, C., Rowhani, P. and Ramankutty, N. (2016) Influence of extreme weather disasters on global crop production. *Nature* 529, 84–87.

Liou, Y.A. and Kar, S.K. (2014) Evapotranspiration estimation with remote sensing and various surface energy balance algorithms - A review. *Energies* 7, 2821–2849.

Li, X., Liu, B., Zheng, G., Ren, Y., Zhang, S. *et al.* (2020) Deep-learning-based information mining from ocean remote-sensing imagery. *National Science Review* 7(10), 1584–1605.

Lobell, D.B., Schlenker, W. and Costa-Roberts, J. (2011) Climate trends and global crop production since 1980. *Science* 333, 616–620.

Lobet, G., Pagès, L. and Draye, X. (2011) A novel image-analysis toolbox enabling quantitative analysis of root system architecture. *Plant Physiology* 157(1), 29–39.

Lynch, J.P. (2013) Steep, cheap and deep: An ideotype to optimize water and n acquisition by maize root systems. *Annals of Botany* 112, 347–357.

Mahlein, A.-K., Kuska, M.T., Thomas, S., Wahabzada, M., Behmann, J. *et al.* (2019) Quantitative and qualitative phenotyping of disease resistance of crops by hyperspectral sensors: Seamless interlocking of phytopathology, sensors, and machine learning is needed. *Current Opinion in Plant Biology* 50, 156–162.

Maimaitijiang, M., Sagan, V., Sidike, P., Maimaitiyiming, M., Hartling, S. *et al.* (2019) Vegetation index weighted canopy volume model (CVMVI) for soybean biomass estimation from unmanned aerial system-based RGB imagery. *ISPRS Journal of Photogrammetry and Remote Sensing* 151, 27–41.

Mairhofer, S., Sturrock, C.J., Bennett, M.J., Mooney, S.J. and Pridmore, T.P. (2015) Extracting multiple interacting root systems using X-ray microcomputed tomography. *The Plant Journal* 84(5), 1034–1043.

Mishra, A.C., Das, J. and Awtar, R. (2024) An emerging era of research in agriculture using AI. *Journal of Scientific Research and Technology* 2(4), 1–7.

Mokry, F.B., Higa, R.H., Mudadu, M., de-Lima, A.O., Meirelles, S.L.C. *et al.* (2013) Genome-wide association study for backfat thickness in Canchim beef cattle using random forest approach. *BMC Genetics* 14, 47. DOI: 10.1186/1471-2156-14-47.

Montesinos-Lopez, O.A., Montesinos-Lopez, A., Crossa, J., Gianola, D., Hernandez-Suarez, C.M. and Martin-Vallejo, J. (2018) Multi-trait, multi-environment deep learning modeling for genomic-enabled prediction of plant traits. G3 Genes Genomes Genetics 8, 3829–3840.

Munns, R. (2010) Approaches to identifying genes for salinity tolerance and the importance of timescale. In: *Plant Stress Tolerance: Methods and Protocols*. Humana Press, Totowa, New Jersey, pp. 25–38.

Nagel, K.A., Putz, A., Gilmer, F., Heinz, K., Fischbach, A. *et al.* (2012) GROWSCREEN-Rhizo is a novel phenotyping robot enabling simultaneous measurements of root and shoot growth for plants grown in soil-filled rhizotrons. *Functional Plant Biology* 39(11), 891–904.

Neilson, E.H., Edwards, A.M., Blomstedt, C.K., Berger, B., Møller, B.L. *et al.* (2015) Utilization of a high-throughput shoot imaging system to examine the dynamic phenotypic responses of a C4 cereal crop plant to nitrogen and water deficiency over time. *Journal of Experimental Botany* 66(7), 1817–1832.

Pal, S.K. and Mitra, S. (1992) Multilayer perceptron, fuzzy sets, classification. *IEEE Transactions on Neural Networks* 3, 683–697. DOI: 10.1109/72.159058.

Pasala, P. and Pandey, B.B. (2020) Plant phenomics: High-throughput technology for accelerating genomics. *Journal of Biosciences* 45, 111. DOI: 10.1007/s12038-020-00083-w.

Perea, R.G., Poyato, E.C., Montesinos, P. and Diaz, J.R. (2018) Prediction of applied irrigation depths at farm level using artificial intelligence techniques. *Agricultural Water Management* 206, 229–240.

Pound, M.P., French, A.P., Atkinson, J.A., Wells, D.M., Bennett, M.J. *et al.* (2013) RootNav: Navigating images of complex root architectures. *Plant Physiology* 162(4), 1802–1814.

Priya, R. and Ramesh, D.J.S.C.I. (2020) ML based sustainable precision agriculture: A future generation perspective. *Sustainable Computing: Informatics and Systems* 28, 100439.

Qian, B., Zhang, X., Smith, W., Grant, B., Jing, Q. *et al.* (2019) Climate change impacts on Canadian yields of spring wheat, canola and maize for global warming levels of 1.5 °C, 2.0 °C, 2.5°C and 3.0°C. *Environmental Research Letters* 14, 074005.

Rai, K.K. (2022) Integrating speed breeding with artificial intelligence for developing climate-smart crops. *Molecular Biology Reports* 49, 11385–11402. DOI: 10.1007/s11033-022-07769-4.

Reinoso-Pelaez, E.L., Gianola, D. and Gonzalez-Recio, O. (2022) Genome-enabled prediction methods based on machine learning. In: Ahmadi, N. and Bartholomé, J. (eds) *Genomic Prediction of Complex Traits. Methods in Molecular Biology*, Vol. 2467. Humana, New York. DOI: 10.1007/978-1-0716-2205-6_7.

Reuzeau, C., Pen, J., Frankard, V., de Wolf, J., Peerbolte, R. *et al.* (2005) TraitMill™: A discovery engine for identifying yield enhancement genes in cereals. *Mol. Plant Breed* 5, 753–759.

Reuzeau, C., Pen, J., Frankard, V., Wolf, J., Peerbolte, R. *et al.* (2010) TraitMill: A discovery engine for identifying yield-enhancement genes in cereals. *Plant Gene and Trait* 1.

Reynolds, M., Sukumaran, S., Pinto, F. and Molero, G. (2020) Physiological breeding for climate change. In: *Population, Agriculture, and Biodiversity: Problems and Prospects*. University of Missouri Press, Columbia, Missouri, p. 181.

Rocha, A., Groen, T., Skidmore, A., Darvishzadeh, R. and Willemen, L. (2018) Machine learning using hyperspectral data inaccurately predicts plant traits under spatial dependency. *Remote Sensing* 10, 1263. DOI: 10.3390/rs10081263.

Rutkoski, J., Poland, J., Mondal, S., Autrique, E., Perez, L.Z *et al.* (2016) Canopy temperature and vegetation indices from high-throughput phenotyping improve accuracy of pedigree and genomic selection for grain yield in wheat. *G3 Genes* 6, 2799–2808. DOI: 10.1534/g3.116.032888.

Saad, N.S.M., Neik, T.X., Thomas, W.J.W., Amas, J.C., Cantila, A.Y. *et al.* (2022) Advancing designer crops for climate resilience through an integrated genomics approach. *Current Opinion in Plant Biology* 67, 102220. DOI: 10.1016/j.pbi.2022.102220.

Sadeghi-Tehran, P., Sabermanesh, K., Virlet, N. and Hawkesford, M.J. (2017) Automated method to determine two critical growth stages of wheat: Heading and flowering. *Frontiers in Plant Science* 8, 252.

Schweiger, A.K., Cavender-Bares, J., Townsend, P.A., Hobbie, S.E., Madritch, M.D. *et al.* (2018) Plant spectral diversity integrates functional and phylogenetic components of biodiversity and predicts ecosystem function. *Nature Ecology and Evolution* 2, 976–982. DOI: 10.1038/s41559-018-0551-1.

Selvaraj, M.G., Valderrama, M., Guzman, D., Valencia, M., Ruiz, H. *et al.* (2020) Machine learning for high-throughput field phenotyping and image processing provides insight into the association of above and below-ground traits in cassava (*Manihotesculenta crantz*). *Plant Methods* 16, 87.

Shahzad, Z., Kellermeier, F., Armstrong, E.M., Rogers, S., Lobet, G. *et al.* (2018) EZ-root-VIS: A software pipeline for the rapid analysis and visual reconstruction of root system architecture. *Plant Physiology* 177(4), 1368–1381.

Sheikh, M., Iqra, F., Ambreen, H., Pravin, K.A., Ikra, M. *et al.* (2024) Integrating artificial intelligence and high-throughput phenotyping for crop improvement. *Journal of Integrative Agriculture* 23(6), 1787–1802.

Singh, B., Mishra, S., Bohra, A., Joshi, R. and Siddique, K.H. (2018) Crop phenomics for abiotic stress tolerance in crop plants. In: *Biochemical, Physiological and Molecular Avenues for Combating Abiotic Stress Tolerance in Plants*. Academic Press, pp. 277–296.

Singh, R.P., Roy, S. and Kogan, F. (2003) Vegetation and temperature condition indices from NOAA AVHRR data for drought monitoring over India. *International Journal of Remote Sensing* 24, 4393–4402.

Song, L., Liu, S., Kustas, W.P., Zhou, J. and Ma, Y. (2015) Using the surface temperature-albedo space to separate regional soil and vegetation temperatures from aster data. *Remote Sensing* 7, 5828–5848.

Su, Q., Lu, W., Du, D., Chen, F., Niu, B. *et al.* (2017) Prediction of the aquatic toxicity of aromatic compounds to tetrahymena pyriformis through support vector regression. *Oncotarget* 8, 49359–49369. DOI: 10.18632/oncotarget.17210.

Talaviya, T., Shah, D., Patel, N., Yagnik, H. and Shah, M. (2020) Implementation of artificial intelligence in agriculture for optimisation of irrigation and application of pesticides and herbicides. *Artificial Intelligence in Agriculture* 4, 58–73.

Tardieu, F., Cabrera-Bosquet, L., Pridmore, T. and Bennett, M. (2017) Plant phenomics, from sensors to knowledge. *Current Biology* 27, R770–R783. DOI: 10.1016/j.cub.2017.05.055.

Thomas, S., Behmann, J., Steier, A., Kraska, T., Muller, O. *et al.* (2018) Quantitative assessment of disease severity and rating of barley cultivars based on hyperspectral imaging in a non-invasive, automated phenotyping platform. *Plant Methods* 14(1), 45.

Tisné, S., Serrand, Y., Bach, L., Gilbault, E., Ben Ameur, R. *et al.* (2013) Phenoscope: An automated large-scale phenotyping platform offering high spatial homogeneity. *The Plant Journal* 74(3), 534–544.

Trachsel, S., Kaeppler, S.M., Brown, K.M. and Lynch, J.P. (2011) Shovelomics: High throughput phenotyping of maize (*Zea mays* L.) root architecture in the field. *Plant and Soil* 341(1), 75–87.

Vapnik, V. (2000) *The Nature of Statistical Learning Theory*. Springer Verlag, New York. DOI: 10.1007/978-1-4757-3264-1.

Verma, A.K. and Deepti, S. (2016) Abiotic stress and crop improvement: Current scenario. *Advances in Plants & Agriculture Research* 4(4), 345–346.

Walter, A., Scharr, H., Gilmer, F., Zierer, R., Nagel, K.A. *et al.* (2007) Dynamics of seedling growth acclimation towards altered light conditions can be quantified via GROWSCREEN: A setup and procedure designed for rapid optical phenotyping of different plant species. *New Phytologist* 174(2), 447–455.

Wungrampha, S., Joshi, R., Rathore, R.S., Singla-Pareek, S.L. and Govindjee, G. (2019) CO2 uptake and chlorophyll a fluorescence of *Suaeda fruticosa* grown under diurnal rhythm and after transfer to continuous dark. *Photosynthesis Research* 142(2), 211–227.

Xiong, Q., Tang, G., Zhong, L., He, H. and Chen, X. (2018) Response to nitrogen deficiency and compensation on physiological characteristics, yield formation, and nitrogen utilization of rice. *Frontiers in Plant Science* 9, 1075. DOI: 10.3389/fpls.2018.01075.

Xiong, Q., Hu, J., Wei, H., Zhang, H. and Zhu, J. (2021) Relationship between plant roots, rhizosphere microorganisms, and nitrogen and its special focus on rice. *Agriculture* 11(3), 234.

Xu, X., Gao, P., Zhu, X., Guo, W., Ding, J. *et al.* (2019) Design of an integrated climatic assessment indicator (ICAI) for wheat production: A case study in Jiangsu Province, China. *Ecological Indicators* 101, 943–953. DOI: 10.1016/j.ecolind.2019.01.059.

Yang, Q., Shi, L., Han, J., Yu, J. and Huang, K. (2020) A near real-time deep learning approach for detecting rice phenology based on UAV images. *Agricultural and Forest Meteorology* 287, 107938.

Yang, Y. and Siau, K.L. (2018) A qualitative research on marketing and sales in the artificial intelligence age. In: *Proceedings of the Thirteenth Midwest Association for Information Systems Conference*. Saint Louis, MO, pp. 1–6.

Yazdanbakhsh, N. and Fisahn, J. (2009) High throughput phenotyping of root growth dynamics, lateral root formation, root architecture and root hair development enabled by PlaRoM. *Functional Plant Biology* 36(11), 938–946.

Xu, Y., Liu, X., Cao, X., Huang, C., Liu, E. *et al.* (2021) Artificial intelligence: A powerful paradigm for scientific research. *Innovation* 2, 100179.

York, L.M. and Lynch, J.P. (2015) Intensive field phenotyping of maize (*Zea mays* L.) root crowns identifies phenes and phene integration associated with plant growth and nitrogen acquisition. *Journal of Experimental Botany* 66(18), 5493–5505.

Zargar, S.M., Raatz, B., Sonah, H., Bhat, J.A., Dar, Z.A. *et al.* (2015) Recent advances in molecular marker techniques: Insight into QTL mapping, GWAS and genomic selection in plants. *Journal of Crop Science and Biotechnology* 18, 293–308.

Zhang, X., Huang, C., Wu, D., Qiao, F., Li, W. *et al.* (2017) High-throughput phenotyping and QTL mapping reveals the genetic architecture of maize plant growth. *Plant Physiology* 173, 1554–1564. DOI: 10.1104/pp.16.01516.

Zhou, W., Zhang, J., Zou, M., Liu, X., Du, X, *et al.* (2019) Prediction of cadmium concentration in brown rice before harvest by hyperspectral remote sensing. *Environmental Science and Pollution Research* 26(2), 1848–1856.

Zhu, J.Q., Werf, W., Anten, N.P.R., Vos, J. and Evers, J.B. (2015) The contribution of phenotypic plasticity to complementary light capture in plant mixtures. *New Phytologist* 207, 1213–1222. DOI: 10.1111/nph.13416.

Zohry, A.E.-H. and Ouda, S. (2022) Climate-resilient crops. In: *Climate-Smart Agriculture: Reducing Food Insecurity*. Springer International Publishing, Cham, pp. 115–135. DOI: 10.1007/978-3-030-93111-7_6.

6 Multiomics Approaches with Advanced Breeding Tools for Crop Improvement

Garima Aggarwal[1], Shalu Kumari[2], Namata Kumari[1], Kajal Mehra[1], Mayank Tiwari[1], Mithilesh Kumar Singh[3] and Bishawajit Kumar[1]*

[1]Department of Genetics and Plant Breeding, G. B. Pant University of Agriculture & Technology, Pantnagar (Udham Singh Nagar), Uttarakhand, India; [2]Department of Genetics and Plant Breeding, I.A.S, BHU, Varanasi, Uttar Pradesh, India; [3]Department of Genetics and Plant Breeding, Dr. Rajendra Prasad Central Agricultural University, Pusa (Samastipur), Bihar, India

Abstract

The escalating challenges posed by climate change, environmental stresses, and rapid population growth call for urgent advancements in crop strategies to ensure global food security. Central to this mission is the development of resilient crop varieties through higher genetic gains, achieved by leveraging modern molecular breeding techniques such as marker-assisted and genomic selection. Multiomics approaches encompassing genomics, transcriptomics, proteomics, and metabolomics offer valuable insights into the intricate biological mechanisms governing crop traits and their interactions with environmental stressors. High-throughput sequencing technologies further enable the identification of essential genes and metabolic pathways, facilitating the development of high-yielding, stress-tolerant varieties. Tools like genome-wide association studies and quantitative trait loci mapping have refined gene discovery, accelerating crop improvement. In combination with genome editing technologies like CRISPR/Cas9, multiomics empowers breeders to create climate-resilient and nutrient-enhanced crops. This chapter delves into the transformative potential of multiomics in plant breeding, emphasizing its role in linking genotype and phenotype through systems biology. Additionally, it addresses challenges and future directions for integrating these advanced methods into agricultural practices, ensuring sustainable solutions in a rapidly changing global environment. The collaboration between multiomics and modern breeding technologies underscores the importance of interdisciplinary research in tackling the pressing demands of food security.

6.1 Introduction

The global food scenario underscores the urgent need for improved crop strategies, particularly in light of challenges like climate change, drought, and population growth. These factors out a strain on food production, highlighting the importance of developing resilient crop varieties that can sustain agricultural output despite environmental stressors (Saleem *et al.*, 2024). To meet these demands, genetic gains in crop improvement are essential, focusing on developing high-yielding varieties that can thrive under changing climatic conditions. While progress has been made, there are still gaps in achieving optimal resilience and productivity, necessitating further research.

Since the beginning of agriculture, ensuring food security has been a fundamental

*Corresponding author: bishawajitpbg@gmail.com

 From Gene Discovery to Climate-Resilient Crops: Omics in Crop Improvement (eds S. Pandey *et al.*)
DOI: 10.1079/9781800628717.0006

concern. Plant breeding, one of the oldest agricultural practices, has evolved alongside human civilization, initially focusing on selecting crops with desirable traits like taste, yield, and resistance to environmental conditions (Lemke *et al.*, 2022). The discovery of Mendel's laws in 1866 and the identification of molecular markers in the 1980s were pivotal in advancing plant breeding, paving the way for molecular breeding techniques like marker-assisted selection (MAS) and genetically modified (GM) crops (Ricroch, 2021). These breakthroughs transitioned plant breeding from relying solely on phenotype to integrating both genotype and phenotype. Modern techniques such as MAS and genomic selection further enhance genetic gain, accelerating improvements in crop for yield, quality, stress tolerance and resilience (Anilkumar, *et al.*, 2022).

The challenge of feeding a rapidly growing global population is compounded by climate change and the decline of fertile land. Staple crops like rice, corn (maize), and wheat are increasingly threatened by extreme weather, nutrient imbalances, and biotic stressors, leading to declining food productivity. In response, modern plant breeding has embraced technological innovations, particularly multiomics approaches, which integrate genomics, transcriptomics, proteomics, metabolomics, ionomics, and phenomics (Hina *et al.*, 2024). As shown in Fig. 6.1, multiomics starts from the DNA level: the genome represents the entire genetic material, which is studied through genomics. The next level is the transcriptome, which includes all RNA transcripts produced by the genome, analyzed via transcriptomics. Moving further, the proteome encompasses all the proteins expressed, and proteomics is the discipline that studies it. The metabolome represents the small molecules involved in metabolism, studied by metabolomics. The ionome refers to the elemental composition of an organism, with ionomics as the field of study. Finally, the phenome represents the observable traits and characteristics, which are analyzed through phenomics. Each of these levels provides a comprehensive understanding of biological systems (Marshall *et al.*, 2022). It is enabled by next-generation sequencing (NGS) and high-throughput technologies; these methods have advanced our understanding of crop growth, yield, and stress responses. Applied to key crops like rice, wheat, and maize, multiomics reveals connections between genomes and phenotypes under varying conditions, helping to develop more resilient crop varieties.

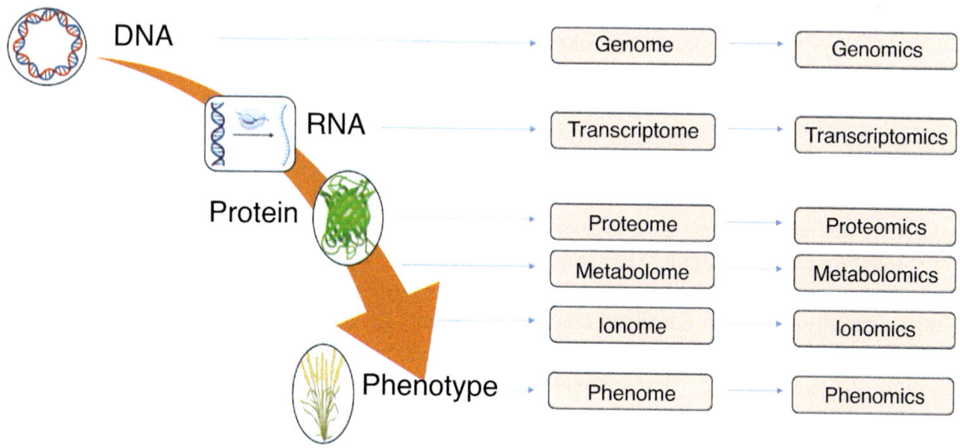

Fig. 6.1. The hierarchical layers of biological information and their associated omics fields. It begins with the genome, studied by genomics, followed by the transcriptome (RNA transcripts) analyzed by transcriptomics. Next is the proteome (all proteins) studied in proteomics, and the metabolome examined through metabolomics. The ionome (elemental composition) is studied by ionomics, and the phenome (observable traits) is analyzed through phenomics. Each level offers insights into biological systems. Author's own image.

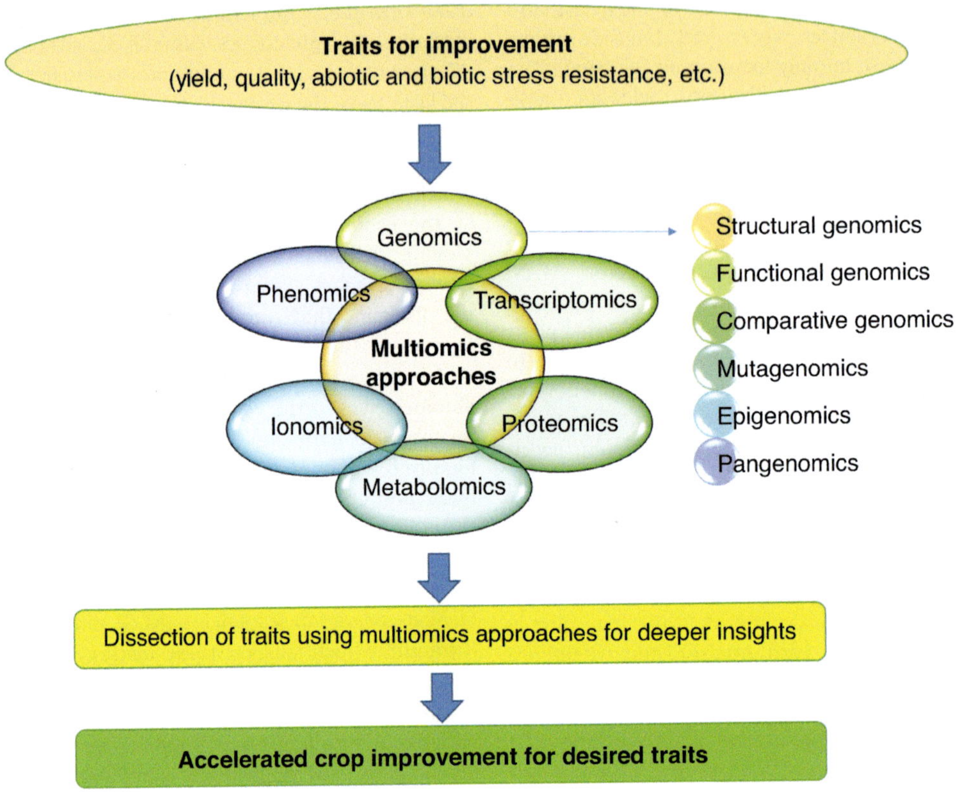

Fig. 6.2. Multiomics approaches used for crop improvement. Author's own image.

Multiomics technologies offer detailed insights into the complex biological mechanisms underlying crop traits, identifying how these traits interact with environmental stressors such as drought and disease. Integration of different multiomics approaches (Fig. 6.2) enables researchers to pinpoint key genetic loci and novel targets for crop improvement (Scossa *et al.*, 2021; Mahmood *et al.*, 2022). Furthermore, combining multiomics with advanced breeding techniques like CRISPR and marker-assisted selection accelerates the development of climate-resilient varieties, essential for meeting the challenges of rising global food demand and changing climatic conditions (Saad, *et al.*, 2022). By optimizing plant breeding strategies and leveraging beneficial traits from crop wild relatives, multiomics offers innovative solutions for sustainable agriculture, ultimately bolstering global food security.

6.2 Genomics

Genomics is the study of an organism's complete set of nucleic acids (DNA or RNA), encompassing all its genes. The term was introduced by Thomas Roderick in 1986. This field covers various areas such as structure, function, evolution, mapping, epigenomics, mutagenomics, and genome editing (Kulski, 2016). Recent advancements in molecular biology, particularly NGS, have revolutionized high-throughput genome sequencing, gene expression analysis, and genomic characterization. NGS involves isolating genomic DNA, amplifying it using the polymerase chain reaction (PCR), sequencing, and assessing the sequence's integrity. The sequencing and assembly of DNA, followed by structural and functional gene annotation, enable large-scale studies on gene activities and their interactions at both cellular and organismal levels (Satam *et al.*, 2023).

Structural genomics focuses on sequence polymorphism and chromosomal organization, aiding in the construction of physical and genetic maps to identify traits of interest, particularly in crop breeding. Functional genomics explores gene functions in trait regulation. Epigenomics examines genomic-level changes, such as histone modifications and DNA or RNA methylations, which affect gene expression. Mutagenomics focuses on genetic changes from mutations, while pangenomics analyzes the complete set of genes within a species. These emerging areas offer valuable insights for crop science and genetic improvement, playing a crucial role in enhancing crop breeding efficiency by elucidating genetic variation and improving desirable traits in crop species (Mahmood *et al.*, 2022).

6.2.1 Structural genomics

Structural genomics, which aims to understand the physical structure of genomes, relies heavily on molecular markers for tagging and mapping genes of interest. These markers are critical in crop breeding programs for improving traits such as yield, stress tolerance, and disease resistance. Techniques like restriction fragment length polymorphisms (RFLP), a non-PCR-based method, detect DNA polymorphisms by hybridizing labeled probes to digested DNA fragments. PCR-based techniques, such as random amplified polymorphic DNA (RAPD), amplified fragment length polymorphisms (AFLP), and single nucleotide polymorphisms (SNPs), have also become essential tools for precise gene mapping (Akpınar *et al.*, 2013).

The advent of NGS technologies has revolutionized the identification and application of SNPs, making them more efficient to use in crop genomics. With NGS, structural genomics can generate large-scale sequence data, facilitating advanced approaches like quantitative trait loci (QTL) mapping and genome-wide association studies (GWAS). QTL mapping links complex phenotypes to specific genotypes, helping breeders focus on genes associated with critical traits. Many crop varieties have been developed through the introgression of QTLs for traits such as quality, and resistance to biotic and abiotic stresses, using genomics-assisted breeding, as shown in Table 6.1. GWAS, on the other hand, identifies genetic variants linked to specific traits across populations, providing a broader understanding of trait inheritance (Edwards *et al.*, 2013). These methods, when combined with genomic selection (GS), enable breeders to accelerate crop improvement by predicting the performance of different genetic variants, thus optimizing breeding outcomes. Through the integration of molecular markers, NGS, QTL mapping, and GWAS, structural genomics plays a crucial role in enhancing crop breeding programs, ensuring the development of high-yielding and stress-tolerant varieties for sustainable agriculture.

6.2.2 Genomic selection

Genomic selection has emerged as a powerful strategy in breeding programs, driven by advances in crop genome sequencing and the availability of high-density genome-wide marker systems. Unlike traditional marker-assisted selection (MAS), GS relies on genome-estimated breeding values (GEBVs), allowing breeders to select genotypes based on information from marker alleles spanning the entire genome. This approach significantly enhances selection efficiency (Krishnappa *et al.*, 2021). The concept of GS was first proposed by Meuwissen *et al.* (2001) for dairy cattle breeding, where it has been highly effective, resulting in genetic gains per generation that are double those of traditional breeding methods. Over time, the application of GS has expanded to crop plants and tree species due to its potential for faster and more precise breeding advancements. Fig. 6.3 illustrates the basic methodology involved in genomic selection, highlighting the key steps from data collection to trait prediction and selection. In wheat, GS has been applied to predict agro-morphological and quality traits. Ward *et al.* (2019) analyzed 329 genotypes using 7748 SNP markers, utilizing the G-BLUP model. Their study achieved a prediction accuracy ranging from 0.29 to 0.67, demonstrating moderate reliability in predicting wheat traits

Table 6.1. Crop varieties developed in India through genomic approaches. Author's own table.

S/N	Trait for improvement	Variety/ genotype	QTL/gene introgressed	Molecular markers used
Rice				
1	Bacterial blight resistance	PR 121	Xa4, xa13 and Xa21	SSR
2	Bacterial blight resistance	Pusa Basmati 1728	PR 121	SSR and STS
3	Bacterial blight resistance	CR Dhan 800	xa5, xa13 and Xa21	STS
4	Bacterial blight resistance	PR 127	Xa45(t)	STS
5	Bacterial blight resistance	DRR Dhan 59	Xa33	SSR
6	Blast resistance	Pusa 6 (Pusa 1612)	Pi2 and Pi54	SSR
7	Blast resistance	Pusa Basmati 1637	Pi9	SSR
8	Blast resistance	Pusa Samba 1850	Pi1, Pi54 and Pita	SSR and STS
9	Bacterial blight and blast resistance	DRR Dhan 62	Xa21, xa13, xa5, Pi2 and Pi-54	SSR
10	Blast resistance	Pusa Basmati 1885	Xa21, xa13, Pi2 and Pi54	SSR
11	Blast resistance	Pusa Basmati 1886	Xa21, xa13, Pi2 and Pi54	SSR
12	Bacterial blight resistance and seedling stage salinity tolerance	DRR Dhan 58	Xa21, xa13, xa5 and qSaltol	SSR
13	Bacterial blight resistance and low soil phosphorous tolerance	DRR Dhan 6	Xa21, xa13, xa5 and qPup1	SSR
14	Submergence tolerance	Swarna Sub 1	qSub1	SSR
15	Submergence tolerance	IR 64 Sub1	qSub1	SSR
16	Submergence tolerance	CO 43 Sub 1	qSub1	InDel
17	Drought tolerance	IR 64 Drt1 (DRR Dhan 42)	qDTY2.2 and qDTY4.1	SSR
18	Submergence and drought tolerance	DRR Dhan 50	qSub1, qDTY2.1 and qDTY3.1	SSR
19	Submergence and drought tolerance	CR Dhan 802 (Subhash)	qSub1, qDTY1.1 and qDTY2.1	SSR
20	Herbicide (Imazethapyr) tolerance	Pusa Basmati 1979	AHAS	SSR
21	Herbicide (Imazethapyr) tolerance	Pusa Basmati 1985	AHAS	SSR
Wheat				
22	Stripe rust resistance	PBW 761 (Unnat PBW 550)	Yr15	SSR
23	Stripe rust resistance	PBW 752	Yr10	
24	Stripe & leaf rust resistance	PBW 723 (Unnat PBW 343)	Yr17, Yr40, Lr37 and Lr57	SSR and CAPS
25	Stripe & leaf rust resistance	PBW 771	Yr40 and Lr57	CAPS and SSR

Continued

Table 6.1. Continued

S/N	Trait for improvement	Variety/ genotype	QTL/gene introgressed	Molecular markers used
Maize				
26	Lysine and tryptophan	Vivek QPM9	opaque2	SSR
27	Lysine and tryptophan	Pusa HM4 Improved	opaque2	SSR
28	Lysine and tryptophan	Pusa HM8 Improved	opaque2	SSR
29	Lysine and tryptophan	Pusa HM9 Improved	opaque2	SSR
30	Provitamin-A	Pusa Vivek QPM9 Improved	crtRB1	InDel
31	Provitamin-A	Pusa Vivek Hybrid-27 Improved	crtRB1	InDel
32	Provitamin-A	Pusa HQPM-5 Improved	crtRB1 and lcyE	InDel
33	Provitamin-A	Pusa HQPM-7 Improved	crtRB1 and lcyE	InDel
34	Provitamin-A	Pusa HQPM-1 Improved	crtRB1 and lcyE	InDel
35	Provitamin-A	Pusa Biofortified Maize Hybrid-1	crtRB1 and lcyE	InDel
Pearl Millet				
36	Downy mildew resistance	HHB 67 Improved	qRSg1 and qRSg4	RFLP
37	Downy mildew resistance	HHB 67 Improved 2	qRSg3.1, qRSg4.2 and qRSg6.1	SSR
Chickpea				
38	Fusarium wilt resistance	Super Annigeri-1	foc4	SSR
39	Fusarium wilt resistance	Pusa Chickpea 20,211	foc1, foc3, foc4 and foc5	SSR
40	Fusarium wilt resistance	IPCMB 19–3 (Samriddhi)	foc2	SSR
41	Drought tolerance	Pusa Chickpea 10,216	QTL hotspot on LG4	SSR
42	Drought tolerance	Pusa Chickpea 4005	QTL hotspot on LG4	SSR
43	Drought tolerance	IPCL4-14	QTL hotspot on LG4	SSR
Soybean				
44	Kunitz trypsin inhibitor free	NRC 127	Null allele of KTi3	STS and SSR
45	Kunitz trypsin inhibitor free	MACSNRC 1667	Null allele of KTi3	STS and SSR
46	Less beany flavour	NRC 132	Null allele of lox2	STS and SSR
47	Kunitz trypsin inhibitor free & less beany flavour	NRC 142	Null allele of KTi3 and lox2	STS and SSR
48	YMV resistance	NRCSL 1	Rymv	SSR
49	Early maturity	NRC 138	Null allele of E1	STS and SSR
Groundnut				
50	Oleic acid	Girnar 4	ahFAD2a and ahFAD2b	AS-PCR and CAPS

Continued

Table 6.1. Continued

S/N	Trait for improvement	Variety/ genotype	QTL/gene introgressed	Molecular markers used
51	Oleic acid	Girnar 5	ahFAD2a and ahFAD2b	AS-PCR and CAPS

SSR, simple sequence repeats; STS, sequence tagged site; InDel, insertion-deletion marker; CAPS, cleaved amplified polymorphic sequence; RFLP, restriction fragment length polymorphism; AS-PCR, allele-specific PCR.

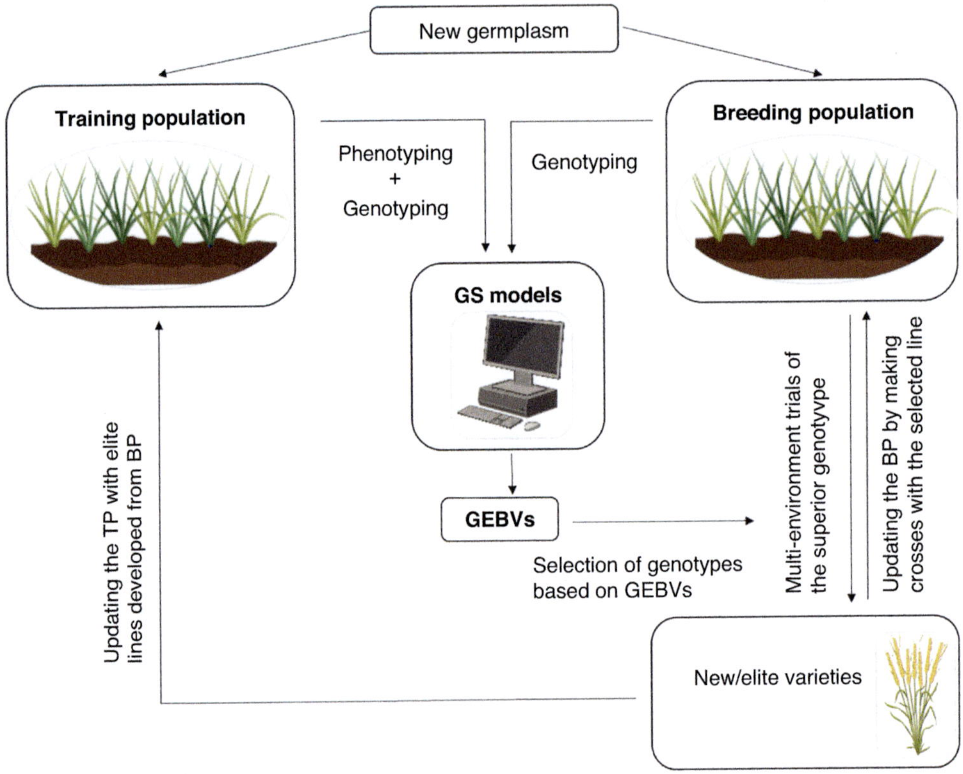

Fig. 6.3. Schematic representation of the methodology involved in genomic selection (GS). BP, breeding population; GEBVs, genome-estimated breeding values; TP, training population. Author's own image.

based on genetic data. Similarly, in rice, Spindel *et al.* (2015) employed GS to predict grain yield, days to 50% flowering, and plant height in 369 elite lines using 73,147 SNPs. The RR-BLUP model yielded prediction accuracies between 0.31 and 0.63, offering valuable insights for improving these key agronomic traits in rice. Maize breeding has also benefited from GS. Massman *et al.* (2013) evaluated 479 maize hybrids using 669 SNP markers, applying both BLUP and RR-BLUP models to predict traits such as grain yield, moisture content, root lodging, and stalk lodging. With prediction accuracies ranging from 0.66 to 0.90, the study highlighted the high reliability of GS for maize, especially in enhancing grain yield and lodging resistance. These studies underscore the significant potential of GS across different crops, where the combination of genetic markers and predictive models can substantially improve the accuracy and efficiency of selection in breeding programs.

6.2.3 Genome-wide association studies

In genome-wide association studies (GWAS), SNPs markers used for genotyping are distributed across the entire genome, allowing researchers to evaluate all loci involved in controlling traits with observable variation in a single analysis. Originally developed for human disease genetics, GWAS has proven highly successful in this field. In plants, GWAS initially focused on single feature polymorphism, recombination, and linkage disequilibrium (LD). However, its application has since expanded to various crop species (Tibbs Cortes *et al.*, 2021). GWAS identifies marker-trait associations by leveraging LD between markers and the genes or QTLs of interest. The main objective of GWAS is to link genomic variants to phenotypic traits. This can be done using a variety of mathematical approaches, and numerous software packages have been developed to conduct GWAS and interpret the results. Despite its utility, GWAS has faced criticism, including concerns that it could implicate the entire genome in disease predisposition and that many association signals may reflect variants or genes with little direct biological relevance to the traits under study. However, GWAS has been widely used in crops to gain insights into the genetic architecture of important traits (Sebastiani *et al.*, 2009).

6.2.4 Functional genomics

Functional genomics aims to understand gene function, structure, and regulation by linking genotypes with phenotypes. Using high-throughput techniques, it explores gene expression and interactions, providing insights for crop improvement. Key approaches include both sequence-based and hybridization-based methods (Wang *et al.*, 2023). Sequencing-based approaches like expressed sequence tags (ESTs) allow rapid, cost-effective gene discovery in crops such as maize, rice, and wheat. Despite some limitations, EST sequencing identifies differentially expressed genes under various conditions. Serial analysis of gene expression (SAGE) and its variants like SuperSAGE have been used to quantify transcript abundance, while massively parallel signature sequencing (MPSS) offers higher specificity and throughput, making it especially useful for species without fully sequenced genomes (Akpınar *et al.*, 2013).

Hybridization-based approaches involve microarray techniques to assess gene expression through pre-designed probes. While widely applied to study stress responses in crops like wheat and barley, limitations such as cross-hybridization and noise can affect data accuracy. These methods also struggle to capture subtle transcriptional changes in complex tissues. Overall, functional genomics provides essential tools for linking gene function with traits of interest, aiding in the discovery of stress-tolerant genes, the development of functional markers, and the identification of genetic variants for crop improvement (Roychowdhury *et al.*, 2023).

6.2.5 Comparative genomics

Comparative genomics is a valuable tool for studying genome organization and evolution by comparing genetic sequences across species, particularly in plants. It has revealed extensive genome conservation, especially in important crops like grasses, despite variations in genome size (Akpınar *et al.*, 2013). Techniques like the 'genome zipper' predict gene order in less-studied species by comparing them with fully sequenced genomes such as rice and sorghum. While it uncovers evolutionary relationships and large-scale genomic changes, some species-specific features, like recent genetic rearrangements, require complete reference genomes. Efforts to sequence such genomes, particularly for crops like wheat and barley, are ongoing to enhance understanding and crop improvement (Roychowdhury *et al.*, 2023).

6.2.6 Mutagenomics

Mutagenomics is an emerging field that focuses on studying mutational events within genomes using advanced genomics technologies. Techniques like targeted induced local lesions in genomes (TILLING), high-resolution melt (HRM), and microarray analysis allow researchers to identify mutations and their effects on gene functions. Originally developed for functional genomics, TILLING has become a key tool in crop breeding, enabling the identification of mutants with desirable traits, such as improved oil composition in soybean and mutations in crops like rice, wheat, and tomato (Yang *et al.*, 2021).

This approach enables both forward and reverse genetic studies, offering insights into gene functions and crop improvement by inducing targeted mutations. TILLING by Sequencing (TbyS) has further enhanced mutation detection through next-generation sequencing, making it a high-throughput method for screening crop varieties. MutMap and other techniques have also advanced the study of mutagenesis by identifying key mutations without the need for outcrossing (Kumar *et al.*, 2017).

Applications of mutagenomics extend to trait improvement in crops, such as increasing oleic acid content in soybean or improving starch quality in potatoes. The CRISPR/Cas9 system has revolutionized mutagenesis, enabling precise genetic modifications, such as enhancing resistant starch in rice. Overall, mutagenomics is a powerful tool for enhancing crop quality, nutritional content, and resistance to environmental stresses (Penna and Jain, 2017).

6.2.7 Epigenomics

Epigenomics investigates heritable changes in gene expression not caused by changes in the DNA sequence. This field integrates environmental influences on genetic regulation, allowing for the evaluation of epigenetic modifications related to stress responses. Techniques such as bisulfite sequencing and chromatin immunoprecipitation sequencing (ChIP-Seq) are employed to study DNA methylation and histone modifications, which are critical for understanding plant resilience to abiotic and biotic stress (Varotto *et al.*, 2022).

6.2.8 Pangenomics

Pangenomics refers to the complete genomic makeup of a species, including core and dispensable genes. Recent advancements in sequencing technology have facilitated the identification of these genes in various crop species, highlighting their roles in maintaining genetic diversity and improving quality traits. Pangenomic studies have revealed that the dispensable genome is often enriched with genes associated with disease resistance and abiotic stress tolerance, indicating its importance in crop breeding programs (Tay Fernandez *et al.*, 2022).

In summary, genomics provides essential tools and methodologies for crop improvement through enhanced understanding of genetic architectures, functional roles of genes, and the intricate relationships between genotype and phenotype. The integration of structural and functional genomics, alongside advanced techniques in epigenomics and pangenomics, offers promising avenues to address the challenges of sustainable agriculture in a changing environment.

6.3 Transcriptomics

Transcriptomics is the study of the complete set of RNA transcripts (the transcriptome) produced by an organism's genome in a cell or tissue. It includes both messenger RNAs (mRNAs) and non-coding RNAs (ncRNAs) such as microRNAs (miRNAs), long non-coding RNAs (lncRNAs), and circular RNAs (circRNAs). These RNAs play key roles in gene expression and regulation, making transcriptomics a crucial tool in understanding how plants respond to various conditions. Initially, transcriptome analysis relied on techniques like cDNA-AFLP and differential display-PCR (DD-PCR), but advancements in high-throughput sequencing, including RNA-Seq, have significantly improved the accuracy of transcriptome profiling (Wang *et al.*, 2009). RNA-Seq has been instrumental in identifying genes related to stress responses in crops. For example, in rice and wheat, transcriptomics has uncovered the role of transcription factors like WRKY in regulating responses to drought and heat stress (Chen *et al.*, 2017). Non-coding RNAs have also become promising targets for crop improvement. miRNAs are involved in post-transcriptional modifications, while lncRNAs and circRNAs function as regulators of splicing, transcription, and gene expression.

The advent of single-cell RNA sequencing (scRNA-Seq) has enabled researchers to study gene expression at the cellular level, uncovering heterogeneity in plant cell types. This technique has been used to explore polyploidization events and reproductive biology in *Arabidopsis*, providing new insights into plant genetics and breeding (Song *et al.*, 2020).

Recent innovations like spatially resolved transcriptomics and alternative splicing (AS) transcriptomics are further expanding our understanding of gene expression. For instance, AS transcriptomics in rice, maize, and sorghum has identified splicing factors that control responses to abiotic stresses (Bashir *et al.*, 2019). These approaches show great potential for improving crops by enhancing their resilience to environmental stressors.

6.4 Proteomics

Proteomics is the study of the entire set of proteins (proteome) expressed by a cell, organ, or organism at a specific time. Since proteins are key regulators of stress responses, understanding changes in plant proteomes is crucial. Proteins are influenced by factors such as development, environmental conditions, and stress, which result in different protein sets being expressed. This makes certain proteins potential biomarkers for specific stressors like drought or salinity (Barkla, 2016).

Proteomic research enables the detection, characterization, and study of proteins' activity profiles, post-translational modifications (PTMs), and protein–protein interactions. For example, in cereal crops like wheat, rice, and maize, proteomic analyses have revealed changes in protein expression under drought stress, providing insight into the plants' stress adaptation mechanisms (Mustafa and Komatsu, 2021). Common techniques include protein electrophoresis (such as 2-DE and DIGE) and mass spectrometry (MS), with MS methods like LC-MS/MS and MALDI-MS being widely used for protein identification (Wittmann-Liebold *et al.*, 2006).

Proteomics also includes the study of post-translational modifications like phosphorylation and glycosylation, which are vital in stress response (Yang, 2022). Techniques such as phosphoproteomics have identified key phosphorylation sites, helping to understand protein function under stress conditions. For instance, under salt stress, comparative proteome analyses in *Medicago* species revealed the expression of differential root proteins associated stress response (Long *et al.*, 2016).

Advanced techniques like SILAC and iTRAQ enable quantitative proteome analysis, helping to measure protein expression changes under environmental stress (Neilson *et al.*, 2011). For example, quantitative proteomics has identified salt-alkali responsive proteins involved in metabolic processes and reactive oxygen species (ROS) homeostasis in crops (Ling *et al.*, 2022). These insights are valuable for improving crop resilience to environmental stressors through breeding programs.

6.5 Metabolomics

Metabolomics is a powerful omics technology that focuses on studying the metabolites, small molecules essential for the plant's physiological activities. Plants produce a wide range of metabolites, including primary ones involved in basic functions like growth, and secondary ones that help them cope with environmental stresses (Manickam *et al.*, 2023). Primary metabolites, such as sugars and amino acids, are critical for normal plant development, while secondary metabolites, like flavonoids and phenolics, are involved in stress responses.

The diversity of metabolites in the plant kingdom is vast, with at least 200,000 unique compounds identified from 391,000 known plant species worldwide. Their concentrations and types vary widely across species and environmental conditions (Yeshi *et al.*, 2022). By studying the metabolic profiles of plants, researchers gain insight into how plants adapt to stress, such as drought, salinity, or heat. For example, trehalose, a sugar, helps plants maintain membrane integrity under drought, while polyamines like spermidine and spermine aid in stress tolerance.

Metabolomics provides an advantage over genomics, transcriptomics, and proteomics because it represents the final products of gene and protein activity, giving a more direct link to the plant's phenotype. Techniques such as liquid chromatography-mass spectrometry (LC–MS) and nuclear magnetic resonance (NMR) are commonly used for analyzing plant metabolites. These methods allow researchers to detect stress-responsive metabolites and understand how plants regulate their metabolism under adverse conditions (Jorge *et al.*, 2016).

Examples of crop improvement using metabolomics include studies on rice, maize, and sesame. In maize, metabolites like phenolics and phenylpropanoids have been linked to biotic and abiotic stress resistance. Similarly, in rice and barley, drought stress has been shown to increase the production of specific secondary metabolites, enhancing the plants' survival (Yadav *et al.*, 2021). In sesame, metabolomic and transcriptomic analyses identified amino acid and ABA metabolism pathways that help in drought tolerance. Such insights are essential for breeding crops better suited to changing environmental conditions.

By integrating metabolomics with other omics platforms, such as genomics and proteomics, scientists can develop crops with improved stress resilience and nutritional value. This approach allows for a more comprehensive understanding of plant biology, providing valuable information for future crop development.

6.6 Phenomics

Phenomics involves the comprehensive study of phenotypes, which are the observable characteristics of an organism influenced by genes and environmental factors. This field employs high-throughput methods to analyze an organism's morphological, physiological, and biochemical traits (Houle *et al.*, 2010). In plants, phenomics connects growth, performance, and composition with genetic, epigenetic, and environmental conditions, providing insights into how these factors produce desirable traits. By integrating phenomics with other omics technologies, researchers can uncover cellular networks that shape final phenotypic outcomes.

Plant phenotypic traits, which result from gene–environment interactions (G × E), can be analyzed across various environmental conditions, revealing relationships between traits and abiotic stresses. Two approaches in phenomics include forward phenomics, which uses high-throughput automated tools to identify valuable traits, and reverse phenomics, which investigates traits in detail to uncover underlying mechanisms (Mir *et al.*, 2019).

Phenomics can be applied at multiple scales, from the cellular level to entire ecosystems, with recent advances focusing on large-scale, non-invasive imaging technologies for precise and rapid phenotyping. High-throughput phenotyping (HTP) includes methods such as visible light imaging, hyperspectral imaging, infrared imaging, and X-ray tomography (Li *et al.*, 2021). These techniques quantify plant traits like chlorophyll fluorescence, water content, and geographical parameters, and are supported by advanced software systems for data analysis.

In addition to 2D imaging, 3D techniques are increasingly used to measure plant height, dry matter, and crop structure. Other methods include fluorescence imaging for detecting early stress responses, and pulse amplitude modulation (PAM) for measuring fluorescence parameters under stress conditions. Advanced spectroscopy and remote sensing techniques, like unmanned aerial vehicles (UAVs) and hyperspectral satellite data, are used for field phenotyping to assess stress responses, including drought and salinity tolerance. The application of RGB imaging (red, green, and blue) is commonly used to estimate canopy and root systems, while techniques like infrared thermography help evaluate stomatal conductance under stress (Humplík *et al.*, 2015). Additionally, boron toxicity and disease symptoms in crops have been analyzed using digital imaging methods. Phenomics has become an essential tool for crop breeding, especially when integrated with other omics. By automating phenotyping through technologies such as Li-Cor 6400, which measures gas exchange and photosynthesis, and the use of UAVs for remote sensing, researchers are able to improve stress tolerance and crop management.

Challenges in plant phenomics include the need to organize large datasets for future analysis. However, advances like the Internet of Things (IoT)-based CropSight system and ontology-driven information systems (PHIS) are making it easier to store, share, and analyze phenotyping data across different scales. The integration of next-generation sensors and computational tools like artificial neural networks (ANNs) promises further improvements in crop breeding and stress management.

In summary, phenomics plays a crucial role in advancing crop breeding by providing a detailed understanding of how plants respond to environmental stresses. This field's integration with other omics and new technologies is key to developing more resilient and productive crop varieties.

6.7 Ionomics

Ionomics studies the ionome, the total composition of mineral nutrients and trace elements in an organism, focusing on inorganic components in plant systems. It measures elemental profiles to understand how they change due to genetic modifications, physiological stimuli, or environmental conditions. This field helps uncover the role of ions in plant growth, development, and stress responses, offering insights into gene function and networks (Salt *et al.*, 2008). Key analytical tools in ionomics include ICP-MS, ICP-OES, NAA, and XRF. These high-throughput techniques have been applied to study ionomes in crops like rice and maize, identifying QTLs linked to mineral content and stress adaptation. Ionomics has also helped reveal mechanisms of ion uptake and regulation under abiotic stresses, such as the role of sodium (Na^+) in salinity stress and calcium's (Ca^{2+}) mitigating effects (Ali *et al.*, 2021).

Research has used ionomics to explore how genetic and environmental factors influence elemental accumulation in plants, with applications in crops like maize, barley, soybean, and fruit species. Integrating ionomics with genomics and metabolomics enables the discovery of genes and pathways that enhance plant resistance to environmental stress, contributing to crop improvement efforts.

6.8 Integration of Multiomics Approaches

The integration of genomics, transcriptomics, and other omics technologies has revolutionized the study of complex traits in crop science. Traditional approaches like GWAS and QTL mapping often yield large candidate gene intervals, making it difficult to identify key genes (Scossa *et al.*, 2021). By incorporating transcriptomics, researchers can analyze gene expression data to refine these intervals, enabling the discovery of critical genes for traits such as drought resistance in rice and seed germination in maize. This integrated approach enhances the precision of gene identification, expediting crop improvement.

Advancements in sequencing technologies have further propelled crop research. High-quality reference genomes for crops like rice, maize, and cotton, combined with tools like bulked segregant analysis sequencing (BSA-seq) and genotyping-by-sequencing (GBS), have enabled the identification of genetic variations associated with important agronomic traits (Riaz *et al.*, 2021). While traditional methods have limitations, the integration of genetic and phenotypic data—facilitated by high-throughput phenotyping platforms—has led to more accurate dissection of complex traits. Technologies such as UAVs and hyperspectral imaging have improved phenotypic data collection, accelerating the discovery of genes linked to traits like drought tolerance and plant growth.

The integration of genomics, transcriptomics, and metabolomics has emerged as a powerful tool for crop improvement. Metabolomics sheds light on the metabolic pathways and genetic variations that influence complex traits (Litvinov *et al.*, 2021). By combining metabolomics with transcriptomics, researchers have identified genes involved in key metabolic processes, such as drought tolerance in maize and seed coat color in *Brassica rapa*. This integrated approach has also revealed metabolic changes in tomatoes during breeding, offering insights into how trait selection alters metabolite profiles, further advancing the understanding of crop quality and yield (Zhang *et al.*, 2022).

Systems biology complements these multiomics approaches by modeling gene, protein, and metabolite interactions to predict phenotypic responses. This approach has been applied to study stress tolerance and yield-related traits in crops like rice and maize (Roychowdhury *et al.*, 2023). Systems biology enables both top-down (phenotype to genotype) and bottom-up (genotype to phenotype) modeling, providing a comprehensive understanding of how genes and proteins function under various conditions. Integrating these models with omics data has led to the identification of stress tolerance networks, aiding the development of more resilient crop varieties (Kumar *et al.*, 2015).

Overall, the combination of genomics, transcriptomics, metabolomics, and systems biology creates a robust framework for understanding complex traits in crops. This multiomics approach not only uncovers the molecular mechanisms behind important traits but also accelerates the development of high-performing and resilient crop varieties, essential for addressing global food security challenges.

6.9 Bioinformatics Tools for Multiomics and Their Integration

Omics technologies have transformed our understanding of crop plants by revealing the genetic and molecular mechanisms behind plant growth, development, and responses to environmental stresses. With advancements in high-throughput sequencing and decreasing costs, vast amounts of molecular data are being generated, leading to the establishment of major public databases like the National Genomics Data Center (NGDC) and the European Bioinformatics Institute (EBI). Table 6.2 represents various databases used for multiomics and omics approaches in different crops. These databases store genomic, epigenomic, transcriptomic, proteomic, and metabolomic data, which are invaluable for crop improvement. Genomic data, for example, help in identifying genes responsible for desirable traits like disease resistance and high yield, while transcriptomic data offer insights into gene expression patterns across different tissues and conditions (Chao *et al.*, 2024).

Epigenomic data add another layer of complexity by revealing how changes in chromatin structure influence gene expression, providing deeper insights into the regulation of traits that are important for crop breeding. Integrating epigenomic data with other omics datasets, despite its challenges, allows researchers to identify key regulatory genes and pathways. This information is crucial for developing crop varieties with enhanced traits such as stress tolerance, higher yield, and improved nutritional content. Technologies like ATAC-Seq and ChIP-seq have made the collection of epigenomic data more accessible, while databases such as RiceENCODE and FruitENCODE are instrumental in studying crop-specific regulatory mechanisms.

In addition to genomic and epigenomic data, proteomics and metabolomics play a vital role in understanding crop traits at the protein and metabolite levels. Proteomic data help researchers identify proteins involved in essential metabolic pathways or associated with traits like stress response, flavor, and nutritional value. Similarly, metabolomic analysis provides insights into the biochemical cascades that link genome, transcriptome, and proteome to phenotype, aiding the development of crops with superior nutritional content or stress tolerance (Yan *et al.*, 2022). Databases such as the Plant Proteome Database and the Plant Metabolic Network serve as key resources for proteomics and metabolomics research, offering a comprehensive view of crop plant metabolism.

The integration of omics data through bioinformatics tools has become indispensable for crop research. Bioinformatics facilitates the storage, analysis, and interpretation of multiomics data, providing essential insights into crop biology. Tools like Gramene, Plant Reactome, and PlantTFDB offer multiomics-based integration, while specialized software packages like STRING and Gromacs enable the analysis of protein interactions and simulations. By leveraging these tools and databases, researchers can accelerate the development of new crop varieties that are more productive, resilient, and nutritionally valuable, thus contributing to global food security.

Table 6.2. Databases for multi-omics and various omics approaches in different crops. Author's own table.

S/N	Crop	Database	Omics type	Link	References
1	*Brassica napus*	BnIR	Multiomics	http://yanglab.hzau.edu.cn/BnIR	Yang et al., 2023a
2	Cotton	CottonMD	Multiomics	http://yanglab.hzau.edu.cn/CottonMD/	Yang et al., 2023b
3	Cotton	COTTONOMICS	Multiomics	http://cotton.zju.edu.cn/	Dai at al., 2022
4	Wheat	WheatOmics	Multiomics	http://wheatomics.sdau.edu.cn/	Ma et al., 2021
5	Maize	ZEAMAP	Multiomics	http://www.zeamap.com/	Gui et al., 2020
6	Millets	Milletdb	Multiomics	http://milletdb.novogene.com/	Sun et al., 2023
7	Soybean	SoyOmics	Multiomics	https://ngdc.cncb.ac.cn/soyomics/index	Liu et al., 2023
8	Sesame	Sinbase 2.0	Multiomics	http://www.sesame-bioinfo.org/Sinbase2.0	Wang et al., 2021
9	Bottle gourd	GourdBase	Multiomics	http://www.gourdbase.cn/	Wang et al., 2018
10	Rice	CARMO	Multiomics	http://bioinfo.sibs.ac.cn/carmo	Wang et al., 2015
11	Rice	UniVIO	Multiomics	http://univio.psc.riken.jp/	Kudo et al., 2013
12	Rice	Oryzabase	Genomics	https://shigen.nig.ac.jp/rice/oryzabase/	Kurata et al., 2006
13	Rice	OryGenesDB	Genomics	https://orygenesdb.cirad.fr/index.html	Droc et al., 2006
14	Rice	RAP-DB	Genomics	https://rapdb.dna.affrc.go.jp/	Sakai et al., 2013
15	Rice	PLAZA 4.0	Functional, evolutionary and comparative genomics	https://bioinformatics.psb.ugent.be/plaza	Van Bel et al., 2018
16	Legume	LegumeIP	Comparative genomics and transcriptomics	http://plantgrn.noble.org/LegumeIP/	Li et al., 2012
17	Carrot	CarrotDB	Genomic and transcriptomic	http://apiaceae.njau.edu.cn/carrotdb	Xu et al., 2014
18	Rice	RiceXPro	Transcriptomics	https://ricexpro.dna.affrc.go.jp/	Sato et al., 2010
19	Brassica	BrassicaEDB	Transcriptomics	https://brassica.biodb.org/	Chao et al., 2020
20	Little Millet	LMT[db]	Transcriptomics	https://igkv.ac.in/xenom/index.aspx	Shekhar et al., 2023

Continued

Table 6.2. Continued

S/N	Crop	Database	Omics type	Link	References
21	Rice	OMA browser	Proteomics	https://omabrowser.org/oma/home/	Schneider et al., 2007
22	Rice	Panther	Proteomics	http://www.pantherdb.org/	Thomas et al., 2003
23	Multi-species	CropMetabolome	Metabolomics	https://www.cropmetabolome.com/	Shi et al., 2024
24	Soybean	SoyMetDB	Metabolomics	http://soymetdb.org.	Joshi et al., 2010
25	Multi-species	KNApSAcK	Metabolomics	http://www.kazusa.or.jp/komics/en/	Nakamura et al., 2014

6.10 Conclusions and Future Prospects

The advancement of NGS technologies has produced vast amounts of data, requiring an interdisciplinary approach to utilize it effectively for crop improvement. Integrating multiomics technologies like genomics, transcriptomics, proteomics, and metabolomics links genotype to phenotype, helping to identify key genes for crop improvement. Multiomics data provides a systematic framework for crop genetic improvement by considering 'phenotype to genotype,' 'genotype to phenotype,' and genotype–environment interactions. Epigenetic diversity is also important in breeding, while de novo domestication of wild crops can help retain genetic diversity and elite alleles, addressing future agricultural challenges (Kakoulidou *et al.*, 2021). Despite the promise of multiomics, integrating disparate datasets and improving phenotypic prediction remain challenges. Advanced algorithms, including machine learning and deep learning, are needed to predict complex phenotypes and heterosis. To facilitate this, robust data visualization tools, bioinformatics, and cloud computing should be integrated (Farooq *et al.*, 2024).

Single-cell multiomics is an emerging frontier, enabling simultaneous analysis of genomic, transcriptomic, metabolomic, and epigenomic data from individual cells. This technology could revolutionize our understanding of genotype–phenotype transitions at the cellular level (Baysoy *et al.*, 2023). Recent advances in sequencing and phenotyping have improved the identification of valuable genetic variations from diverse germplasm for plant breeding. Omics techniques have been key in identifying genes, proteins, and metabolites associated with critical traits and stress responses. Machine learning enhances phenotypic predictions from GWAS, further supporting crop breeding efforts (Sun *et al.*, 2024).

The integration of multiomics with systems biology and GWAS can improve our understanding of abiotic stress tolerance and enhance crop quality, yield, and resilience. Despite limited research on major crops, these techniques hold great promise for addressing environmental challenges and food security.

In conclusion, multiomics research is shaping the future of crop breeding, providing insights for developing resilient, high-yield varieties. Collaborative efforts among scientists, computational biologists, and breeding companies will drive innovations that meet global food security needs amidst environmental stress.

References

Akpınar, B.A., Lucas, S.J. and Budak, H. (2013) Genomics approaches for crop improvement against abiotic stress. *The Scientific World Journal* 2013(1), 361921. DOI: 10.1155/2013/361921.

Ali, S., Tyagi, A. and Bae, H. (2021) Ionomic approaches for discovery of novel stress-resilient genes in plants. *International Journal of Molecular Sciences* 22(13), 7182. DOI: 10.3390/ijms22137182.

Anilkumar, C., Sunitha, N.C., Harikrishna, S., Devate, N.B. and Ramesh, S. (2022) Advances in integrated genomic selection for rapid genetic gain in crop improvement: A review. *Planta* 256(5), 87. DOI: 10.1007/s00425-022-03996-y.

Barkla, B.J. (2016) Identification of abiotic stress protein biomarkers by proteomic screening of crop cultivar diversity. *Proteomes* 4(3), 26. DOI: 10.3390/proteomes4030026.

Bashir, K., Matsui, A., Rasheed, S. and Seki, M. (2019) Recent advances in the characterization of plant transcriptomes in response to drought, salinity, heat, and cold stress. *F1000Research* 8. DOI: 10.12688/f1000research.18424.1.

Baysoy, A., Bai, Z., Satija, R. and Fan, R. (2023) The technological landscape and applications of single-cell multi-omics. *Nature Reviews Molecular Cell Biology* 24(10), 695–713. DOI: 10.1038/s41580-023-00615-w.

Chao, H., Li, T., Luo, C., Huang, H., Ruan, Y. *et al.* (2020) BrassicaEDB: A gene expression database for Brassica crops. *International Journal of Molecular Sciences* 21(16), 5831. DOI: 10.3390/ijms21165831.

Chao, H., Zhang, S., Hu, Y., Ni, Q., Xin, S. *et al.* (2024) Integrating omics databases for enhanced crop breeding. *Journal of Integrative Bioinformatics* 20(4), 20230012. DOI: 10.1515/jib-2023-0012.

Chen, F., Hu, Y., Vannozzi, A., Wu, K., Cai, H. *et al.* (2017) The WRKY transcription factor family in model plants and crops. *Critical Reviews in Plant Sciences* 36(5–6), 311–335. DOI: 10.1080/07352689.2018.1441103.

Dai, F., Chen, J., Zhang, Z., Liu, F., Li, J. *et al.* (2022) COTTONOMICS: A comprehensive cotton multi-omics database. *Database* baac080. DOI: 10.1093/database/baac080.

Droc, G., Ruiz, M., Larmande, P., Pereira, A., Piffanelli, P. *et al.* (2006) OryGenesDB: A database for rice reverse genetics. *Nucleic Acids Research* 34(suppl_1), D736–D740. DOI: 10.1093/nar/gkj012.

Edwards, D., Batley, J. and Snowdon, R.J. (2013) Accessing complex crop genomes with next-generation sequencing. *Theoretical and Applied Genetics* 126, 1–11. DOI: 10.1007/s00122-012-1964-x.

Farooq, M.A., Gao, S., Hassan, M.A., Huang, Z., Rasheed, A. *et al.* (2024) Artificial intelligence in plant breeding. *Trends in Genetics* 40(10), 891–908. DOI: 10.1016/j.tig.2024.07.001.

Gui, S., Yang, L., Li, J., Luo, J., Xu, X. *et al.* (2020) ZEAMAP, a comprehensive database adapted to the maize multi-omics era. *IScience* 23(6). DOI: 10.1016/j.isci.2020.101241.

Hina, A., Abbasi, A., Arshad, M., Imtiaz, S., Shahid, S. *et al.* (2024) Utilization of multi-omics approaches for crop improvement. *OMICs-based Techniques for Global Food Security* 91–121. DOI: 10.1002/9781394209156.ch5.

Houle, D., Govindaraju, D.R. and Omholt, S. (2010) Phenomics: The next challenge. *Nature Reviews Genetics* 11(12), 855–866. DOI: 10.1038/nrg2897.

Humplík, J.F., Lazár, D., Husičková, A. and Spíchal, L. (2015) Automated phenotyping of plant shoots using imaging methods for analysis of plant stress responses–A review. *Plant Methods* 11, 1–10. DOI: 10.1186/s13007-015-0072-8.

Jorge, T.F., Rodrigues, J.A., Caldana, C., Schmidt, R., van Dongen, J.T. *et al.* (2016) Mass spectrometry-based plant metabolomics: Metabolite responses to abiotic stress. *Mass Spectrometry Reviews* 35(5), 620–649. DOI: 10.1002/mas.21449.

Joshi, T., Yao, Q., Levi, D.F., Brechenmacher, L., Valliyodan, B. *et al.* (2010) SoyMetDB: The soybean metabolome database. In: *2010 IEEE International Conference on Bioinformatics and Biomedicine (BIBM)*, IEEE, pp. 203–208. DOI: 10.1109/BIBM.2010.5706563.

Kakoulidou, I., Avramidou, E.V., Baránek, M., Brunel-Muguet, S., Farrona, S. *et al.* (2021) Epigenetics for crop improvement in times of global change. *Biology* 10(8), 766. DOI: 10.3390/biology10080766.

Krishnappa, G., Savadi, S., Tyagi, B.S., Singh, S.K., Masthigowda, M.H. *et al.* (2021) Integrated genomic selection for rapid improvement of crops. *Genomics* 113, 1070–1086. DOI: 10.1016/j.ygeno.2021.02.007.

Kudo, T., Akiyama, K., Kojima, M., Makita, N., Sakurai, T. *et al.* (2013) UniVIO: A multiple omics database with hormonome and transcriptome data from rice. *Plant and Cell Physiology* 54(2), e9–e9. DOI: 10.1093/pcp/pct003.

Kulski, J.K. (2016) Next-generation sequencing—an overview of the history, tools, and "Omic" applications. *Next Generation Sequencing - Advances, Applications and Challenges* 10, 61964. DOI: 10.5772/61964.

Kumar, A., Pathak, R.K., Gupta, S.M., Gaur, V.S. and Pandey, D. (2015) Systems biology for smart crops and agricultural innovation: Filling the gaps between genotype and phenotype for complex traits linked with robust agricultural productivity and sustainability. *Omics: A Journal of Integrative Biology* 19(10), 581–601. DOI: 10.1089/omi.2015.0106.

Kumar, A.P., McKeown, P.C., Boualem, A., Ryder, P., Brychkova, G. *et al.* (2017) TILLING by Sequencing (TbyS) for targeted genome mutagenesis in crops. *Molecular Breeding* 37, 1–12. DOI: 10.1007/s11032-017-0620-1.

Kurata, N. and Yamazaki, Y. (2006) Oryzabase. An integrated biological and genome information database for rice. *Plant Physiology* 140(1), 12–17. DOI: 10.1093/nar/gkr939.

Lemke, S., Tao, X. and Kushner, G.J. (2022) Assuring the food safety of crops developed through breeding. *ACS Agricultural Science & Technology* 2(6), 1151–1165. DOI: 10.1021/acsagscitech.2c00268.

Li, D., Quan, C., Song, Z., Li, X., Yu, G. *et al.* (2021) High-throughput plant phenotyping platform (HT3P) as a novel tool for estimating agronomic traits from the lab to the field. *Frontiers in Bioengineering and Biotechnology* 8, 623705. DOI: 10.3389/fbioe.2020.623705.

Li, J., Dai, X., Liu, T. and Zhao, P.X. (2012) LegumeIP: An integrative database for comparative genomics and transcriptomics of model legumes. *Nucleic Acids Research* 40(D1), D1221–D1229. Available at: DOI: 10.1093/nar/gkr939.

Ling, L., An, Y., Wang, D., Tang, L., Du, B. *et al.* (2022) Proteomic analysis reveals responsive mechanisms for Saline-Alkali stress in Alfalfa. *Plant Physiology and Biochemistry* 170, 146–159. DOI: 10.1016/j.plaphy.2021.12.003.

Litvinov, D.Y., Karlov, G.I. and Divashuk, M.G. (2021) Metabolomics for crop breeding: General considerations. *Genes* 12(10), 1602. DOI: 10.3390/genes12101602.

Liu, Y., Zhang, Y., Liu, X., Shen, Y., Tian, D. *et al.* (2023) SoyOmics: A deeply integrated database on soybean multi-omics. *Molecular Plant* 16(5), 794–797. DOI: 10.1016/j.molp.2023.03.011.

Ma, S., Wang, M., Wu, J., Guo, W., Chen, Y. *et al.* (2021) WheatOmics: A platform combining multiple omics data to accelerate functional genomics studies in wheat. *Molecular Plant* 14(12), 1965–1968. DOI: 10.1016/j.molp.2021.10.006.

Long, R., Li, M., Zhang, T., Kang, J., Sun, Y. *et al.* (2016) Comparative proteomic analysis reveals differential root proteins in *Medicago sativa* and *Medicago truncatula* in response to salt stress. *Frontiers in Plant Science* 7, 424. DOI: 10.3389/fpls.2016.00424.

Mahmood, U., Li, X., Fan, Y., Chang, W., Niu, Y. *et al.* (2022) Multi-omics revolution to promote plant breeding efficiency. *Frontiers in Plant Science* 13, 1062952. DOI: 10.3389/fpls.2022.1062952.

Manickam, S., Rajagopalan, V.R., Kambale, R., Rajasekaran, R., Kanagarajan, S. *et al.* (2023) Plant metabolomics: Current initiatives and future prospects. *Current Issues in Molecular Biology* 45(11), 8894–8906. DOI: 10.3390/cimb45110558.

Marshall, J.L., Peshkin, B.N., Yoshino, T., Vowinckel, J., Danielsen, H.E. *et al.* (2022) The essentials of multiomics. *The Oncologist* 27(4), 272–284.

Massman, J.M., Gordillo, A., Lorenzana, R.E. and Bernardo, R. (2013) Genomewide predictions from maize single-cross data. *Theoretical and Applied Genetics* 126(1), 13–22. DOI: 10.1007/s00122-012-1955-y.

Meuwissen, T.H., Hayes, B.J. and Goddard, M. (2001) Prediction of total genetic value using genome-wide dense marker maps. *Genetics* 157(4), 1819–1829. DOI: 10.1093/genetics/157.4.1819.

Mir, R.R., Reynolds, M., Pinto, F., Khan, M.A. and Bhat, M.A. (2019) High-throughput phenotyping for crop improvement in the genomics era. *Plant Science* 282, 60–72. DOI: 10.1016/j.plantsci.2019.01.007.

Mustafa, G. and Komatsu, S. (2021) Plant proteomic research for improvement of food crops under stresses: A review. *Molecular Omics* 17(6), 860–880. DOI: 10.1039/D1MO00151E.

Nakamura, Y., Mochamad Afendi, F., Kawsar Parvin, A., Ono, N., Tanaka, K. *et al.* (2014) KNApSAcK metabolite activity database for retrieving the relationships between metabolites and biological activities. *Plant and Cell Physiology* 55(1), e7–e7. DOI: 10.1093/pcp/pct176.

Neilson, K.A., Mariani, M. and Haynes, P.A. (2011) Quantitative proteomic analysis of cold-responsive proteins in rice. *Proteomics* 11(9), 1696–1706. DOI: 10.1002/pmic.201000727.

Penna, S. and Jain, S.M. (2017) Mutant resources and mutagenomics in crop plants. *Emirates Journal of Food & Agriculture (EJFA)* 29(9), 651–657. DOI: 10.9755/ejfa.2017.v29.i9.86.

Riaz, A., Kanwal, F., Börner, A., Pillen, K., Dai, F. *et al.* (2021) Advances in genomics-based breeding of barley: Molecular tools and genomic databases. *Agronomy* 11(5), 894. DOI: 10.3390/agronomy11050894.

Ricroch, A. (2021) The evolution of agriculture and tools for plant innovation. *Plant Biotechnology: Experience and Future Prospects* 3–15. DOI: 10.1007/978-3-030-68345-0_1.

Roychowdhury, R., Das, S.P., Gupta, A., Parihar, P., Chandrasekhar, K. *et al.* (2023) Multi-omics pipeline and omics-integration approach to decipher plant's abiotic stress tolerance responses. *Genes* 14(6), 1281. DOI: 10.3390/genes14061281.

Saad, N.S.M., Neik, T.X., Thomas, W.J., Amas, J.C., Cantila, A.Y. *et al.* (2022) Advancing designer crops for climate resilience through an integrated genomics approach. *Current Opinion in Plant Biology* 67, 102220. DOI: 10.1016/j.pbi.2022.102220.

Sakai, H., Lee, S.S., Tanaka, T., Numa, H., Kim, J. *et al.* (2013) Rice Annotation Project Database (RAP-DB): An integrative and interactive database for rice genomics. *Plant and Cell Physiology* 54(2), e6–e6. DOI: 10.1093/pcp/pcs183.

Saleem, A., Anwar, S., Nawaz, T., Fahad, S., Saud, S. *et al.* (2024) Securing a sustainable future: The climate change threat to agriculture, food security, and sustainable development goals. *Journal of Umm Al-Qura University for Applied Sciences* 1–17. DOI: 10.1007/s43994-024-00177-3.

Salt, D.E., Baxter, I. and Lahner, B. (2008) Ionomics and the study of the plant ionome. *Annual Review of Plant Biology* 59(1), 709–733. DOI: 10.1146/annurev.arplant.59.032607.092942.

Satam, H., Joshi, K., Mangrolia, U., Waghoo, S., Zaidi, G. *et al.* (2023) Next-generation sequencing technology: Current trends and advancements. *Biology* 12(7), 997. DOI: 10.3390/biology12070997.

Sato, Y., Antonio, B.A., Namiki, N., Takehisa, H., Minami, H. *et al.* (2010) RiceXPro: A platform for monitoring gene expression in Japonica rice grown under natural field conditions. *Nucleic Acids Research* 39(suppl_1), D1141–D1148. DOI: 10.1093/nar/gkq1085.

Schneider, A., Dessimoz, C. and Gonnet, G.H. (2007) OMA Browser—exploring orthologous relations across 352 complete genomes. *Bioinformatics* 23(16), 2180–2182. DOI: 10.1093/bioinformatics/btm295.

Scossa, F., Alseekh, S. and Fernie, A.R. (2021) Integrating multi-omics data for crop improvement. *Journal of Plant Physiology* 257, 153352. DOI: 10.1016/j.jplph.2020.153352.

Sebastiani, P., Timofeev, N., Dworkis, D.A., Perls, T.T. and Steinberg, M.H. (2009) Genome-wide association studies and the genetic dissection of complex traits. *American Journal of Hematology* 84(8), 504–515. DOI: 10.1002/ajh.21440.

Shekhar, S., Prasad, A.S., Banjare, K., Kaushik, A., Mannade, A.K. *et al.* (2023) LMTdb: A comprehensive transcriptome database for climate-resilient, nutritionally rich little millet (*Panicum sumatrense*). *Frontiers in Plant Science* 14, 1106104. DOI: 10.3389/fpls.2023.1106104.

Shi, H., Zhu, Y., Wu, X., Jiang, T., Li, X. *et al.* (2024) CropMetabolome: A comprehensive metabolome database for major crops cross eight categories. *The Plant Journal* 119, 1613–1626. DOI: 10.1111/tpj.16858.

Song, Q., Ando, A., Jiang, N., Ikeda, Y. and Chen, Z.J. (2020) Single-cell RNA-seq analysis reveals ploidy-dependent and cell-specific transcriptome changes in Arabidopsis female gametophytes. *Genome Biology* 21, 1–18. DOI: 10.1186/s13059-020-02094-0.

Spindel, J., Begum, H., Akdemir, D., Virk, P., Collard, B. *et al.* (2015) Genomic selection and association mapping in rice (*Oryza sativa*): Effect of trait genetic architecture, training population composition, marker number and statistical model on accuracy of rice genomic selection in elite, tropical rice breeding lines. *PLoS Genetics* 11(2), e1004982. DOI: 10.1371/journal.pgen.1004982.

Sun, L., Lai, M., Ghouri, F., Nawaz, M.A., Ali, F. *et al.* (2024) Modern plant breeding techniques in crop improvement and genetic diversity: From molecular markers and gene editing to artificial intelligence—a critical review. *Plants* 13(19), 2676. DOI: 10.3390/plants13192676.

Sun, M., Yan, H., Zhang, A., Jin, Y., Lin, C. *et al.* (2023) Milletdb: A multi-omics database to accelerate the research of functional genomics and molecular breeding of millets. *Plant Biotechnology Journal* 21(11), 2348–2357. DOI: 10.1111/pbi.14136.

Tay Fernandez, C.G., Nestor, B.J., Danilevicz, M.F., Gill, M., Petereit, J. *et al.* (2022) Pangenomes as a resource to accelerate breeding of under-utilised crop species. *International Journal of Molecular Sciences* 23(5), 2671. DOI: 10.3390/ijms23052671.

Thomas, P.D., Campbell, M.J., Kejariwal, A., Mi, H., Karlak, B. *et al.* (2003) PANTHER: A library of protein families and subfamilies indexed by function. *Genome Research* 13(9), 2129–2141. DOI: 10.1101/gr.772403.

Tibbs Cortes, L., Zhang, Z. and Yu, J. (2021) Status and prospects of genome-wide association studies in plants. *The Plant Genome* 14(1), e20077. DOI: 10.1002/tpg2.20077.

Van Bel, M., Diels, T., Vancaester, E., Kreft, L., Botzki, A. *et al.* (2018) PLAZA 4.0: An integrative resource for functional, evolutionary and comparative plant genomics. *Nucleic Acids Research* 46(D1), D1190–D1196. DOI: 10.1093/nar/gkx1002.

Varotto, S., Krugman, T., Aiese Cigliano, R., Kashkush, K., Kondić-Špika, A. *et al.* (2022) Exploitation of epigenetic variation of crop wild relatives for crop improvement and agrobiodiversity preservation. *Theoretical and Applied Genetics* 135(11), 3987–4003. DOI: 10.1007/s00122-022-04122-y.

Wang, J., Chen, Y. and Zou, Q. (2023) Comparative genomics and functional genomics analysis in plants. *International Journal of Molecular Sciences* 24(7), 6539. DOI: 10.3390/ijms24076539.

Wang, L., Yu, J., Zhang, Y., You, J., Zhang, X. and Wang, L. (2021) Sinbase 2.0: An updated database to study multi-omics in *Sesamum indicum*. *Plants* 10(2), 272. DOI: 10.3390/plants10020272.

Wang, Y., Xu, P., Wu, X., Wu, X., Wang, B. *et al.* (2018) GourdBase: A genome-centered multi-omics database for the bottle gourd (*Lagenaria siceraria*), an economically important cucurbit crop. *Scientific Reports* 8(1), 3604. DOI: 10.1038/s41598-018-22007-3.

Wang, J., Qi, M., Liu, J. and Zhang, Y. (2015) CARMO: A comprehensive annotation platform for functional exploration of rice multi-omics data. *The Plant Journal* 83(2), 359–374. DOI: 10.1111/tpj.12894.

Wang, Z., Gerstein, M. and Snyder, M. (2009) RNA-Seq: A revolutionary tool for transcriptomics. *Nature Reviews Genetics* 10(1), 57–63. DOI: 10.1038/nrg2484.

Ward, B.P., Brown-Guedira, G., Tyagi, P., Kolb, F.L., Van Sanford, D.A. *et al.* (2019) Multienvironment and multitrait genomic selection models in unbalanced early-generation wheat yield trials. *Crop Science* 59, 491–507. DOI: 10.2135/cropsci2018.03.0189.

Wittmann-Liebold, B., Graack, H.R. and Pohl, T. (2006) Two-dimensional gel electrophoresis as tool for proteomics studies in combination with protein identification by mass spectrometry. *Proteomics* 6(17), 4688–4703. DOI: 10.1002/pmic.200500874.

Xu, Z.S., Tan, H.W., Wang, F., Hou, X.L. and Xiong, A.S. (2014) CarrotDB: A genomic and transcriptomic database for carrot. *Database* bau096. DOI: 10.1093/database/bau096.

Yadav, B., Jogawat, A., Rahman, M.S. and Narayan, O.P. (2021) Secondary metabolites in the drought stress tolerance of crop plants: A review. *Gene Reports* 23, 101040. DOI: 10.1016/j.genrep.2021.101040.

Yan, S., Bhawal, R., Yin, Z., Thannhauser, T.W. and Zhang, S. (2022) Recent advances in proteomics and metabolomics in plants. *Molecular Horticulture* 2(1), 17. DOI: 10.1186/s43897-022-00038-9.

Yang, P. (2022) Proteomics of post-translational modifications in plants. *Frontiers in Plant Science* 13, 894314. DOI: 10.3389/fpls.2022.894314.

Yang, Y., Saand, M.A., Huang, L., Abdelaal, W.B., Zhang, J. *et al.* (2021) Applications of multi-omics technologies for crop improvement. *Frontiers in Plant Science* 12, 563953. DOI: 10.3389/fpls.2021.563953.

Yang, Z., Wang, S., Wei, L., Huang, Y., Liu, D. *et al.* (2023a) BnIR: A multi-omics database with various tools for *Brassica napus* research and breeding. *Molecular Plant* 16(4), 775–789. DOI: 10.1016/j.molp.2023.03.007.

Yang, Z., Wang, J., Huang, Y., Wang, S., Wei, L. *et al.* (2023b) CottonMD: A multi-omics database for cotton biological study. *Nucleic Acids Research* 51(D1), D1446–D1456. DOI: 10.1093/nar/gkac863.

Yeshi, K., Crayn, D., Ritmejerytė, E. and Wangchuk, P. (2022) Plant secondary metabolites produced in response to abiotic stresses has potential application in pharmaceutical product development. *Molecules* 27(1), 313. DOI: 10.3390/molecules27010313.

Zhang, R., Zhang, C., Yu, C., Dong, J. and Hu, J. (2022) Integration of multi-omics technologies for crop improvement: Status and prospects. *Frontiers in Bioinformatics* 2, 1027457. DOI: 10.3389/fbinf.2022.1027457.

7 Plant-specific Bioinformatic Tools and Resources for Omics in Crop Improvement

Saurabh Pandey[1]*, Ashutosh Singh[2] and Mithilesh Kumar Singh[3]
[1]Department of Molecular Biology and Biotechnology, Indira Gandhi Agricultural University, Chhattisgarh, India; [2]Centre for Advanced Studies on Climate Change, Dr. Rajendra Prasad Central Agricultural University, Pusa, Samastipur, Bihar, India; [3]Department of Genetics and Plant Breeding, Dr. Rajendra Prasad Central Agricultural University, Pusa, Samastipur, India

Abstract
The advent of sequencing in the last century revolutionized functional biology, resulting in the generation of enormous datasets. The interpretation of these datasets required computer programs, which gave rise to bioinformatics. Earlier, the scope of data interpretation with fewer programs and software was limited and focused on animal systems. However, the concentrated efforts of plant biologists for rice genome sequencing led to plant-specific databases, software, and tools. Almost all the model plant species have been sequenced, or their sequencing project has started. Plant genomes are more complex due to different ploidy levels, repeats, and other genomic elements. These complexities led to differential expression regulation and metabolic networks requiring specific tools and resources. The evolution of sequencing technologies led to the generation of multiomics data in plants, giving insights into their growth, development, and stress responses. In this chapter, we will cover the plant-specific databases, advanced omics tools used in crop improvement, and tools giving insights into plant evolution. This chapter will improve our understanding of plant-specific bioinformatics tools that could be integrated with functional biology, leading to climate-resilient crops developed in a shorter duration.

7.1 Introduction

Around 70% of the world's population is poor and resides in rural areas, and they frequently experience hunger due to climatic changes (Pandey and Singh, 2024). Therefore, continuous improvement in agriculture is needed to feed the deprived population (Sanchez and Swaminathan, 2005). The production of many crops is steadily growing due to advances in biological and agricultural research; however, crop characteristics must be improved via biotechnology techniques such as genomics and bioinformatics (Mishra *et al.*, 2024). Although the Bioinformatics and Genome Research conference series began in 1990, bioinformatics has only lately gained significant traction and established itself as the norm for academic researchers in agriculture. The developments in molecular biology and high-throughput sequencing all contributed to the explosion in bioinformatics and genomic research. The revolution in bioinformatics is a result of the quick development of computer technology. This involves gathering biological data previously scattered across filing cabinets and

*Corresponding author: pandey.saurabh784@gmail.com, saurabh.pandey@igkv.ac.in

 From Gene Discovery to Climate-Resilient Crops: Omics in Crop Improvement (eds S. Pandey *et al.*)
DOI: 10.1079/9781800628717.0007

cubbyholes. These data now find their way into enormous databases that, because of their comprehensiveness and instant cross-accessibility, turn the information into a commodity with greater value than the sum of its parts. This has resulted in significant changes to the nature of bioinformatics; it is now primary information and a primary asset in and of itself, rather than merely an instrument through which we gather and manage other information. The agricultural sector has thus become highly data-intensive (Lim and Butt, 1998).

High-throughput next-generation sequencing technologies from platforms such as Roche, IlluminaMiSeq, and 454 Pyro sequencers are used for automated DNA sequencing. Molecular modeling and molecular simulation methods are used to predict protein structures. The Chou Fasmann method, secondary structure prediction using GOR, and quaternary structure analysis have all been simplified. Advanced tools like Gromos, Proline, T-Coffee, and others are frequently used in *in silico* analysis for structure prediction (Mitra *et al.*, 2022). Under the broader heading of bioinformatics research, investigations in transcriptomics, meta-transcriptomics, proteomics, metabolomics, and phylogenomics are carried out.

Since bioinformatics encompasses everything from protein sequencing to structural and functional prediction to developing and manufacturing medicinal compounds to treat various illnesses, it is a very significant area of study. Furthermore, bioinformatics alone has enabled transcriptomics, meta-transcriptomics, phylogenomics, metabolomics, and proteomics research to be conducted more accurately and in a shorter amount of time. On a massive scale, biological databases and tools are essential repositories for storing and retrieving complicated information. Various data sources, such as web servers for managing, analyzing, and verifying accessibility, are being developed. Therefore, the method is typically used to obtain and access the data gathered from several sources in bioinformatics research.

Considering this background, this chapter discusses the role of bioinformatics in advancing crop genomics.

7.2 Different Plant Databases

Numerous plant biological databases encompass information from various research areas such as metabolomics, proteomics, genomics, microarray gene expression, and phylogeny. The data within these biological databases pertains to gene functions, molecular composition, chromosomal positioning, the clinical effects of mutations, and the resemblance of biological sequences and traits (Sood *et al.*, 2021). Several databases have been created across various platforms for different uses in genomic data. Comprising the DNA Data Bank of Japan (DDBJ) at The National Institute of Genetics in Mishima, Japan, the International Nucleotide Sequence Database Collaboration (INSDC) is the biggest genome database. GenBank is at the National Center of Biotechnology Information (NCBI) in Bethesda, USA, and the European Molecular Biology Laboratory (EMBL) Nucleotide Sequence Database is housed at the European Bioinformatics Institute (EBI) in the UK. Coordinated worldwide coverage is ensured by these groups' daily data exchange (Sterk *et al.*, 2007; Sugawara *et al.*, 2008; Wheeler *et al.*, 2008; Karsch-Mizrachi *et al.*, 2012). The issue of storing and analyzing these data has grown since the advent of sophisticated next-generation sequencing techniques. Custom databases such as TAGdb have made it possible to look through a large amount of data. Still, more and more crop-improvement apps use assembled and annotated genomic data (Edwards and Batley, 2010). The two primary general web-based tools for viewing plant genomes are Ensembl and GBrowse. Due to their widespread use, it is not unusual to encounter comparable genomic data maintained on both platforms. Although having all publicly available nucleic acid sequences in one location is beneficial, the resource's bulk restricts how this data may be seen. Genome viewers offer more significant context-oriented data inquiry by situating genomic data within the framework of partly or fully sequenced genomes (Donlin, 2007; Arnaoudova *et al.*, 2009). Different plant databases are presented in Table 7.1.

Table 7.1. Different crop databases. Author's own table.

S.N.	Database	Use	Web link	References
1	RicyerDB	Rice biology	http://server.malab.cn/Ricyer/index.html	Jiang *et al*., 2018
2	PlantRGDB	38,997 retrocopies with their categorization and 49 plant species	http://aegilops.wheat.ucdavis.edu/plantrgdb	Wang, 2017
3	Functional trait databases	Functional trait database Mediterranean Basin vascular plants	https://www.try-db.org/TryWeb/Home.php	Tavşanoğlu and Pausas, 2018
4	ERISdb	Splice site in different plants	http://lemur.amu.edu.pl/share/ERISdb/	Szcześniak *et al*., 2013
5	PlantProm	Plant promoter repository	http://www.softberry.com/plantprom2016	Shahmuradov *et al*., 2003
6	CTDB	Chickpea transcriptome information	http://nipgr.res.in/ctdb.html	Verma *et al*., 2015
7	CicArMiSatDB	Chickpea Microsatellite Database for genetic research support	http://cicarmisatdb.icrisat.org	Doddamani *et al*., 2014
8	MSDB	Accessing and analyzing SSR information in various species of plants	http://tdb.ccmb.res.in/msdb	Avvaru *et al*., 2017
9	Kazusa Marker DataBase	SSR and SNP markers database	http://marker.kazusa.or.jp	Shirasawa *et al*., 2014
10	Plant snoRNA database	Plant species database on small nucleolar RNAs	http://www.scri.sari.ac.uk/plant_snoRNA/	Brown *et al*., 2003
11	PlantRNA	tRNA database	http://plantrna.ibmp.cnrs.fr/	Cognat *et al*., 2012
12	PRGdb	Plant resistance gene database	http://prgdb.org	Osuna-Cruz *et al*., 2018
13	Ricebase	Genomic database of rice	http://ricebase.org	Edwards *et al*., 2016
14	ReprOlive	Olive tree reproductive transcriptome database	https://doi.org/10.3389/fpls.2015.00625	Carmona *et al*., 2015
15	Planteome database	Species-specific ontologies for crops and genetic and phenotypic annotation	http://www.planteome.org	Cooper *et al*., 2018

7.3 Advanced Omics Technologies (Software or Algorithm) for Crop Improvement

The advent of omics technologies has revolutionized plant breeding and crop improvement. Omics refers to high-throughput approaches such as genomics, transcriptomics, proteomics, metabolomics, and phenomics that allow for the comprehensive analysis of biological systems (Srivastava *et al*., 2023). Different software and tools were used for these plant-specific analyses, as mentioned in Table 7.2. Integrating bioinformatics, artificial intelligence (AI), and machine learning (ML) with omics data has led to the development of novel software and algorithms tailored for crop improvement. These advancements help decipher complex genetic networks, improve marker-assisted selection (MAS), and enable precision breeding for enhanced crop productivity and resilience (Kumawat *et al*., 2020; Afzal *et al*., 2023). AI and ML have played

Table 7.2. Omics tools for crop improvement applications. Author's own table.

Tool/algorithm name	Omics category	Application	Reference
TASSEL	Genomics	Genome-wide association studies (GWAS) for trait mapping	Bradbury *et al.*, 2007
GATK	Genomics	Variant discovery and genotyping	McKenna *et al.*, 2010
Cufflinks	Transcriptomics	Transcript assembly and differential expression analysis	Trapnell *et al.*, 2012
HISAT2	Transcriptomics	Fast and sensitive alignment of RNA-Seq reads	Kim *et al.*, 2019
MaxQuant	Proteomics	Label-free and label-based quantitative proteomics	Cox and Mann, 2008
Progenesis QI	Proteomics	Data processing and normalization for LC–MS proteomics	Chen *et al.*, 2016
XCMS	Metabolomics	Metabolite feature detection and peak alignment	Smith *et al.*, 2006
MetaboAnalyst	Metabolomics	Statistical and functional analysis of metabolomic data	Xia *et al.*, 2009
KBase	Multi-omics	Integrative analysis of genomics, transcriptomics, and metabolomics	Arkin *et al.*, 2018
PlantRegMap	Regulatory omics	Prediction of cis-regulatory elements in plant genomes	Tian *et al.*, 2020
WGCNA	Systems biology	Gene co-expression network analysis for trait discovery	Langfelder and Horvath, 2008
RASP	transcriptomics	RNA secondary structure probing data	Li *et al.*, 2021
DeepVariant	AI-based genomics and epigenomics	Deep learning-based variant calling	Fishman *et al.*, 2023
Deep Metabolome	AI-based metabolomics	Machine learning-assisted metabolite identification	Pomyen *et al.*, 2020

a pivotal role in improving crop improvement strategies. Machine learning algorithms such as Random Forest, Support Vector Machines (SVM), and deep learning models are used for genotype–phenotype predictions (Grinberg *et al.*, 2020). DeepVariant, an AI-powered tool developed by Google, improves variant calling accuracy in crop genomes (Abdelwahab *et al.*, 2023). Similarly, DeepGene and PlantCV offer automated phenotypic analysis, helping breeders make data-driven decisions (Pierz *et al.*, 2023). Predictive modeling of crop traits using AI enhances genomic selection (GS) by identifying genes linked to yield, disease resistance, and abiotic stress tolerance.

Next-generation sequencing (NGS) and third-generation sequencing (TGS) technologies have dramatically enhanced crop genome analysis. Platforms such as Illumina, PacBio, and Oxford Nanopore Technologies provide massive sequencing data that require efficient computational tools for analysis (Wee *et al.*, 2019). Bioinformatics pipelines such as CLC Genomics Workbench, Galaxy, and TASSEL are widely used to handle large-scale sequencing datasets (Rachappanavar *et al.*, 2022). These tools facilitate genome assembly, variant calling, and annotation, allowing breeders to identify key genetic loci associated with agronomic traits.

Genomics forms the foundation of multiomics research, providing insights into the genetic architecture of agronomic traits. It serves as a source of information for future genetic manipulation in crops, for example, tomato (Sharma *et al.*, 2017; Pandey and Prasad, 2024). Genome-sequencing technologies like Illumina, PacBio, and Oxford Nanopore enable whole-genome assembly and variant discovery (Espinosa *et al.*, 2024). Genome-wide association studies (GWAS) and quantitative trait locus (QTL) mapping are commonly used to link genetic variations to phenotypic traits. For example, a GWAS study in rice identified significant single nucleotide polymorphisms (SNPs) related to yield under salinity stress (Era *et al.*, 2024). Similarly, genome-wide identification of trait-specific gene families is possible through bioinformatics tools that mine the genomic data available (Raturi *et al.*, 2024). Software tools such as PLINK, TASSEL, and GAPIT facilitate large-scale GWAS analyses in crops (Chavhan *et al.*, 2024). GWAS have become fundamental in identifying genetic variants linked to desirable crop traits. Computational tools such as PLINK, GAPIT, and GWASpoly provide robust statistical frameworks for detecting associations between genetic markers and phenotypic characteristics. GWAS-based software utilizes mixed linear models (MLMs) and Bayesian inference techniques to analyze large-scale SNP datasets, aiding in discovering key candidate genes (Pandey *et al.*, 2019; Jain *et al.*, 2024). Pangenomics approaches analyze the entire gene repertoire within a species, identifying core and dispensable genes contributing to crop adaptation. Tools such as PanTools, Roary, and PanX facilitate the construction of pangenomes, allowing for comparative genomic studies among cultivars (Sarawad *et al.*, 2025). Structural variants such as inversions, duplications, and translocations influence crop trait variations. Computational pipelines like Sniffles and SyRI help detect and annotate structural variants in crop genomes (Yuan *et al.*, 2021). This analysis helps detect structural variants at the pangenome level. Functional genomics approaches, such as CRISPR-Cas9 and RNA interference (RNAi), have transformed crop genetic engineering. Computational tools such as CRISPResso2, CHOPCHOP, and Cas-OFFinder aid in designing genome editing experiments with high specificity (Pandey *et al.*, 2024). These tools assist in optimizing gene knockout and knock-in strategies for crop trait improvement. For example, CRISPR has been utilized against plant viruses to provide resistance/tolerance (Pandey and Mishra, 2023).

RNA sequencing (RNA-Seq) enables studying transcriptomic changes in response to environmental and developmental cues (Mani *et al.*, 2021). Software tools such as HISAT2, STAR, and DESeq2 are widely used for differential gene expression analysis (Perelo *et al.*, 2024). Weighted gene co-expression network analysis (WGCNA) aids in identifying key regulatory genes controlling agronomic traits (Yu *et al.*, 2020). Single-cell transcriptomics, facilitated by tools like Seurat and Scanpy, provides insights into cell-type-specific gene expression in crops (Singh *et al.*, 2024a). Transcriptomics studies gene expression patterns across different developmental stages and environmental conditions. RNA-Seq technologies, including Illumina HiSeq and Nanopore direct RNA sequencing, generate high-resolution transcriptome profiles. These tools also help identify small RNA molecules that are regulatory and essential in plant defense (Pandey *et al.*, 2023).

Proteomics studies complement genomics by analyzing the functional proteins involved in crop development and stress responses. Software such as MaxQuant, Proteome Discoverer, and Perseus enable quantitative proteomic analyses (Shah *et al.*, 2019). Proteomics complements transcriptomics by providing insights into protein abundance, modifications, and interactions. Mass spectrometry (MS)-based platforms such as LC–MS/MS and MALDI-TOF are widely used in crop proteomics. Understanding post-translational modifications (PTMs) through tools such as PTM-Prophet enhances knowledge of protein function and regulation in crop development (Geiszler *et al.*, 2021).

Metabolomics tools like XCMS, MetaboAnalyst, and MassHunter facilitate metabolic profiling, identifying bioactive compounds associated with stress tolerance and crop nutritional quality (Razzaq *et al.*, 2019). Integrating proteomics and metabolomics with genomic data enhances systems biology approaches in crop improvement. Metabolomics analyzes small-molecule metabolites influencing agronomic traits such as stress response, flavor, and nutritional quality. Chromatography–mass spectrometry (GC–MS, LC–MS) and nuclear magnetic resonance (NMR) spectroscopy enable metabolite profiling (McCullagh and Probert, 2024). Integrating metabolomics with genomics and transcriptomics aids pathway reconstruction and biomarker discovery for crop improvement. It also includes lipidomics, which exploits the metabolic pathway candidates essential for plant function and stress responses (Pandey *et al.*, 2025). Epigenomics studies DNA methylation, histone modifications, and chromatin accessibility that regulate gene expression without altering DNA sequences. Whole-genome bisulfite sequencing (WGBS), ATAC-seq and ChIP-seq techniques help analyze crop epigenetic patterns and provide high-resolution epigenomic maps (Singh *et al.*, 2024b). Bismark, DeepTools, and MethylKit allow methylation analysis and chromatin state predictions (Kumar *et al.*, 2024). Epigenomic insights enable the development of epigenome-based breeding strategies for stress resilience and improved crop yields. Understanding epigenetic regulation enhances the development of climate-resilient and high-yielding crops (Dinkar *et al.*, 2024).

Multiomics approaches have transformed crop research by providing comprehensive insights into the molecular basis of agronomic traits. Multiomics approaches integrate genomics, transcriptomics, proteomics, metabolomics, and epigenomics to understand the molecular mechanisms underlying crop agronomic traits (Satrio *et al.*, 2024). These approaches enable the identification of genes regulating yield, disease resistance, abiotic stress tolerance, and nutritional quality. With advancements in computational tools and high-throughput technologies, multiomics strategies have become essential for systems biology and precision breeding. Integrating multiomics datasets enhances the understanding of complex biological interactions in crops. Platforms such as iOmicsPASS, MixOmics, and Cytoscape facilitate network-based analysis of multiomics data (Kumar and Gupta, 2024). ML-driven multiomics integration tools, including MOFA (Multi-Omics Factor Analysis) and DIABLO (Data Integration Analysis for Biomarker Discovery), help identify biomarkers associated with crop performance traits (Satya *et al.*, 2024). Integrating multiomics data is crucial for understanding complex trait regulation. Multiomics approaches have been successfully applied to improve critical agronomic traits in crops. For example, Integrative omics studies have identified key genes and metabolites associated with drought resistance and grain quality (Zhao *et al.*, 2023). Similarly, GWAS combined with transcriptomics has uncovered regulatory networks controlling kernel size and nitrogen-use efficiency in maize (corn) (Liu *et al.*, 2023). Epigenomic modifications have been linked to heat stress tolerance, aiding in selecting stress-resilient cultivars in wheat (Zhang *et al.*, 2023). Despite significant progress, multiomics research faces challenges such as data standardization, computational complexity, and integration bottlenecks. Future directions should focus on developing interoperable databases and pipelines for multiomics data sharing and advancing AI-driven predictive modeling for trait selection with expanding cloud-based computing frameworks to enhance accessibility and scalability.

Digital agriculture leverages omics technologies for precision farming, optimizing crop management and resource utilization. Remote sensing, unmanned aerial vehicle (UAV)-based phenotyping, and Internet of Things (IoT)-driven sensors generate large-scale agricultural datasets (Shamshiri *et al.*, 2024). AI tools have been applied extensively in agriculture; for example, AI tools were used for seed quality analysis for automation and high-throughput screening (Singh *et al.*, 2025). AI-powered platforms such as FieldView, DroneDeploy, and PhenoAI enable real-time monitoring of crop health, disease outbreaks, and environmental stress factors (Coelho *et al.*, 2022).

7.4 Tools Revealing the Origin and Evolution of Crops

Understanding the origin and evolution of crops is essential for plant breeding, conservation, and sustainable agriculture. Bioinformatics tools have significantly contributed to deciphering the evolutionary pathways of domesticated plants by analyzing genomic data, detecting selection signatures, and reconstructing phylogenetic relationships. This section explores bioinformatics tools used in studying crop evolution, their applications, and recent advancements in the field.

Advances in high-throughput sequencing have facilitated the study of crop domestication and evolution (Dev *et al.*, 2024). Whole-genome sequencing (WGS), resequencing, and comparative genomics enable researchers to trace genetic

Table 7.3. Bioinformatics tools for studying crop evolution. Author's own table.

Tool name	Application	Reference
STRUCTURE	Population structure analysis	Pritchard *et al.*, 2000
BEAST	Bayesian phylogenetic inference	Drummond and Rambaut, 2007
TreeMix	Detecting gene flow and migration events	Pickrell and Pritchard, 2012
ADMIXTURE	Estimating population ancestry	Alexander *et al.*, 2009
SNPhylo	SNP-based phylogenetic tree construction	Lee *et al.*, 2014
ANGSD	Population genomics analysis	Korneliussen *et al.*, 2014
fastSTRUCTURE	Scalable population structure inference	Raj *et al.*, 2014
VCFtools	Variant call format (VCF) data processing	Danecek *et al.*, 2011
PLINK	GWAS and population genetic analysis	Purcell *et al.*, 2007
Dsuite	Detecting introgression in genomic datasets	Malinsky *et al.*, 2021
GATK	Variant discovery and genotyping	McKenna *et al.*, 2010
MEGA	Molecular evolutionary genetics analysis	Tamura *et al.*, 2021
RAxML	Maximum likelihood-based phylogenetic inference	Stamatakis, 2015
OrthoFinder	Ortholog identification for comparative genomics	Emms and Kelly, 2019
iTOL	Interactive phylogenetic tree visualization	Letunic and Bork, 2021

variations, gene duplications, and structural variations contributing to crop adaptation (Feuillet *et al.*, 2011). Techniques such as SNP analysis, GWAS, and pangenomics provide insights into the genetic diversity of cultivated and wild crop relatives. Several computational tools and databases aid in analyzing crop origin and evolutionary patterns (Table 7.3). These tools perform phylogenetic analysis, population genomics, selection scanning, and gene flow estimation.

Bioinformatics tools such as BEAST, RAxML, and MEGA facilitate the construction of phylogenetic trees, allowing researchers to trace the evolutionary history of crops (Drummond and Rambaut, 2007; Stamatakis, 2015; Tamura *et al.*, 2021). Phylogenetic studies using BEAST and RAxML have traced the independent domestication events of *indica* and *japonica* rice subspecies (Choi *et al.*, 2017). Comparative genomic studies using OrthoFinder help identify conserved genes and evolutionary divergence among species (Emms and Kelly, 2019). Whole-genome sequencing and comparative genomics have unraveled the complex polyploidization events leading to modern bread wheat varieties (IWGSC *et al.*, 2018). Population genetics tools such as STRUCTURE, ADMIXTURE, and fastSTRUCTURE enable the identification of ancestral relationships, admixture patterns, and genetic clusters in crop populations. These tools have been extensively applied in maize, wheat, and rice domestication studies. For example, genomic analysis using STRUCTURE and GWAS identified key genes responsible for maize domestication from its wild ancestor, teosinte (Hufford *et al.*, 2012). Selection signatures in crop genomes are identified using tools like ANGSD, Dsuite, and TreeMix (Table 7.3). These analyses reveal genes under selection during domestication and adaptation to environmental stressors. Pangenomics approaches utilizing tools like PLINK and VCFtools provide insights into genome-wide variations across multiple crop cultivars (Purcell *et al.*, 2007; Danecek *et al.*, 2011). This information is crucial for identifying beneficial alleles in wild relatives and landraces.

Bioinformatics tools have significantly advanced our understanding of crop origins and evolution by enabling large-scale genomic analysis, population structure assessment, and phylogenetic inference. These insights are crucial for crop improvement and conservation efforts. Future research should focus on integrating AI and cloud computing for enhanced computational power and accessibility in evolutionary genomics.

7.5 Meta-Analysis and Comparative Analysis of Crop Omics Datasets

Meta-analysis and comparative analysis of crop omics datasets have become essential approaches for extracting meaningful biological insights from high-throughput genomics, transcriptomics, proteomics, and metabolomics studies (Satrio *et al.*, 2024). These approaches enable researchers to integrate multiple datasets, identify conserved patterns, and draw robust conclusions about crop traits, gene functions, and evolutionary relationships. With the rapid accumulation of omics datasets in public repositories, meta-analysis allows researchers to enhance statistical power by integrating multiple datasets, increasing the reliability of conclusions. Comparative analysis also helps identify conserved and divergent patterns by highlighting key regulatory elements and trait-associated genes across different studies. Its use improved reproducibility by cross-validation of findings using independent datasets and strengthened research outcomes. Also, novel biomarkers could be identified for genetic and metabolic signatures linked to agronomic traits.

For meta-analysis and comparative analysis, data collection and preprocessing are first required. Datasets are retrieved from repositories such as NCBI Gene Expression Omnibus (GEO), European Nucleotide Archive (ENA), and the Sequence Read Archive (SRA). After that, standardization of raw data formats, removal of batch effects, and normalization are performed using tools like Limma and ComBat. Then statistical meta-analysis methods such as Hedge's g and Fisher's combined probability test are used to quantify the significance of results across studies. Furthermore, tools like DESeq2 and edgeR enable comparative transcriptomics by identifying consistently upregulated or downregulated genes. After that, network and pathway integration is conducted through weighted gene co-expression network analysis (WGCNA), which helps identify functional gene modules. Researchers applied DESeq2 and WGCNA to integrate transcriptomic datasets from different environmental conditions, identifying key genes involved in drought response (Baldoni *et al.*, 2021). Meta-analysis of transcriptomic and proteomic data identified novel regulators of heat stress tolerance, improving genomic selection strategies (Zhang *et al.*, 2025). The Kyoto Encyclopedia of Genes and Genomes (KEGG) performs the pathway enrichment analysis, and gene ontology (GO) terms provide insights into biological functions and processes. Finally, comparative genomics and evolutionary analysis are done. Comparative genomics relies on pangenomic studies and ortholog detection tools such as OrthoMCL and Roary to identify conserved and lineage-specific genes. Comparative genomic analysis using Roary revealed structural variations contributing to nitrogen-use efficiency in different rice cultivars (Hou *et al.*, 2021). All these tools are mentioned in Table 7.4.

Meta-analysis and comparative analysis of crop omics datasets provide potent insights into plant biology, trait selection, and evolutionary adaptation. Advances in computational tools, statistical methodologies, and ML models will further accelerate discoveries in crop research. As omics data grows, integrating multiomics datasets will enhance our understanding of complex agronomic traits and facilitate crop improvement strategies.

7.6 Conclusions

Developing novel omics technologies, software, and algorithms has revolutionized crop improvement by facilitating precision breeding, functional gene annotation, and stress resilience studies (Banshidhar *et al.*, 2023). For example, it could be used for precision breeding where GWAS and genomic selection (TASSEL and GATK) help identify key genes linked to desirable agronomic traits. Similarly, it could be used for stress tolerance analysis where transcriptomics tools such as Cufflinks and HISAT2 enable the identification of stress-responsive genes under drought, salinity, and temperature stress conditions. Omics tools have also been used for nutritional enhancement. For example, metabolomics tools like XCMS and MetaboAnalyst help identify bioactive compounds that improve crop

Table 7.4. An overview of widely used software for meta-analysis and comparative analysis of crop omics datasets. Author's own table.

Tool name	Application	Reference
GEO2R	Differential expression analysis of GEO datasets	Barrett *et al.*, 2013
MetaOmics	Integrative omics meta-analysis	Ma *et al.*, 2019
WGCNA	Gene co-expression network analysis	Langfelder and Horvath, 2008
DESeq2	RNA-Seq differential expression analysis	Love *et al.*, 2014
edgeR	Differential expression in RNA-Seq data	Robinson *et al.*, 2010
Limma	Microarray and RNA-Seq normalization	Ritchie *et al.*, 2015
ComBat	Batch effect correction in omics data	Johnson *et al.*, 2007
Roary	Pangenome analysis in prokaryotes and plants	Page *et al.*, 2015
OrthoMCL	Ortholog detection across species	Li *et al.*, 2003
KEGG	Pathway annotation and functional enrichment	Kanehisa *et al.*, 2017
Cytoscape	Visualization of molecular interaction networks	Shannon *et al.*, 2003
STRING	Protein–protein interaction analysis	Szklarczyk *et al.*, 2023
iDEP	Interactive omics data exploration	Ge, 2021

nutritional quality. Also, Regulatory Network Mapping with PlantRegMap and WGCNA facilitates understanding of gene regulatory networks involved in plant growth and development. AI-driven analytics, high-throughput sequencing, and an integrative multiomics approach enhance our understanding of complex crop traits. For example, integrating ML with the help of AI-based approaches using DeepVariant and DeepMetabolomics enhances the predictive modeling of complex crop traits. Integrating multiomics data and ML approaches will further accelerate discoveries in plant research, enabling the development of resilient, high-yielding, and nutritionally enhanced crops. New omics technologies, encompassing genomics, transcriptomics, proteomics, and metabolomics, have revolutionized crop improvement strategies. As computational tools evolve, their application in digital agriculture and genome editing will further accelerate the development of climate-resilient and high-yielding crops.

Advancements in omics technologies and computational tools have significantly contributed to crop improvement by enabling precision breeding, functional gene annotation, and evolutionary insights. Integrating AI, big data analytics, and high-throughput sequencing technologies continues to enhance the predictive power of omics-based studies. Multiomics approaches provide deeper insights into the molecular basis of agronomic traits, and comparative analysis techniques refine our understanding of crop genomics. Future research should focus on improving interoperability among datasets, refining AI-driven models, and expanding public omics databases to accelerate crop improvement efforts.

References

Abdelwahab, O., Belzile, F. and Torkamaneh, D. (2023) Performance analysis of conventional and AI-based variant callers using short and long reads. *BMC Bioinformatics* 24(1), 472.

Afzal, S., Mubeen, M., Hussain, S., Ali, M., Javeed, H.M.R. *et al.* (2023) Modern breeding approaches for climate change. In: *Climate Change Impacts on Agriculture: Concepts, Issues and Policies for Developing Countries*. Springer International Publishing, Cham, pp. 299–313.

Alexander, D.H., Novembre, J. and Lange, K. (2009) Fast model-based estimation of ancestry in unrelated individuals. *Genome Research* 19(9), 1655–1664.

Arkin, A.P., Cottingham, R.W., Henry, C.S., Harris, N.L., Stevens, R.L. *et al.* (2018) KBase: The United States department of energy systems biology knowledgebase. *Nature Biotechnology* 36(7), 566–569.

Arnaoudova, E.G., Bowens, P.J., Chui, R.G., Dinkins, R.D., Hesse, U. *et al.* (2009) Visualizing and sharing results in bioinformatics projects: GBrowse and GenBank exports. *BMC Bioinformatics* 10, 1–2.

Avvaru, A.K., Saxena, S., Sowpati, D.T. and Mishra, R.K. (2017) MSDB: A comprehensive database of simple sequence repeats. *Genome Biology and Evolution* 9(6), 1797–1802.

Baldoni, E., Frugis, G., Martinelli, F., Benny, J., Paffetti, D. *et al.* (2021) A comparative transcriptomic meta-analysis revealed conserved key genes and regulatory networks involved in drought tolerance in cereal crops. *International Journal of Molecular Sciences* 22(23), 13062.

Banshidhar, S., Pandey, S., Singh, A., Jaiswal, P., Singh, M.K. *et al.* (2023) The potentialities of omics resources for millet improvement. *Functional & Integrative Genomics* 23(3), 210.

Barrett, T., Wilhite, S.E., Ledoux, P., Evangelista, C., Kim, I.F. *et al.* (2013) NCBI GEO: Archive for functional genomics data sets—update. *Nucleic Acids Research* 41(D1), D991–D995.

Bradbury, P.J., Zhang, Z., Kroon, D.E., Casstevens, T.M., Ramdoss, Y. *et al.* (2007) TASSEL: Software for association mapping of complex traits in diverse samples. *Bioinformatics* 23(19), 2633–2635.

Brown, J.W., Echeverria, M., Qu, L.H., Lowe, T.M., Bachellerie, J.P. *et al.* (2003) Plant snoRNA database. *Nucleic Acids Research* 31(1), 432–435.

Carmona, R., Zafra, A., Seoane, P., Castro, A.J., Guerrero-Fernández, D. *et al.* (2015) ReprOlive: A database with linked data for the olive tree (*Olea Europaea* L.) reproductive transcriptome. *Frontiers in Plant Science* 6, 625.

Chavhan, R.L., Hinge, V.R., Wankhade, D.J., Deshmukh, A.S., Mahajan, N. *et al.* (2024) Bioinformatics for molecular breeding and enhanced crop performance: Applications and perspectives. In: Chen, J.T. (ed.) *Bioinformatics for Plant Research and Crop Breeding*. Wiley and Sons, Hoboken, New Jersey, pp. 21–74.

Chen, Y., Wang, F., Xu, F. and Yang, T. (2016) Mass spectrometry-based protein quantification. In: Mirzaei, H. and Carrasco, M. (eds) *Modern Proteomics – Sample Preparation, Analysis and Practical Applications. Advances in Experimental Medicine and Biology*, Vol. 919. Springer, Cham, pp. 255–279. DOI: 10.1007/978-3-319-41448-5_15.

Choi, J.Y., Platts, A.E., Fuller, D.Q., Hsing, Y.I., Wing, R.A. *et al.* (2017) The rice paradox: Multiple origins but single domestication in Asian rice. *Molecular Biology and Evolution* 34(4), 969–979.

Coelho, A.L.D.F., Oliveira, T.F. and Netto, M.N. (2022) Platforms, applications, and software. In: *Digital Agriculture*. Springer International Publishing, Cham, pp. 259–272.

Cognat, V., Pawlak, G., Duchêne, A., Daujat, M., Gigant, A. *et al.* (2012) PlantRNA, a database for tRNAs of photosynthetic eukaryotes. *Nucleic Acids Research* 41(D1), D273–D279. DOI: 10.1093/nar/gks935.

Cooper, L., Meier, A., Laporte, M.A., Elser, J.L., Mungall, C. *et al.* (2018) The planteome database: An integrated resource for reference ontologies, plant genomics and phenomics. *Nucleic Acids Research* 46(D1), D1168–D1180.

Cox, J. and Mann, M. (2008) MaxQuant enables high peptide identification rates, individualized p. p.b.-range mass accuracies, and proteome-wide protein quantification. *Nature Biotechnology* 26(12), 1367–1372.

Danecek, P., Auton, A., Abecasis, G., Albers, C.A., Banks, E. *et al.* (2011) The variant call format and VCFtools. *Bioinformatics* 27(15), 2156–2158.

Dev, D., Antre, S.H., Nandni, S., Pandey., S., Singh, M.K. *et al.* (2024) Accelerated crop domestication: Candidates, advanced tools and prospects. *South African Journal of Botany* 171, 237–244.

Dinkar, V., Pandey, S., Kumar, A., Shiv, A. and Lal, D. (2024) Epigenetic regulations under plant stress: A cereals perspective. *Environmental and Experimental Botany* 220, 105688.

Doddamani, D., Katta, M.A., Khan, A.W., Agarwal, G., Shah, T.M. *et al.* (2014) CicArMiSatDB: The chickpea microsatellite database. *BMC Bioinformatics* 15, 1–7.

Donlin, M.J. (2007) Using the generic genome browser (GBrowse). *Current Protocols in Bioinformatics* 28(1), 9–9.

Drummond, A.J. and Rambaut, A. (2007) BEAST: Bayesian evolutionary analysis by sampling trees. *BMC Evolutionary Biology* 7, 214.

Edwards, D. and Batley, J. (2010) Plant genome sequencing: Applications for crop improvement. *Plant Biotechnology Journal* 8(1), 2–9.

Edwards, J.D., Baldo, A.M. and Mueller, L.A. (2016) Ricebase: A breeding and genetics platform for rice, integrating individual molecular markers, pedigrees and whole-genome-based data. *Database* baw107.

Emms, D.M. and Kelly, S. (2019) OrthoFinder: Phylogenetic orthology inference for comparative genomics. *Genome Biology* 20, 1–14.

Era, F.M., Raihan, M.S., Jahan, N., Pandey, S., Alalawy, A.I. *et al.* (2024) Identification of significant SNPs for yield-related salt tolerant traits in rice through genome-wide association analysis. *Cellular and Molecular Biology* 70(12), 18–25.

Espinosa, E., Bautista, R., Larrosa, R. and Plata, O. (2024) Advancements in long-read genome sequencing technologies and algorithms. *Genomics* 110842.

Feuillet, C., Leach, J.E., Rogers, J., Schnable, P.S. and Eversole, K. (2011) Crop genome sequencing: Lessons and rationales. *Trends in Plant Science* 16(2), 77–88.

Fishman, V., Sindeeva, M., Chekanov, N., Shashkova, T., Ivanisenko, N. *et al.* (2023) AI in genomics and epigenomics. In: *Artificial Intelligence for Healthy Longevity*. Springer International Publishing, Cham, pp. 217–243.

Ge, X. (2021) IDEP web application for RNA-Seq data analysis. *RNA Bioinformatics* 2284, 417–443.

Geiszler, D.J., Kong, A.T., Avtonomov, D.M., Yu, F., da Veiga Leprevost, F. *et al.* (2021) PTM-Shepherd: Analysis and summarization of post-translational and chemical modifications from open search results. *Molecular & Cellular Proteomics* 20, 100018.

Grinberg, N.F., Orhobor, O.I. and King, R.D. (2020) An evaluation of machine-learning for predicting phenotype: Studies in yeast, rice, and wheat. *Machine Learning* 109(2), 251–277.

Hou, M., Yu, M., Li, Z., Ai, Z. and Chen, J. (2021) Molecular regulatory networks for improving nitrogen use efficiency in rice. *International Journal of Molecular Sciences* 22(16), 9040.

Hufford, M.B., Xu, X., Van Heerwaarden, J., Pyhäjärvi, T., Chia, J.M. *et al.* (2012) Comparative population genomics of maize domestication and improvement. *Nature Genetics* 44(7), 808–811.

International Wheat Genome Sequencing Consortium (IWGSC), Appels, R., Eversole, K., Stein, N., Feuillet, C. *et al.* (2018) Shifting the limits in wheat research and breeding using a fully annotated reference genome. *Science* 361(6403), eaar7191.

Jain, P., Dutta, B. and Sevanthi, A.M. (2024) Trait based association mapping in plants. In: *Genomics Data Analysis for Crop Improvement*. Springer Nature, Singapore, pp. 159–190.

Jiang, J., Xing, F., Zeng, X. and Zou, Q. (2018) RicyerDB: A database for collecting rice yield-related genes with biological analysis. *International Journal of Biological Sciences* 14(8), 965–970.

Johnson, W.E., Li, C. and Rabinovic, A. (2007) Adjusting batch effects in microarray expression data using empirical Bayes methods. *Biostatistics* 8(1), 118–127.

Kanehisa, M., Furumichi, M., Tanabe, M., Sato, Y. and Morishima, K. (2017) KEGG: New perspectives on genomes, pathways, diseases and drugs. *Nucleic Acids Research* 45(D1), D353–D361.

Karsch-Mizrachi, I., Nakamura, Y. and Cochrane, G. (2012) The international nucleotide sequence database collaboration. *Nucleic Acids Research* 40(D1), D33–D37.

Kim, D., Paggi, J.M., Park, C., Bennett, C. and Salzberg, S.L. (2019) Graph-based genome alignment and genotyping with HISAT2 and HISAT-genotype. *Nature Biotechnology* 37(8), 907–915.

Korneliussen, T.S., Albrechtsen, A. and Nielsen, R. (2014) ANGSD: Analysis of next generation sequencing data. *BMC Bioinformatics* 15, 1–13.

Kumar, S. and Gupta, M.K. (2024) Data management in cross-omics. In: *Integrative Omics*. Academic Press, pp. 201–223.

Kumar, V., Gangani, S., Shukla, R., Prajapati, S.K., Shekhar, H. *et al.* (2024) Next-generation sequencing and genomic data analysis. In: *Unraveling New Frontiers and Advances in Bioinformatics*. Springer Nature, Singapore, pp. 49–76.

Kumawat, G., Kumawat, C.K., Chandra, K., Pandey, S., Chand, S. *et al.* (2020) Insights into marker assisted selection and its applications in plant breeding. In: *Plant Breeding - Current and Future Views*. Intechopen. DOI: 10.5772/intechopen.95004.

Langfelder, P. and Horvath, S. (2008) WGCNA: An R package for weighted correlation network analysis. *BMC Bioinformatics* 9(1), 559.

Lee, T.H., Guo, H., Wang, X., Kim, C. and Paterson, A.H. (2014) SNPhylo: A pipeline to construct a phylogenetic tree from huge SNP data. *BMC Genomics* 15, 1–6.

Letunic, I. and Bork, P. (2021) Interactive Tree Of Life (iTOL) v5: An online tool for phylogenetic tree display and annotation. *Nucleic Acids Research* 49(W1), W293–W296.

Li, L., Stoeckert, C.J. and Roos, D.S. (2003) OrthoMCL: Identification of ortholog groups for eukaryotic genomes. *Genome Research* 13(9), 2178–2189.

Li, P., Zhou, X., Xu, K. and Zhang, Q.C. (2021) RASP: An atlas of transcriptome-wide RNA secondary structure probing data. *Nucleic Acids Research* 49(D1), D183–D191.

Lim, H.A. and Butt, T.R. (1998) Bioinformatics takes charge. *Trends in Biotechnology* 16(3), 104–107.

Liu, X., He, K., Ali, F., Li, D., Cai, H. *et al.* (2023) Genetic dissection of N use efficiency using maize inbred lines and testcrosses. *The Crop Journal* 11(4), 1242–1250. DOI: 10.1016/j.cj.2023.03.010.

Love, M.I., Huber, W. and Anders, S. (2014) Moderated estimation of fold change and dispersion for RNA-Seq data with DESeq2. *Genome Biology* 15, 1–21.

Mani, S.D., Pandey, S., Govindan, M., Muthamilarasan, M. and Nagarathnam, R. (2021) Transcriptome dynamics underlying elicitor-induced defense responses against septoria leaf spot disease of tomato (*Solanum lycopersicum* L.). *Physiology and Molecular Biology of Plants* 27, 873–888.

Ma, T., Huo, Z., Kuo, A., Zhu, L., Fang, Z. *et al.* (2019) MetaOmics: Analysis pipeline and browser-based software suite for transcriptomic meta-analysis. *Bioinformatics (Oxford, England)* 35(9), 1597–1599.

Malinsky, M., Matschiner, M. and Svardal, H. (2021) Dsuite-Fast D-statistics and related admixture evidence from VCF files. *Molecular Ecology Resources* 21(2), 584–595.

McCullagh, J. and Probert, F. (2024) New analytical methods focusing on polar metabolite analysis in mass spectrometry and NMR-based metabolomics. *Current Opinion in Chemical Biology* 80, 102466.

McKenna, A., Hanna, M., Banks, E., Sivachenko, A., Cibulskis, K. *et al.* (2010) The Genome Analysis Toolkit: A MapReduce framework for analyzing next-generation DNA sequencing data. *Genome Research* 20(9), 1297–1303.

Mishra, R., Karada, M.S. and Agnihotri, D. (2024) Bioinformatics in crop improvement and agricultural genomics. In: *Unraveling New Frontiers and Advances in Bioinformatics*. Springer Nature, Singapore, pp. 293–313.

Mitra, D., Mitra, D., Bensaad, M.S., Sinha, S., Pant, K. *et al.* (2022) Evolution of bioinformatics and its impact on modern bio-science in the twenty-first century: Special attention to pharmacology, plant science and drug discovery. *Computational Toxicology* 24, 100248.

Osuna-Cruz, C.M., Paytuvi-Gallart, A., Di Donato, A., Sundesha, V., Andolfo, G. *et al.* (2018) PRGdb 3.0: A comprehensive platform for prediction and analysis of plant disease resistance genes. *Nucleic Acids Research* 46(D1), D1197–D1201.

Page, A.J., Cummins, C.A., Hunt, M., Wong, V.K., Reuter, S. *et al.* (2015) Roary: Rapid large-scale prokaryote pan genome analysis. *Bioinformatics* 31(22), 3691–3693.

Pandey, S. and Mishra, S. (2023) CRISPR/Cas-based strategies for resistance against plant viruses. In: *CRISPR/Cas-Mediated Genome Editing in Plants*. Apple Academic Press, pp. 391–414.

Pandey, S. and Prasad, M. (2024) Recent advances in tomato genetic engineering for virus resistance: Achievements and limitations. In: *Genetic Engineering of Vegetable Crops*. CABI, Walllingford, UK, pp. 1–22.

Pandey, S. and Singh, A. (2024) Climate change: Effects and need of climate-resilient agriculture. In: *Climate-Resilient Agriculture*. Apple Academic Press, pp. 3–8.

Pandey, S., Mishra, S. and Chandra, K. (2019) Tools and resources for SNP mining in crop plants. *International Journal of Current Microbiology and Applied Sciences* 8(1), 2819–2829.

Pandey, S., Antre, S.H., Kumari, S. and Singh, A. (2023) MicroRNA-mediated regulation of plant viral disease development. In: *Plant MicroRNAs and Stress Response*. CRC Press, pp. 242–256.

Pandey, S., Divakar, S. and Singh, A. (2024) Genome editing prospects for heat stress tolerance in cereal crops. *Plant Physiology and Biochemistry* 108989.

Pandey, S., Mishra, S. and Dwivedi, K.K. (2025) High-temperature stress: Effects on plant metabolism specific to lipidome. In: *Impact of High Night Temperature on Plant Biology*. Apple Academic Press, pp. 113–131.

Perelo, L.W., Gabernet, G., Straub, D. and Nahnsen, S. (2024) How tool combinations in different pipeline versions affect the outcome in RNA-Seq analysis. *NAR Genomics and Bioinformatics* 6(1), lqae020.

Pickrell, J.K. and Pritchard, J.K. (2012) Inference of population splits and mixtures from genome-wide allele frequency data. *PLoS Genetics* 8(11), e1002967.

Pierz, L.D., Heslinga, D.R., Buell, C.R. and Haus, M.J. (2023) An image-based technique for automated root disease severity assessment using PlantCV. *Applications in Plant Sciences* 11(1), e11507.

Pomyen, Y., Wanichthanarak, K., Poungsombat, P., Fahrmann, J., Grapov, D. *et al.* (2020) Deep metabolome: applications of deep learning in metabolomics. *Computational and Structural Biotechnology Journal* 18, 2818–2825.

Pritchard, J.K., Stephens, M. and Donnelly, P. (2000) Inference of population structure using multilocus genotype data. *Genetics* 155(2), 945–959.

Purcell, S., Neale, B., Todd-Brown, K., Thomas, L., Ferreira, M.A. *et al.* (2007) PLINK: A tool set for whole-genome association and population-based linkage analyses. *The American Journal of Human Genetics* 81(3), 559–575.

Rachappanavar, V., Padiyal, A., Sharma, J.K. and Negi, N. (2022) Analytical pipelines for the GBS analysis. In: Sonah, H., Goyal, V., Shivaraj, S.M. and Deshmukh, R.K. (eds) *Genotyping by Sequencing for Crop Improvement*. John Wiley & Sons, pp. 161–187.

Raj, A., Stephens, M. and Pritchard, J.K. (2014) fastSTRUCTURE: Variational inference of population structure in large SNP data sets. *Genetics* 197(2), 573–589.

Raturi, A., Shekhar, S., Jha, R.K., Chauhan, D., Pandey, S. *et al.* (2024) Genome-wide comparative analysis of photosynthetic enzymatic genes provides novel insights into foxtail millet and other cereals. *Frontiers in Genetics* 15, 1449113.

Razzaq, A., Sadia, B., Raza, A., Khalid Hameed, M. and Saleem, F. (2019) Metabolomics: A way forward for crop improvement. *Metabolites* 9(12), 303.

Ritchie, M.E., Phipson, B., Wu, D.I., Hu, Y., Law, C.W. *et al.* (2015) *Limma* powers differential expression analyses for RNA-sequencing and microarray studies. *Nucleic Acids Research* 43(7), e47–e47.

Robinson, M.D., McCarthy, D.J. and Smyth, G.K. (2010) EdgeR: A Bioconductor package for differential expression analysis of digital gene expression data. *Bioinformatics* 26(1), 139–140.

Sanchez, P.A. and Swaminathan, M.S. (2005) Cutting world hunger in half. *Science* 307(5708), 357–359.

Sarawad, A., Hosagoudar, S. and Parvatikar, P. (2025) Pan-genomics: Insight into the functional genome, applications, advancements, and challenges. *Current Genomics* 26(1), 2–14.

Satrio, R.D., Fendiyanto, M.H. and Miftahudin, M. (2024) Tools and techniques used at global scale through genomics, transcriptomics, proteomics, and metabolomics to investigate plant stress responses at the molecular level. In: *Molecular Dynamics of Plant Stress and Its Management*. Springer Nature, Singapore, pp. 555–607.

Satya, S., Sharma, S., Choudhary, G. and Kaushik, G. (2024) Advances in environmental microbiology: A multi-omic perspective. In: *Microbial Omics in Environment and Health*. Springer Nature, Singapore, pp. 175–204.

Shah, A.D., Goode, R.J., Huang, C., Powell, D.R. and Schittenhelm, R.B. (2019) LFQ-analyst: An easy-to-use interactive web platform to analyze and visualize label-free proteomics data preprocessed with MaxQuant. *Journal of Proteome Research* 19(1), 204–211.

Shahmuradov, I.A., Gammerman, A.J., Hancock, J.M., Bramley, P.M. and Solovyev, V.V. (2003) PlantProm: A database of plant promoter sequences. *Nucleic Acids Research* 31(1), 114–117.

Shamshiri, R.R., Sturm, B., Weltzien, C., Fulton, J., Khosla, R. *et al.* (2024) Digitalization of agriculture for sustainable crop production: A use-case review. *Frontiers in Environmental Science* 12, 1375193.

Shannon, P., Markiel, A., Ozier, O., Baliga, N.S., Wang, J.T. *et al.* (2003) Cytoscape: A software environment for integrated models of biomolecular interaction networks. *Genome Research* 13(11), 2498–2504.

Sharma, S., Pandey, S., Muthamilarasan, M., Chaudhry, V., Dulani, P. *et al.* (2017) Genomics resources for abiotic stress tolerance in Solanaceae crops. In: Kumar Chakrabarti, S., Xie, C. and Kumar Tiwari, J. (eds) *The Potato Genome*. Springer Nature, Cham, pp. 195–216.

Shirasawa, K., Isobe, S., Tabata, S. and Hirakawa, H. (2014) Kazusa Marker DataBase: A database for genomics, genetics, and molecular breeding in plants. *Breeding Science* 64(3), 264–271.

Singh, K., Balyan, H.S. and Gupta, P.K. (2024b) Bioinformatics for Analyzing the Role of Epigenetics in Plant Disease Resistance. *Bioinformatics for Plant Research and Crop Breeding*, pp. 351–390.

Singh, S., Praveen, A., Dudha, N., Sharma, V.K. and Bhadrecha, P. (2024a) Single-cell transcriptomics: A new frontier in plant biotechnology research. *Plant Cell Reports* 43(12), 294.

Singh, S.K., Jha, R., Pandey, S., Mohan, C., Ghosh, S. *et al.* (2025) Artificial intelligence-based tools for next-generation seed quality analysis. *Crop Design* 4(1), 100094. DOI: 10.1016/j.cropd.2024.100094.

Smith, C.A., Want, E.J., O'Maille, G., Abagyan, R. and Siuzdak, G. (2006) XCMS: Processing mass spectrometry data for metabolite profiling using nonlinear peak alignment, matching, and identification. *Analytical Chemistry* 78(3), 779–787.

Sood, P., Pandey, S. and Prasad, M. (2021) Biotechnological advances and advanced mutation breeding techniques for crop plants. In: Kumar, P. and Kumar Thakur, A. (eds) *Crop Improvement*. CRC Press, Boca Raton, pp. 139–165.

Srivastava, U., Kanchan, S., Kesheri, M., Gupta, M.K. and Singh, S. (2023) Types of omics data: Genomics, metagenomics, epigenomics, transcriptomics, proteomics, metabolomics, and phenomics. In: Kumar Gupta, M., Katara, P., Mondal, S. and Ram Lakhan Singh, R. (eds) *Integrative Omics*. Academic Press, pp. 13–34. DOI: 10.1016/B978-0-443-16092-9.00002-3.

Stamatakis, A. (2015) Using RAxML to infer phylogenies. *Current Protocols in Bioinformatics* 51(1), 6–14.

Sterk, P., Kulikova, T., Kersey, P. and Apweiler, R. (2007) The EMBL nucleotide sequence and genome reviews databases. *Plant Bioinformatics: Methods and Protocols* 1–21.

Sugawara, H., Ogasawara, O., Okubo, K., Gojobori, T. and Tateno, Y. (2008) DDBJ with new system and face. *Nucleic Acids Research* 36(suppl_1), D22–D24.

Szcześniak, M.W., Kabza, M., Pokrzywa, R., Gudyś, A. and Makałowska, I. (2013) ERISdb: A database of plant splice sites and splicing signals. *Plant and Cell Physiology* 54(2), e10.

Szklarczyk, D., Kirsch, R., Koutrouli, M., Nastou, K., Mehryary, F. *et al.* (2023) The STRING database in 2023: Protein–protein association networks and functional enrichment analyses for any sequenced genome of interest. *Nucleic Acids Research* 51(D1), D638–D646.

Tamura, K., Stecher, G. and Kumar, S. (2021) MEGA11: Molecular evolutionary genetics analysis version 11. *Molecular Biology and Evolution* 38(7), 3022–3027.

Tavşanoğlu, Ç. and Pausas, J.G. (2018) A functional trait database for Mediterranean Basin plants. *Scientific Data* 5(1), 1–18.

Tian, F., Yang, D.C., Meng, Y.Q., Jin, J. and Gao, G. (2020) PlantRegMap: Charting functional regulatory maps in plants. *Nucleic Acids Research* 48(D1), D1104–D1113.

Trapnell, C., Roberts, A., Goff, L., Pertea, G., Kim, D. *et al.* (2012) Differential gene and transcript expression analysis of RNA-Seq experiments with TopHat and Cufflinks. *Nature Protocols* 7(3), 562–578.

Verma, M., Kumar, V., Patel, R.K., Garg, R. and Jain, M. (2015) CTDB: An integrated chickpea transcriptome database for functional and applied genomics. *PloS One* 10(8), e0136880.

Wang, Y. (2017) PlantRGDB: A database of plant retrocopied genes. *Plant and Cell Physiology* 58(1), e2–e2.

Wee, Y., Bhyan, S.B., Liu, Y., Lu, J., Li, X. *et al.* (2019) The bioinformatics tools for the genome assembly and analysis based on third-generation sequencing. *Briefings in Functional Genomics* 18(1), 1–12.

Wheeler, D.L., Barrett, T., Benson, D.A., Bryant, S.H., Canese, K. *et al.* (2008) Database resources of the national center for biotechnology information. *Nucleic Acids Research* 36(suppl_1), D13–D21.

Xia, J., Psychogios, N., Young, N. and Wishart, D.S. (2009) MetaboAnalyst: A web server for metabolomic data analysis and interpretation. *Nucleic Acids Research* 37(suppl_2), W652–W660.

Yu, B., Liu, J., Wu, D., Liu, Y., Cen, W. *et al.* (2020) Weighted gene coexpression network analysis-based identification of key modules and hub genes associated with drought sensitivity in rice. *BMC Plant Biology* 20, 1–21.

Yuan, Y., Bayer, P.E., Batley, J. and Edwards, D. (2021) Current status of structural variation studies in plants. *Plant Biotechnology Journal* 19(11), 2153–2163.

Zhang, H., Jin, Z., Cui, F., Zhao, L., Zhang, X. *et al.* (2023) Epigenetic modifications regulate cultivar-specific root development and metabolic adaptation to nitrogen availability in wheat. *Nature Communications* 14(1), 8238.

Zhang, N., Tang, L., Li, S., Liu, L., Gao, M. *et al.* (2025) Integration of multi-omics data accelerates molecular analysis of common wheat traits. *Nature Communications* 16(1), 2200.

Zhao, M., Du, C., Zeng, J., Gao, Z., Zhu, Y. *et al.* (2023) Integrated omic analysis provides insights into the molecular regulation of stress tolerance by partial root-zone drying in rice. *Frontiers in Plant Science* 14, 1156514. DOI: 10.3389/fpls.2023.1156514.

8 Multiomics for Dissecting Complex Traits

Temsuramok Jamir[1], Amarjeet Kumar[2*], Alka Bharati[3], Ajay Kumar Chandra[4], Vishal Dinkar[5], Bhupendra Choudhary[1], Narendra Kumar[2], Aditya Pratap Singh[6] and Ashutosh Singh[7]

[1]CPGS-AS, CAU(I), Umiam, Meghalaya, India; [2]College of Agriculture, CAU(I), Kyrdemkulai, Meghalaya, India; [3]ICAR – Research Complex for NEH region, Umiam, Meghalaya, India; [4]IIT, Guwahati, Assam, India; [5]ICAR–Indian Institute of Pulses Research, Kanpur, India; [6]Department of Genetics and Plant Breeding, Bidhan Chandra Krishi Vishwavidyalaya, Mohanpur, West Bengal, India; [7]Dr. Rajendra Prasad Central Agricultural University, Pusa, Bihar, India

Abstract

Understanding the intricate relationship between phenotype and genotype in complex traits remains a significant challenge for researchers. However, advancements in omics technologies have greatly improved the precision of trait analysis. Omic technologies including genomics, transcriptomics, proteomics, metabolomics, ionomics and their integration with high-throughput technologies and advanced computational tools has made the application of these omics-driven approaches more feasible and widespread in crop science. Key candidate genes and metabolic pathways identified through these methods have been successfully incorporated into breeding strategies such as genomics-assisted and marker-assisted selection, as well as molecular breeding and genetic engineering. Furthermore, high-throughput phenotyping integrated with metabolomics has reinforced the impact of omics approaches in crop improvement efforts. By combining multiomics methodologies with advanced plant breeding techniques, high-throughput experimentation, sophisticated bioinformatics, computational modeling, and systems biology, researchers can gain deeper insights into complex agricultural traits. These include yield potential, resistance to environmental stressors, and improvements in quality attributes. Ultimately, this integrative approach paves the way for the development of climate-resilient crop varieties, ensuring sustainable agricultural productivity.

8.1 Introduction

The rising global demand for food crops, coupled with the threats of climate change, presents a key challenge to sustainable food production. While fertilizers and chemicals have been used in the past to boost crop production by managing weeds and pests, these stretegies are unsustainable and depend on limited resources (Scheben *et al.*, 2017). Humanity faces substantial challenges in addressing global climate change, meeting nutritional needs, and ensuring an adequate energy supply. Cereal crops, which belongs to the Poaceae family grown for their edible grains, play a crucial role in providing food energy for both humans and livestock,

*Corresponding author: amarjeetgpb@gmail.com

DOI: 10.1079/9781800628717.0008

making them the most extensively produced crop species. Wheat, rice and maize (corn) are the leading cereal crops worldwide in terms of production (Goron and Raizada, 2015). While rice, wheat, and maize provide a significant portion (nearly 40%) of our daily calories, they require more resources to cultivate than climate-smart 'minor' cereals, such as pearl millet. Other cereal crops include barley, oat, rye, sorghum, foxtail millet, and finger millet (Sood *et al.*, 2019; Dinkar *et al.*, 2024). However, the ongoing production of wheat rice and maize, and other key cereal crops, are under threat from climate change, which jeopardizes global food security.

Over time, climatic conditions are anticipated to worsen, causing significant risks to natural ecosystems, as well as agriculture, and ultimately to the global economy (Pretty *et al.*, 2018). With the unpredictable consequences of climate change, production of enough food to meet the needs of the growing global population is particularly daunting for agriculture. Climate change has the potential to disrupt weather patterns, alter precipitation cycles, shift temperatures, and modify carbon dioxide levels in specific regions (Kole *et al.*, 2015). As a result of these changes, crops may experience more stress from the environment, a rise in pest and disease outbreaks, and a general decline in crop production.

The primary goal of climate-smart agriculture is to sustainably increase food production while minimizing environmental impact. To address the issue of the crop production deficit arising from climate change, transitioning from conventional agriculture to climate-smart practices is essential (Steenwerth *et al.*, 2014). Addressing these challenges requires a collaborative effort utilizing interdisciplinary and transdisciplinary approaches. Different omics tools can be utilized independently or in integration to elucidate the biological functions associated with the genetic information, ultimately aiding in the development of adaptive strategies to tackle climate change. Omics refers to high-throughput technologies offering a comprehensive perspective of biological systems. Omics-based strategies have demonstrated their effectiveness in uncovering the genetic and molecular basis of crop development by analyzing alterations in DNA, gene expression, proteins, metabolites, and mineral nutrients in response to environmental and physiological stresses. (Muthamilarasan *et al.*, 2019; Kumar *et al.*, 2024).

8.2 Climate Change and Agriculture

Scientists today face significant challenges in securing food production by enhancing agricultural productivity as well as maintaining environmental sustainability while adapting to climate change. A rise in global temperatures, unpredictable pattern of rainfall, and expansion of drought- and flood-prone regions are some of the results of the ongoing shift in climate, which adversely affects plant growth and crop yields (Lamaoui *et al.*, 2018). Severity in abiotic stresses, such as drought, heat, salinity, cold, and submergence, is likely to be heightened by climate-related changes (Pandey *et al.*, 2017). Additionally, climate change intensifies both abiotic stresses (e.g. drought, heat, salinity, cold, and flooding) and biotic stresses, leading to increased pest infestations, the spread of plant diseases, the emergence of invasive weeds, and declines in beneficial pollinators and soil microbes (Kole *et al.*, 2015; Raza *et al.*, 2022).

To counter these challenges, sustainable agricultural practices must be adopted to enhance crop resilience in changing climatic conditions (Vaughan *et al.*, 2018). The development of crop varieties resilient to climate change, followed by efficient resource utilization, integrated ecosystem management, and diversification of agricultural systems will be the key strategies for mitigating the effect of climate change (Gil *et al.*, 2017; Dhankher and Foyer, 2018; Evans and Lawson, 2020). Exploring wild relatives of crops and underutilized plant species can provide essential genetic diversity for breeding climate-resilient varieties (Chandra *et al.*, 2024).

8.2.1 Current challenges in crop production

The impact of climate change on agricultural production is now widely acknowledged as an unavoidable threat. Climate change presents an undeniable threat to global agriculture, particularly affecting staple cereals. By 2050, food production must increase by 60–100% to support the projected global population of nine billion people (FAO, 2013).

The major stress factors with the potential to significantly impact crop yield are high temperature and drought (Bharati *et al.*, 2024). To enhance yield safety, future breeding efforts should be informed by new knowledge about the processes that govern the development of plant and its response to environmental stress. It is crucial to investigate the interactions between plant structure, function, and the environment at different stages of plant development, including organismal, cellular, and molecular levels, to gain a comprehensive understanding. Economically important crop species, such as cereals, are frequently constrained in their growth and productivity by abiotic stresses like extreme temperatures and low water availability (Barnabás *et al.*, 2008). Cereal breeding should prioritize the development of cultivars that can withstand both high temperatures and low water supply, as they often occur together (Tester and Bacic, 2005). Heat stress during the growth phase leads to various metabolic and physiological changes, including disruptions in hormone balance. Change in hormone level, its regulation, interaction and cross talk are responsible for cellular-, organ- and plant-level change due to heat stress (Maestri *et al.*, 2002).

The yield of cereal crops is significantly affected by environmental stresses. The impact of heat and drought stress on crop yield is multifaceted, influencing processes such as gametogenesis, fertilization, and grain development (Sehgal *et al.*, 2018). Despite recent advancements, our understanding of these processes remains incomplete, necessitating the application of modern biotechnological tools to decode the intricate molecular networks regulating plant responses to environmental stress.

8.3 Omics Technologies in Complex Trait for Crop Improvement

Accurate phenotyping has significant role in trait improvement. Therefore, phenotypic and genotypic characterization of complex traits is a challenging task for a breeder and biotechnologist. The advancement and inclusion of multiomics technologies like genomics, transcriptomics, proteomics, metabolomics have refined our understanding and knowledge of the complex traits of crop plants (Fig. 8.1). The genetic dissection of phenotypically complex traits plays a crucial role in understanding their mechanism as well as in identifying the corresponding variations.

Candidate allele mining is a critical step in the dissection of a complex traits (Mahmood *et al.*, 2022). Presently, many crops have fully sequenced genome information that helps us in understanding the genotypic and phenotypic association of complex traits precisely. The most common method used for identifying genes or quantitative trait loci (QTLs) for complex traits is linkage mapping for biparental populations like the F2 population, recombinant inbred lines (RILs), and backcross population. In addition, the exploitation of genetic variation at the whole-genomic level in natural populations can be also used through genome-wide association studies (GWAS). Joint linkage–LD (linkage disequilibrium) mapping for a naturally collected plant population and family-based population is also a very powerful tool for mapping the genomic region responsible for complex traits.

In order to develop climate-smart crops with improved tolerance to environmental and biological stresses, higher nutritional value, and important agronomic traits, it is essential to employ novel techniques for crop enhancement. These techniques should enable the utilization of wild crop relatives and lesser-known species that possess the necessary natural genetic variation (Sood *et al.*, 2019; Kumar *et al.*, 2024). Among these methods, the use and integration of 'omics' technologies plays a crucial role. These technologies have been instrumental in identifying key genes, proteins, and metabolic pathways associated with various important agronomic traits (Table 8.1). The identified genes and QTLs are further exploited for breeding crops, mainly cereal crops, through marker-assisted backcross breeding.

Advancements in genome sequencing, pangenomics, and GWAS are helping identify species-wide genetic variations. Meanwhile,

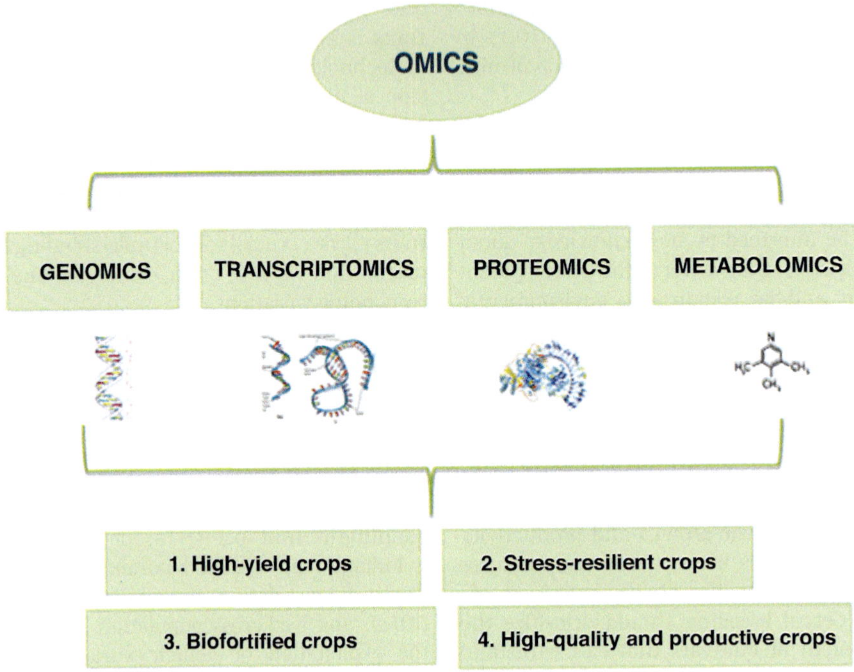

Fig. 8.1. Major omics strategies for crop improvement and climate resilience. Author's own image.

Table 8.1. Examples of significant genes contributing to complex climate-resilience traits in crops. Author's own table.

Crop	Target gene	Trait observed	Reference
Rice	Sub1A, Sub1B and Sub1C	Submergence tolerance	Singh *et al.*, 2020
	RR22	Enhanced salt tolerance	Zhang *et al.*, 2019
	HIS1	Herbicide resistance	Maeda *et al.*, 2019
Wheat	TdCBF-A14	Frost tolerance	Sieber *et al.*, 2016
	DREB2 and ERF3	Improved drought tolerance	Kim *et al.*, 2018
Maize	ZmTIP1	Root hair elongation and drought tolerance	Zhang *et al.*, 2020
	ARGOS8	Improved grain yield under drought stress tolerance	Shi *et al.*, 2017
Barley	HvCKX2.1	Drought stress response	Czajkowska *et al.*, 2019
Foxtail millet	SiDREB2	Dehydration tolerance	Lata *et al.*, 2011

transcriptomics, particularly RNA sequencing, is uncovering complex plant responses to abiotic and biotic stresses (Zenda *et al.*, 2021). Similarly, improvements in proteomics technologies are enabling detailed analyses of protein interactions, metabolic pathways, and post-translational modifications, providing deeper insights into stress adaptation mechanisms (Abdelrahman *et al.*, 2018). With high-throughput phenomics, researchers can now explore new avenues such as the study of root systems (e.g. root system architecture) and also the microbiome associated with the root system.

8.3.1 Genomics

The foundation of crop genetic and genomic research lies in the genome sequences of reference crops, which offer valuable insights into the genetic composition, variability in the genome, and the genetic basis of agronomic traits (Scheben *et al.*, 2017). With the widespread adoption of genome sequencing technologies, there has been a growing number of assembled plant genomes, including cultivated crop species as well as their wild relatives. Structural genomics has revealed the complexity of plant genomes. Plant genomes present a greater challenge for assembly than animal genomes due to their large size, repetitive sequences, and polyploidy. The assembly of the repetitive genomic region using short read-sequences poses obstacles. The use of long-read sequencing like SMRT (single-molecule real-time) along with optical mapping is aimed at addressing this issue and enhancing plant genome assemblies.

Structural genomics investigates genome architecture. It characterizes the genome by the construction of gene maps and sequences, as well comparative gene mapping and sequence analysis, aiding gene discovery and characterization (Mohanta *et al.*, 2017). With the advent of next-generation sequencing, whole-genome information is now available for many crop species, but future research needs to focus on elucidating gene function from genomic structure. Functional proteomic libraries can facilitate this process, allowing for the identification of protein functions through structural similarity comparisons.

8.3.2 Transcriptomics

Transcriptome profiling provides a comprehensive overview of all RNA molecules present in a cell, tissue, organ, or entire organism at a specific time and condition, including 'mRNA, tRNA, rRNA, sRNA, and other non-coding RNA', which cannot be examined at the genomic level (Weckwerth *et al.*, 2020; Chaturvedi *et al.*, 2021). Understanding the transcriptome is essential to deciphering the functional element of genome. The transcriptome uncovers the molecular constituents of the specific tissue or cells, as well as an understanding of how cells respond to changes induced by developmental and environmental stimuli. In contrast to the stable genome, the transcriptome varies under different spatial and temporal condition as well as in different tissue types. Therefore, transcriptomic studies help in unraveling the molecular-level stress responses of an organism (Kukurba and Montgomery, 2015; Escandón *et al.*, 2018).

The analysis of the entire set of RNA transcripts in an organism is crucial for understanding the genome's functional components, gene expression, and how it is controlled (Lowe *et al.*, 2017). The use of short nucleotide signatures and hybridization-based high-throughput methods (e.g. microarray) were a cost-effective approach for analyzing thousands to millions of transcripts concurrently, enabling comprehensive transcriptome studies. However, these methods were cumbersome and time consuming. Sequencing based transcriptome study can be traced back to the sequencing of expressed sequence tags (ESTs) through the Sanger sequencing platform. Following this, the development of technologies such as serial analysis of gene expression (SAGE), massively parallel signature sequencing (MPSS), and cap analysis of gene expression (CAGE) offered precise information of expression data (Chaitankar *et al.*, 2016). Modern transcriptomic technologies, including RNA sequencing, have replaced older methods such as the sequencing of ESTs and hybridization-based microarrays. These high-throughput approaches provide precise digital gene expression data, enabling the identification of genes responsible for environmental stress, as well as their regulatory network.

8.3.3 Proteomics

The study of the complete protein set within an organism, tissue, or cell under particular circumstances is known as proteomics. Unlike transcriptomics, which analyzes gene expression at the RNA level, proteomics provides insights into protein abundance, isoforms, interactions, localization, and post-translational modifications (Yang *et al.*, 2015). The transcriptome composition of a tissue or cell does not always accurately predict the phenotype of plant due to several post-transcriptional modifications such as alternate splicing, trans splicing, RNA degradation, etc. Therefore, transcribed RNA may not represent the downstrean translated protein pool of the tissue. Proteomics provides a crucial link between transcriptomics and metabolomics, as proteins are the direct drivers of plant responses to developmental and environmental cues (Tan *et al.*, 2017; Labuschagne, 2018).

The changing transcript levels may not always match with their corresponding proteins, making proteomic studies crucial. Analyzing the proteome also offers insights into the subcellular localization of a protein, post-translational modifications, three-dimensional protein structure and protein–protein interactions in different cellular or environmental conditions (Yang *et al.*, 2015). The processes of post-translational modification and protein interaction reveal the complexity of how genetic information is ultimately expressed as functional proteins. Studying these mechanisms provides valuable insights into their biological roles, developmental significance, and adaptive functions.

8.3.4 Metabolomics

Plants produce an array of metabolites to counter environmental stresses, including antioxidants, osmoprotectants, and phytohormones through different biochemical, and molecular mechanisms (Sabagh *et al.*, 2021). These responses against environmental and pathogenic stresses include the production of a wide range of metabolites, the activation of antioxidant enzymes, the uptake and transportation of ions, the accumulation of osmoprotectants (especially proline), and the release of phytohormones, among other processes (Pandey *et al.*, 2015; Singhal *et al.*, 2021). Metabolites, which include both intermediate and final products of metabolism, play a crucial role as signaling molecules that facilitate communication between organelles and the cytoplasm through retrograde signaling (Bobik and Burch-Smith, 2015). To decipher the intricate molecular pathways that control physiological processes, a comprehensive qualitative and quantitative analysis of metabolites is necessary across various developmental stages and environmental conditions. Such analyses have been widely applied in comparing different genotypes, studying mutants and transgenic organisms, and assessing plant responses to developmental phases, environmental stressors, nutrient availability, and toxicological effects.

Metabolites encompass a diverse range of primary and secondary compounds, including organic acids, sugar alcohols, hormones, allelochemicals, ketones, amino acids, and steroids (Razzaq *et al.*, 2019). Notably, plants continuously adapt their metabolic processes to withstand adverse conditions by synthesizing protective molecules such as antioxidants, osmoprotectants, and stress-responsive proteins (Ramalingam *et al.*, 2015). In this context, metabolomics serves as a powerful tool for systematically identifying, quantifying, and analyzing small-molecule metabolites (collectively referred to as the metabolome) within cells, tissues, or entire organisms. This field of study provides critical insights into metabolic adaptations during various developmental stages and in response to environmental challenges, ultimately enhancing our understanding of plant resilience and stress tolerance (Fiehn, 2002; Arbona *et al.*, 2013).

8.4 Genetics-to-Genomics for Climate Resilience

The development of high-throughput sequencing technologies, along with decreases in cost and improvements in the algorithms used to assemble sequence data and the platforms used to survey genomes, has made genomic tools and resources widely available for a large number of crop species (Jayakodi *et al.*, 2021). Among the essential resources now available are high-quality reference genomes, DNA markers, and genetic maps, which are fundamental to both genomic research (functional and comparative) and applied efforts in molecular breeding and crop improvement. (Zhang and Hao, 2020).

Reference genomes are available for several key agricultural crops enabling the resequencing of maximum genotypes and accessions of these crop species. High-throughput resequencing has further enhanced our ability to identify and map the rarest candidate genes responsible for important agronomic traits. In addition, the identification of regulatory regions and variation in these genomic regions across genotypes enables us to predict more refined gene models. This, in turn, has provided deeper insights into gene functions and their roles in plant development and stress responses (Morrell *et al.*, 2012; Schreiber *et al.*, 2018). The continuous advancements in genomic research now enable scientists to conduct large-scale genome-wide analyses to identify genes governing critical agronomic traits, thereby facilitating precision breeding strategies to improve crop resilience, productivity, and nutritional value (Zhang *et al.*, 2019; Nakano *et al.*, 2020; Chandra *et al.*, 2024).

8.4.1 Genetic diversity and climate adaptation

Given the urgent need to diversify our food sources, boost agricultural productivity, and improve nutritional security in a changing climate, we must move beyond conventional adaptation strategies and embrace more holistic and transformative solutions. Developing crop varieties that can maintain high yields and productivity in the face of increasingly unpredictable climate conditions is essential. A crucial part of this effort involves harnessing the genetic potential of crop wild relatives (CWRs) and underutilized crop species (UCS), which possess naturally evolved traits for resilience and adaptability (Kole *et al.*, 2015).

Plant breeders can now significantly accelerate crop improvement by leveraging recent advances in plant genomics to pinpoint key genes for important agricultural traits, and by integrating these genomic technologies with approaches such as gene editing and genomic selection (Pourkheirandish *et al.*, 2020). These cutting-edge techniques not only shorten breeding cycles and reduce costs but also enhance critical traits, enabling crops to withstand environmental stresses while improving their nutritional value.

The discovery of genes associated with essential adaptive traits hinges on the development of novel germplasm resources that exhibit unique allelic diversity within valuable genetic backgrounds. One of the most promising strategies for creating climate-resilient crop varieties is the de novo domestication of CWRs. Over millennia of natural selection and evolutionary pressure, CWRs have accumulated a diverse array of genetic adaptations that enable them to thrive in extreme biotic and abiotic environments (Ananda *et al.*, 2020). Modern elite cultivars have been specifically bred for agricultural systems that rely on high inputs and minimize environmental challenges. In contrast, CWRs possess a diverse set of physical and physiological traits that enable them to thrive in harsh and challenging environments. CWRs and traditional landraces are thus highly valuable genetic resources for traits such as tolerance to drought, heat, salinity, and cold. By incorporating these beneficial alleles through conventional breeding or molecular techniques, researchers can develop robust, climate-adaptive crop varieties that ensure food security in an era of escalating climate challenges.

8.4.2 Functional genomics and mutagenomics

The wealth of information and resources from structural genomics will eventually be applied in functional genomics studies to give us a more complete picture of how genes function. Functional genomics focuses on developing systematic experimental approaches to analyze gene activity, interactions, and regulatory mechanisms (Sood *et al.*, 2019). Sophisticated biotechnological tools are available for gene identification, isolation, cloning, characterization, and functional analysis (via overexpression or knockout studies). The advent of crop genome sequencing has significantly facilitated the identification of genes responsible for critical traits such as disease resistance, stress tolerance, and yield optimization.

In addition, precision genome engineering, which utilizes advanced genome-editing technologies such as the CRISPR/Cas9 system and TALENs, has emerged as a powerful approach

Table 8.2. Examples of improved cereal crops produced by induced mutations. Author's own table.

Species name	Mutant name	Type of mutation used	The trait improved
Oryza sativa L.	Sinar 1	Gamma radiation	Higher yield and higher aromatic value
Triticum aestivum L.	Binagom-1	Direct use of an induced mutant	Has higher salinity tolerance, higher yield
Horderum vulgare L.	Phenix	Hybridization with mutant	Improved drought tolerance
Zea mays L.	Longfuyu	Direct use of an induced mutant	Improved resistance to bacterial diseases
Sorghum bicolor L.	Fambe	Direct use of an induced mutant	Resistance to lodging

Source: IAEA program of Nuclear Techniques in Agriculture (www.iaea.org).

for crop improvement (Rinaldo and Ayliffe, 2015). These genome-editing technologies, those without foreign DNA insertion, offer a promising way to boost crop yields by creating pest- and disease-resistant genetically modified crops. A notable example of their application is the successful production of a mutant bread wheat variety with enhanced resistance to powdery mildew through targeted modification of the *Mildew resistance locus O* (*TaMlo*) using TALEN and CRISPR/Cas9 technologies (Wang *et al.*, 2016).

A contemporary omics-driven approach, mutagenomics, enables the investigation of genetic modifications by analyzing mutational events and their impact on phenotypic traits. High-throughput genomic technologies such as serial analysis of gene expression (SAGE), high-resolution melt (HRM) analysis, targeted induced local lesions in genomes (TILLING), and microarray-based studies facilitate the characterization of these genetic alterations. Reverse genetic approaches in mutagenomics provide powerful tools for functional gene analysis by silencing or disrupting specific candidate genes (Table 8.2). Techniques like RNA interference (RNAi) and virus-induced gene silencing (VIGS) are particularly valuable for inducing targeted mutations in crops where naturally occurring mutant alleles are unavailable. These approaches also enable gene knockdown or silencing, allowing researchers to dissect gene function and its role in plant development and stress adaptation (Talukdar and Sinjushin, 2015).

8.4.3 Marker-assisted selection (MAS) and genomic selection

Genotyping by sequencing (GBS) has revolutionized the genotyping of crop by providing an efficient and cost-effective means of identifying genetic variations associated with key agronomic traits. The growing adoption of these techniques has led to single nucleotide polymorphisms (SNPs) emerging as the preferred molecular markers due to their high heritability, broad distribution across the genome, and ability to precisely detect causal variations at a single-base resolution (Sood *et al.*, 2019). GBS approaches primarily include whole genome resequencing (WGR) and reduced representation sequencing (RRS). While WGR provides comprehensive genomic coverage, it remains cost-prohibitive for large-genome crops such as wheat (Mohanta *et al.*, 2017).

In contrast, RRS reduces sequencing costs by selectively targeting a subset of the genome through restriction enzyme digestion. This approach significantly lowers per-sample sequencing expenses, making it more feasible for large-genome crops. RRS is limited by its lower SNP density due to the frequent occurrence of substantial missing data, caused by polymorphisms in the restriction site and also the stochastic nature of sequence sampling. Despite these challenges, advances in computational imputation methods are improving the accuracy and completeness of SNP datasets, enhancing the utility of RRS for crop genotyping. Given its ability to generate informative SNP markers,

GBS has become a valuable tool for molecular breeding, facilitating marker-assisted selection and genomic studies. As sequencing costs continue to decline, GBS will become increasingly accessible for both major staple crops and underutilized cereal species, providing critical resources for breeding climate-resilient crop varieties (Kole *et al.*, 2015).

Integrating genomic selection (GS) into breeding programs has further sped up crop improvement by allowing breeders to bypass the need for extensive phenotypic assessments of individual loci. GS predicts genomic estimated breeding values (GEBVs) for breeding lines based on a reference training population that has been both phenotyped and genotyped. This predictive approach allows breeders to efficiently select and propagate superior individuals across multiple generations without the labor-intensive phenotypic screening traditionally required.

GBS has played a crucial role in empirical GS studies, particularly in wheat, where it has been successfully applied to predict key traits such as yield and disease resistance. Notably, Poland and Rife (2012) and Rutkoski *et al.* (2011) employed GBS-based GS models in elite wheat breeding lines, achieving high prediction accuracies for yield and stem rust resistance, respectively. Furthermore, the use of genomic prediction with GBS data has been successfully extended to other crops, including maize, demonstrating its potential to capture genetic variation in breeding populations. Furthermore, GBS has been shown to perform comparably to high-density SNP arrays while offering a more cost-effective alternative due to its higher marker density. These advancements highlight the transformative potential of GS in breeding programs, enabling breeders to enhance crop productivity, stress resilience, and overall genetic gain more efficiently than traditional selection methods (Varshney *et al.*, 2014).

8.4.4 Transcriptomics in stress response

Transcriptomics aims to analyze the entire collection of RNA transcripts (the transcriptome) present in a particular cell or tissue of an organism (Raza *et al.*, 2022). This field provides a comprehensive approach to understand expression patterns of genes in response to different environmental and physiological stimuli over a given temporal frame (Duque *et al.*, 2013). By profiling the transcriptome, researchers can analyze the differential expression of genes under specific conditions, gaining insights into their functional roles.

Techniques such as cDNA-AFLP (amplified fragment length polymorphism), differential display-PCR (DD-PCR), and suppression subtractive hybridization (SSH) were initially employed for transcriptome profiling, but these methods lacked both the necessary resolution and sensitivity for detailed analysis (Nataraja *et al.*, 2017). The emergence of advanced molecular technologies significantly enhanced transcriptomic analysis, with methods such as microarrays, digital gene expression profiling, next-generation sequencing (NGS), RNA sequencing (RNA-Seq), and SAGE enabling more precise and high-throughput gene expression studies.

8.4.4.1 Trancriptomic profiling under stress

Valuable insights into how gene expression is regulated under stress have been gained through microarray analyses. Research on soybean and barley has demonstrated that gene expression patterns vary across developmental and reproductive stages during drought stress (Le *et al.*, 2012; Guo *et al.*, 2009). Similarly, transcriptome profiling using an Affymetrix GeneChip array revealed differentially expressed genes (DEGs) in soybean under dehydration stress (Khan *et al.*, 2020).

Transcription factors (TFs) play a crucial role in stress responses by modulating gene expression. Two important TFs in rice, C2H2-ZF and DST, have been identified as regulators of stomatal opening, thereby affecting the plant's ability to tolerate salt and drought stress (Huang *et al.*, 2009). Likewise, WRKY transcription factors have been shown to contribute to wheat's abiotic stress response (Okay *et al.*, 2014).

Beyond TFs, other regulatory molecules, including, non-coding RNAs, phytohormones and small peptides, have been shown to play important roles in controlling gene expression when crops like Arabidopsis, rice, wheat, and tomato are exposed to abiotic stress conditions (Matsui *et al.*, 2008; Chekanova, 2015; Bashir *et al.*, 2019). Some TFs, like DRE-binding protein

2 (DREB2) and dehydration-responsive element/C-repeat (DRE/CRT), regulate phytohormone-independent abiotic stress responses in rice (Todaka *et al.*, 2015).

Transcriptomic analyses conducted in sorghum have identified DEGs that are involved in the plant's responses to several environmental stresses, including drought, heat, and osmotic stress, as well as responses to hormone treatments (Dugas *et al.*, 2011; Johnson *et al.*, 2015). A similar pattern of differential gene expression, specifically in the *OsMADS* gene family, has been observed in developing rice plants subjected to drought stress (Jin *et al.*, 2013). The data generated by these transcriptomic studies are valuable for further discovering and validating the role of candidates in the stress response. These candidate gene could be further used crop improvement.

8.4.4.2 *RNA sequencing*

Recent advancements in transcriptomics have been driven by RNA sequencing (RNA-Seq), a high-throughput technique that provides an in-depth analysis of gene expression. Further advancement in technology also allows us to perform RNA sequencing directly within cells or tissues (*in-situ* RNA-Seq), preserving spatial gene expression patterns (Ke *et al.*, 2013). Spatially resolved transcriptomics further refines this approach by mapping gene expression with spatial specificity, providing a detailed molecular understanding of cellular and physiological processes (Burgess, 2015).

RNA-Seq studies have played a vital role in identifying how gene expression patterns differ across various tissues in response to both abiotic and biotic stress in crops like foxtail millet and sweet potato (*Ipomoea batatas* L.). Moreover, total RNA-Seq analyses performed during cowpea (*Vigna unguiculata* L.) development have identified DEGs and simple sequence repeat (SSR) markers, which facilitates marker-assisted selection for crop improvement studies (Qi *et al.*, 2013; Bonthala *et al.*, 2016; Chen *et al.*, 2017).

RNA-Seq technology has proven highly effective in identifying the genes that are activated or repressed in response to drought stress in important crops, including maize, wheat, rice, and rapeseed (*Brassica napus*). Several research projects have utilized this technology to explore the changes that occur in the transcriptome under drought conditions, which has resulted in the identification of candidate genes that may contribute to a plant's ability to tolerate drought (Kakumanu *et al.*, 2012; Huang *et al.*, 2014). The findings discussed here demonstrate the significance of RNA-Seq technology within the field of functional genomics, as it provides valuable information that can be used to develop crop varieties that are more resilient to stress through both traditional breeding techniques that are highly targeted and more advanced biotechnological interventions.

8.4.5 Proteomics for climate resilience

Proteomics is the comprehensive study of all proteins expressed within an organism, categorized into sequence proteomics, structural proteomics, functional proteomics, and expression proteomics (Mosa *et al.*, 2017). Each of these subfields provides unique insights into the role and behavior of proteins in biological systems.

- Sequence proteomics focuses on identifying amino acid sequences within proteins, utilizing high-performance liquid chromatography (HPLC) as a primary method for sequencing (Primrose and Twyman, 2013).
- Structural proteomics aims to determine the three-dimensional configurations of proteins, which is crucial for understanding their functional mechanisms. Techniques such as computer-based modeling, nuclear magnetic resonance (NMR), crystallization, electron microscopy, and X-ray diffraction of protein crystals are commonly employed in this field (Fiser and Sali, 2003; Wood and Woolfson, 2018).
- Functional proteomics investigates the specific roles of proteins in cellular processes through techniques like yeast one-hybrid and two-hybrid assays as well as protein microarray profiling (Leung *et al.*, 2014).

In the last 20 years, plant proteomics has made substantial progress, largely due to improvements in traditional experimental techniques as well as the development of cutting-edge high-throughput and high-resolution technologies. Advancements in these technologies have positively impacted various stages of proteomic research, spanning from sample preparation and protein extraction to protein fractionation, quantification, and data analysis, ultimately leading to a more profound understanding of plant biology and how plants adapt to stressful conditions.

Analyzing plant responses to abiotic and biotic stresses often involves the use of next-generation quantitative proteomics in both comparative and descriptive studies. For instance, an isobaric tag for a relative and absolute quantification (iTRAQ)-based comparative proteomics study in rice identified 368 and 491 upregulated proteins in the salt-tolerant genotype LYP9 when grown under high and moderate salt concentration in growth media, during the maximum tillering stage. These findings provided valuable insights into salinity-responsive proteins and their associated metabolic pathways, thereby contributing to a better understanding of stress adaptation mechanisms in crops (Ahmad *et al.*, 2016; Mustafa and Komatsu, 2021).

These advancements in proteomics not only expand our understanding of plant physiology but also pave the way for developing stress-resilient crop varieties, which are essential for sustainable agriculture in the face of climate change.

8.4.5.1 *Proteomics under stress conditions*

Researchers frequently use comparative transcriptomic analysis to investigate how gene expression patterns change in plants when exposed to both stressful and non-stressful conditions. This approach has provided valuable insights into how plants respond to environmental challenges, such as salinity and heat stress, and has been applied across many different crop species. An example of comparative transcriptomic study in maize highlighted its response to salinity stress, as the salt-tolerant genotype exhibited the upregulation of several key functional genes associated with salt tolerance. These upregulated genes included thioredoxins (*Zm00001d018461*, *Zm00001d041804* and *Zm00001d018238*), salt stress-induced protein (*Zm00001d023516*) and CBL-interacting kinase (*Zm00001d044642*). Additionally, genes involved in defense mechanisms, such as leucine-rich repeat protein (*Zm00001d035756*) and pathogenesis-related protein (*Zm00001d018324*), were also upregulated. Furthermore, multiple TFs from the MYB (*Zm00001d053220*), WRKY (*Zm00001d005622*), and bZIP (*Zm00001d043992*) families, primarily involved in the abscisic acid (ABA) signaling pathway, were significantly expressed. These findings align with previous research that has linked these genes to salt- and drought-stress tolerance (Chen *et al.*, 2014; Zhao *et al.*, 2018; Zenda *et al.*, 2019; Zhang *et al.*, 2023).

When the B73 maize genotype was exposed to heat stress, a study revealed that several families of TFs were significantly more abundant in the protein processing in the endoplasmic reticulum (PPER) pathway, which is known to be a vital mechanism for how plants respond to heat stress. These included AP2-EREBP (*GRMZM2G010555*, etc.), bZIP (*GRMZM2G479760*, etc.), bHLH (*GRMZM2G001930*, etc.), and WRKY (*GRMZM2G324999*, *GRMZM2G071907*, etc.). Additionally, heat shock proteins (HSPs) such as *GRMZM2G069651*, *GRMZM2G366532*, and *GRMZM2G149647* were found to play crucial roles in heat tolerance by assisting in protein folding and stability under high-temperature conditions (Qian *et al.*, 2019). These findings highlight the intricate molecular mechanisms underlying plant responses to abiotic stress and demonstrate the importance of proteomic studies in enhancing stress tolerance in crops.

8.4.5.2 *Functional protein networks*

With the vast amount of publicly available transcriptomic and proteomic data, meta-analysis has become a powerful tool for integrating large-scale datasets to study plant responses to both abiotic and biotic stresses. By analyzing gene expression and co-expression patterns from multiple studies, researchers can obtain more comprehensive and reliable insights into stress adaptation mechanisms (Cohen and Leach, 2019; Tahmasebi *et al.*, 2019).

In a meta-analysis of 391 microarray samples from 23 independent experiments studying stress responses in tomato, 2336 DEGs associated with various stress conditions were identified. Of these, 1862 DEGs were linked to biotic stress responses, while 835 DEGs were related to abiotic stress adaptation. Notably, 4.2% of these DEGs were identified as transcription factor genes, emphasizing their role in stress response regulation (Ashrafi-Dehkordi *et al.*, 2018). Key transcription factors identified in this study included Jasmonate Ethylene Response Factor 1 (*JERF1*), *MYB48*, *EIL2*, *EIL3*, *LATE ELONGATED HYPOCOTYL (LHY)*, and *SlGRAS6*, all of which play significant roles in mediating plant responses to both biotic and abiotic stresses (Ashrafi-Dehkordi *et al.*, 2018). These findings underscore the potential of a meta-analysis for identifying candidate genes that can serve as targets for genetic engineering and molecular breeding, ultimately facilitating the development of crop varieties with enhanced stress tolerance.

The combination of meta-analysis and studies of functional protein networks offers a powerful approach to accelerate the process of identifying genes that are critical for plant adaptation to stress. This accelerated discovery of stress-adaptive genes can then be used to develop more efficient and precisely targeted crop improvement strategies, ultimately aimed at making agriculture more resilient and sustainable as the climate changes.

8.4.6 Metabolomics in stress adapdation

Metabolomics provides a comprehensive analysis of metabolites, the small molecules that participate in the diverse biochemical processes occurring within a biological system. The entire collection of metabolites produced by a plant's metabolic pathways is known as the metabolome (Baharum and Azizan, 2018). Metabolites, being the ultimate products of gene expression, provide insights into a cell's biochemical state and how it physiologically responds to different circumstances (Weckwerth and Fiehn, 2002). While proteomics focuses on the study of proteins, metabolomics offers a broader perspective by assessing protein expression at the metabolic level and identifying the underlying biochemical pathways essential for gene function (Weckwerth and Morgenthal, 2005).

The diverse chemical and physical properties of metabolites necessitate the use of specialized separation and analytical techniques to develop comprehensive metabolic profiles of plant samples (Jogaiah *et al.*, 2013). A variety of analytical platforms are available for quantifying metabolites in plants, including.

- Thin-layer chromatography (TLC)
- Gas chromatography–mass spectrometry (GC–MS)
- Liquid chromatography–mass spectrometry (LC–MS)
- Liquid chromatography–electrochemistry–mass spectrometry (LC–EC–MS)
- Nuclear magnetic resonance (NMR)
- Direct infusion mass spectrometry (DIMS)
- Fourier-transform infrared spectroscopy (FT-IR)
- Capillary electrophoresis–mass spectro metry (CE–MS)

These advanced methodologies enable the accurate identification and quantification of metabolites, offering valuable insights into plant responses to stress conditions

8.4.6.1 Stress-responsive metabolites

Metabolomics is essential for understanding plant systems, partly because plants synthesize a much larger number of metabolites than microbes or animals. Secondary metabolites, in particular, play a crucial role in how plants adapt to environmental challenges. The emerging field of environmental metabolomics focuses on studying metabolic alterations in plants exposed to various abiotic stresses such as drought, salinity, extreme temperatures, and oxidative stress (Brunetti *et al.*, 2013; Viant and Sommer, 2013). For instance, gas chromatography–time of flight mass spectrometry (GC–TOF-MS) technique has been used to detect polyamine metabolites in drought-stressed rice (Do *et al.*, 2013). Researchers have also used metabolomics to screen for salt tolerance in rice and to investigate metabolic changes associated with drought, cold, and heat stress in important crops including wheat, maize, tomato, and soybean. These studies demonstrate the potential of metabolomics in enhancing stress tolerance and improving crop resilience under changing climatic conditions.

8.4.6.2 Metabolic pathways in stress tolerance

The study of metabolic pathways has revealed that plants reconfigure their metabolic networks in response to stress. Techniques such as LC–MS, GC–MS, GC/EI–TOF-MS, HPLC, and NMR have been extensively applied to study abiotic (drought, salinity, oxidative stress, temperature fluctuations) as well as biotic stress responses in crops like rice, tomato, maize, and soybean (Ghatak *et al.*, 2018).

Metabolic engineering has been employed to enhance nutritional value and stress tolerance in crops. For example:

- The genetic modification of rice resulted in β-carotene (pro-vitamin A) accumulation in the endosperm, improving its nutritional quality (Paine *et al.*, 2005).
- Metabolic engineering was used to boost the production of anthocyanin, an important secondary metabolite, in tomato, resulting in better antioxidant properties (Butelli *et al.*, 2008).

The integration of field of metabolomics with other omics fields, such as genomics (the study of genes), transcriptomics (the study of RNA transcripts), and proteomics (the study of proteins), enables a more comprehensive and multi-faceted understanding of plant functions. This approach allows researchers to connect information from the genome all the way to the metabolome (the complete set of metabolites), and to link these molecular processes to the plant's observable characteristics (phenotypic traits). Strong correlations have been observed between these different omics layers, particularly in the context of how crops respond to abiotic stress, providing a strong foundation for developing crop varieties that are better able to withstand environmental challenges.

8.5 Integrating Genetics-to-Genomics for Resistance Breeding of Major Cereal Crops

Climate change-induced factors, such as rising global temperatures and elevated CO_2 levels, are negatively affecting the nutritional quality of staple cereal crops. These declines in essential nutrient density exacerbate the already severe issue of malnutrition, particularly in low-income regions where billions of people depend on cereals as their primary food source (Myers *et al.*, 2014; Macdiarmid and Whybrow, 2019).

Crops provide key nutrients—vitamins, minerals, and dietary fiber—making them essential for human health. Therefore, insufficient access to plant-based diets can lead to:

- Increased risks of non-communicable diseases; and
- Severe micronutrient deficiencies that may not be easily corrected through dietary diversification (Scheelbeek *et al.*, 2018).

In order to overcome these challenges, efforts should be directed towards improving the nutritional content of crops through:

- Breeding programs;
- Agronomic interventions; and
- Genetic engineering and transgenic approaches.

This is essential for sustaining food security and human health in the face of climate change. Successfully improving the nutritional content of crops depends on precisely identifying the key QTLs, genes, and metabolic pathways involved in how plants acquire nutrients. To achieve this, genomic and multiomics approaches have been widely applied to investigate nutrition-related traits, providing critical insights for genomics-assisted breeding (GAB) programs (Singh *et al.*, 2020; Roorkiwal *et al.*, 2022). Recent studies have identified major QTLs associated with improved nutrient content in cereals. Puranik *et al.* (2020) utilized a combination of GBS and GWAS in a population of 190 finger millet (*Eleusine coracana* L.) genotypes. This approach successfully mapped genomic regions responsible for nutritional quality and content in grains, including protein, iron (Fe), zinc (Zn), calcium (Ca), magnesium (Mg), potassium (K), and sodium (Na) levels. These findings highlight the potential of integrating genomic technologies with traditional breeding strategies to develop nutrient-rich, climate-resilient cereal crops, ultimately contributing to global food security and improved human health (Fig. 8.2).

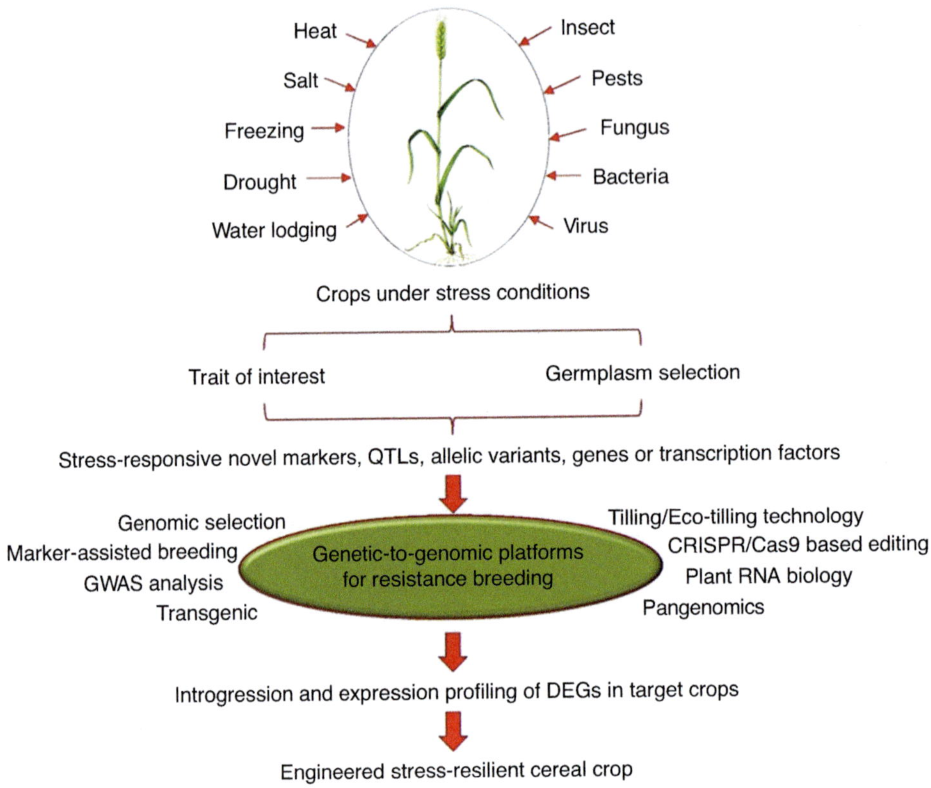

Fig. 8.2. Detailed overview of genetics-to-genomics approaches for designing stress-resilient food crops. Author's own image.

8.6 Conclusions and Future Prospects

The field of crop science has greatly benefited from the integration of various omics disciplines, including genomics, transcriptomics, proteomics, metabolomics, ionomics, and phenomics. These scientific fields utilize high-throughput technologies to investigate the functional roles, underlying molecular mechanisms, and interacting gene networks that are responsible for complex traits observed in crop plants. Integrating genome-wide association studies (GWAS) with metabolomics, transcriptomics, and proteomics has provided researchers with valuable insights into the biochemical pathways and adaptive responses to abiotic stress in model crops. This synergistic approach has been highly effective in identifying key candidate genes and metabolic pathways associated with improved stress tolerance, yield, and nutritional content. Advancements in high-throughput sequencing technologies, computational tools, and bioinformatics pipelines have significantly enhanced the feasibility of integrating multiomics approaches into crop sciences. The candidate genes and metabolic pathways identified through these methods are being incorporated into genomics-assisted breeding (GAB) and marker-assisted selection (MAS), as well as molecular breeding and genetic engineering strategies aimed at developing improved crop varieties. Among the various omics fields, metabolomics has garnered increasing attention due to the critical role of metabolites as metabolic intermediates and their direct association with phenotypic traits. The progress in metabolomics and high-throughput phenotyping platforms has further reinforced the significance of genomics, transcriptomics, and proteomics, expanding the scope of research in crop sciences. Emerging high-throughput phenomics technologies are opening new frontiers in plant science, including the detailed analysis of root system architecture and the study of plant root-associated microbial communities to improve crop health and enhance resilience to climate change. In parallel, significant advancements have been made in gene-editing technologies (e.g. CRISPR-Cas9), bioinformatics tools, machine-learning algorithms, and predictive modeling, all of which are revolutionizing omics research. Looking ahead, the integration of multiomics approaches with modern plant breeding techniques, high-throughput experimental methodologies, advanced computational modeling, and systems biology frameworks is expected to drive the development of climate-resilient, high-yielding, and nutritionally enhanced crops. This interdisciplinary approach will play a crucial role in ensuring sustainable agricultural productivity and global food security in the face of climate change.

References

Abdelrahman, M., El-Sayed, M.A., Hashem, A., Abd_Allah, E.F., Alqarawi, A.A. *et al.* (2018) Metabolomics and transcriptomics in legumes under phosphate deficiency in relation to nitrogen fixation by root nodules. *Frontiers in Plant Science* 9, 922.

Ahmad, P., Latef, A.A.A., Rasool, S., Akram, N.A., Ashraf, M. *et al.* (2016) Role of proteomics in crop stress tolerance. *Frontiers in Plant Science* 7, 1336.

Ananda, G.K., Myrans, H., Norton, S.L., Gleadow, R., Furtado, A. *et al.* (2020) Wild sorghum as a promising resource for crop improvement. *Frontiers in Plant Science* 11, 1108.

Arbona, V., Manzi, M., de Ollas, C. and Gómez-Cadenas, A. (2013) Metabolomics as a tool to investigate abiotic stress tolerance in plants. *International Journal of Molecular Sciences* 14(3), 4885–4911. DOI: 10.3390/ijms14034885.

Ashrafi-Dehkordi, E., Alemzadeh, A., Tanaka, N. and Razi, H. (2018) Meta-analysis of transcriptomic responses to biotic and abiotic stress in tomato. *PeerJ* 6, e4631.

Baharum, S.N. and Azizan, K.A. (2018) Metabolomics in systems biology. *Omics Applications for Systems Biology* 51–68.

Barnabás, B., Jäger, K. and Fehér, A. (2008) The effect of drought and heat stress on reproductive processes in cereals. *Plant, Cell & Environment* 31(1), 11–38.

Bashir, K., Matsui, A., Rasheed, S. and Seki, M. (2019) Recent advances in the characterization of plant transcriptomes in response to drought, salinity, heat, and cold stress. *F1000Research* 8. DOI: 10.12688/f1000research.18424.1.

Bharati, A., Kumar, A., Chandra, A.K., Phogat, S., Poudel, A. *et al.* (2024) Breeding and molecular approaches for heat-resilient crops. In: *Climate-Resilient Agriculture*. Apple Academic Press, pp. 193–226.

Bobik, K. and Burch-Smith, T.M. (2015) Chloroplast signaling within, between and beyond cells. *Frontier Plant Science* 6, 781.

Bonthala, V.S., Mayes, K., Moreton, J., Blythe, M., Wright, V. *et al.* (2016) Identification of gene modules associated with low temperatures response in bambara groundnut by network-based analysis. *PLoS One* 11(2), e0148771.

Brunetti, C., George, R.M., Tattini, M., Field, K. and Davey, M.P. (2013) Metabolomics in plant environmental physiology. *Journal of Experimental Botany* 64(13), 4011–4020.

Burgess, D.J. (2015) Putting transcriptomics in its place. *Nature Reviews Genetics* 16(6), 319–319.

Butelli, E., Titta, L., Giorgio, M., Mock, H.-P., Matros, A. *et al.* (2008) Enrichment of tomato fruit with health-promoting anthocyanins by expression of select transcription factors. *Nature Biotechnology* 26(11), 1301–1308.

Chaitankar, V., Karakülah, G., Ratnapriya, R., Giuste, F.O., Brooks, M.J. *et al.* (2016) Next generation sequencing technology and genomewide data analysis: Perspectives for retinal research. *Progress in Retinal and Eye Research* 55, 1–31.

Chandra, A.K., Kumar, A., Bharati, A., Kumar, L. and Singh, A. (2024) From genetics to genomics: Connecting conventional to modern approaches for designing climate-resilient underutilized orphan crops. In: *Climate-Resilient Agriculture*. Apple Academic Press, pp. 281–300.

Chaturvedi, P., Wiese, A.J., Ghatak, A., Zaveska Drabkova, L., Weckwerth, W. *et al.* (2021) Heat stress response mechanisms in pollen development. *New Phytologist* 231(2), 571–585.

Chekanova, J.A. (2015) Long non-coding RNAs and their functions in plants current opinion in plant biology 27, 207–216.

Chen, H., Wang, L., Liu, X., Hu, L., Wang, S. *et al.* (2017) De novo transcriptomic analysis of cowpea (*Vigna unguiculata* L. Walp.) for genic SSR marker development. *BMC Genetics* 18, 1–12.

Chen, X., Wang, Y., Lv, B., Li, J., Luo, L. *et al.* (2014) The NAC family transcription factor OsNAP confers abiotic stress response through the ABA pathway. *Plant and Cell Physiology* 55(3), 604–619.

Cohen, S.P. and Leach, J.E. (2019) Abiotic and biotic stresses induce a core transcriptome response in rice. *Scientific Reports* 9(1), 6273.

Czajkowska, B.I., Finlay, C.M., Jones, G. and Brown, T.A. (2019) Diversity of a cytokinin dehydrogenase gene in wild and cultivated barley. *PLoS One* 14(12), e0225899.

Dhankher, O.P. and Foyer, C.H. (2018) Climate resilient crops for improving global food security and safety. *Plant, Cell & Environment* 41(5), 877–884.

Dinkar, V., Pandey, S., Kumar, A., Shiv, A., Lal, D. *et al.* (2024) Epigenetic regulations under plant stress: A cereals perspective. *Environmental and Experimental Botany* 220, 105688.

Do, P.T., Degenkolbe, T., Erban, A., Heyer, A.G., Kopka, J. *et al.* (2013) Dissecting rice polyamine metabolism under controlled long-term drought stress. *PLoS One* 8(4), e60325.

Dugas, D.V., Monaco, M.K., Olson, A., Klein, R.R., Kumari, S. *et al.* (2011) Functional annotation of the transcriptome of *Sorghum bicolor* in response to osmotic stress and abscisic acid. *BMC Genomics* 12, 1–21.

Duque, A.S., de Almeida, A.M., da Silva, A.B., da Silva, J.M., Farinha, A.P. *et al.* (2013) Abiotic stress responses in plants: Unraveling the complexity of genes and networks to survive. In: *Abiotic Stress-Plant Responses and Applications in Agriculture*. IntechOpen.

Escandón, M., Meijón, M., Valledor, L., Pascual, J., Pinto, G. *et al.* (2018) Metabolome integrated analysis of high-temperature response in *Pinus radiata*. *Frontiers in Plant Science* 9, 485.

Evans, J.R. and Lawson, T. (2020) From green to gold: Agricultural revolution for food security. *Journal of Experimental Botany* 71(7), 2211–2215.

FAO (2013) The state of food and agriculture: Food systems for better nutrition. In: *Food and Agriculture Organization of the United Nations*. Rome.

Fiehn, O. (2002) Metabolomics – The link between genotypes and phenotypes (2002). *Plant Molecular Biology* 48, 155–171. DOI: 10.1023/A:1013713905833.

Fiser, A. and Sali, A. (2003) ModLoop: Automated modeling of loops in protein structures. *Bioinformatics* 19(18), 2500–2501.

Ghatak, A., Chaturvedi, P. and Weckwerth, W. (2018) Metabolomics in plant stress physiology. *Plant Genetics and Molecular Biology* 187–236.

Gil, J.D., Cohn, A.S., Duncan, J., Newton, P. and Vermeulen, S. (2017) The resilience of integrated agricultural systems to climate change. *Wiley Interdisciplinary Reviews: Climate Change* 8(4), e461.

Goron, T.L. and Raizada, M.N. (2015) Genetic diversity and genomic resources available for the small millet crops to accelerate a new green revolution. *Frontiers in Plant Science* 6, 157.

Guo, P., Baum, M., Grando, S., Ceccarelli, S., Bai, G. *et al.* (2009) Differentially expressed genes between drought-tolerant and drought-sensitive barley genotypes in response to drought stress during the reproductive stage. *Journal of Experimental Botany* 60(12), 3531–3544.

Huang, J., Sun, S.-J., Xu, D.-Q., Yang, X., Bao, Y.-M. *et al.* (2009) Increased tolerance of rice to cold, drought and oxidative stresses mediated by the overexpression of a gene that encodes the zinc finger protein ZFP245. *Biochemical and Biophysical Research Communications* 389(3), 556–561.

Huang, L., Zhang, F., Zhang, F., Wang, W., Zhou, Y. *et al.* (2014) Comparative transcriptome sequencing of tolerant rice introgression line and its parents in response to drought stress. *BMC Genomics* 15(1), 1026.

Jayakodi, M., Schreiber, M., Stein, N. and Mascher, M. (2021) Building pan-genome infrastructures for crop plants and their use in association genetics. *DNA Research* 28(1), dsaa030. DOI: 10.1093/dnares/dsaa030.

Jin, Y., Yang, H., Wei, Z., Ma, H. and Ge, X. (2013) Rice male development under drought stress: Phenotypic changes and stage-dependent transcriptomic reprogramming. *Molecular Plant* 6, 1630–1645.

Jogaiah, S., Govind, S.R. and Tran, L.-S.P. (2013) Systems biology-based approaches toward understanding drought tolerance in food crops. *Critical Reviews in Biotechnology* 33(1), 23–39.

Johnson, K.R., Hindmarch, C.C.T., Salinas, Y.D., Shi, Y., Greenwood, M. *et al.* (2015) A RNA-seq analysis of the rat supraoptic nucleus transcriptome: Effects of salt loading on gene expression. *Plos One* 10(4), e0124523.

Kakumanu, A., Ambavaram, M.M., Klumas, C., Krishnan, A., Batlang, U. *et al.* (2012) Effects of drought on gene expression in maize reproductive and leaf meristem tissue revealed by RNA-seq. *Plant Physiology* 160(2), 846–867.

Ke, R., Mignardi, M., Pacureanu, A., Svedlund, J., Botling, J, *et al.* (2013) *In situ* sequencing for RNA analysis in preserved tissue and cells. *Nature Methods* 10(9), 857–860.

Khan, M.A., Asaf, S., Khan, A.L., Jan, R., Kang, S.-M. *et al.* (2020) Thermotolerance effect of plant growth-promoting *Bacillus cereus* SA1 on soybean during heat stress BMC microbiology 20, 1–14.

Kim, D., Alptekin, B. and Budak, H. (2018) CRISPR/Cas9 genome editing in wheat. *Funct Integr Genomics* 18, 31–41. DOI: 10.1007/s10142-017-0572-x.

Kole, C., Muthamilarasan, M., Henry, R., Edwards, D., Sharma, R. *et al.* (2015) Application of genomics-assisted breeding for generation of climate resilient crops: Progress and prospects. *Frontiers in Plant Science* 6(2015), 563.

Kukurba, K.R. and Montgomery, S.B. (2015) *RNA sequencing and analysis. Cold Spring Harbor Protocols* 11, pdb-top084970.

Kumar, A., Maheshwari, C., Phogat, S., Poudel, A., Arya, A. *et al.* (2024) Impact of biotic and abiotic interaction on crop plants. In: *Climate-Resilient Agriculture*. Apple Academic Press, pp. 107–134.

Labuschagne, M.T. (2018) A review of cereal grain proteomics and its potential for sorghum improvement. *Journal of Cereal Science* 84, 151–158.

Lamaoui, M., Jemo, M., Datla, R. and Bekkaoui, F. (2018) Heat and drought stresses in crops and approaches for their mitigation. *Frontiers in Chemistry* 6, 26.

Lata, C., Bhutty, S., Bahadur, R.P., Majee, M. and Prasad, M. (2011) Association of an SNP in a novel DREB2-like gene SiDREB2 with stress tolerance in foxtail millet [*Setaria italica* (L.)]. *Journal of Experimental Botany* 62(10), 3387–3401.

Le, D.T., Nishiyama, R., Watanabe, Y., Tanaka, M., Seki, M, *et al.* (2012) Differential gene expression in soybean leaf tissues at late developmental stages under drought stress revealed by genome-wide transcriptome analysis. *PloS One* 7(11), e49522.

Leung, C.K., Hasegawa, K., Wang, Y., Deonarine, A., Tang, L. *et al.* (2014) Direct interaction between the WD40 repeat protein WDR-23 and SKN-1/Nrf inhibits binding to target DNA. *Molecular and Cellular Biology* 34(16), 3156–3167.

Lowe, R., Shirley, N., Bleackley, M., Dolan, S. and Shafee, T. (2017) Transcriptomics technologies. *PLoS Computational Biology* 13(5), e1005457.

Macdiarmid, J.I. and Whybrow, S. (2019) Nutrition from a climate change perspective. *Proceedings of the Nutrition Society* 78(3), 380–387.

Maeda, H., Murata, K., Sakuma, N., Takei, S., Yamazaki, A. *et al.* (2019) A rice gene that confers broad-spectrum resistance to β-triketone herbicides. *Science* 365(6451), 393–396.

Maestri, E., Klueva, N., Perrotta, C., Gulli, M., Nguyen, H.T. *et al.* (2002) Molecular genetics of heat tolerance and heat shock proteins in cereals. *Plant Molecular Biology* 48, 667–681.

Mahmood, U., Li, X., Fan, Y., Chang, W., Niu, Y. *et al.* (2022) Multi-omics revolution to promote plant breeding efficiency. *Frontier Plant Science* 13, 1062952. DOI: 10.3389/fpls.2022.1062952.

Matsui, K., Umemura, Y. and Ohme-Takagi, M. (2008) AtMYBL2, a protein with a single MYB domain, acts as a negative regulator of anthocyanin biosynthesis in *Arabidopsis*. *The Plant Journal* 55(6), 954–967.

Mohanta, T.K., Kumar, P. and Bae, H. (2017) Genomics and evolutionary aspect of calcium signaling event in calmodulin and calmodulin-like proteins in plants. *BMC Plant Biology* 17(1), 38.

Morrell, P., Buckler, E. and Ross-Ibarra, J. (2012) Crop genomics: Advances and applications. *Nature Reviews Genetics* 13, 85–96. DOI: 10.1038/nrg3097.

Mosa, K.A., Ismail, A. and Helmy, M. (2017) Plant stress tolerance: An integrated omics approach. *Springer* 24, 978–3319593777.

Mustafa, G. and Komatsu, S. (2021) Plant proteomic research for improvement of food crops under stresses: A review. *Molecular Omics* 17(6), 860–880.

Muthamilarasan, M., Singh, N.K. and Prasad, M. (2019) Multi-omics approaches for strategic improvement of stress tolerance in underutilized crop species: A climate change perspective. *Advances in Genetics* 103, 1–38.

Myers, S.S., Zanobetti, A., Kloog, I., Huybers, P., Leakey, A.D. *et al.* (2014) Increasing CO2 threatens human nutrition. *Nature* 510(7503), 139–142.

Nakano, Y., Kusunoki, K., Maruyama, H., Enomoto, T., Tokizawa, M. *et al.* (2020) A single-population GWAS identified AtMATE expression level polymorphism caused by promoter variants is associated with variation in aluminum tolerance in a local *Arabidopsis* population. *Plant Direct,* 2024 Aug 4(8), e00250.

Nataraja, K.N., Madhura, B.G. and Parvathi, M.S. (2017) Omics: Modern tools for precise understanding of drought adaptation in plants. *Plant OMICS and Crop Breeding* 263–294.

Okay, S., Derelli, E. and Unver, T. (2014) Transcriptome-wide identification of bread wheat WRKY transcription factors in response to drought stress. *Molecular Genetics and Genomics* 289(5), 765–781.

Paine, J.A., Shipton, C.A., Chaggar, S., Howells, R.M., Kennedy, M.J. *et al.* (2005) Improving the nutritional value of golden rice through increased pro-vitamin A content. *Nature Biotechnology* 23(4), 482–487.

Pandey, S., Patel, M.K., Mishra, A. and Jha, B. (2015) Physio-biochemical composition and untargeted metabolomics of cumin (*Cuminum cyminum* L.) make it promising functional food and help in mitigating salinity stress. *PloS One* 10(12), e0144469.

Pandey, S.K., Vinoj, V., Landu, K. and Babu, S.S. (2017) Declining pre-monsoon dust loading over South Asia: Signature of a changing regional climate. *Scientific Reports* 7(1), 16062.

Poland, J.A. and Rife, T.W. (2012) Genotyping-by-sequencing for plant breeding and genetics. *The Plant Genome* 5(3).

Pourkheirandish, M., Golicz, A.A., Bhalla, P.L. and Singh, M.B. (2020) Global role of crop genomics in the face of climate change. *Frontiers in Plant Science* 11, 922.

Pretty, J., Benton, T.G., Bharucha, Z.P., Dicks, L.V., Cornelia Butler Flora, H.C.J.G. *et al.* (2018) Global assessment of agricultural system redesign for sustainable intensification. *Nature Sustainability* 1(8), 441–446.

Primrose, S.B. and Twyman, R. (2013) *Principles of gene manipulation and genomics*. John Wiley & Sons.

Puranik, S., Sahu, P.P., Beynon, S., Srivastava, R.K., Sehgal, D. *et al.* (2020) Genome-wide association mapping and comparative genomics identifies genomic regions governing grain nutritional traits in finger millet (*Eleusine coracana* L). *Plants, People, Planet* 2(6), 649–662.

Qi, L.S., Larson, M.H., Gilbert, L.A., Doudna, J.A., Weissman, J.S. *et al.* (2013) Repurposing CRISPR as an RNA-guided platform for sequence-specific control of gene expression. *Cell* 152(5), 1173–1183.

Qian, Y., Ren, Q., Zhang, J. and Chen, L. (2019) Transcriptomic analysis of the maize (*Zea mays* L.) inbred line B73 response to heat stress at the seedling stage. *Gene* 692, 68–78.

Ramalingam, A., Himabindu, R., Himabindu, K., Kudapa1Lekha, T., PazhamalaLekha, T. *et al.* (2015) Proteomics and metabolomics: Two emerging areas for legume improvement. *Frontiers in Plant Science* 6, 1116. DOI: 10.3389/fpls.2015.01116.

Raza, A., Mubarik, M.S., Sharif, R., Habib, M., Jabeen, W. *et al.* (2022) Developing drought-smart, ready-to-grow future crops. *The Plant Genome*. DOI: 10.1002/tpg2.20279.

Razzaq, A., Sadia, B., Raza, A., Hameed, M.K. and Saleem, F. (2019) A way forward for crop improvement by metabolites. *Metabolomics* 9(12), 303. DOI: 10.3390/metabo9120303.

Rinaldo, A.R. and Ayliffe, M. (2015) Gene targeting and editing in crop plants: A new era of precision opportunities. *Molecular Breeding* 35, 40. DOI: 10.1007/s11032-015-0210-z.

Roorkiwal, M., Bhandari, A., Barmukh, R., Bajaj, P., Valluri, V.K. *et al.* (2022) Genome-wide association mapping of nutritional traits for designing superior chickpea varieties. *Frontiers in Plant Science* 13, 843911.

Rutkoski, J.E., Heffner, E.L. and Sorrells, M.E. (2011) Genomic selection for durable stem rust resistance in wheat. *Euphytica* 179(1), 161–173.

Sabagh, A.E., Mbarki, S., Hossain, A., Iqbal, M.A., Islam, M.S. *et al.* (2021) Potential role of plant growth regulators in administering crucial processes against abiotic stresses. *Frontiers in Agronomy* 3, 648694.

Scheben, A., Wolter, F., Batley, J., Puchta, H. and Edwards, D. (2017) Towards CRISPR/Cas crops – bringing together genomics and genome editing. *New Phytologist* 216(3), 682–698.

Scheelbeek, P.F., Bird, F.A., Tuomisto, H.L., Green, R., Harris, F.B. *et al.* (2018) Effect of environmental changes on vegetable and legume yields and nutritional quality. *Proceedings of the National Academy of Sciences* 115(26), 6804–6809.

Schreiber, M., Stein, N. and Mascher, M. (2018) Genomic approaches for studying crop evolution. *Genome Biology* 19(1), 140.

Sehgal, A., Sita, K., Siddique, K.H., Kumar, R., Bhogireddy, S. *et al.* (2018) Drought or/and heat-stress effects on seed filling in food crops: Impacts on functional biochemistry, seed yields, and nutritional quality. *Frontiers in Plant Science* 9, 1705.

Shi, J., Gao, H., Wang, H., Lafitte, H.R., Archibald, R.L. *et al.* (2017) ARGOS 8 variants generated by CRISPR-Cas9 improve maize grain yield under field drought stress conditions. *Plant Biotechnology Journal* 15(2), 207–216.

Sieber, A.N., Longin, C.F., Leiser, W.L. and Würschum, T. (2016) Copy number variation of CBF-A14 at the Fr-A2 locus determines frost tolerance in winter durum wheat. *Theoretical and Applied Genetics* 129(6), 1087–1097. DOI: 10.1007/s00122-016-2685-3.

Singhal, R.K., Saha, D., Skalicky, M., Mishra, U.N., Chauhan, J. *et al.* (2021) Crucial cell signaling compounds crosstalk and integrative multi-omics techniques for salinity stress tolerance in plants. *Frontiers in Plant Science* 12, 670369.

Singh, R.K., Prasad, A., Muthamilarasan, M., Parida, S.K. and Prasad, M. (2020) Breeding and biotechnological interventions for trait improvement: Status and prospects. *Planta* 252(4), 54.

Sood, S., Joshi, D.C., Chandra, A.K. and Kumar, A. (2019) Phenomics and genomics of finger millet: Current status and future prospects. *Planta* 250(3), 731–751.

Steenwerth, K.L., Hodson, A.K., Bloom, A.J., Carter, M.R., Cattaneo, A. *et al.* (2014) Climate-smart agriculture global research agenda: Scientific basis for action. *Agriculture & Food Security* 3, 1–39.

Tahmasebi, A., Ashrafi-Dehkordi, E., Shahriari, A.G., Mazloomi, S.M. and Ebrahimie, E. (2019) Integrative meta-analysis of transcriptomic responses to abiotic stress in cotton. *Progress in Biophysics and Molecular Biology* 146, 112–122.

Talukdar, D. and Sinjushin, A. (2015) Cytogenomics and mutagenomics in plant functional biology and breeding. *PlantOmics: The Omics of Plant Science* 14(3), 113–156.

Tan, B.C., Lim, Y.S. and Lau, S.E. (2017) Proteomics in commercial crops: An overview. *Journal of Proteomics* 3(169), 176–188. DOI: 10.1016/j.jprot.2017.05.018.

Tester, M. and Bacic, A. (2005) Abiotic stress tolerance in grasses. From model plants to crop plants. *Plant Physiology* 137(3), 791–793.

Todaka, D., Shinozaki, K. and Yamaguchi-Shinozaki, K. (2015) Recent advances in the dissection of drought-stress regulatory networks and strategies for development of drought-tolerant transgenic rice plants. *Frontiers in Plant Science* 6, 84.

Varshney, R.K., Terauchi, R. and McCouch, S.R. (2014) Harvesting the promising fruits of genomics: Applying genome sequencing technologies to crop breeding. *PLoS Biology* 12(6), e1001883.

Vaughan, M.M., Block, A., Christensen, S.A., Allen, L.H. and Schmelz, E.A. (2018) The effects of climate change associated abiotic stresses on maize phytochemical defenses. *Phytochemistry Reviews* 17, 37–49.

Viant, M.R. and Sommer, U. (2013) Mass spectrometry based environmental metabolomics: A primer and review. *Metabolomics* 9, 144–158.

Wang, Y., Wang, C., Quan, W., Jia, X., Fu, Y. *et al.* (2016) Identification and mapping of PmSE5785, a new recessive powdery mildew resistance locus, in synthetic hexaploid wheat. *Euphytica* 207, 619–626.

Weckwerth, W. and Morgenthal, K. (2005) Metabolomics: From pattern recognition to biological interpretation. *Drug Discovery Today* 10(22), 1551–1558.

Weckwerth, W. and Fiehn, O. (2002) Can we discover novel pathways using metabolomic analysis? *Current Opinion in Biotechnology* 13(2), 156–160.

Weckwerth, W., Ghatak, A., Bellaire, A., Chaturvedi, P. and Varshney, R.K. (2020) *Panomics meets germplasm*. *Plant Biotechnology Journal* 18(7), 1507–1525.

Wood, C.W. and Woolfson, D.N. (2018) C CB uilder 2.0: Powerful and accessible coiled-coil modeling. *Protein Science* 27(1), 103–111.

Yang, Y., Hu, M., Yu, K., Zeng, X. and Liu, X. (2015) Mass spectrometry-based proteomic approaches to study pathogenic bacteria-host interactions. *Protein & Cell* 6(4), 265–274.

Zenda, T., Liu, S., Wang, X., Liu, G., Jin, H. *et al.* (2019) Key maize drought-responsive genes and pathways revealed by comparative transcriptome and physiological analyses of contrasting inbred lines. *International Journal of Molecular Sciences* 20(6), 1268.

Zenda, T., Liu, S., Dong, A., Li, J., Wang, Y. *et al.* (2021) Omics-facilitated crop improvement for climate resilience and superior nutritive value. *Frontiers in Plant Science* 12, 774994.

Zhang, A., Liu, Y., Wang, F., Li, T., Chen, Z. *et al.* (2019) Enhanced rice salinity tolerance via CRISPR/Cas9-targeted mutagenesis of the OsRR22 gene. *Molecular Breeding* 39(3), 47.

Zhang, C. and Hao, Y.-J. (2020) Advances in genomic, transcriptomic, and metabolomic analyses of fruit quality in fruit crops. *Horticultural Plant Journal* 6(6).

Zhang, C., Chen, B., Zhang, P., Han, Q., Zhao, G. *et al.* (2023) Comparative transcriptome analysis reveals the underlying response mechanism to salt stress in maize seedling roots. *Metabolites* 13(11), 1155.

Zhang, P., Zhong, K., Zhong, Z. and Tong, H. (2019) Genome-wide association study of important agronomic traits within a core collection of rice (*Oryza sativa* L.). *BMC Plant Biology* 19(1), 259.
Zhang, X., Mi, Y., Mao, H., Liu, S., Chen, L. *et al*. (2020) Genetic variation in ZmTIP1 contributes to root hair elongation and drought tolerance in maize. *Plant Biotechnology Journal* 18(5), 1271–1283.
Zhao, Y., Cheng, X., Liu, X., Wu, H., Bi, H. *et al*. (2018) The wheat MYB transcription factor TaMYB31 is involved in drought stress responses in *Arabidopsis*. *Frontiers in Plant Science* 9, 1426.

9 Omics Approaches for Biotic and Abiotic Stress Tolerance

Divya Chaudhary[1], Anand Singh Jeena[2*], Basu Sudhakar Reddy[2], Rohit [2] and Garima Aggarwal[2]

[1]ICAR-National Bureau of Plant Genetic Resources, New Delhi, India; [2]Department of Genetics and Plant Breeding, G. B. Pant University of Agriculture and Technology, Pantnagar, Uttrakhand, India

Abstract

The growing need to enhance agricultural productivity and resilience has led to an increased focus on understanding and improving plant tolerance to biotic and abiotic stresses. Omics approaches, encompassing genomics, transcriptomics, proteomics, metabolomics, and phenomics, have emerged as powerful tools in this quest. These high-throughput technologies enable comprehensive profiling and analysis of biological molecules, offering insights into the complex networks and pathways involved in stress responses. This chapter delves into the application of omics approaches for enhancing biotic and abiotic stress tolerance in plants. It covers the identification of key genes, proteins, and metabolites associated with stress resistance, the development of biomarkers for early stress detection, and the integration of multiomics data to construct holistic models of stress tolerance. Furthermore, the chapter discusses the translation of omics findings into practical strategies for crop improvement, including genetic engineering, marker-assisted selection, and breeding programs. By leveraging the wealth of data generated through omics technologies, researchers can develop more resilient crop varieties, contributing to sustainable agriculture and food security in the face of a changing climate and increasing global population. The chapter addresses the challenges and limitations of omics technologies, emphasizing the need for large-scale data integration. Ultimately, integrating omics approaches into traditional breeding programs promises to enhance crop resilience and support sustainable agriculture.

9.1 Introduction

As global climate conditions become more erratic and extreme, the agricultural sector faces increasing challenges in maintaining crop productivity. Biotic stresses from pests, pathogens, and harmful microorganisms, along with abiotic stresses such as drought, salinity, heat, and nutrient deficiencies, are major threats to food security worldwide. The complexity of plant responses to these stressors has spurred the development of innovative research approaches, with omics technologies playing a crucial role in understanding and enhancing plant stress tolerance. These technologies, which include genomics, transcriptomics, proteomics, metabolomics, and epigenomics, have provided unprecedented insights into the molecular mechanisms underlying stress responses (Fig. 9.1), enabling more effective strategies for crop improvement (Qi *et al.*, 2023). Genomics has been foundational in identifying stress-responsive genes, alleles,

*Corresponding author: dr.asjeena@gmail.com

DOI: 10.1079/9781800628717.0009

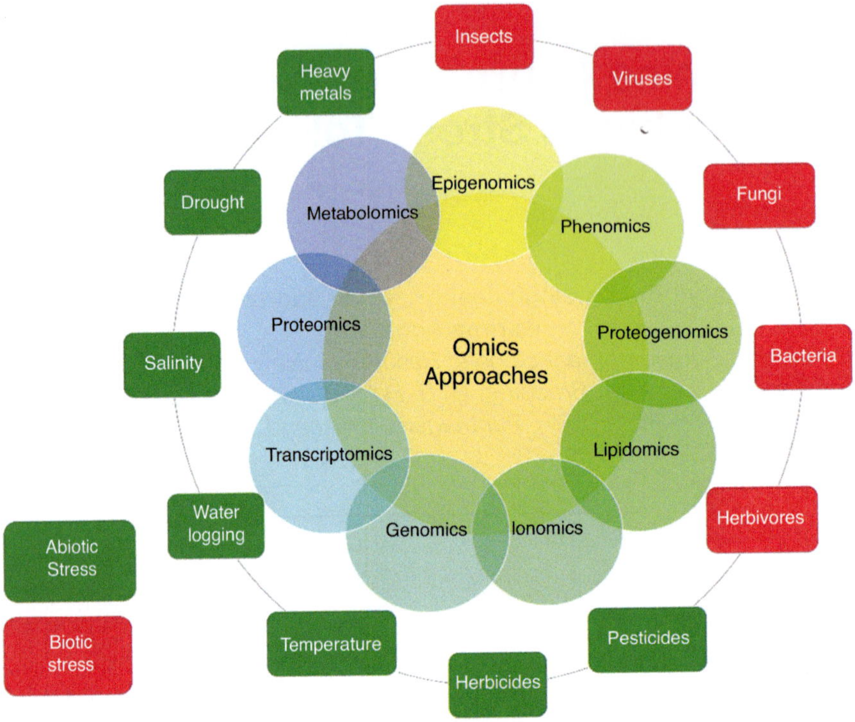

Fig. 9.1. Different omics approaches for tolerance to biotic and abiotic stresses. Author's own image.

and regulatory elements that confer tolerance to both biotic and abiotic stresses. Advances such as genome-wide association studies (GWAS) and high-throughput sequencing technologies allow researchers to map quantitative trait loci (QTLs) linked to stress tolerance traits in different crops (Bhatta *et al.*, 2019). These genomic data are essential for developing marker-assisted selection (MAS) programs, accelerating the breeding of resilient crop varieties (Xu *et al.*, 2020).

Transcriptomics, particularly RNA sequencing (RNA-Seq), provides critical insights into how plants dynamically adjust their gene expression profiles in response to stress. For instance, differential gene expression analysis under drought or pathogen attack can identify key transcription factors and regulatory RNAs (Lee *et al.*, 2022). Proteomics and metabolomics complement these insights by examining the downstream effects of gene expression at the protein and metabolite levels. Proteomics focuses on stress-induced changes in protein abundance, post-translational modifications, and protein–protein interactions, while metabolomics assesses the fluctuations in primary and secondary metabolites that help plants mitigate stress. Together, these approaches reveal the biochemical and physiological adaptations plants undergo during stress, such as osmoprotectant accumulation and antioxidant activity, which are essential for survival in challenging environments (Qi *et al.*, 2023). Epigenomics is another omic technique, which explores modifications such as DNA methylation and histone acetylation, adding another layer of complexity to plant stress responses (Liu *et al.*, 2018). Epigenetic changes can modulate gene expression in response to environmental stressors, providing plants with a mechanism to 'remember' previous exposures and adapt more efficiently in future stress events (Ramakrishnan *et al.*, 2021). This epigenetic memory is particularly valuable in perennial crops and trees, which face recurrent stress conditions over long lifespans.

The integration of multiomics data through systems biology approaches allows researchers to build comprehensive models of plant stress responses. These models facilitate the identification of regulatory hubs and key metabolic pathways that are crucial for enhancing stress tolerance. For instance, integrating genomic, transcriptomic, and metabolomic data has uncovered novel genes and pathways that contribute to drought and salinity tolerance in various crops (Sun *et al.*, 2022). In this chapter, we explore the latest advancements in omics technologies and their applications in improving plant stress tolerance. The discussion focuses on how these approaches are reshaping our understanding of biotic and abiotic stress responses, as well as the potential for translating omics-based discoveries into practical strategies for developing more resilient crop varieties. With global food security at stake, leveraging omics tools in crop improvement is essential to meet the challenges of a changing world.

9.2 Genomics Approaches

Genomics approaches are crucial in deciphering plant responses to biotic (pathogens, pests) and abiotic (drought, heat, salinity) stresses. Through whole-genome sequencing (WGS) and other techniques, researchers can identify stress-related genes and explore genetic diversity linked to tolerance mechanisms. Next-generation sequencing (NGS) technologies, such as Illumina and PacBio, provide rapid insights into the entire genome, allowing the discovery of genes responsible for stress tolerance traits like drought resistance and pathogen defense (Ashraf *et al.*, 2022). Genomics enables the identification of genetic variations and mutations, facilitating breeding for stress tolerance through methods such as genetic engineering, marker-assisted selection (MAS), and genomic selection (GS). Integrating genomics into breeding programs accelerates the development of resilient crop varieties, reducing susceptibility to environmental stressors.

Genome sequencing is a powerful tool for identifying genes involved in stress responses. High-throughput sequencing of stress-exposed plants has uncovered numerous stress-responsive genes involved in various plant physiological processes like water uptake, root development, and hormone regulation. For example, sequencing studies on rice and wheat identified genes such as DREB (dehydration-responsive element-binding proteins) that enhance drought resistance by modulating gene expression during water deficit (Hirayama and Shinozaki, 2010). Crop reference genome sequences form the cornerstone of genetic and genomic studies, providing critical insights into gene composition, genomic variability, and the genetic foundations of tolerance to both biotic and abiotic stresses.

9.2.1 Genome-wide association studies (GWAS)

GWAS reveal various QTLs associated with stress tolerance traits across plant genomes. By analyzing populations of plants under stress conditions, GWAS identifies single nucleotide polymorphisms (SNPs) that correlate with traits like drought and salt tolerance. The research on maize and sorghum has shown that GWAS can uncover significant loci associated with root architecture and osmotic adjustment under drought stress (Li *et al.*, 2019). For instance, a GWAS study on rice recognized important QTLs associated with salt tolerance by analyzing SNP data across diverse rice genotypes, leading to the discovery of candidate genes involved in ion transport and salt exclusion (Kong *et al.*, 2021). Table 9.1 depicts various GWAS studies for biotic and abiotic stresses in different crops. Such findings help breeders select varieties that are more resilient to environmental challenges.

9.2.2 Marker-assisted selection (MAS) for stress tolerance

MAS accelerates plant breeding by using molecular markers linked to desirable traits. MAS has been instrumental in breeding programs aimed at enhancing stress tolerance traits such as drought resistance, salinity tolerance, and pest resilience. Markers associated with QTLs for stress tolerance are identified through techniques like GWAS, and these markers are used to select plants with favorable alleles. For

Table 9.1. Recent genome-wide association study (GWAS) conducted on different crop species to identify QTLs responsible for various biotic and abiotic stresses. Author's own table.

S.no.	Crop	Stress	Candidate genes/QTLs	References
1	Wheat	Various biotic stress: wheat rusts, crown rot, cereal cyst nematode, and Hessian fly	124 significant markers trait association	Bhatta *et al.*, 2019
2	Sesame	Drought and salt stress	9 QTLs for drought and 15 for salt responses	Li *et al.*, 2018
3	Cotton	Drought and salt stress	53 candidate genes	Abdelraheem *et al.*, 2021
4	*Camelina sativa*	Salt stress	17 significant salinity stress associated SNPs	Luo *et al.*, 2020
5	Cotton	Biotic and abiotic stresses: salt, cold and *Verticillium* wilt	223 genes were observed in a candidate gene interval	Sun *et al.*, 2019
6	Barley	Cold stress	52 significant quantitative trait nucleotides located on chromosomes 1 hour, 3 hour, and 6 hour for cold tolerance	Elakhdar *et al.*, 2023
7	Tomato	Cold stress	685 candidate genes	Shah *et al.*, 2024

example, in rice breeding, MAS has been used to select for traits like submergence tolerance, where markers linked to the Sub1A gene, which confers flood tolerance, were used to breed stress-tolerant varieties (Neeraja *et al.*, 2007). MAS is especially effective for traits with complex inheritance patterns, such as drought tolerance, where phenotypic selection is challenging. Through MAS, breeders can expedite the selection process, reducing the need for extensive field trials (Song *et al.*, 2023). Breeders can utilize linked markers associated with biotic and abiotic stress to identify specific QTLs or genes of interest and efficiently incorporate them from the donor parent into the desired genetic background of the recurrent parent through MAS. The various improved cultivars developed by this technique are described in Table 9.2.

9.3 Transcriptomics Approaches

The transcriptome encompasses the entire set of RNA molecules produced by a cell or group of cells, including mRNAs, tRNAs, miRNAs, and other noncoding RNAs. Unlike the genome, the transcriptome is highly variable, changing with cell type, developmental stage, and in response to biotic and abiotic conditions. Transcriptome variability arises from two primary factors: differences in the sequences being transcribed and variations in the levels of gene expression (Wang *et al.*, 2009). While multiple techniques have been developed to analyze transcriptomes, it is the advent of next-generation sequencing (RNA-Seq) that has enabled researchers to examine both types of variability simultaneously and with high resolution (McGettigan, 2013).

RNA-Seq has revealed that both the sequences being transcribed and their expression levels change significantly in response to various biotic and abiotic stresses. These variations include processes like alternative splicing, alternative transcription start and stop sites, and RNA editing, which are widely observed in the transcriptomes of extensively studied species (Trapnell *et al.*, 2010). Additionally, RNA-Seq has led to the discovery of numerous previously unidentified noncoding RNAs. Small RNA sequencing (miRNA-Seq) has now become just as common as traditional mRNA sequencing (Mironova *et al.*, 2015). RNA-Seq is particularly valuable for species without fully sequenced genomes or those with poorly annotated genomes. This technique uses NGS technologies to convert RNA into millions of short reads. These sequencing platforms, based on sequencing-by-synthesis, use either DNA

Table 9.2. Crop varieties developed by MAS for stress tolerance in India. Author's own table.

Crop	Varieties	Improved trait	Recurrent parent	Donor parent	Introgressed gene	Year of variety release
Rice	DRR Dhan 59	Bacterial blight resistance	Akshyadhan	FBR1-13	*Xa33*	2021
	DRR Dhan 62	Bacterial blight and blast resistance	Improved Samba Mahsuri	C101A51 and Tetep	*Xa21*, *xa13*, *xa5*, *Pi2*, and *Pi-54*	2021
	DRR Dhan 58	Bacterial blight resistance and seedling stage salinity tolerance	Improved Samba Mahsuri	FL478	*Xa21*, *xa13*, *xa5*, and *qSaltol*	2021
	IR 64 Sub1	Submergence tolerance	IR 64	IR 49830-7-1-2-3 (FR13A)	*qSub1*	2020
	CR Dhan 803 (Trilochan)	Submergence tolerance	Pooja	Swarna Sub1	*qSub1*	2021
	CR Dhan 802 (Subhash)	Submergence and drought tolerance	Swarna Sub1	IR 81896-B-B-195	*qSub1*, *qDTY1.1*, and *qDTY2.1*	2019
	Pusa Basmati 1985	Herbicide (Imazethapyr) tolerance	Pusa Basmati 1509	Robin	*AHAS*	2021
Wheat	PBW 761 (Unnat PBW 550)	Stripe rust resistance	PBW 550	Avocet + Yr15	*Yr15*	2019
	PBW 757	Stripe rust resistance	PBW 550	Avocet + Yr15	*Yr15*	2019
	PBW 771	Stripe and leaf rust resistance	DBW 17	*Ae. umbellulata* (Accession no. 3732)	*Yr40* and *Lr57*	2020
Pearl Millet	HHB 67 Improved 2 (Hybrid)	Downy mildew resistance	H 77/833-2-202 (male parent of hybrid, HHB 67)	P1449 and 863B	*qRSg3.1*, *qRSg4.2*, and *qRSg6.1*	2021
Chickpea	Super Annigeri-1	*Fusarium* wilt resistance	Annigeri-1	WR-315 (ICC 8933)	*foc4*	2020
	Pusa Chickpea 20,211	*Fusarium* wilt resistance	Pusa 391	WR 315	*foc1*, *foc3*, *foc4*, and *foc5*	2021
	Pusa Chickpea 4005	Drought tolerance	Pusa 362	ICC 4958	QTL hotspot on *LG4*	2021
Soybean	NRCSL 1	Yellow Mosaic Virus (YMV) resistance	JS 335	SL525	*Rymv*	2021

Table 9.3. Studies on gene expression responses to biotic and abiotic stresses using the RNA-Seq technique. Author's own table.

Crop	Stress	Differentially expressed genes (DEGs) / other events during stress condition	References
Soybean	Drought and flooding	2724 genes in drought and 2498 genes under flood	Chen *et al.*, 2016
Populus ussuriensis	Cold stress	49 genes are involved in signal transduction pathways related to cold stress	Li *et al.*, 2022
Sweet Potato	Drought stress	Thousands of upregulated and downregulated genes	Arisha *et al.*, 2020
Gossypium davidsonii (Cotton wild species)	Salt stress	4744 in roots and 5337 in leaves	Zhang *et al.*, 2016
Rice	Salt stress	447 genes upregulated under stress condition	Chandran *et al.*, 2019
Gossypium hirsutum	Salt stress	DEGs identified related to physiological and metabolic processes	Wang *et al.*, 2020
Wheat	Heat Stress	Thousands of DEGs	Lee *et al.*, 2022
Soybean	Drought stress	RNA-Seq identified many alternative splicing events under stress conditions	Song *et al.*, 2020
Maize	Drought stress	Flowering-time-related genes *ID1*, *Hd3a*, and *ZCN8*	Kim *et al.*, 2021

polymerase (e.g. Roche 454, Illumina, Helicos, Pacific Biosciences) or a DNA ligase (e.g. Life Technologies SOLiD, Complete Genomics) as the core enzyme. These platforms can be further categorized into single-molecule-based technologies, such as Helicos and Pacific Biosciences, or ensemble-based systems like Illumina and SOLiD (Metzker, 2010). Among these, Illumina and SOLiD stand out for their low error rates, below 1% per base, which is particularly important for analyzing miRNAs. Their high sequencing capacity also makes them well-suited for measuring low-abundance transcripts (Metzker, 2010). Different studies on gene expression responses to biotic and abiotic stresses using the RNA-Seq approach are shown in Table 9.3. This technology provides insights into stress-induced gene expression, enabling the identification of novel genes, alternative splicing events, and even non-coding RNAs involved in stress responses (Wang *et al.*, 2009; Stark *et al.*, 2019). Under abiotic stresses such as drought or salinity, RNA-Seq can reveal key genes involved in osmotic regulation, antioxidant defenses, and hormone signaling, while under biotic stresses like pathogen attacks, RNA-Seq highlights defense-related genes such as those encoding pathogenesis-related (PR) proteins and receptor-like kinases (Fàbregas *et al.*, 2020; Yu *et al.*, 2023). The high sensitivity of RNA-Seq enables the detection of even low-abundance transcripts, making it ideal for identifying stress-responsive genes (Zhao *et al.*, 2021).

RNA-Seq is particularly effective in understanding the transcriptional regulation that plants undergo during biotic and abiotic stresses. When detecting stress-induced gene expression, RNA-Seq requires pipelines that analyze the relative abundance of RNA between different sample groups. These pipelines take normalized RNA abundance data as input, compute test statistics for each transcript, and rank them based on their statistical significance. Some widely used software for identifying differentially expressed genes (DEGs) includes baySeq (Hardcastle and Kelly, 2010), DESeq (Anders and Huber, 2010), edgeR (Robinson *et al.*, 2010), and PoissonSeq (Li *et al.*, 2012).

Gene co-expression network analysis is a valuable approach to uncover the complex relationships between genes that respond similarly to stress. Co-expression networks are constructed based on the correlation of gene expression levels across different conditions, allowing the identification of gene modules that are co-regulated during stress responses (Langfelder and Horvath, 2008; Serin *et al.*, 2016). These networks can help identify key regulatory or hub genes that act as central players in stress tolerance mechanisms (Zhou *et al.*, 2022). Functional analysis of co-expressed genes through enrichment analysis can highlight critical biological processes, such as photosynthesis, hormone signaling, and protein folding, that are affected during stress (You *et al.*, 2016). By integrating RNA-Seq, DGE analysis, and co-expression networks, transcriptomics offers a robust framework for understanding plant stress responses at the molecular level. This knowledge contributes to the development of crop varieties with enhanced tolerance to environmental stresses.

9.4 Proteomics Approaches

Proteomics is fundamental for identifying proteins that show differential expression in response to environmental challenges, including drought, heat, salinity, and pathogen infection. These stress-responsive proteins often serve as crucial markers for tolerance and adaptation, with the identification of master regulator proteins being vital for enhancing abiotic and biotic stress resistance through genetic engineering and crop modeling. A commonly used technique in proteomics is comparative analysis, where the protein profiles of plants under stress are compared to those in optimal conditions. This helps to identify proteins involved in stress responses. Additionally, subcellular proteomics offers insights into localized cellular responses and communication between subcellular compartments during both plant development and stress conditions (Komatsu, 2019). Several techniques, such as 2D gel electrophoresis (2-DE) and mass spectrometry (MS), are employed for stress-responsive protein identification (Table 9.4). Both label-free techniques like liquid chromatography–mass spectrometry (LC–MS) and tag-based methods such as isotope-coded affinity tag (ICAT) and isobaric tag for relative and absolute quantitation (iTRAQ) are frequently used to quantify and compare protein abundance under stress. However, due to the complexity and dynamic nature of plant proteomes, no single method can fully capture this diversity. As a result, multiple advanced approaches have been developed to increase resolution and coverage of plant proteomes. This has led to the development of high-throughput techniques that allow for the identification of protein–protein interactions and post-translational modifications (PTMs), which are key to understanding stress adaptation at the molecular level. Laser-capture microdissection (LCM) has been applied for identifying tissue- and cell-specific proteins crucial for stress responses in plants (Vats *et al.*, 2022). These proteins often undergo PTMs, such as phosphorylation, glycosylation, and ubiquitination, which regulate stress-related processes. Phosphoproteomics, an MS-based technique, is particularly valuable for assessing *in vivo* kinase activity under stress conditions (Roychowdhury *et al.*, 2023). Phosphorylation and other PTMs play central roles in regulating abscisic acid (ABA) signaling, which helps crops adapt to harsh environments (Wang and Komatsu, 2016). A recent study on wheat under drought stress using N-glycoproteomic analysis identified N-glycosylation sites that were concentrated at specific motifs, contributing to enhanced protein stability and stress resilience (Chang *et al.*, 2021). Similarly, phosphoproteomic analysis in maize (corn) showed that phosphorylation regulates the transition from cell proliferation to differentiation during de-etiolation, particularly affecting gene transcription and signal transduction processes (Arefian *et al.*, 2021). Advanced techniques like immobilized metal affinity chromatography (IMAC) and immunoprecipitation using phospho-specific antibodies have led to the discovery of numerous *in vivo* phosphorylation sites (Yan *et al.*, 2022). Moreover, sophisticated techniques such as stable isotope labeling with amino acids in cell culture (SILAC) and iTRAQ provide deeper insights into kinase domain dynamics. Recent developments, such as sequential window acquisition of all theoretical mass spectra (SWATH-MS), offer label-free alternatives for high-quality proteomic analysis (Chen *et al.*, 2021). To achieve comprehensive protein coverage, optimized protein extraction methods, like sodium deoxycholate extraction, are employed across various plant

Table 9.4. Overview of proteomic techniques and total protein counts in various crop species under biotic and abiotic stresses. Author's own table.

Crop species	Cultivar/ variety	Pathogen/ stress factor	Database search software	Technique	Protein count	References
Biotic stress						
Rice	Qingliu, TN1	*Cnaphalocrocis medinalis*	PeakView	SWATH–MS	505	Cheah *et al.*, 2020
Rice	Gangyuan8 (GY8), Lijiangxintuanheigu (LTH	*Magnaporthe oryzae*	Mascot	iTRAQ	865	Ma *et al.*, 2020
Wheat	UC1110, PI610750	*Fusarium pseudograminearum*	MaxQuant	TMT-based LC–MS/MS	163	Qiao *et al.*, 2021
Maize	B73	*Fusarium verticillioides*	MaxQuant	LC–MS/MS analysis	742	Hafiz *et al.*, 2022
Maize	African tall	*Spodoptera litura*	Proteome Discoverer	LC–MS/MS	345	Singh *et al.*, 2024
Sorghum	ICSV700, IS2205	*Chilo partellus*	ProteinLynx Global Server	Label-free LC-MS/MS	967	Tamhane *et al.*, 2021
Cabbage	YCR, CT-18	*Plasmodiophora brassicae*	MASCOT	MALDI-TOF/TOF MS	24	Moon *et al.*, 2020
Chickpea	JG 11, ICC 506-EB	*Helicoverpa armigera*	MASCOT	2-DE, MALDI-TOF/TOF MS	600	Bhattacharjee *et al.*, 2020
Pigeonpea	TTB7, ICPW 068	*Helicoverpa armigera*	MaxQuant	MS/MS	3900	Rathinam *et al.*, 2020
Sugarcane	ROC22, MT11–610	*Acidovorax avenae*	Proteome Discoverer	iTRAQ-based LC–MS/MS	1027	Zhou *et al.*, 2021
Sugarcane	LCP 85–384, ROC20	*Xanthomonas albilineans*	Proteome Discoverer	iTRAQ-based LC–MS/MS	4295	Meng *et al.*, 2020
Abiotic stress						
Rice	Wuyujing 3	Heat stress	MaxQuant	Data independent acquisition mass spectrometry (DIA– MS)	840	Liu *et al.*, 2021
Wheat	Chinese Spring	Drought, salinity, submergence	Proteome Discoverer	Label-free nanoLC–ESI–MS/MS	2273	Yan *et al.*, 2021
Wheat	Jing411	Cold stress	Proteome Discoverer	iTRAQ-based LC–MS/MS	669	Xu *et al.*, 2022
Wheat	XM 55, YM 158	Waterlogging	Mascot	iTRAQ	7710	Yang *et al.*, 2022
Maize	SD609, SD902	Drought stress	Mascot	iTRAQ-based LC–MS/MS	300	Li *et al.*, 2021

Continued

Table 9.4. Continued

Crop species	Cultivar/ variety	Pathogen/ stress factor	Database search software	Technique	Protein count	References
Sorghum	BT × 623	Drought stress	Mascot	MALDI-TOF-TOF MS	65	Li *et al*., 2020
Rapeseed	C18, C6	Cold stress	Proteome Discoverer	iTRAQ, Nano-LC-MS/MS	241	Mehmood *et al*., 2021
Chickpea	P-362	Drought	Mascot	2-DE, MALDI-TOF-TOF MS	110	Gupta *et al*., 2020
Sugarcane	Co 99004, Co 0315	Heat stress	MASCOT	MALDI-TOF/MS	352	Gomathi *et al*., 2024

ESI, electrospray ionization; MALDI, matrix-assisted laser desorption ionization; TOF, time of flight.

tissues. SWATH-MS has been successfully applied in rice to identify defense-related enzymes like phenylalanine ammonia lyase, which accumulate in resistant varieties under herbivory stress (Cheah *et al.*, 2020).

Abiotic stress responses are being extensively studied across various crops, with proteomic analyses playing a crucial role in unraveling the molecular mechanisms involved (Barkla *et al.*, 2016; Komatsu *et al.*, 2022; Mancini *et al.*, 2022). For instance, proteomic analyses in maize have revealed the involvement of stress-responsive proteins in pathways such as the TCA cycle, glycolysis, reactive oxygen species (ROS) homeostasis, and sucrose metabolism under salt-alkali stress (Zhu *et al.*, 2020). Furthermore, proteome dynamics during heat acclimation have been analyzed to identify the factors determining plant survival or death in response to heat. Proteins related to ROS regulation and protein translation were found to be associated with acclimation, while changes in secondary metabolism and defense mechanisms were linked to cell death (Echevarría-Zomeño *et al.*, 2016). Biotic stress, caused by pests and pathogens like bacteria, viruses, fungi, nematodes, and insects, also triggers complex defensive responses in plants. A common defense mechanism includes the amplification of ROS production and phytohormone signaling (Scheler *et al.*, 2013). For instance, a comparative iTRAQ proteomics analysis of wheat stripe rust resistance in the Suwon11 cultivar identified proteins involved in ROS metabolism, such as peptidyl-prolyl isomerases (PPIases), RNA-binding proteins (RBPs), and chaperonins, which play key roles in the plant's defense against *Puccinia striiformis* f. sp. tritici (Pst) infection (Yang YuHeng *et al.*, 2016). Additionally, a proteomic study examining the interaction between resistant and susceptible tomato cultivars infected with tomato yellow leaf curl virus (TYLCV) identified 86 differentially expressed proteins involved in processes such as photosynthesis, protein metabolism, carbohydrate metabolism, signal transduction, and antioxidation (Huang Ying *et al.*, 2016). In terms of bioinformatics, databases like PRIDE (del Toro, *et al.*, 2014, July) and PeptideAtlas (Kusebauch *et al.*, 2014) are essential resources for proteomic data. Tools like STRING (Szklarczyk *et al.*, 2023) provide comprehensive protein–protein interaction networks by integrating data from multiple sources to map the associations between proteins, further elucidating their roles in stress responses. Despite considerable progress, computational tools processing large proteomics datasets are still in development and often lack intuitive interfaces for users without technical expertise. The integration of the different layers of biological data toward building a unifying model of plant stress response is apparently challenging due to sheer volume and complexity of information involved apart from its cost and associated technical requirements. The identification of low-abundance proteins, which may play important roles in mediating stress tolerance, remains difficult, potentially resulting in the omission of key proteins. In addition, the proteomic responses to stress are highly variable across plant tissues, organs, or developmental stages, challenging the interpretation of data and limiting the generalizability of results across whole-plant systems or species.

9.5 Metabolomics Approaches

Metabolic profiling, also known as metabolomics, is a powerful tool that allows researchers to examine a vast array of small molecules, known as metabolites, within biological systems like cells, tissues, or whole organisms (Fig. 9.2). These low-molecular-weight compounds, which are typically less than 1 kDa, play a crucial role in the functioning of living organisms. In plants, metabolomics is particularly important under stress conditions like drought, salinity, extreme temperatures, or pathogen attacks where these stresses trigger significant changes in metabolic processes necessary for adaptation and survival. The term 'metabolome' was first coined by Oliver and colleagues in 1998, referring to the complete set of metabolites in an organism. Given the complexity of the metabolome, it encompasses a chemically diverse group of compounds, including lipids, organic acids, carbohydrates, amino acids, nucleotides, and steroids. Interestingly, many of these compounds are still being studied, with some yet to be fully validated and reported. By analyzing these metabolic shifts, researchers gain a deeper understanding of how plants adjust their metabolic pathways in response to stress, revealing the intricate processes that underpin plant stress physiology.

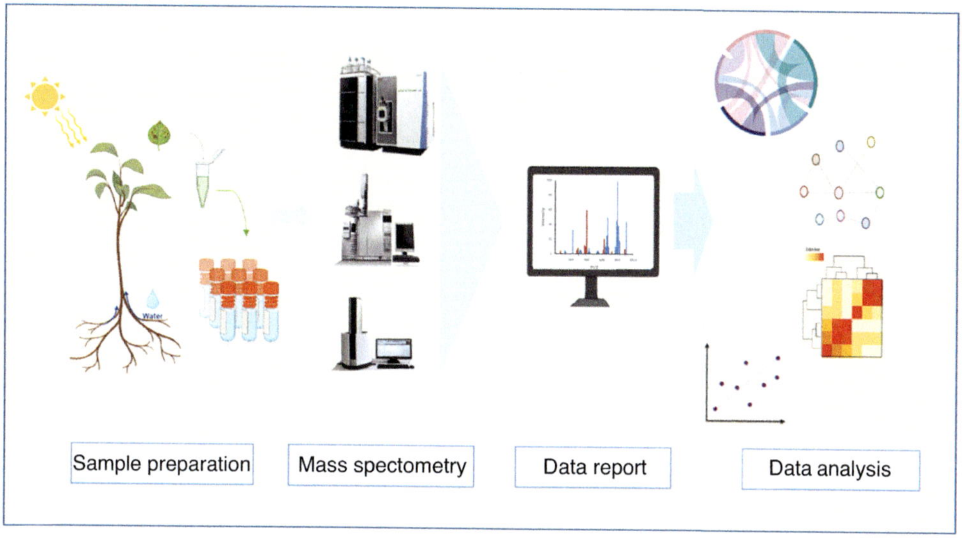

Fig. 9.2. The workflow for metabolite extraction, identification, and quantification using mass spectrometry techniques. Author's own image.

Thus, metabolomics has emerged as a vital tool to develop stress-resistant crop varieties.

9.5.1 Metabolic profiling under stress conditions

Recent advancements in plant biotechnology, particularly in metabolomics, have transformed our understanding of metabolic regulation within plants. Mass spectrometry (MS)-based technologies, including GC–MS, LC–MS, nuclear magnetic resonance (NMR), matrix-assisted laser desorption ionization (MALDI), and capillary electrophoresis MS, are key to exploring the plant metabolome on a large scale. MS is especially valuable due to its high sensitivity and its ability to uncover novel metabolic biomarkers and molecules, while platforms like NMR are more suitable for studying the physical properties of metabolites and protein–ligand interactions. GC–MS and LC–MS are powerful analytical techniques used to explore a wide range of biomolecules. GC–MS excels in analyzing less polar compounds, such as alkylsilyl derivatives, eicosanoids, essential oils, esters, perfumes, terpenes, waxes, volatiles, carotenoids, flavonoids, and lipids, making it particularly effective for detecting volatile primary and secondary metabolites. In contrast, LC–MS is designed for more polar biomolecules, including organic acids, organic amines, nucleosides, ionic species, nucleotides, and polyamines, and it offers the advantage of not requiring derivatization. While GC–MS focuses on the volatility and separation of non-polar compounds, LC–MS utilizes techniques like electrospray ionization (ESI) and atmospheric pressure chemical ionization (APCI) to provide a robust platform for both targeted and untargeted metabolomic studies. Together, these methods provide valuable insights into the biochemical makeup of biological systems. Capillary electrophoresis MS provides high-resolution separation of various analyte groups.

9.5.2 Identification of key metabolites for stress tolerance

Metabolomics plays a key role in understanding the regulation of abiotic and biotic stress tolerance in plants. Under abiotic stresses such as drought, salinity, and nutrient deficiency, plants experience significant metabolic changes. For example, amino acids like proline and secondary metabolites like flavonoids accumulate, acting as osmoprotectants

Table 9.5. Overview of metabolites identified under a stress response by using different analytical techniques. Author's own table.

Crop	Stress type	Key metabolites	Technique used	Reference
Wheat	Drought	Amino acids (proline, valine), sugars, phenylpropanoids	GC–MS, LC–MS	Bowne *et al.*, 2012; Yadav *et al.*, 2019
Rice	Salinity	Leucine, isoleucine, shikimate, quinate	GC–MS, jasmonate pathway	Gupta and De, 2017 Chang *et al.*, 2019 Kurotani *et al.*, 2015
Barley	Nutrient stress	Organic acids, amino acids	UHPLC	Ghosson *et al.*, 2018 Huang *et al.*, 2008
Maize	Drought, biotic	Lipids, glutathione, benzoxazinoids	RP-UPLC–MS, LC–MS	Skirycz and Inzé, 2010
Tomato	Heat stress, Biotic	Polyphenols, lignin, flavonoids	FT-IR, NMR	Chebrolu *et al.*, 2016
Rice	Pathogen stress	Glyceric acid, jasmonic acid, mucic acid	CE/TOF-MS	Suharti *et al.*, 2016
Horse gram	Viral infection	Octadecanoic acid, cucurbitacin B	GC–MS	Sudhagar *et al.*, 2022; Rajaprakasam *et al.*, 2023
Maize	Fungal resistance	Smiglaside, smilaside	LC–MS, GC–MS	Zhou *et al.*, 2019

FT-IR, Fourier transform infrared spectroscopy; UHPLC, ultra high-performance liquid chromatography.

and antioxidants. Studies on wheat and maize have shown that drought and heat stress lead to elevated levels of carbohydrates and amino acids (Table 9.5). Similarly, jasmonate pathways in rice have been linked to salt stress tolerance. Under biotic stress, metabolites like benzoxazinoids in maize and other secondary metabolites are crucial for plant defense. Comparative metabolomics of pathogen-infected and healthy plants have revealed higher accumulations of defense-related metabolites like phenylalanine, jasmonic acid, and flavonoids. In rice and maize, metabolites like polyphenols, smiglaside, and smilaside were found to be associated with resistance to pathogens like *Fusarium graminearum* and *Magnaporthe grisea*.

Metabolomics, though smaller in scale compared to the proteome and genome, provides an efficient platform for data analysis and processing. Around 3000 metabolites play key roles in major metabolic pathways. The study of metabolic quantitative trait loci (mQTLs) offers crucial insights into the genetic control of metabolite networks, bridging the gap between genotype and phenotype. mQTL mapping identifies SNPs linked to key metabolites, which are vital for stress responses. Advanced tools like NGS have enabled ultra-high-density maps to find candidate genes for traits such as secondary metabolite production. Studies in rice, barley, tomato, and other crops have revealed significant mQTLs linked to stress tolerance and metabolite biosynthesis (Table 9.6). For example, mQTL studies in barley under drought stress identified stress-responsive metabolites like sinapic acid and flavones, while in *Brassica napus*, 105 mQTLs were linked to glucosinolate biosynthesis. Combining metabolomics with genomics, transcriptomics, and proteomics leads to a deeper understanding of plant stress responses, while integrating metabolomics data into systems biology models facilitates predictive modeling of plant behavior.

9.5.3 Pathway analysis and metabolite quantification

Advancements in metabolomics have greatly improved the identification and quantification of metabolites, especially under stress conditions. Tools such as MET-COFEA, ChromaTOF,

Table 9.6. Identified mQTLs associated with abiotic stress tolerance in various crops. Author's own table.

Crop	mQTLs Identified	Key metabolites	Stress/condition studied	Reference
Rice	19 mQTLs on 23 loci	Secondary metabolites (phenolics, flavonoids)	Comparative metabolomic study	Gong *et al.*, 2013
Barley	98 mQTLs	Sinapic acid, ferulic acid, flavones	Drought stress	Piasecka *et al.*, 2017
Brassica napus	105 mQTLs	Glucosinolates	Glucosinolate biosynthesis	Feng *et al.*, 2012
Tomato	679 mQTLs	Secondary metabolites (carotenoids, alkaloids)	Environmental stress tolerance	Alseekh *et al.*, 2015
Wheat	558 mQTLs	Phenylpropanoids, flavonoids, alkaloids	Double-haploid wheat lines	Hill *et al.*, 2015
Maize	Multiple mQTLs	Amino acids, carbohydrates	Yield and stress traits	Wen *et al.*, 2015

and METLIN allow researchers to process high-throughput data, while web-based platforms like MetaboAnalyst and the Plant Metabolic Network (PMN) enhance data sharing and analysis. These tools facilitate the discovery of metabolic markers linked to traits of interest, thereby offering insights into crop physiology, stress adaptation, and strategies for improvement. MetaboAnalyst, for instance, provides comprehensive capabilities for data processing, analysis, and interpretation through techniques such as Metabolite Set Enrichment Analysis (MSEA) and Metabolic Pathway Analysis (MetPA). Pathway enrichment tools like MetPA and Mummichog further contextualize experimental data within relevant biological pathways. A variety of databases, including HMDB, METLIN, and LIPID MAPS, support metabolite identification, while pathway databases like KEGG, MetaCyc, and Reactome assist in mapping these metabolites to specific pathways. Together, these tools enable a deeper understanding of metabolic processes and their regulation under various conditions.

9.6 Epigenomics Approaches

Epigenomics examines heritable changes in gene expression that do not involve changes to the DNA sequence itself. These changes are crucial in how plants adapt to stress, as they allow rapid and reversible alterations in gene activity.

9.6.1 DNA methylation patterns in stress responses

DNA methylation is a fundamental epigenomic mechanism that significantly influences plant responses to environmental stressors (Fig. 9.3). This modification involves the addition of methyl groups to cytosine residues, primarily in the contexts of CG, CHG, and CHH (where H is A, T, or C). Methylation results in gene silencing through the compaction of chromatin structure, which prevents transcriptional machinery from accessing DNA. Under stress conditions, plants exhibit dynamic changes in DNA methylation patterns, allowing them to activate or repress stress-responsive genes swiftly. For example, during drought stress, certain regions of the plant genome undergo hypomethylation, which facilitates the activation of genes associated with water-use efficiency and drought tolerance. Conversely, hypermethylation in other regions can lead to the silencing of genes that are deemed unnecessary or potentially harmful under drought conditions. These reversible alterations in DNA methylation not only enable immediate responses to environmental challenges but also contribute to a form of 'stress memory,' allowing plants to 'remember' previous stress exposures. This memory enhances their ability to mount more effective responses when faced with similar stresses in the future.

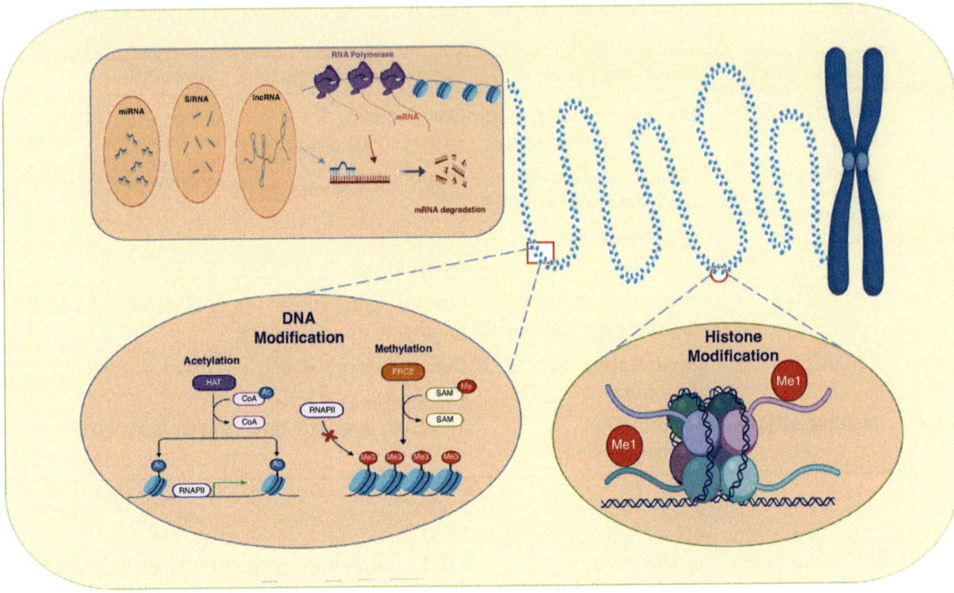

Fig. 9.3. Epigenetic modifications associated with plant stress responses. Author's own image.

Transposable elements (TEs) are another critical aspect of DNA methylation dynamics. TEs are often linked to differentially methylated regions (DMRs) within the genome, influencing the expression of nearby protein-coding genes, particularly during stress responses (Liu *et al.*, 2018; Ramakrishnan *et al.*, 2021). The activity of these TEs can be modulated by changes in methylation status, which can impact genomic stability and gene regulation. Small RNAs, particularly small interfering RNAs (siRNAs), are involved in maintaining hypermethylation at these loci through the RNA-directed DNA methylation (RdDM) pathway. This coordination between DNA methylation and small RNA pathways reinforces the plant's ability to regulate gene expression effectively in response to both biotic (e.g. pathogen attack) and abiotic stresses (e.g. drought and heat) (Erdmann and Picard, 2020; Sun *et al.*, 2022). Active demethylation processes also play a significant role in stress adaptation. Contrary to the earlier belief that methylated cytosine simply reverted to unmethylated cytosine, active demethylation involves complex mechanisms such as oxidation and deamination. Enzymes like DNA glycosylases are integral to this active demethylation process, helping to reset the methylation status of specific genes in response to environmental cues. Recent advancements in sequencing technologies, including nanopore and single-molecule real-time sequencing, have significantly enhanced our understanding of transcriptomes and epigenetic alterations, elucidating the critical roles of DNA methylation in plant adaptation to stress (Fan *et al.*, 2024).

9.6.2 Histone modifications in biotic and abiotic stress tolerance

Histone modifications represent another crucial layer of epigenetic regulation that influences gene expression in response to environmental stresses. Various post-translational modifications (PTMs) of histones, including acetylation, methylation, phosphorylation, and ubiquitination, alter chromatin structure and accessibility. These modifications can either promote or repress gene expression, allowing plants to finely tune their transcriptional responses during stress conditions. For instance, histone acetylation, particularly on histone H3, is often associated with gene activation. In response to heat stress, the levels of acetylation on histone H3 increase, leading to the activation of heat-shock proteins (HSPs), which play essential roles in protecting cellular components from heat-induced damage. Similarly, other abiotic stresses like drought, salinity, and UV-B exposure trigger dynamic changes in histone modifications, thereby modulating gene expression to enhance stress tolerance (Stetler *et al.*, 2010; Kuta *et al.*, 2020).

Histone methylation also plays a vital role in plant immunity during biotic stress. Modifications on specific lysine residues, such as H3K4me3 (associated with gene activation) and H3K27me3 (associated with gene repression), are critical for regulating defense-related genes (Zhang *et al.*, 2021a). During pathogen attacks, the patterns of histone methylation can shift, leading to the activation of defense genes and suppression of those that may compromise plant immunity. The interplay of histone modifications with other epigenetic mechanisms, such as small RNAs and chromatin remodelers, forms a complex regulatory network that enables plants to deploy precise, stress-specific transcriptional programs. This dynamic regulation underscores the importance of histone modifications in enhancing plant resilience to both abiotic and biotic stresses.

9.6.3 Non-coding RNAs and stress adaptation

Non-coding RNAs (ncRNAs) play a multifaceted role in regulating plant stress adaptation mechanisms, acting as critical components of the epigenetic and transcriptional machinery. These ncRNAs include microRNAs (miRNAs), small interfering RNAs (siRNAs), and long non-coding RNAs (lncRNAs), each of which influences gene regulation at various levels. miRNAs are particularly well characterized in their roles in plant responses to abiotic stresses (Waititu *et al.*, 2020; Meng *et al.*, 2021; Gill *et al.*, 2024). They operate at the post-transcriptional level by binding to complementary mRNA sequences, resulting in either mRNA degradation or translational inhibition. For example, miR398 regulates the expression of superoxide dismutase (SOD), an enzyme essential for detoxifying ROS generated during oxidative stress (Sunkar *et al.*, 2006). By modulating SOD activity, miR398 helps maintain cellular redox balance, which is critical for cell survival under stress conditions. siRNAs also play a significant role in maintaining genome stability during stress by contributing to RNA-directed DNA methylation (RdDM). This process facilitates the silencing of transposable elements and stress-related genes through epigenetic changes like DNA methylation and chromatin remodeling. siRNAs can guide the methylation of stress-responsive loci, allowing for tight regulation of gene expression in response to environmental fluctuations. Long non-coding RNAs (lncRNAs), while less understood, have been shown to act as molecular scaffolds or decoys that coordinate gene expression during stress responses. By interacting with proteins, other RNAs, or DNA, lncRNAs can regulate chromatin states or modulate the transcriptional activity of stress-responsive genes. Emerging studies suggest that lncRNAs contribute to both immediate stress responses and the establishment of stress memory, enabling plants to recall and respond more efficiently to recurring environmental challenges. Together, ncRNAs serve as integral components of the plant's epigenetic regulatory network, enhancing robustness and protection against various stresses. They not only facilitate immediate adaptation to environmental challenges but also contribute to the potential for transgenerational inheritance of stress-adaptive traits. This interplay among ncRNAs, DNA methylation, and histone modifications represents a critical mechanism for plant survival in challenging environments, highlighting the sophisticated regulatory strategies that plants employ to cope with stress (Bure *et al.*, 2022).

9.7 Integrated Omics Approaches

Humans rely on about 30 domesticated crop species for around 90% of our food needs. These crops are highly susceptible to yield losses from both biotic and abiotic stresses, which can range

Table 9.7. Online platforms to access multiomics data in different crops. Author's own table.

Crop	Omics database	Reference
Maize	ZEAMAP, MaizeNetome	Gui *et al.*, 2020; Feng *et al.*, 2023
Wheat	WheatOmics	Ma *et al.*, 2021
Brassica napus	BnaOmics, BnIR	Cui *et al.*, 2023; Yang *et al.*, 2023
Rice	MBKbase	Peng *et al.*, 2020
Soybean	SoyMD, SoyOmics	Liu *et al.*, 2023; Zhang *et al.*, 2024
Millets	Milletdb	Sun *et al.*, 2023
Horticultural crops	GERDH	Cheng *et al.*, 2023
Cotton	CottonMD, COTTONOMICS	Dai *et al.*, 2022; Yang *et al.*, 2023
Brassica juncea	BjuIR	Zhang *et al.*, 2024

from 50 to 80%. Climate change exacerbates these vulnerabilities, particularly in arid and semi-arid regions, where changing climatic conditions can lead to significant crop losses.

Addressing this complex challenge requires a united effort to develop innovative solutions. This era of large-scale omics data is enhancing our understanding of the genetic factors that influence plants' responses to abiotic and biotic stresses (Mosa *et al.*, 2017). Techniques in omics can facilitate the development of new crop varieties through genome-assisted breeding and genome editing. These approaches help identify crucial genetic pathways responsible for stress responses and the sequence variants that significantly affect how plants respond to challenges (Khan *et al.*, 2023). Integrating multiomics data can achieve a more holistic perspective of how plants respond to abiotic stresses at the molecular level. It allows an understanding from genotype to phenotype. Table 9.7 shows some of the online platforms used to access the multiomics combined data in major crops.

9.7.1 Combining genomics, transcriptomics, proteomics, and metabolomics

Combining genomics, transcriptomics, proteomics, and metabolomics, collectively known as multi-omics, offers a powerful approach to biological research (Fig. 9.4). Each of these fields provides distinct insights into the biological systems being studied, and their integration allows researchers to achieve a more comprehensive understanding of biological processes. The last decade has witnessed remarkable progress in sequencing technologies, resulting in the development of numerous reference genomes. These advancements have significantly deepened our understanding of the pathways that connect genotype to phenotype. This has been particularly beneficial for stress breeding, as it has enabled the identification and characterization of functional genes, allowing for the tracing of gene-to-protein pathways in major crop species.

The process begins with GWAS and SNP analysis, which identify the sources of genetic variation or candidate genes. Integrating this information with transcriptome data enables the analysis of the expression levels of these genes, providing deeper insights into their role. Natural variation in the promoter of DROUGHT1 (DROT1) was identified to confer drought resistance in rice (Sun *et al.*, 2022). *ZmFBL41* gene encoding F-box protein that confirms resistance to sheath blight and banded leaves was reported using GWAS (Li *et al.*, 2019). GWAS combined with transcriptome-wide association study (TWAS) explained more heritable variation. Ma *et al.* (2021) confirmed six causal elements for the heat tolerance (HT) response, three of which overlapped with the GWAS results and are involved in protein kinase activity. The most susceptible gene identified was GhHRK1 in cotton. Zeng *et al.*, 2020 elucidated novel UV-B responsive mechanisms in plants using Barley. Likewise, flavonoid decoration pathway of wheat kernels was deciphered (Chen *et al.*,

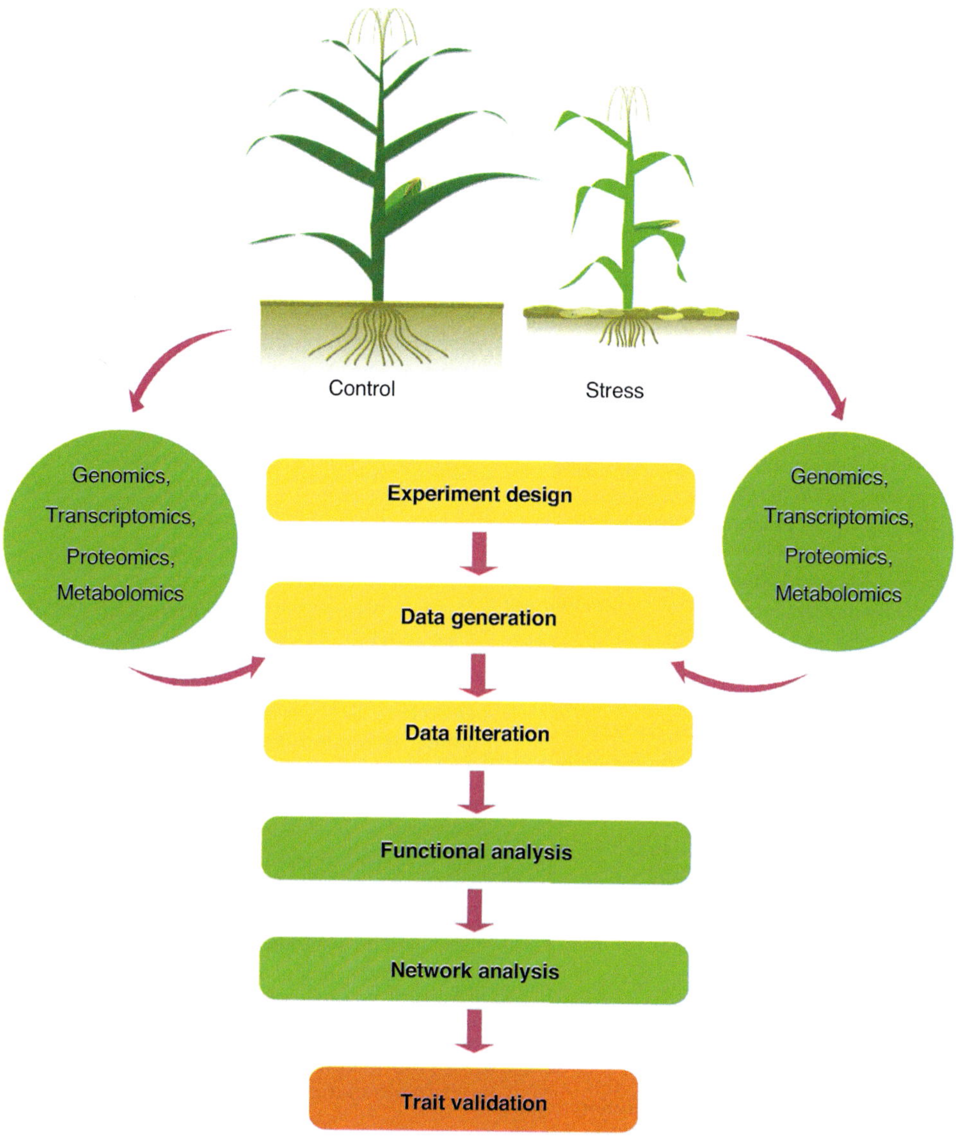

Fig. 9.4. Integrated multiomics scheme for stress tolerance breeding in plants. Author's own image.

2020). Combined metabolomics and transcriptomics analysis established the metabolite–gene relation it in eQTL analysis to identify drought tolerance hub genes (Zhang *et al.*, 2021b). Integrative transcriptomic and proteomic analysis revealed molecular network of glutamine synthetase two corresponding to nitrogen deficiency in rice (Liang *et al.*, 2021).

9.7.2 Systems biology for comprehensive stress response analysis

Coupling of different omics approaches could be useful for identifying putative candidate genes for functional analysis. But plants contain different cellular compartments and tissues that carry complex biological systems. An understanding of this crosstalk between various pathways and their effect on the plant can be attempted via a complete biological system study through modeling. This predicts the behavior of all components and interactions among genes, proteins, and metabolites with respect to external stimuli. There are two system-based approaches, first, top-down or integrative systems biology which hunts traits using data from bioinformatics and second, bottom-up or predictive systems biology here, genes and proteins is used to predict the behavior of systems under different conditions.

Modeling provides a comprehensive understanding of how gene expression influences protein behavior, ultimately driving the production of metabolites in response to stress in plants. Systems modeling relies on biological hypotheses to predict system-wide responses. A biological system is simulated based on the hypothesis, and the results are compared with experimental data to assess and refine the hypothesis, ensuring alignment with observations (Pazhamala *et al.*, 2021). This iterative process is crucial for accurately predicting system behavior under both control and stress conditions over time (Macklin, 2019). Constraint-based modeling, as described by Price *et al.* (2004), has proven especially effective in offering deep insights into cellular metabolism under abiotic stress, such as in rice (Muthuramalingam *et al.*, 2019).

9.7.3 Data integration and network analysis for stress tolerance

The initial steps prior to data integration require 'experimental design' followed by 'data generation' and 'data pre-processing' (Roychowdhury *et al.*, 2023). Data integration is a very vital step that can apply computational methods to integrate the multiomics datasets. Data are primarily subjected to multivariate analysis and strong correlations are searched for among the biological entities. Most of the statistical methodologies include dimensionality reduction and studying the coexpressed clusters among the different data measured on the same samples. Web applications like PaintOmics 4 (Liu *et al.*, 2022) and PANTHER 11 (Mi *et al.*, 2021) simplify data integration using biological pathway maps. 'Network analysis' is a further step to construct biological networks that capture the associations between different molecules identified in the integrated data. Network analysis techniques, such as co-expression networks or protein–protein interaction networks, can help identify key hub genes or proteins involved in stress responses. A detailed study on wheat utilizing an eFP browser revealed 152 hub genes, of which 32% contained the ethylene-responsive element binding factor-associated amphiphilic repression (EAR) transcriptional repression motif (Da Ros *et al.*, 2023). In grapevine, metabolite profiling and network analysis revealed coordinated changes involved in the water-stress response (Hochberg *et al.*, 2023). Coexpression gene network analysis for cold-tolerance was studied in *Solanum commersonii* (Esposito *et al.*, 2021).

9.8 Other Omics Approaches and Their Role in Plant Stress Tolerance

In recent years, the integration of multiple omics technologies has revolutionized plant biology, offering comprehensive insights into the complex molecular responses of plants to environmental stresses. While genomics, transcriptomics, and proteomics are well-established, other advanced omics approaches such as phenomics, proteogenomics, lipidomics, ionomics, and interactomics have gained significant attention for their ability to provide deeper layers of understanding. These approaches collectively enhance our capacity to develop stress-tolerant crops by uncovering new biomarkers, pathways, and molecular interactions.

9.8.1 Phenomics

Phenomics refers to the large-scale study of phenotypes, i.e. observable traits of plants, under various environmental conditions. With the advent of high-throughput phenotyping platforms, phenomics allows for the simultaneous measurement of a wide range of traits, including growth patterns, photosynthetic efficiency, water use, and nutrient uptake. These phenotypes are essential for linking the genotype to the environment, providing critical insights into how plants adapt to stresses like drought, salinity, heat, and cold (Song *et al.*, 2021). In the context of plant stress tolerance, phenomics helps identify traits associated with resilience and productivity under stress. For example, advanced imaging systems combined with automated analysis can monitor changes in leaf morphology, root architecture, and stomatal conductance under drought or heat stress (Haworth *et al.*, 2023). Phenomics data can also be integrated with genomics and transcriptomics to link specific genes to stress-adaptive traits, facilitating the breeding of stress-tolerant varieties (Campbell *et al.*, 2018). Additionally, phenomics can help elucidate plant responses to biotic stresses such as pathogen infections by tracking disease progression and plant immune responses in real time.

9.8.2 Proteogenomics, lipidomics, ionomics, and interactomics

Proteogenomics is an emerging field that combines proteomics and genomics data to provide a comprehensive view of gene function and regulation. This approach leverages protein-level information to improve genome annotations and identify novel protein-coding regions. In the context of plant stress tolerance, proteogenomics enhances our understanding of how genetic variations are translated into functional proteins that mediate stress responses. Proteogenomics can identify stress-responsive proteins and PTMs that are crucial for plant survival under stress conditions. For example, phosphorylation and ubiquitination of key signaling proteins can regulate stress tolerance pathways such as ABA signaling in drought conditions or immune responses during pathogen attack (Qi *et al.*, 2018). By integrating proteogenomics with transcriptomics, researchers can also study how transcriptional changes correspond to protein abundance, revealing the extent of post-transcriptional regulation under stress.

Lipidomics focuses on the comprehensive analysis of lipids in biological systems. Lipids play critical roles in plant responses to environmental stresses by maintaining membrane integrity, signaling, and energy storage. During stress conditions such as drought or cold, plants alter their lipid composition to protect cellular membranes from damage and maintain fluidity (Henschel *et al.*, 2024). In response to abiotic stresses, lipidomics has revealed significant changes in the levels of phospholipids, sphingolipids, and fatty acids, which are involved in membrane remodeling and signaling. For instance, under salt stress, an increase in phosphatidic acid (PA) has been associated with improved stress tolerance by mediating ROS scavenging and membrane stabilization (Han and Yang, 2021). Lipidomics also helps in identifying lipid-based signaling molecules such as jasmonic acid and its derivatives, which play crucial roles in biotic stress responses (Pretorius *et al.*, 2021). By linking lipid profiles with stress tolerance traits, lipidomics contributes to the identification of metabolic pathways that can be targeted for crop improvement.

Ionomics is the large-scale study of elemental composition (ions) in plants, particularly how nutrient and trace element levels are regulated in response to environmental changes. Elements like potassium, calcium, sodium, and magnesium are essential for maintaining osmotic balance, enzyme activity, and signaling under stress conditions (Baxter, 2010). Plants subjected to abiotic stresses such as salinity or drought often experience ion imbalances, leading to toxicity or deficiency. Ionomics helps elucidate how plants regulate ion homeostasis under such conditions. For instance, under salt stress, ionomics can reveal how plants exclude sodium from sensitive tissues and maintain potassium uptake for proper cell function (Shariatipour *et al.*, 2022). By studying the ion profiles of plants, researchers can identify key genes involved in ion transport and signaling pathways that confer stress tolerance. Ionomics also plays a role in understanding the uptake of beneficial elements such as silicon, which can enhance resistance to biotic stresses like fungal infections.

Interactomics is the study of the interactions between proteins, DNA, RNA, and small molecules within a biological system. In the context of plant stress tolerance, interactomics provides insights into how different molecules work together to mediate complex stress responses (Liu *et al.*, 2023). By mapping protein–protein interactions (PPIs), protein–DNA interactions, and signaling networks, interactomics uncovers the molecular crosstalk that governs stress adaptation. For example, under abiotic stress, plants rely on a network of stress-responsive transcription factors, kinases, and signaling molecules that interact to regulate gene expression and physiological responses. Interactomics can identify key hubs in these networks that, when manipulated, enhance stress tolerance. Similarly, in biotic stress responses, interactomics helps unravel the defense signaling networks activated by pathogen recognition receptors (PRRs) and their downstream effectors, providing targets for improving disease resistance in crops.

9.9 Biotechnological Applications

Biotechnology leverages biological processes, living organisms, or entire systems to create innovative products and technologies that enhance human well-being. In recent years, omics technologies have played a pivotal role in advancing agricultural research, particularly in the development of crop varieties with enhanced stress tolerance (Yadav *et al.*, 2024).

9.9.1 Genetic engineering for enhanced stress tolerance

Genetic engineering has significantly advanced the improvement of crops, targeting both biotic and abiotic stress tolerance as well as quality enhancement (Hrbáčková *et al.*, 2020). Key strategies in genetic engineering for enhanced stress tolerance include manipulating genes responsible for osmolyte biosynthesis, which helps in maintaining cellular stability under stress, and modifying genes that encode enzymes involved in scavenging ROS to prevent oxidative damage (Sharma *et al.*, 2014). Additionally, genes encoding late embryogenesis abundant (LEA) proteins are engineered to enhance drought resistance, while genes for heterologous enzymes with varying temperature optima are used to broaden a plant's environmental adaptability. Other approaches involve the genetic engineering of molecular chaperones for protein stability, transcription factors that regulate stress-responsive pathways, and proteins crucial for maintaining ion homeostasis under adverse conditions (Meriç *et al.*, 2020). Researchers are also focusing on selecting stress-inducible promoters for more precise gene expression and identifying new stress-related genes to develop transgenic plants better suited for agricultural use. These innovations are paving the way for more resilient and productive crops. Brini *et al.* (2020) highlighted the interplay between stress signaling and nutritional enhancement in rice by regulating antioxidant, osmolyte, and metabolite levels. OsiLf-2 from *Streptomyces albidoflavus* helped *Camellia oleifera* withstand drought stress (He *et al.*, 2023). Likewise, tobacco lines expressing the *Daucus carota* DcLCYB1 gene exhibited increased DcLCYB1 expression, leading to extensive changes in the tobacco transcriptome (~2000 genes), proteome (~700 proteins), and metabolome (26 metabolites); these changes affected genes associated with metabolic processes, including cell wall formation, lipid metabolism, glycolysis, and secondary metabolism, resulting in enhanced abiotic stress tolerance (Moreno *et al.*, 2021).

9.9.2 CRISPR-Cas9 and other genome editing tools

Currently, three primary methods are widely used for genome editing: transcription activator-like effector nucleases (TALENs), clustered regularly interspaced short palindromic repeats (CRISPR), and zinc finger nucleases (ZFNs). CRISPR-Cas9 is favored over other techniques due to its cost-effectiveness, simplicity, speed, and efficiency in gene knockout. In contrast, ZFNs and TALENs are commonly utilized for a wider range of genetic modifications by creating DNA double-strand breaks (Zhou *et al.*, 2023). CRISPR-Cas9 has revolutionized stress tolerance breeding, becoming the most widely used method to modify tolerant (T) genes or knocking out susceptible (S) genes and engineer them without relying on transgenic approaches (Erdoğan *et al.*, 2023). In omics, genome editing tools contribute immensely to the functional analysis of genes and metabolites. The CRISPR-Cas9 system can target multiple genes simultaneously and overcome the problem posed by functional redundancy of genes in abiotic breeding (Mohamed *et al.*, 2024). Lou *et al.* (2017) conferred the role of ABA sensitivity and tolerance to drought stress in rice on *OsSAPK2* gene using loss of function mutants. For enhanced salt tolerance the expression level of the *OsRAV2* gene was altered using CRISPR/Cas-based targeted mutations in rice (Duan *et al.*, 2016). A novel rice germplasm for salt-tolerance at seedling stage was developed using CRISPR-Cas9 to directionally edit the *OsRR22* gene (Han *et al.*, 2022). CRISPR-Cas9 can modify/eliminate genes for abiotic stresses, e.g. salinity, drought, extreme temperatures, heavy metals, and nutrient deficiencies (Table 9.8).

9.9.3 Breeding for stress tolerance using omics data

In recent years, there has been a substantial surge in research and literature exploring the application of omics technologies in crop breeding (Chao *et al.*, 2024). Omics technologies complement traditional breeding efforts by providing a deeper understanding of plant stress responses at multiple biological levels. This multi-layered approach allows for the identification of specific candidate genes, regulatory elements, and metabolic pathways that are crucial for stress resilience. Once these key genes are identified, they can be targeted for improvement through selective breeding or precise gene-editing techniques like CRISPR (Hrbáčková *et al.*, 2020). Breeding strategy for stress tolerance using omics data requires following steps:

1. Experimental design: growing crop genotype in stress and control environment.
2. Data generation: in the experiment at appropriate crop stage collect the data at required omics level.
3. Data processing: trimming, aligning, annealing, etc. of the large data collected at the respective omics level.
4. Data integration: integrate multiomics datasets for functional analysis and network analysis.
5. Validation of experiment hypotheses: verify the experiment results by hunting the candidate gene/QTL.
6. Application in breeding: apply breeding principles to mutate, edit, backcross, MAB, accumulate the gene/QTL, etc. for improvement of cultivar.

Omics-based analysis can integrate data from diverse platforms, offering a comprehensive, systems-level understanding of abiotic stress tolerance in crops. This approach enables the identification of critical regulatory networks, biomarkers, and candidate genes that can be targeted for breeding, facilitating precision agriculture strategies (Derbyshire *et al.*, 2022). Transcriptomic analysis of sorghum under dry and controlled conditions from seedling to maturity has been detailed for future breeding applications (Varoquaux *et al.*, 2019). Similarly, studies on foxtail millet have explored the temporal network and interaction between drought stress and the Circadian clock (Yi *et al.*, 2022). Both crops serve as valuable references for drought tolerance research. Kudapa *et al.* (2024) used multiomics in chickpea breeding and found proteins involved in activating drought stress response mechanisms.

9.10 Challenges and Future Perspectives

Substantial advancements in omics technology have advanced various promising perspectives regarding plant stress tolerance; however, several constraints prevail, thus preventing their complete efficacy. Although the cost of technology has dramatically decreased in the recent decade, making large-scale genome sequencing less expensive to perform, the expenses associated with generating reference-quality genome assemblies remain high (Schwarze *et al.*, 2018). Integrating multiomics data, such as genomics, transcriptomics, proteomics, and metabolomics, poses a challenge because of the complexity and vast amount of data generated through such analysis. This complexity prevents proper correlation of different layers of biological information. Additionally, most of the data generated by omics studies were performed in controlled lab conditions, which may not fully reflect the complexity of real-world agricultural environments, leading to

Table 9.8. CRISPR-Cas9 and other gene editing tools in stress breeding. Author's own table.

Stress tolerance	Crop	Target gene	Pathway	Mode of action	Reference
Drought	Rice	*OsSAPK2*	Abscisic acid	loss of function	Lou *et al.*, 2017
Drought	Maize	*ARGOS8*	Ethylene sensitivity	Increased expression	Shi *et al.*, 2017
Drought	Tomato	*SlNPR1*	Plant defense response	Gene expression	Li *et al.*, 2019
Drought	Chickpea	*4CL and RVE7*	Phenylpropanoid metabolism and circadian rhythm	Loss of function	Badhan *et al.*, 2021
Drought	Rice	*OsWRKY5*	ABA signaling	Loss of function	Lim *et al.*, 2022
Drought	Rice	*OsSAP*	ROS	Loss of function	Park *et al.*, 2022
Salinity	Rice	*OsRAV2*	RAV	Gene expression	Duan *et al.*, 2016
Drought and salinity	Rice	*DST* gene	Protein–protein interaction	Loss of function	Kumar *et al.*, 2020
Salinity	Tomato	*SlARF4*	Auxin response factors	gene knockout	Bouzroud *et al.*, 2020
Salinity	Rice	OsRR22	Regulatory protein	gene knockout	Han *et al.*, 2022
Salinity	Wheat	*TaDREB2*	Ethylene responsive	Decreased expression	Kim *et al.*, 2018
Salinity	Barley	*HvITPK1*	Phosphate storage	Loss of function	Vlcko and Ohnoutkova, 2020
Salinity	Soybean	*GmAITR*	ABA signaling	Increased expression	Wang *et al.*, 2021
Salinity	Wheat	*TaHAG1*	ROS	Gene knockout	Zheng *et al.*, 2021
Salinity	Maize	*ZmSTL1*	Transpiration	Loss of function	Wang *et al.*, 2022
Cold	Tomato	*SlCBF1*	C-repeat binding factors	Gene knockout	Li *et al.*, 2018
Cold	Rice	*OsPIN5b, GS3, and OsMYB30*		Loss of function	Zeng *et al.*, 2020
Drought and cold	*Fragaria vesca*	*FvICE1*	ABA signaling	Increased expression	Han *et al.*, 2023
Cold	Rice	*OsCS511*	ROS	Increased expression	Park *et al.*, 2024
Cold	Tomato	*SlNAC3*	Ethylene responsive	Increased expression	Wang *et al.*, 2024
Heat stress	Maize	*ZmTMS5*	Male-sterile	Gene knockout	Li *et al.*, 2017

discrepancies in crop responses to stress. Sample size, statistical accuracy, and the number of replicates are further challenges, especially when dealing with rare or expensive samples. Ensuring stability and accuracy in results while reducing false positives (FP) and false negatives (FN) is important. Sequencing-based omics exhibit inherently higher reproducibility than MS-based omics (proteomics and metabolomics) because the results are reliant on the type of mass spectrometer and sample processing methods. Such intrinsic heterogeneity may influence the accuracy and reproducibility of the data obtained using specific omics. The cells that are naturally present in tissues themselves are heterogeneous and omics technologies can only capture a brief overview of their constantly changing states, complicating the understanding of their role in stress responses. In fact, the phenotyping of such traits becomes significantly complex under diverse conditions of stress, making it a little more complicated regarding the production of stress-resilient crops (Raza *et al.*, 2021). Concomitantly, some bottlenecks also lie in the current bioinformatics and computational approaches that lag behind the omics technologies trends. Lack of standardization in data collection, storage, and analysis hinders the reproducibility and cross-comparison of studies. These constraints together highlight the need for more refined systems, good statistical models, and superior computational tools for making efficient use of the potentiality of omics in any stress tolerance-related research.

The steady development of omics technologies seems to give a promising aspect towards future studies on stress tolerance in agricultural settings. Advances in high-throughput sequencing, mass spectrometry, and bioinformatics now further develop the ability to interrogate complex datasets along all biological dimensions. New approaches, such as single-cell omics, have vast potential since cell-specific stress reactions that may be masked by bulk tissue-based analyses could be discovered. Similarly, spatial omics technologies are likely to enhance our understanding by allowing researchers to sketch out the molecular changes due to stress at the tissue or subcellular levels, hence giving a better view of the plant response to stress (Yu *et al.*, 2023). Machine learning (ML) and artificial intelligence (AI) are becoming pivotal in handling the vast datasets generated by omics technologies (Isewon *et al.*, 2022). The nexus of these resources with the Internet of Things (IoT) will link complex biological data with phenotypic outcomes (Chaudhary *et al.*, 2023). This integration can overcome the bottlenecks of current data interpretation, hence enhancing the predictive power of omics in breeding for stress-resilient crops. AI-driven approaches could be explored to identify novel biomarkers and regulatory networks critical to stress tolerance; therefore, the process of breeding crops sensitive to environmental fluctuations can be streamlined. New disciplines like peptidomics, phosphoproteomics, and redox proteomics are emerging thus allowing far better insight into the molecular interactions and functionalities of proteins important for plant stress responses (Mondal *et al.*, 2022). Single-cell RNA sequencing (scRNA-Seq) which has already been used in crops such as rice (Zhang *et al.*, 2021b) will eventually revolutionize cell-level examination of transcriptional responses to stress. Future research will likely focus on combining omics technologies with precision agriculture and phenomics, facilitating the development of predictive models of crop performance under stress conditions, thereby providing a better route for breeding programs and climate-resilient crop varieties that come up more expeditiously. More intense use of precise genes or regulatory elements defined by the data available through these studies would be made using technologies such as CRISPR/Cas9 for an improvement in crop tolerance towards an expanded range of environmental stresses. The intersection of omics technologies with cutting-edge practices like speed breeding and genome editing can significantly accelerate agricultural innovation. By fostering faster plant growth and enabling the production of crops suited to specific environments, these technologies could help meet the growing global food demand sustainably. Over the next few decades, metagenomics will likely contribute a lot towards a promising future in the plant microbiome concerning definitive beneficial microbial interactions with positive impact on stress tolerance. These advances must be accompanied by policy-supporting frameworks to advance these developments into practical, real-world solutions essential for integrating omics technologies into sustainable agricultural practices.

9.11 Conclusions

The integration of omics approaches represents a paradigm shift in our efforts to enhance plant stress tolerance. By leveraging genomics, transcriptomics, proteomics, metabolomics, and epigenomics, researchers have been able to dissect the intricate molecular mechanisms underlying plant responses to various stressors. These technologies have paved the way for identifying stress-responsive genes, proteins, metabolites, and regulatory networks, offering novel targets for genetic engineering and breeding programs aimed at improving crop resilience.

As omics technologies continue to evolve, the potential for discovering new pathways and regulatory elements involved in stress tolerance will expand, providing even more opportunities for innovative solutions in agriculture. However, the challenge lies in effectively integrating the vast amounts of data generated by these approaches and translating this knowledge into practical applications. Future research will need to focus on improving data integration and developing more sophisticated models to fully exploit the power of omics technologies. In this context, the chapter underscores the pivotal role of omics in driving the future of plant stress tolerance research and its implications for global food security.

References

Abdelraheem, A., Thyssen, G.N., Fang, D.D., Jenkins, J.N., McCarty, J.C. *et al.* (2021) GWAS reveals consistent QTL for drought and salt tolerance in a MAGIC population of 550 lines derived from intermating of 11 upland cotton (*Gossypium hirsutum*) parents. *Molecular Genetics and Genomics* 296, 119–129.

Alseekh, S., Tohge, T., Wendenberg, R., Scossa, F., Omranian, N. *et al.* (2015) Identification and mode of inheritance of quantitative trait loci for secondary metabolite abundance in tomato. *The Plant Cell* 27(3), 485–512. DOI: 10.1105/tpc.114.132266.

Anders, S. and Huber, W. (2010) Differential expression analysis for sequence count data. *Nature Precedings* 1–1.

Arefian, M., Bhagya, N. and Prasad, T.K. (2021) Phosphorylation-mediated signalling in flowering: Prospects and retrospects of phosphoproteomics in crops. *Biological Reviews* 96(5), 2164–2191.

Arisha, M.H., Ahmad, M.Q., Tang, W., Liu, Y., Yan, H. *et al.* (2020) RNA-sequencing analysis revealed genes associated drought stress responses of different durations in hexaploid sweet potato. *Scientific Reports* 10(1), 12573.

Ashraf, M.F., Hou, D., Hussain, Q., Imran, M., Pei, J. *et al.* (2022) Entailing the next-generation sequencing and metabolome for sustainable agriculture by improving plant tolerance. *International Journal of Molecular Sciences* 23(2), 651.

Badhan, S., Ball, A.S. and Mantri, N. (2021) First report of CRISPR/Cas9 mediated DNA-free editing of 4CL and RVE7 genes in chickpea protoplasts. *International Journal of Molecular Sciences* 22(1), 396.

Barkla, B.J., Vera-Estrella, R. and Raymond, C. (2016) Single-cell-type quantitative proteomic and ionomic analysis of epidermal bladder cells from the halophyte model plant *Mesembryanthemum crystallinum* to identify salt-responsive proteins. *BMC Plant Biology* 16(1), 110.

Baxter, I. (2010) Ionomics: The functional genomics of elements. *Briefings in Functional Genomics* 9(2), 149–156. DOI: 10.1093/bfgp/elp055.

Bhatta, M., Morgounov, A., Belamkar, V., Wegulo, S.N., Dababat, A.A. *et al.* (2019) Genome-wide association study for multiple biotic stress resistance in synthetic hexaploid wheat. *International Journal of Molecular Sciences* 20(15), 3667.

Bhattacharjee, M., Dhar, S., Handique, P.J., Acharjee, S. and Sarmah, B.K. (2020) Defense response in chickpea pod wall due to simulated herbivory unfolds differential proteome profile. *The Protein Journal* 39, 240–257.

Bouzroud, S., Gasparini, K., Hu, G., Barbosa, M.A.M., Rosa, B.L. *et al.* (2020) Down regulation and loss of auxin response factor 4 function using CRISPR/Cas9 alters plant growth, stomatal function and improves tomato tolerance to salinity and osmotic stress. *Genes* 11(3), 272.

Bowne, J.B., Erwin, T.A., Juttner, J., Schnurbusch, T., Langridge, P. *et al.* (2012) Drought responses of leaf tissues from wheat cultivars of differing drought tolerance at the metabolite level. *Molecular Plant* 5(2), 418–429.

Brini, F., Yakoubi, I. and Saibi, W. (2020) Genomics and genetic engineering of rice elucidating cross talk between stress signaling and nutrition enhancement via regulation of antioxidant, osmolyte, and metabolite levels. In: *Rice Research for Quality Improvement: Genomics and Genetic Engineering: Volume 2: Nutrient Biofortification and Herbicide and Biotic Stress Resistance in Rice*. pp. 409–432.

Bure, I.V., Nemtsova, M.V. and Kuznetsova, E.B. (2022) Histone modifications and non-coding RNAs: Mutual epigenetic regulation and role in pathogenesis. *International Journal of Molecular Sciences* 23(10), 5801.

Campbell, Z.C., Acosta-Gamboa, L.M., Nepal, N. and Lorence, A. (2018) Engineering plants for tomorrow: How high-throughput phenotyping is contributing to the development of better crops. *Phytochemistry Reviews* 17, 1329–1343.

Chandran, A.K.N., Kim, J.W., Yoo, Y.H., Park, H.L., Kim, Y.J. *et al.* (2019) Transcriptome analysis of rice-seedling roots under soil–salt stress using RNA-seq method. *Plant Biotechnology Reports* 13, 567–578.

Chang, J., Cheong, B.E., Natera, S. and Roessner, U. (2019) Morphological and metabolic responses to salt stress of rice (*Oryza sativa* L.) cultivars which differ in salinity tolerance. *Plant Physiology and Biochemistry* 144, 427–435.

Chang, Y., Zhu, D., Duan, W., Deng, X., Zhang, J. *et al.* (2021) Plasma membrane N-glycoproteome analysis of wheat seedling leaves under drought stress. *International Journal of Biological Macromolecules* 193, 1541–1550.

Chao, H., Zhang, S., Hu, Y., Ni, Q., Xin, S. *et al.* (2024) Integrating omics databases for enhanced crop breeding. *Journal of Integrative Bioinformatics* 20(4), 20230012.

Chaudhary, D., Sharma, S., Khar, S. and Srivastava, R.K. (2023) Internet of Things (IoT) and big data: Future of farming. *Agricultural Engineering Today* 47(1), 59–62.

Cheah, B.H., Lin, H.H., Chien, H.J., Liao, C.T., Liu, L.Y.D. *et al.* (2020) SWATH-MS-based quantitative proteomics reveals a uniquely intricate defense response in *Cnaphalocrocis medinalis*-resistant rice. *Scientific Reports* 10(1), 6597.

Chebrolu, K.K., Fritschi, F.B., Ye, S., Krishnan, H.B., Smith, J.R. *et al.* (2016) Impact of heat stress during seed development on soybean seed metabolome. *Metabolomics* 12, 1–14.

Cheng, H., Zhang, H., Song, J., Jiang, J., Chen, S. *et al.* (2023) GERDH: An interactive multi-omics database for cross-species data mining in horticultural crops. *The Plant Journal* 116(4), 1018–1029.

Chen, J., Hu, X., Shi, T., Yin, H., Sun, D. *et al.* (2020) Metabolite-based genome-wide association study enables dissection of the flavonoid decoration pathway of wheat kernels. *Plant Biotechnology Journal* 18(8), 1722–1735.

Chen, M.X., Zhang, Y., Fernie, A.R., Liu, Y.G. and Zhu, F.Y. (2021) SWATH-MS-based proteomics: Strategies and applications in plants. *Trends in Biotechnology* 39(5), 433–437.

Chen, W., Yao, Q., Patil, G.B., Agarwal, G., Deshmukh, R.K. *et al.* (2016) Identification and comparative analysis of differential gene expression in soybean leaf tissue under drought and flooding stress revealed by RNA-Seq. *Frontiers in Plant Science* 7, 1044.

Da Ros, L., Bollina, V., Soolanayakanahally, R., Pahari, S., Elferjani, R. *et al.* (2023) Multi-omics atlas of combinatorial abiotic stress responses in wheat. *The Plant Journal* 116(4), 1118–1135.

Cui, X., Hu, M., Yao, S., Zhang, Y., Tang, M. *et al.* (2023) BnaOmics: A comprehensive platform combining pan-genome and multi-omics data from *Brassica napus*. *Plant Communications* 4(5).

Dai, F., Chen, J., Zhang, Z., Liu, F., Li, J. *et al.* (2022) COTTONOMICS: A comprehensive cotton multi-omics database. *Database* baac080.

del Toro, N., Reisinger, F., Foster, J.M., Contell, J., Fabregat, A. *et al.* (2014, July) PRIDE proteomes: A condensed view of the plethora of public proteomics data available in the PRIDE repository. In: *DILS*, Vol. 2014. pp. 21–24.

Derbyshire, M.C., Batley, J. and Edwards, D. (2022) Use of multiple 'omics techniques to accelerate the breeding of abiotic stress tolerant crops. *Current Plant Biology* 32, 100262.

Duan, Y.B., Li, J., Qin, R.Y., Xu, R.F., Li, H. *et al.* (2016) Identification of a regulatory element responsible for salt induction of rice OsRAV2 through *ex situ* and *in situ* promoter analysis. *Plant Molecular Biology* 90, 49–62.

Echevarría-Zomeño, S., Fernández-Calvino, L., Castro-Sanz, A.B., López, J.A., Vázquez, J. *et al.* (2016) Dissecting the proteome dynamics of the early heat stress response leading to plant survival or death in *Arabidopsis*. *Plant, cell & environment* 39(6), 1264–1278.

Elakhdar, A., Slaski, J.J., Kubo, T., Hamwieh, A., Hernandez Ramirez, G. *et al.* (2023) Genome-wide association analysis provides insights into the genetic basis of photosynthetic responses to low-temperature stress in spring barley. *Frontiers in Plant Science* 14, 1159016.

Erdmann, R.M. and Picard, C.L. (2020) RNA-directed DNA methylation. *PLoS Genetics* 16(10), e1009034.

Erdoğan, İ., Cevher-Keskin, B., Bilir, Ö., Hong, Y. and Tör, M. (2023) Recent developments in CRISPR/Cas9 genome-editing technology related to plant disease resistance and abiotic stress tolerance. *Biology* 12(7), 1037.
Esposito, S., Aversano, R., Bradeen, J., D'Amelia, V., Villano, C. *et al.* (2021) Coexpression gene network analysis of cold-tolerant *Solanum commersonii* reveals new insights in response to low temperatures. *Crop Science* 61(5), 3538–3550.
Fàbregas, N., Yoshida, T. and Fernie, A.R. (2020) Role of Raf-like kinases in SnRK2 activation and osmotic stress response in plants. *Nature Communications* 11(1), 6184.
Fan, Y., Sun, C., Yan, K., Li, P., Hein, I. *et al.* (2024) Recent advances in studies of genomic DNA methylation and its involvement in regulating drought stress response in crops. *Plants* 13(10), 1400.
Feng, J., Long, Y., Shi, L., Shi, J., Barker, G. *et al.* (2012) Characterization of metabolite quantitative trait loci and metabolic networks that control glucosinolate concentration in the seeds and leaves of *Brassica napus*. *New Phytologist* 193(1), 96–108. DOI: 10.1111/j.1469-8137.2011.03890.x.
Feng, J.W., Han, L., Liu, H., Xie, W.Z., Liu, H. *et al.* (2023) MaizeNetome: A multi-omics network database for functional genomics in maize. *Molecular Plant* 16(8), 1229–1231.
Ghosson, H., Schwarzenberg, A., Jamois, F. and Yvin, J.C. (2018) Simultaneous untargeted and targeted metabolomics profiling of underivatized primary metabolites in sulfur-deficient barley by ultra-high performance liquid chromatography-quadrupole/time-of-flight mass spectrometry. *Plant Methods* 14, 1–17.
Gill, S.S., Khan, N.A., Agarwala, N., Singh, K., Sunkar, R. *et al.* (2024) NcRNAs in plant development and stress responses. *Plant Physiology and Biochemistry* 108950.
Gomathi, R., Kohila, S., Viswanathan, R., Krishnapriya, V., Appunu, C. *et al.* (2024) Comparative proteomic analysis of high-temperature response in sugarcane (*Saccharum spp.*). *Sugar Tech* 1–15.
Gong, L., Chen, W., Gao, Y., Liu, X., Zhang, H. *et al.* (2013) Genetic analysis of the metabolome exemplified using a rice population. *Proceedings of the National Academy of Sciences* 110(50), 20320–20325.
Gui, S., Yang, L., Li, J., Luo, J., Xu, X. *et al.* (2020) ZEAMAP, a comprehensive database adapted to the maize multi-omics era. *IScience* 23(6).
Gupta, P. and De, B. (2017) Metabolomics analysis of rice responses to salinity stress revealed elevation of serotonin, and gentisic acid levels in leaves of tolerant varieties. *Plant Signaling & Behavior* 12(7), e1335845.
Gupta, S., Mishra, S.K., Misra, S., Pandey, V., Agrawal, L. *et al.* (2020) Revealing the complexity of protein abundance in chickpea root under drought-stress using a comparative proteomics approach. *Plant Physiology and Biochemistry* 151, 88–102.
Hafiz, A.H., Jun, Z.H.A.N.G., Yu-Shuang, G., Mei-Xu, G. and Wei, G.U.O. (2022) Proteomic analysis of pathogen-responsive proteins from maize stem apoplast triggered by *Fusarium verticillioides*. *Journal of Integrative Agriculture* 21(2), 446–459.
Han, J., Li, X., Li, W., Yang, Q., Li, Z. *et al.* (2023) Isolation and preliminary functional analysis of FvICE1, involved in cold and drought tolerance in *Fragaria vesca* through overexpression and CRISPR/Cas9 technologies. *Plant Physiology and Biochemistry* 196, 270–280.
Han, X. and Yang, Y. (2021) Phospholipids in salt stress response. *Plants* 10(10), 2204.
Han, X., Chen, Z., Li, P., Xu, H., Liu, K. *et al.* (2022) Development of novel rice germplasm for salt-tolerance at seedling stage using CRISPR-Cas9. *Sustainability* 14(5), 2621.
Hardcastle, T.J. and Kelly, K.A. (2010) BaySeq: Empirical Bayesian methods for identifying differential expression in sequence count data. *BMC Bioinformatics* 11, 1–14.
Haworth, M., Marino, G., Atzori, G., Fabbri, A., Daccache, A. *et al.* (2023) Plant physiological analysis to overcome limitations to plant phenotyping. *Plants* 12(23), 4015. DOI: 10.3390/plants12234015.
Henschel, J.M., Andrade, A.N.D., dos Santos, J.B.L., da Silva, R.R., da Mata, D.A. *et al.* (2024) Lipidomics in plants under abiotic stress conditions: An overview. *Agronomy* 14(8), 1670.
He, Z., Cui, K., Wang, R., Xu, T., Zhang, Z. *et al.* (2023) Multi-omics joint analysis reveals how Streptomyces albidoflavus OsiLf-2 assists *Camellia oleifera* to resist drought stress and improve fruit quality. *Frontiers in Microbiology* 14, 1152632.
Hill, C.B., Taylor, J.D., Edwards, J., Mather, D., Langridge, P. *et al.* (2015) Detection of QTL for metabolic and agronomic traits in wheat with adjustments for variation at genetic loci that affect plant phenology. *Plant Science* 233, 143–154. DOI: 10.1016/j.plantsci.2015.01.008.
Hirayama, T. and Shinozaki, K. (2010) Research on plant abiotic stress responses in the post-genome era: Past, present and future. *The Plant Journal* 61(6), 1041–1052.

Hochberg, U., Perry, A., Rachmilevitch, S., Ben-Gal, A. and Sperling, O. (2023) Instantaneous and lasting effects of drought on grapevine water use. *Agricultural and Forest Meteorology* 338, 109521.

Hrbáčková, M., Dvořák, P., Takáč, T., Tichá, M., Luptovčiak, I. *et al.* (2020) Biotechnological perspectives of omics and genetic engineering methods in Alfalfa. *Frontiers in Plant Science* 11, 592.

Huang, C.Y., Roessner, U., Eickmeier, I., Genc, Y., Callahan, D.L. *et al.* (2008) Metabolite profiling reveals distinct changes in carbon and nitrogen metabolism in phosphate-deficient barley plants (*Hordeum vulgare* L.). *Plant and Cell Physiology* 49(5), 691–703.

Huang Ying, H.Y., Ma HongYu, M.H., Huang Wei, H.W., Wang Feng, W.F., Xu ZhiSheng, X.Z. *et al.* (2016) Comparative proteomic analysis provides novel insight into the interaction between resistant vs susceptible tomato cultivars and TYLCV infection.

Isewon, I., Apata, O., Oluwamuyiwa, F., Aromolaran, O. and Oyelade, J. (2022) Machine learning algorithms: Their applications in plant omics and agronomic traits' improvement. *F1000Research* 11, 1256.

Khan, M.K.R., Ditta, A., Wang, B., Fang, L., Anwar, Z. *et al.* (2023) The intervention of multi-omics approaches for developing abiotic stress resistance in cotton crop under climate change. In: *Sustainable Agriculture in the Era of the OMICs Revolution*. Springer International Publishing, Cham, Switzerland, pp. 37–82.

Kim, D., Alptekin, B. and Budak, H. (2018) CRISPR/Cas9 genome editing in wheat. *Functional & Integrative Genomics* 18, 31–41.

Kim, K.H., Song, K., Park, J.M., Kim, J.Y. and Lee, B.M. (2021) RNA-Seq analysis of gene expression changes related to delay of flowering time under drought stress in tropical maize. *Applied Sciences* 11(9), 4273.

Komatsu, S. (2019) Plant proteomic research 2.0: Trends and perspectives. *International Journal of Molecular Sciences* 20(10), 2495.

Komatsu, S., Yamaguchi, H., Hitachi, K., Tsuchida, K., Rehman, S.U. *et al.* (2022) Morphological, biochemical, and proteomic analyses to understand the promotive effects of plant-derived smoke solution on wheat growth under flooding stress. *Plants* 11(11), 1508.

Kong, W., Zhang, C., Zhang, S., Qiang, Y., Zhang, Y. *et al.* (2021) Uncovering the novel QTLs and candidate genes of salt tolerance in rice with linkage mapping, RTM-GWAS, and RNA-Seq. *Rice* 14, 1–12.

Kudapa, H., Ghatak, A., Barmukh, R., Chaturvedi, P., Khan, A. *et al.* (2024) Integrated multi-omics analysis reveals drought stress response mechanism in chickpea (*Cicer arietinum* L.). *The Plant Genome* 17(1), e20337.

Kumar, S., Verma, V.V., Yadav, R.K., Yadav, S.K., Watts, P. *et al.* (2020) CRISPR-Cas9 mediated genome editing of drought and salt tolerance (OsDST) gene in Indica mega rice cultivar MTU1010. *Physiology and Molecular Biology of Plants* 26, 1099–1110.

Kurotani, K.I., Hayashi, K., Hatanaka, S., Toda, Y., Ogawa, D. *et al.* (2015) Elevated levels of CYP94 family gene expression alleviate the jasmonate response and enhance salt tolerance in rice. *Plant and Cell Physiology* 56(4), 779–789.

Kusebauch, U., Deutsch, E.W., Campbell, D.S., Sun, Z., Farrah, T. *et al.* (2014) Using PeptideAtlas, SRMAtlas, and PASSEL: Comprehensive resources for discovery and targeted proteomics. *Current Protocols in Bioinformatics* 46(1), 13–25.

Kuta, R., Larochelle, N., Fernandez, M., Pal, A., Minotti, S. *et al.* (2020) Depending on the stress, histone deacetylase inhibitors act as heat shock protein co-inducers in motor neurons and potentiate arimoclomol, exerting neuroprotection through multiple mechanisms in ALS models. *Cell Stress and Chaperones* 25(1), 173–191.

Langfelder, P. and Horvath, S. (2008) WGCNA: An R package for weighted correlation network analysis. *BMC Bioinformatics* 9, 1–13.

Lee, M.H., Kim, K.M., Sang, W.G., Kang, C.S. and Choi, C. (2022) Comparison of gene expression changes in three wheat varieties with different susceptibilities to heat stress using RNA-Seq analysis. *International Journal of Molecular Sciences* 23(18), 10734.

Liang, T., Yuan, Z., Fu, L., Zhu, M., Luo, X. *et al.* (2021) Integrative transcriptomic and proteomic analysis reveals an alternative molecular network of glutamine synthetase 2 corresponding to nitrogen deficiency in rice (*Oryza sativa* L.). *International Journal of Molecular Sciences* 22(14), 7674.

Li, D., Dossa, K., Zhang, Y., Wei, X., Wang, L. *et al.* (2018) GWAS uncovers differential genetic bases for drought and salt tolerances in sesame at the germination stage. *Genes* 9(2), 87.

Li, H., Li, Y., Ke, Q., Kwak, S.S., Zhang, S. *et al*. (2020) Physiological and differential proteomic analyses of imitation drought stress response in *Sorghum bicolor* root at the seedling stage. *International Journal of Molecular Sciences* 21(23), 9174.

Li, H., Yang, M., Zhao, C., Wang, Y. and Zhang, R. (2021) Physiological and proteomic analyses revealed the response mechanisms of two different drought-resistant maize varieties. *BMC Plant Biology* 21, 1–15.

Li, J., Witten, D.M., Johnstone, I.M. and Tibshirani, R. (2012) Normalization, testing, and false discovery rate estimation for RNA-Sequencing data. *Biostatistics* 13(3), 523–538. DOI: 10.1093/biostatistics/kxr031.

Li, J., Zhang, H., Si, X., Tian, Y., Chen, K. *et al*. (2017) Generation of thermosensitive male-sterile maize by targeted knockout of the ZmTMS5 gene. *Journal of Genetics and Genomics= Yi Chuan Xue Bao* 44(9), 465–468.

Li, R., Liu, C., Zhao, R., Wang, L., Chen, L. *et al*. (2019) CRISPR/Cas9-Mediated SlNPR1 mutagenesis reduces tomato plant drought tolerance. *BMC Plant Biology* 19(3), 1–13.

Liu, C., Mentzelopoulou, A., Muhammad, A., Volkov, A., Weijers, D. *et al*. (2023) An actin remodeling role for *Arabidopsis* processing bodies revealed by their proximity interactome. *The EMBO Journal* 42(9), e111885.

Lim, C., Kang, K., Shim, Y., Yoo, S.C. and Paek, N.C. (2022) Inactivating transcription factor OsWRKY5 enhances drought tolerance through abscisic acid signaling pathways. *Plant Physiology* 188(4), 1900–1916.

Liu, G., Xia, Y., Liu, T., Dai, S. and Hou, X. (2018) The DNA methylome and association of differentially methylated regions with differential gene expression during heat stress in *Brassica rapa*. *International Journal of Molecular Sciences* 19(5), 1414.

Liu, T., Salguero, P., Petek, M., Martinez-Mira, C., Balzano-Nogueira, L. *et al*. (2022) PaintOmics 4: New tools for the integrative analysis of multi-omics datasets supported by multiple pathway databases. *Nucleic Acids Research* 50(W1), W551–W559.

Liu, W., Yin, T., Zhao, Y., Wang, X., Wang, K. *et al*. (2021) Effects of high temperature on rice grain development and quality formation based on proteomics comparative analysis under field warming. *Frontiers in Plant Science* 12, 746180.

Li, W., Fu, Y., Lv, W., Zhao, S., Feng, H. *et al*. (2022) Characterization of the early gene expression profile in *Populus ussuriensis* under cold stress using PacBio SMRT sequencing integrated with RNA-Seq reads. *Tree Physiology* 42(3), 646–663.

Lou, D., Wang, H., Liang, G. and Yu, D. (2017) OsSAPK2 confers abscisic acid sensitivity and tolerance to drought stress in rice. *Frontiers in Plant Science* 8, 993.

Luo, Z., Szczepanek, A. and Abdel-Haleem, H. (2020) Genome-Wide Association Study (GWAS) analysis of camelina seedling germination under salt stress condition. *Agronomy* 10(9), 1444.

Mancini, I., Domingo, G., Bracale, M., Loreto, F. and Pollastri, S. (2022) Isoprene emission influences the proteomic profile of *Arabidopsis* plants under well-watered and drought-stress conditions. *International Journal of Molecular Sciences* 23(7), 3836.

Ma, S., Wang, M., Wu, J., Guo, W., Chen, Y. *et al*. (2021) WheatOmics: A platform combining multiple omics data to accelerate functional genomics studies in wheat. *Molecular Plant* 14(12), 1965–1968.

Ma, Z., Wang, L., Zhao, M., Gu, S., Wang, C. *et al*. (2020) ITRAQ proteomics reveals the regulatory response to *Magnaporthe oryzae* in durable resistant vs. susceptible rice genotypes. *PLoS One* 15(1), e0227470.

Macklin, P. (2019) Key challenges facing data-driven multicellular systems biology. *Gigascience* 8(10), giz127.

McGettigan, P.A. (2013) Transcriptomics in the RNA-Seq era. *Current Opinion in Chemical Biology* 17(1), 4–11. DOI: 10.1016/j.cbpa.2012.12.008.

Mehmood, S.S., Lu, G., Luo, D., Hussain, M.A., Raza, A. *et al*. (2021) Integrated analysis of transcriptomics and proteomics provides insights into the molecular regulation of cold response in *Brassica napus*. *Environmental and Experimental Botany* 187, 104480.

Meng, J.Y., Ntambo, M.S., Rott, P.C., Fu, H.Y., Huang, M.T. *et al*. (2020) Identification of differentially expressed proteins in sugarcane in response to infection by *Xanthomonas albilineans* using iTRAQ quantitative proteomics. *Microorganisms* 8(1), 76.

Meng, X., Li, A., Yu, B. and Li, S. (2021) Interplay between miRNAs and lncRNAs: Mode of action and biological roles in plant development and stress adaptation. *Computational and Structural Biotechnology Journal* 19, 2567–2574.

Meriç, S., Ayan, A. and Atak, Ç. (2020) Molecular abiotic stress tolerans strategies: From genetic engineering to genome editing era. *Abiotic Stress in Plants* 91.

Metzker, M.L. (2010) Sequencing technologies—the next generation. *Nature Reviews Genetics* 11(1), 31–46.

Mi, H., Ebert, D., Muruganujan, A., Mills, C., Albou, L.P. *et al.* (2021) PANTHER version 16: A revised family classification, tree-based classification tool, enhancer regions and extensive API. *Nucleic Acids Research* 49(D1), D394–D403.

Mironova, V.V., Weinholdt, C. and Grosse, I. (2015) RNA-Seq data analysis for studying abiotic stress in horticultural plants. In: *Abiotic Stress Biology in Horticultural Plants*. pp. 197–220.

Mohamed, H.I., Khan, A. and Basit, A. (2024) CRISPR-Cas9 system mediated genome editing technology: An ultimate tool to enhance abiotic stress in crop plants. *Journal of Soil Science and Plant Nutrition* 1–24.

Mondal, S., Song, H., Zhang, L. and Cao, Y. (2022) Multi-omics and computational biology in horticultural plants: From genotype to phenotype. *Frontiers in Plant Science* 13, 1073266.

Moon, J.Y., Kim, S.T., Choi, G.J., Kwon, S.Y., Cho, H.S. *et al.* (2020) Comparative proteomic analysis of host responses to *Plasmodiophora brassicae* infection in susceptible and resistant *Brassica oleracea*. *Plant Biotechnology Reports* 14, 263–274.

Moreno, J.C., Martinez-Jaime, S., Kosmacz, M., Sokolowska, E.M., Schulz, P. *et al.* (2021) A multi-OMICs approach sheds light on the higher yield phenotype and enhanced abiotic stress tolerance in tobacco lines expressing the carrot lycopene β-cyclase1 gene. *Frontiers in Plant Science* 12, 624365.

Mosa, K.A., Ismail, A. and Helmy, M. (2017) Omics and system biology approaches in plant stress research. In: *Plant Stress Tolerance: An Integrated Omics Approach*. Springer International Publishing, Cham, Switzerland, pp. 21–34.

Muthuramalingam, P., Jeyasri, R., Krishnan, S.R., Pandian, S.T.K., Sathishkumar, R. *et al.* (2019) Integrating the bioinformatics and omics tools for systems analysis of abiotic stress tolerance in *Oryza sativa* (L.). In: *Advances in Plant Transgenics: Methods and Applications*. Springer Singapore, Singapore, pp. 59–77.

Neeraja, C.N., Maghirang-Rodriguez, R., Pamplona, A., Heuer, S., Collard, B.C. *et al.* (2007) A marker-assisted backcross approach for developing submergence-tolerant rice cultivars. *Theoretical and Applied Genetics* 115, 767–776.

Park, J.R., Kim, E.G., Jang, Y.H., Jan, R., Farooq, M. *et al.* (2022) Applications of CRISPR/Cas9 as new strategies for short breeding to drought gene in rice. *Frontiers in Plant Science* 13, 850441.

Park, J.R., Kim, E.G., Jang, Y.H., Jan, R., Farooq, M. *et al.* (2024) CRISPR/Cas9-mediated genome editing of OsCS511 enhances cold tolerance in *Oryza sativa* L. *Environmental and Experimental Botany* 226, 105932.

Pazhamala, L.T., Kudapa, H., Weckwerth, W., Millar, A.H. and Varshney, R.K. (2021) Systems biology for crop improvement. *The Plant Genome* 14(2), e20098.

Peng, H., Wang, K., Chen, Z., Cao, Y., Gao, Q. *et al.* (2020) MBKbase for rice: An integrated omics knowledgebase for molecular breeding in rice. *Nucleic Acids Research* 48, D1085–D1092.

Piasecka, A., Sawikowska, A., Kuczyńska, A., Ogrodowicz, P., Mikołajczak, K. *et al.* (2017) Drought-related secondary metabolites of barley (*Hordeum vulgare* L.) leaves and their metabolomic quantitative trait loci. *The Plant Journal* 89(5), 898–913. DOI: 10.1111/tpj.13430.

Pretorius, C. J., Zeiss, D. R., & Dubery, I. A. (2021) The presence of oxygenated lipids in plant defense in response to biotic stress: A metabolomics appraisal. *Plant Signaling & Behavior*, 16(12), 1989215. DOI: 10.1080/15592324.2021.1989215.

Price, N.D., Reed, J.L. and Palsson, B.Ø. (2004) Genome-scale models of microbial cells: Evaluating the consequences of constraints. *Nature Reviews Microbiology* 2(11), 886–897.

Qiao, F., Yang, X., Xu, F., Huang, Y., Zhang, J. *et al.* (2021) TMT-based quantitative proteomic analysis reveals defense mechanism of wheat against the crown rot pathogen *Fusarium pseudograminearum*. *BMC Plant Biology* 21, 1–17.

Qi, J., Song, C. P., Wang, B., Zhou, J., Kangasjärvi, J., Zhu, J. K., & Gong, Z. (2018) Reactive oxygen species signaling and stomatal movement in plant responses to drought stress and pathogen attack. *Journal of integrative plant biology*, 60(9), 805-826.

Qi, W., Chen, J., Han, Y., Li, Z., Su, X., & Yeo, F. K. S. (2023) Omics-driven crop improvement for stress tolerance. *Frontiers in Plant Science*, 14, 1172228.

Rajaprakasam, S., Shanmugavel, P., Chockalingam, V., Jegadeesan, S., Latha, T.K.S. *et al.* (2023) Comparative metabolomic profiling of horse gram (*Macrotyloma uniflorum* (Lam.) Verdc.) genotypes for horse gram yellow mosaic virus resistance. *Metabolites* 13(2), 165.

Ramakrishnan, M., Satish, L., Kalendar, R., Narayanan, M., Kandasamy, S. *et al.* (2021) The dynamism of transposon methylation for plant development and stress adaptation. *International Journal of Molecular Sciences* 22(21), 11387.

Rathinam, M., Roschitzki, B., Grossmann, J., Mishra, P., Kunz, L. *et al.* (2020) Unraveling the proteomic changes involved in the resistance response of *Cajanus platycarpus* to herbivory by *Helicoverpa armigera*. *Applied Microbiology and Biotechnology* 104, 7603–7618.

Raza, A., Tabassum, J., Kudapa, H. and Varshney, R.K. (2021) Can omics deliver temperature resilient ready-to-grow crops? *Critical Reviews in Biotechnology* 41(8), 1209–1232.

Robinson, M.D., McCarthy, D.J. and Smyth, G.K. (2010) EdgeR: A bioconductor package for differential expression analysis of digital gene expression data. *Bioinformatics* 26(1), 139–140.

Roychowdhury, R., Das, S.P., Gupta, A., Parihar, P., Chandrasekhar, K. *et al.* (2023) Multi-omics pipeline and omics-integration approach to decipher plant's abiotic stress tolerance responses. *Genes* 14(6), 1281.

Scheler, C., Durner, J. and Astier, J. (2013) Nitric oxide and reactive oxygen species in plant biotic interactions. *Current Opinion in Plant Biology* 16(4), 534–539.

Schwarze, K., Buchanan, J., Taylor, J.C. and Wordsworth, S. (2018) Are whole-exome and whole-genome sequencing approaches cost-effective? A systematic review of the literature. *Genetics in Medicine* 20(10), 1122–1130.

Serin, E.A., Nijveen, H., Hilhorst, H.W. and Ligterink, W. (2016) Learning from co-expression networks: Possibilities and challenges. *Frontiers in Plant Science* 7, 444. DOI: 10.3389/fpls.2016.00444.

Shah, L.R., Ahmed, N., Hussain, K., Mansoor, S., Khan, T. *et al.* (2024) Mapping phenotypic performance and novel snps for cold tolerance in tomato (*Solanum lycopersicum*) genotypes through GWAS and population genetics. *BMC Genomic Data* 25(1), 9.

Shariatipour, N., Shams, Z. and Heidari, B. (2022) The role of ionomics as functional genomics of elements in plant abiotic stress tolerance. In: *Omics Analysis of Plants under Abiotic Stress*. Apple Academic Press, pp. 111–152.

Sharma, R., Bhardwaj, R., Thukral, A.K., Handa, N., Kaur, R. *et al.* (2014) Osmolyte dynamics: New strategies for crop tolerance to abiotic stress signals. In: *Emerging Technologies and Management of Crop Stress Tolerance*. Academic Press, pp. 405–430.

Shi, J., Gao, H., Wang, H., Lafitte, H.R., Archibald, R.L. *et al.* (2017) ARGOS 8 variants generated by CRISPR-Cas9 improve maize grain yield under field drought stress conditions. *Plant Biotechnology Journal* 15(2), 207–216.

Singh, S., Singh, I.K. and Singh, A. (2024) Comparative proteome analysis of *Spodoptera litura*-infested *Zea mays* reveals a robust defense strategy targeting insect peritrophic membrane. *Plant Physiology and Biochemistry* 108835.

Skirycz, A. and Inzé, D. (2010) More from less: Plant growth under limited water. *Current Opinion in Biotechnology* 21(2), 197–203.

Song, L., Pan, Z., Chen, L., Dai, Y., Wan, J. *et al.* (2020) Analysis of whole transcriptome RNA-Seq data reveals many alternative splicing events in soybean roots under drought stress conditions. *Genes* 11(12), 1520.

Song, L., Wang, R., Yang, X., Zhang, A. and Liu, D. (2023) Molecular markers and their applications in Marker-Assisted Selection (MAS) in bread wheat (*Triticum aestivum* L.). *Agriculture* 13(3), 642.

Song, P., Wang, J., Guo, X., Yang, W. and Zhao, C. (2021) High-throughput phenotyping: Breaking through the bottleneck in future crop breeding. *The Crop Journal* 9(3), 633–645.

Stark, R., Grzelak, M. and Hadfield, J. (2019) RNA sequencing: The teenage years. *Nature Reviews Genetics* 20(11), 631–656.

Stetler, R.A., Gan, Y., Zhang, W., Liou, A.K., Gao, Y. *et al.* (2010) Heat shock proteins: Cellular and molecular mechanisms in the central nervous system. *Progress in Neurobiology* 92(2), 184–211.

Sudhagar, R., Priyanka, S., Chockalingam, V., Sendhilvel, V., Souframanien, J. *et al.* (2022) Development and metabolic characterization of horse gram (*Macrotyloma uniflorum* Lam. (Verdc.)) mutants for powdery mildew resistance. *Agronomy* 12(4), 800.

Suharti, W.S., Nose, A. and Zheng, S.H. (2016) Metabolomic study of two rice lines infected by *Rhizoctonia solani* in negative ion mode by CE/TOF-MS. *Journal of Plant Physiology* 206, 13–24.

Sun, H., Meng, M., Yan, Z., Lin, Z., Nie, X. *et al.* (2019) Genome-wide association mapping of stress-tolerance traits in cotton. *The Crop Journal* 7(1), 77–88.
Sun, M., Yang, Z., Liu, L. and Duan, L. (2022) DNA methylation in plant responses and adaption to abiotic stresses. *International Journal of Molecular Sciences* 23(13), 6910.
Sun, M., Yan, H., Zhang, A., Jin, Y., Lin, C. *et al.* (2023) Milletdb: A multi-omics database to accelerate the research of functional genomics and molecular breeding of millets. *Plant Biotechnology Journal* 21(11), 2348–2357.
Sunkar, R., Kapoor, A. and Zhu, J.K. (2006) Posttranscriptional induction of two Cu/Zn superoxide dismutase genes in *Arabidopsis* is mediated by downregulation of miR398 and important for oxidative stress tolerance. *The Plant Cell* 18(8), 2051–2065.
Szklarczyk, D., Kirsch, R., Koutrouli, M., Nastou, K., Mehryary, F. *et al.* (2023) The STRING database in 2023: Protein–protein association networks and functional enrichment analyses for any sequenced genome of interest. *Nucleic Acids Research* 51(D1), D638–D646.
Tamhane, V.A., Sant, S.S., Jadhav, A.R., War, A.R., Sharma, H.C. *et al.* (2021) Label-free quantitative proteomics of sorghum bicolor reveals the proteins strengthening plant defense against insect pest *Chilo partellus*. *Proteome Science* 19, 1–25.
Trapnell, C., Williams, B.A., Pertea, G., Mortazavi, A., Kwan, G. *et al.* (2010) Transcript assembly and abundance estimation from RNA-Seq reveals thousands of new transcripts and switching among isoforms. *Nature Biotechnology* 28(5), 511.
Varoquaux, N., Cole, B., Gao, C., Pierroz, G., Baker, C.R. *et al.* (2019) Transcriptomic analysis of field-droughted sorghum from seedling to maturity reveals biotic and metabolic responses. *Proceedings of the National Academy of Sciences* 116(52), 27124–27132.
Vats, S., Bansal, R., Rana, N., Kumawat, S., Bhatt, V. *et al.* (2022) Unexplored nutritive potential of tomato to combat global malnutrition. *Critical Reviews in Food Science and Nutrition* 62(4), 1003–1034.
Vlcko, T. and Ohnoutkova, L. (2020) Allelic variants of CRISPR/Cas9 induced mutation in an inositol trisphosphate 5/6 kinase gene manifest different phenotypes in barley. *Plants* 9(2), 195.
Waititu, J.K., Zhang, C., Liu, J. and Wang, H. (2020) Plant non-coding RNAs: Origin, biogenesis, mode of action and their roles in abiotic stress. *International Journal of Molecular Sciences* 21(21), 8401.
Wang, D., Lu, X., Chen, X., Wang, S., Wang, J. *et al.* (2020) Temporal salt stress-induced transcriptome alterations and regulatory mechanisms revealed by PacBio long-reads RNA sequencing in *Gossypium hirsutum*. *BMC Genomics* 21, 1–15.
Wang, X. and Komatsu, S. (2016) Plant subcellular proteomics: Application for exploring optimal cell function in soybean. *Journal of Proteomics* 143, 45–56.
Wang, T., Xun, H., Wang, W., Ding, X., Tian, H. *et al.* (2021) Mutation of GmAITR genes by CRISPR/Cas9 genome editing results in enhanced salinity stress tolerance in soybean. *Frontiers in Plant Science* 12, 779598.
Wang, T., Ma, X., Chen, Y., Wang, C., Xia, Z. *et al.* (2024) SlNAC3 suppresses cold tolerance in tomatoes by enhancing ethylene biosynthesis. *Plant, Cell & Environment*.
Wang, Y., Cao, Y., Liang, X., Zhuang, J., Wang, X. *et al.* (2022) A dirigent family protein confers variation of Casparian strip thickness and salt tolerance in maize. *Nature Communications* 13(1), 2222.
Wang, Z., Gerstein, M. and Snyder, M. (2009) RNA-Seq: A revolutionary tool for transcriptomics. *Nature Reviews Genetics* 10(1), 57–63.
Wen, W., Li, K., Alseekh, S., Omranian, N., Zhao, L. *et al.* (2015) Genetic determinants of the network of primary metabolism and their relationships to plant performance in a maize recombinant inbred line population. *The Plant Cell* 27(7), 1839–1856. DOI: 10.1105/tpc.15.00208.
Xu, K., Zhao, Y., Gu, J., Zhou, M., Gao, L. *et al.* (2022) Proteomic analysis reveals the molecular mechanism underlying the cold acclimation and freezing tolerance of wheat (*Triticum aestivum* L.). *Plant Science* 318, 111242.
Xu, Z., Hua, J., Wang, F., Cheng, Z., Meng, Q. *et al.* (2020) Marker-assisted selection of qMrdd8 to improve maize resistance to rough dwarf disease. *Breeding Science* 70(2), 183–192.
Yadav, A.K., Carroll, A.J., Estavillo, G.M., Rebetzke, G.J. and Pogson, B.J. (2019) Wheat drought tolerance in the field is predicted by amino acid responses to glasshouse-imposed drought. *Journal of Experimental Botany* 70(18), 4931–4948.
Yadav, A.K., Shree, B., Lakhwani, D. and Singh, A.K. (2024) Omics technologies for crop improvement. In: *Integrative Omics*. Academic Press, pp. 293–313.

Yang, R., Li, M., Harrison, M.T., Fahad, S., Wei, M. *et al.* (2022) ITRAQ proteomic analysis of wheat (*Triticum aestivum* L.) genotypes differing in waterlogging tolerance. *Frontiers in Plant Science* 13, 890083.

Yan, M., Zheng, L., Li, B., Shen, R. and Lan, P. (2021) Comparative proteomics reveals new insights into the endosperm responses to drought, salinity and submergence in germinating wheat seeds. *Plant Molecular Biology* 105, 287–302.

Yan, S., Bhawal, R., Yin, Z., Thannhauser, T.W. and Zhang, S. (2022) Recent advances in proteomics and metabolomics in plants. *Molecular Horticulture* 2(1), 17.

Yang, Z., Wang, S., Wei, L., Huang, Y., Liu, D. *et al.* (2023) BnIR: A multi-omics database with various tools for Brassica napus research and breeding. *Molecular Plant* 16(4), 775–789.

Yang YuHeng, Y.Y., Yu Yang, Y.Y., Bi ChaoWei, B.C. and Kang ZhenSheng, K.Z. (2016) Quantitative proteomics reveals the defense response of wheat against *Puccinia striiformis* f. sp. tritici. *Scientific Reports* 6(1), 34261.

Yi, F., Huo, M., Li, J. and Yu, J. (2022) Time-series transcriptomics reveals a drought-responsive temporal network and crosstalk between drought stress and the circadian clock in foxtail millet. *The Plant Journal* 110(4), 1213–1228.

You, Q., Zhang, L., Yi, X., Zhang, K., Yao, D. *et al.* (2016) Co-expression network analyses identify functional modules associated with development and stress response in *Gossypium arboreum*. *Scientific Reports* 6(1), 38436.

Yu, H., Yang, H., Haridas, S., Hayes, R.D., Lynch, H. *et al.* (2023) Conservation and expansion of transcriptional factor repertoire in the *Fusarium oxysporum* species complex. *Journal of Fungi* 9(3), 359.

Yu, X., Liu, Z. and Sun, X. (2023) Single-cell and spatial multi-omics in the plant sciences: Technical advances, applications, and perspectives. *Plant Communications* 4(3).

Zeng, Y., Wen, J., Zhao, W., Wang, Q. and Huang, W. (2020) Rational improvement of rice yield and cold tolerance by editing the three genes OsPIN5b, GS3, and OsMYB30 with the CRISPR–Cas9 system. *Frontiers in Plant Science* 10, 1663.

Zhang, F., Zhu, G., Du, L., Shang, X., Cheng, C. *et al.* (2016) Genetic regulation of salt stress tolerance revealed by RNA-Seq in cotton diploid wild species, *Gossypium davidsonii*. *Scientific Reports* 6(1), 20582.

Zhang, A., Wei, Y., Shi, Y., Deng, X., Gao, J. *et al.* (2021a) Profiling of H3K4me3 and H3K27me3 and their roles in gene subfunctionalization in allotetraploid cotton. *Frontiers in Plant Science* 12, 761059.

Zhang, T.Q., Chen, Y., Liu, Y., Lin, W.H. and Wang, J.W. (2021b) Single-cell transcriptome atlas and chromatin accessibility landscape reveal differentiation trajectories in the rice root. *Nature Communications* 12(1), 2053.

Zhang, L., Xiao, J., Liang, C., Chen, Y., Yu, C. *et al.* (2024) BjuIR: A multi-omics database with various tools for accelerating functional genomics research in *Brassica juncea*. *Plant Communications* 5(8).

Zhao, M., Ren, Y. and Li, Z. (2021) Transcriptome profiling of Jerusalem artichoke seedlings (*Helianthus tuberosus* L.) under polyethylene glycol-simulated drought stress. *Industrial Crops and Products* 170, 113696.

Zheng, M., Lin, J., Liu, X., Chu, W., Li, J. *et al.* (2021) Histone acetyltransferase TaHAG1 acts as a crucial regulator to strengthen salt tolerance of hexaploid wheat. *Plant Physiology* 186(4), 1951–1969.

Zhou, S., Zhang, Y.K., Kremling, K.A., Ding, Y., Bennett, J.S. *et al.* (2019) Ethylene signaling regulates natural variation in the abundance of antifungal acetylated diferuloylsucroses and *Fusarium graminearum* resistance in maize seedling roots. *New Phytologist* 221(4), 2096–2111.

Zhou, J.R., Sun, H.D., Ali, A., Rott, P.C., Javed, T. *et al.* (2021) Quantitative proteomic analysis of the sugarcane defense responses incited by *Acidovorax avenae* subsp. avenae causing red stripe. *Industrial Crops and Products* 162, 113275.

Zhou, Q., Zhao, M., Xing, F., Mao, G., Wang, Y. *et al.* (2022) Identification and expression analysis of CAMTA genes in tea plant reveal their complex regulatory role in stress responses. *Frontiers in Plant Science* 13, 910768.

Zhou, X., Zhao, Y., Ni, P., Ni, Z., Sun, Q. *et al.* (2023) CRISPR-mediated acceleration of wheat improvement: Advances and perspectives. *Journal of Genetics and Genomics*.

Zhu, Y., Jia, X., Wu, Y., Hu, Y., Cheng, L. *et al.* (2020) Quantitative proteomic analysis of *Malus halliana* exposed to salt-alkali mixed stress reveals alterations in energy metabolism and stress regulation. *Plant Growth Regulation* 90, 205–222.

10 Multiomics Strategies for Enhancement of Quality Traits in Agricultural Crops

Dhanapati Keerthana[1], Kadam Kiran[1], A.D.S.S. Ajay Nikhil[2], Ruchi Asati[1], Girija Choudhary[1], Vijay Kumar Yadav[3], S.K. Chaturvedi[1], Anshuman Singh[1] and Rakesh Choudhary[1*]

[1]Rani Lakshmi Bai Central Agricultural University, Jhansi, India; [2]Sher-e-Kashmir University of Agricultural Sciences and Technology of Kashmir, Srinagar, India; [3]ICAR-Indian Institute of Grassland and Fodder Research Institute, Jhansi, India

Abstract

The major concern of agriculture in the past few decades is achieving food security. With the advent of high-yielding varieties during the green revolution period agricultural production has been increased drastically to meet the growing population demand. But in the current scenario malnutrition is a major issue, and growing consumer preference for high-quality food products that not only meet basic nutritional needs but also offer superior taste, texture, and appearance. Quality breeding focuses on enhancing the nutritive value of crops, which is crucial for addressing global food security challenges. By improving the chemical composition of proteins, vitamins, and minerals in crops, breeders can contribute to better health outcomes for consumers. High-quality crops can fetch better prices and enhance profitability for farmers. To date, crop improvement has achieved three major milestones, namely phenotype-based artificial selection, hybrid breeding, and molecular breeding. With the invention of various omics-based tools like genomics, phenomics, transcriptomics, metabolomics, proteomics, and ionomics, one can precisely identify the candidate genes, associated markers, and regulatory pathways governing the complex traits. Multiomics allows for a more complete understanding of complex biological processes by revealing how various omic layers interact and influence each other. These identified genes can be introgressed into breeding pipelines for the development of resilient varieties.

10.1 Introduction

Enhancing quality traits in agricultural crops is essential for addressing global challenges such as food security, malnutrition, climate resilience, sustainability, and economic stability. With the world population projected to reach nearly 10 billion by 2050, agricultural production must increase by 47% to meet food demands (U.S. Department of Agriculture, n.d.). Nutritional enhancement through biofortification aims to combat malnutrition affecting more than 2 billion people worldwide by enriching staple crops like rice, wheat, and maize (corn) with essential vitamins and minerals. Climate-resilient varieties are critical, as extreme weather and pests currently threaten up to 40% of global crop yields annually (Gaikwad *et al.*, 2020).

*Corresponding author: rakesh.gpb.rlbcau@gmail.com

DOI: 10.1079/9781800628717.0010

Additionally, breeding crops that reduce reliance on chemical inputs can significantly lower agriculture's contribution to the 23% of global greenhouse gas emissions it generates. Improved yields and quality also enhance farmer's profitability, with research showing that every penny invested in crop improvement delivers economically feasible returns (Roychowdhury *et al.*, 2023).

Quality traits in agricultural crops are essential for determining their suitability for various uses, including human feed, animal feed, and industrial applications. These traits can be categorized into several types, each affecting the overall quality and marketability of the crops. Organoleptic traits are related to sensory characters such as taste, aroma, and texture. The aroma of cooked rice is critical for consumer preference (Ankit Kumar *et al.*, 2022; Abu and Yimer, 2023). Nutritional quality refers to content of essential micronutrients and macronutrients in crops, such as proteins, vitamins, and minerals. For example, enhancing lysine content in cereals or methionine in pulses is vital for improving their nutritional value (Wu *et al.*, 2022). Morphological quality traits include physical attributes like size, shape, and color of the crop produce like wheat quality can be assessed through grain size and color (Ankit Kumar *et al.*, 2022; Abu and Yimer, 2023). Biological quality traits are related to the digestibility and nutritional efficiency of crops when consumed by animals or humans. Key metrics include the biological value of proteins and protein efficiency ratios (Ankit Kumar *et al.*, 2022). Safety traits are becoming increasingly important as they ensure food safety, such as the ability to resist heavy metal accumulation or reduce harmful residues in crops (Wu *et al.*, 2022). Quality traits play an important role in ensuring nutritional security, market demand and economic viability. Nutritional security by improving crop quality leads to better health outcomes by addressing micronutrient deficiencies prevalent in many populations. Market demand is affected because consumers prefer high-quality produce that meets specific sensory and nutritional standards, influencing breeding programs; crops with superior quality traits often fetch higher prices in the market, benefiting farmers economically (Wu *et al.*, 2022).

Improving quality traits in crops presents significant challenges due to their complex nature and the interplay between traditional and modern breeding methods. Quality traits are often governed by multiple genes and are highly influenced by external factors like environment, making their improvement a multifaceted challenge. The difficulty in phenotyping these traits for assessing their expression in plants adds another layer of complexity. Unlike yield-related traits that can be measured in a more straightforward manner, quality traits require nuanced evaluation methods to capture subtle differences in flavour, texture, and nutritional value (Wu *et al.*, 2022; Ortiz, 2023). Traditional breeding methods, while foundational in crop improvement, face several limitations that hinder their effectiveness in enhancing quality traits. These methods are highly time-consuming, often taking more than a decade to develop a new cultivar due to their dependence on natural selection and observable traits (phenotypes), which are frequently impacted by environmental conditions, resulting in inconsistent outcomes and the risk of selecting suboptimal traits (Anand *et al.*, 2023). Genetic precision is another challenge because traditional approaches predominantly focus on dominant alleles, frequently overlooking valuable recessive traits that could enhance crop performance. Additionally, these methods often result in genetic bottlenecks by utilizing a limited number of parent lines, reducing genetic diversity and increasing the crop's vulnerability to diseases and environmental changes (Lamichhane and Thapa, 2022). Despite their historical importance, these limitations underscore the need for integrating advanced breeding technologies to overcome these challenges (Breseghello and Coelho, 2013).

In rice, key quality parameters include milling quality, such as head rice rate (HRR), brown rice rate (BRR), white rice rate (WRR); cooking and sensory quality; size and shape as appearance quality; and lysine, B-vitamins, and minerals as nutritional quality (Sultana *et al.*, 2022). In wheat, important factors include gluten content, falling number (alpha-amylase activity), starch content, grain hardness, moisture content, and color (Dhar, 2023). For maize, grain starch, protein, and oil content are essential quality traits (Sun *et al.*, 2023). Sorghum quality is determined by moisture content (max. 14.5%), total defects (limited to 8%), aflatoxin levels (max. 10 μg/kg), and a minimum test weight of 71 kg/hl (Rooney and Murty, 1982). Pearl millet, or bajra, quality includes moisture content (max. 14%), absence of harmful substances like aflatoxins (max. 30 μg/kg) and uric acid (max. 100 mg/kg). Finger millet, or ragi, with 12% moisture content, contains high protein (6.7–12.3g per 100 g), amino acids, and key minerals such as calcium (up to 344.59 mg) and iron (around 5.16 mg), but anti-nutritional factors like phytates and tannins can limit mineral bioavailability (Shobha and Ravishankar, 2017).

Chickpea is an affordable source of protein, carbohydrates, minerals, vitamins (folic acid, tocopherols), dietary fiber, and health-promoting fatty acids (Jukanti *et al.*, 2012). Black gram beans provide 341 calories and 25.21g of protein, with high dietary fiber, B-vitamins, particularly folate, and essential minerals like iron, calcium, and potassium. These beans also contain isoflavones that may reduce the risk of post-menopausal cancers. Green gram is rich in vitamins, calcium, iron, and phosphorus, and offers antioxidant, anti-inflammatory, and anti-microbial properties due to bioactive compounds such as proteins, polyphenols, and phytosterols. Cowpea provides essential nutrients like protein, iron, and zinc, along with folate, fiber, and significant amounts of B-vitamins, calcium, and magnesium, supporting bone health and nerve function (Pandey, 2019).

Groundnuts (peanuts) are rich in healthy fats, particularly monounsaturated fats, and an excellent source of protein (25.8 g per 100 g), containing all essential amino acids. They are also a significant source of dietary fiber and vitamins such as niacin and vitamin E, along with minerals like magnesium, copper, and iron. Additionally, groundnuts contain bioactive compounds such as arginine, resveratrol, phytosterols, phenolic acid, and flavonoids (Arya *et al.*, 2016). In *Brassica* species, minimizing the levels of erucic acid and glucosinolates is important, while in soybean, reducing lipoxygenase 2 and eliminating Kunitz trypsin inhibitors (KTI) is essential. Reducing anti-nutritional factors such as lectins, phytic acid, and KTI, while enhancing micronutrient and vitamin availability, is crucial for improving crop quality. Key quality traits in various field crops are shown in Fig. 10.1.

Next-generation sequencing (NGS) technologies have transformed biological research, making it possible to produce high-quality whole genome sequences and conduct large-scale genomic and transcriptomic analysis. These advances have produced vast datasets, significantly enhancing our understanding of structural and functional genomics (Kulski, 2016). By providing insights into changes in DNA, mRNA, proteins, metabolites, and mineral nutrient content under varied environmental and physiological stress, NGS has paved the way for exploring the genetic and molecular basis of crop development (Muthamilarasan *et al.*, 2019). Building on these advancements, omics-based approaches such as genomics, transcriptomics, metabolomics, proteomics, ionomics, and phenomics have emerged as powerful tools for studying plant biology (Fig. 10.2). These approaches have been invaluable in examining genetic and molecular mechanisms related to stress responses, senescence, quality traits, and yield in economically significant crops (Cao *et al.*, 2022; Li *et al.*, 2022; Zhou and Zheng, 2022). Furthermore, omics techniques help in the discovery and synthesis of secondary metabolites and plant signaling molecules, demonstrating their indispensable role in advancing plant science (Luo *et al.*, 2022).

The integration of various omics technologies has led to the development of multiomics, which combines molecular, system, and computational biology tools to analyze complex biological systems (Fig. 10.3). Multiomics approaches provide extensive data for understanding physiological processes in crop plants (Table 10.1) (Zhou *et al.*, 2022). These integrated approaches have proven particularly valuable for enhancing crop nutrient content and uncovering the genetic, molecular, and physiological mechanisms behind quality nutrient accumulation and distribution in crops (Müller and Hippe, 1987; Zenda *et al.*, 2021). Since mono-omics approaches alone are insufficient for understanding intricate plant responses to environmental conditions, integrating multiomics is necessary to achieve meaningful and actionable results (Roychowdhury *et al.*, 2023). Omics and multiomics techniques collectively represent a multidisciplinary effort to decode plant biology, offering unparalleled opportunities to enhance crop resilience, productivity, crop quality and nutritional value (Table 10.2). These tools provide a keen understanding of fundamental processes and help to find innovative strategies to overcome global challenges in agriculture and food security (Deshmukh *et al.*, 2014;

Fig. 10.1. Different quality traits in agricultural crops. Author's own image.

Khandagale *et al.*, 2020; Muthuramalingam *et al.*, 2022; Raza, 2022).

10.2 Overview of Multiomics in Improving Quality Traits

10.2.1 Genomics is decoding the blueprint of complex traits

Genome is complete set of genes present in a cell or organism. Genomics in plant breeding refers to the application of genomic tools to enhance the efficiency and effectiveness of breeding programs. This approach integrates advanced molecular techniques with traditional breeding methods to develop improved plant varieties that meet agricultural demands. Decoding the genome through high-throughput NGS technology involves isolating genomic DNA, amplifying it via PCR, sequencing the DNA, and evaluating the integrity of the resulting sequence (Xu *et al.*, 2022). Following sequencing, structural and functional gene annotation aids in understanding their roles in trait inheritance and the interactions of gene products at both cellular and organismal levels (Zanini *et al.*, 2022). Recently, numerous advanced tools for quantitative trait loci (QTL) mapping have emerged, leveraging deep sequencing methods such as restriction-site associated DNA sequencing (RAD-Seq), genotyping-by-sequencing (GBS), bulked segregant analysis sequencing (BSA-Seq), and specific locus amplified fragment sequencing (SLAF-Seq) (Bundo *et al.*, 2022).

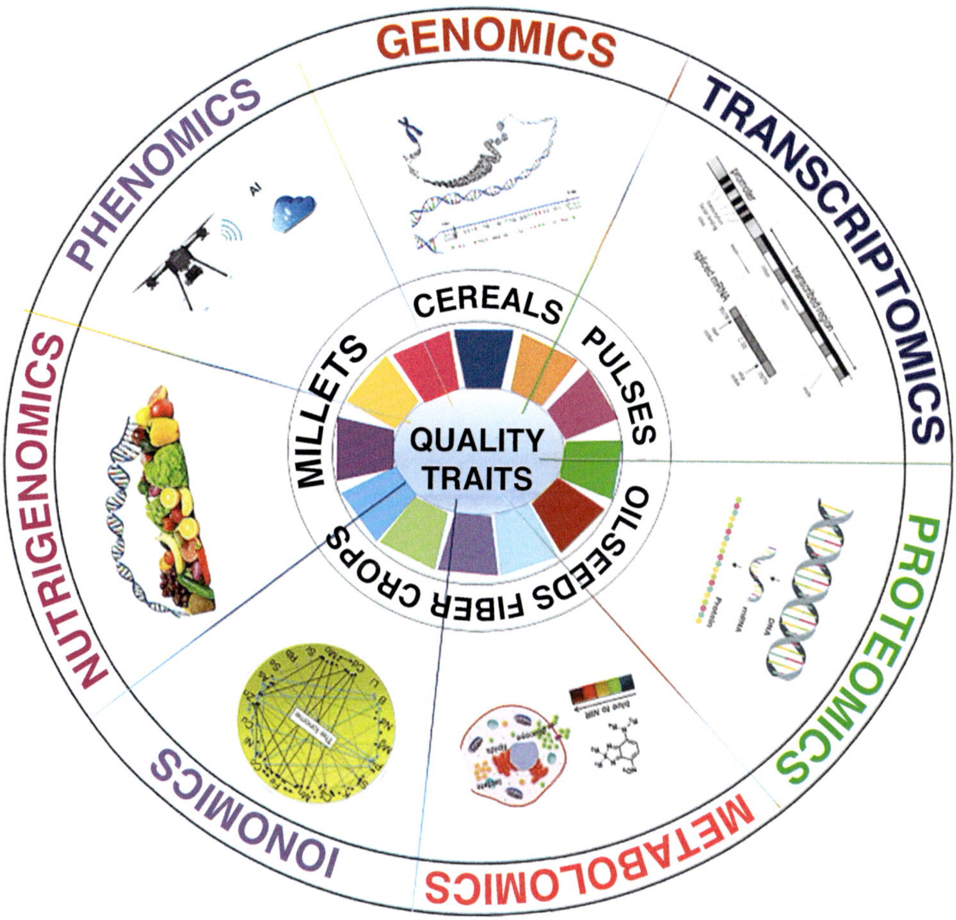

Fig. 10.2. Different cohorts of omics-based tools for enhancement of quality traits in crop breeding. Author's own image.

These mapping techniques rely on genetic maps or a limited number of single nucleotide polymorphisms (SNPs) to identify candidate genomic regions and genes associated with traits, but their precision remains insufficient (Zhang *et al.*, 2022).

Genomics became a boon in agriculture and is mainly classified into three cohorts: structural genomics, functional genomics, and comparative genomics. Functional genomics analyzes the genome sequence data to unravel gene functions and gene interactions, as well as the regulatory networks of genes with the help of various forward and reverse genetic approaches (Sharma *et al.*, 2022). Structural genomics focuses on deciphering the three-dimensional structure of genes to identify their positions, order, and arrangement on chromosomes using tools such as genomic selection, sequencing, mapping, and marker-based resources (Lou *et al.*, 2017). Comparative genomics involves aligning biological sequences to identify conserved regions. These studies have unveiled significant synteny among related species (Kalyana Babu *et al.*, 2022).

Guo *et al.* (2023) identified a genetic marker located on fifth chromosome which is significantly linked to the starch branching enzyme (SBE) gene in maize, which is directly involved in starch synthesis. Likewise, many genes were identified in maize for the increment of protein and oil.

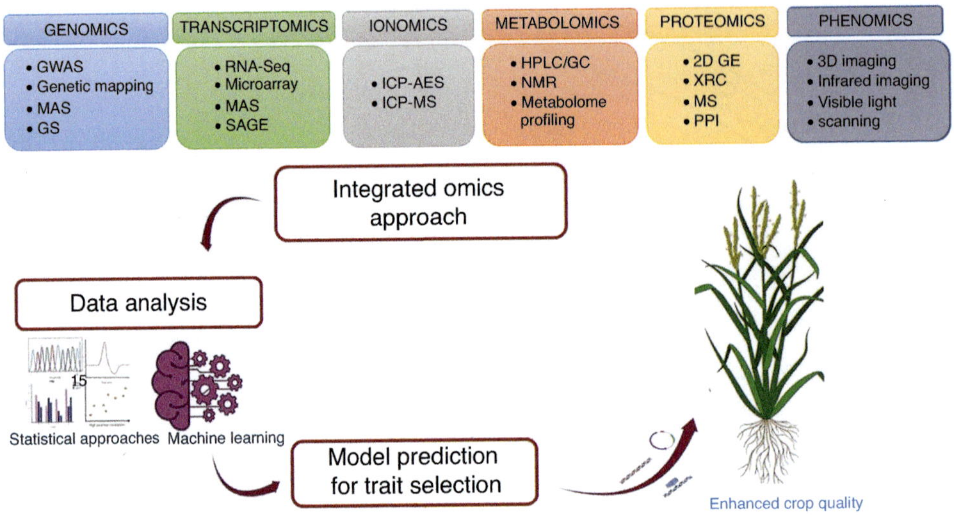

Fig. 10.3. Utilization of an integrated omics approach for the enhancement of quality traits in agricultural crops. Author's own image.

10.2.1.1 Genome-wide association studies

Genome-wide association studies (GWAS) have become a crucial resource in crop improvement, enabling breeders to pinpoint genetic variants linked to key traits. By utilizing the natural genetic variation present in crop populations, GWAS significantly boosts the accuracy and effectiveness of contemporary breeding strategies.

GWAS operates by scanning the genomes of numerous individuals within a population to discover DNA markers, typically SNPs, that are linked with observable characters (phenotypes) such as grain yield, and biotic and abiotic stress tolerance. This method establishes genetic associations, which enables us to pinpoint the underlying genetic basis of complex traits with greater accuracy compared to traditional approaches like QTL mapping (Bentley *et al.*, 2022; Aleem *et al.*, 2024). Unlike QTL mapping, which necessitates the development of specific mapping populations, GWAS can exploit existing germplasm collections, making it more efficient and resource-effective (Nakano and Kobayashi, 2020). Furthermore, GWAS offers higher resolution mapping, allowing for the detection of minor gene effects and complex interactions that are often missed by classical bi-parental mapping methods (Gali *et al.*, 2019; Susmitha *et al.*, 2023).

The practical utility of GWAS for crop breeding has been demonstrated in major crops where it has facilitated the identification of genetic loci associated to critical quantitative traits including yield, early maturity, and stress tolerance. These findings are instrumental in developing crop varieties that can better withstand environmental stresses and contribute to global food security. Moreover, the integration of GWAS with genomic selection (GS) further accelerates genetic improvement by combining marker data with predictive models, enabling the early generation selection of elite genotypes and improving the overall breeding process (Gali *et al.*, 2019; Susmitha *et al.*, 2023).

GWAS faces challenges such as population structure and phenotyping bottlenecks that can impact accuracy and effectiveness. Population structure within study groups can lead to spurious associations, necessitating careful experimental design and statistical corrections to ensure reliable marker-trait relationships (Gali *et al.*, 2019; Nakano and Kobayashi, 2020). Additionally, while genotyping has advanced significantly, phenotyping remains a bottleneck due to its

Table 10.1. Common multiomics techniques and their application in quality-trait improvement (Nayak *et al.*, 2021).

Omics layer	Technique	Applications
Genomics	Genome-wide association studies (GWAS), marker-assisted selection (MAS), genome sequencing, genomic selection (GS)	Identification of genetic markers for key traits
Phenomics	3D imaging, infra-red imaging, visible light scanning	Assessment of physical and biochemical traits under different environments
Transcriptomics	Quantitative real time polymerase chain reaction (qRT-PCR), RNA-sequencing, microarrays	Understanding gene expression patterns under specific conditions
Proteomics	2-Dimensional gel electrophoresis (2D GE), X-ray crystallography (XRC), matrix-assisted laser desorption/ionization-time of flight (MALDI-TOF), liquid chromatography–mass spectrophotometry (LC–MS)	Identification of proteins involved in quality traits
Metabolomics	Gas chromatography–mass spectrophotometry (GC–MS), high performance liquid chromatography (HPLC), ultra performance liquid chromatography (UPLC), nuclear Magnetic resonance (NMR), capillary electrophoresis (CE), super-critical fluid chromatography (SFC)	Profiling metabolites associated with key traits

labor-intensive nature. High-throughput and precise phenotyping techniques are crucial for accurately measuring traits across large populations, ensuring the success of GWAS in crop improvement (Sansa *et al.*, 2024).

10.2.1.2 Genomic selection

Genomic selection (GS) is a transformative approach in crop improvement that utilizes genomic data to enhance the efficiency and effectiveness of breeding programs. By utilizing high-density genetic markers, GS allows breeders to predict the genetic potential of crop lines based on their genomic estimated breeding values (GEBVs). This method significantly reduces the need for extensive phenotyping, which has traditionally been a time-consuming aspect of plant breeding. Instead, GS enables rapid selection of high-value individuals from early-generation populations, facilitating faster genetic gains in traits such as yield, nutritional quality, and resilience to biotic and abiotic stresses.

The GS procedure begins by forming a training population (TP) that includes individuals with both types of information (genotypic and phenotypic), which is used to develop predictive models for estimating GEBVs. These models analyze the relationship between genotype and phenotype, allowing breeders to predict the performance of individuals in a breeding population (BP) based solely on their genetic profiles. Selected candidates are then chosen for further breeding cycles, and the model is continually validated and refined through field trials. One of the primary benefits of genomic selection is its ability to improve the accuracy of trait prediction for complex quantitative traits controlled by multiple genes. This is particularly important for crops where traditional marker-assisted selection has limitations. GS has shown promising results in various cereal crops, including wheat,

Table 10.2. Application of multiomics techniques in identification of genes/transcription factors associated with key quality traits in important agricultural crops. Author's own table.

Crop	Trait	Technique	Key findings/genes involved	Reference
Cereals				
Rice	Aroma	Genomics (QTL mapping/GWAS/GS)	Identification of genes linked to aroma (e.g. 2 AP)	Verma *et al.*, 2020
	Amino acids, flavonoids, phenolamines, terpenoids	HPLC-ESI-QTOF/MS	Identified 143 single nucleotide polymorphisms (SNPs) and 89 metabolites	Matsuda *et al.*, 2015
	Starch metabolism	RNA-Seq; RT-PCR analysis	Higher specific expression level of genes such as transcription factor (TF) genes encoding ADP glucose pyrophosphorylase (AGPase) and granule bound starch synthase I (GBSS I)	Venu *et al.*, 2011
	Cooking temperature	Re-sequencing	ssIIa	Waters *et al.*, 2006
	Fragrance	Re-sequencing	badh2	Bradbury *et al.*, 2005
Wheat	Seed protein content	Proteomics, transcriptomics (RNA-Seq)	Gene identified and outcome of that gene incorporation	Jha *et al.*, 2022
	Amino acids, (poly)amines, flavonoids, organic acids, sugars, vitamins, etc	HPLC-ESI-QTRAP-MS/MS	1098 mGWAS associations were detected with large effects	Chen *et al.*, 2020
	Milling quality	RNA-Seq	Encoding FLA	Nirmal *et al.*, 2017
	Wheat hardness	RNA-Seq	Pin genes	Nirmal *et al.*, 2016
Maize	Provitamin A content	Genomics (GWAS, GS), transcriptomics	Identification of key genes (*crtRB1, lycE*) regulating provitamin A levels	Chauhan *et al.*, 2022
	Carotenoids	HPLC	The GWAS confirmed carotenoid biosynthesis genes *lcyE* and *crtRB1*. In addition, significant novel associations were detected for several transcription factors (e.g. RING zinc finger domain and HLH DNA-binding domain super family proteins). SNP was identified which displayed an association with β-carotene and provitamin A levels	Azmach *et al.*, 2018

Continued

Table 10.2. Continued

Crop	Trait	Technique	Key findings/genes involved	Reference
	Flavonoids	LC-ESI-QqTOF-MS/MS	The function of three candidate genes, including two UDP-glycosyltransferases (*UGT*) and an oxygenase, was revealed via preliminary molecular functional characterization.	Jin *et al*., 2017
Sorghum	Grain quality, Tannin content	Metabolomics, genomics	Identification of genes regulating tannin biosynthesis (*TAN1, TAN2*) for improved digestibility	Zhang *et al*., 2023
	Fe and Zn content	Gene expression analysis	PgIZIP, PgINRAMP, and PgIFER	Mahendrakar *et al*., 2020
	Protein and fat content in sorghum grains	RNA-Seq dataset; GWAS	alpha-amylase 3 gene	Rhodes *et al*., 2017
Barley	Malting quality	Genomics, transcriptomics	Identification of genes related to starch and β-glucan degradation during malting	Panahi *et al*., 2024
Millets				
Pearl millet	Fe and Zn content	Transcriptome profiling	155 and 251 transcripts were found to be up- and downregulated respectively	Satyavathi *et al*., 2022
	Drought tolerance, grain quality	Transcriptomics	Identification of drought-responsive genes (*PgDREB2A, PgNAC21*) linked to yield stability	Shivhare *et al*., 2020
Finger millet	Calcium content, grain quality	Genomics, metabolomics	Identification of genes for high calcium and improved grain texture (CaM regulatory proteins)	Sharma *et al*., 2017
	Ca content	RNA-Seq	Calcium sensor genes (CaM, CaMLs, CBLs, CIPKs, and CRKs)	Singh *et al*., 2014
Foxtail millet	Protein content, drought tolerance	Genomics, transcriptomics	QTLs linked to enhanced protein and water-use efficiency (*SiLEA14, SiARDP*)	Loni *et al*., 2023
	Alkaloids, amino acids, fatty acids, organic acids, phenolamides, polyphenols (i.e. flavonoids, anthocyanins), sugars, vitamins	HPLC-ESI-QTRAP-MS/MS	Cyanidin 3-*O*-glucoside and quercetin *O*-acetylhexside are concentratedly located at 43.55 Mb on chromosome 5 and 26.9 Mb on chromosome 7, and two Lc were mined as candidate genes, respectively.	Wei *et al*., 2021

Continued

Table 10.2. Continued

Crop	Trait	Technique	Key findings/genes involved	Reference
Oilseed crops				
Brassica species	SOC (seed oil content)	GWAS and transcriptome-wide association analysis (TWAS)	Identified two genes, *BnPMT6s*, that negatively regulate SOC	Tang *et al.*, 2021
	Oil quality, glucosinolate content	Lipidomics, genomics	Genes like *Bnlat2* & *Bnlat5* for improved oil quality and reduced anti-nutritional glucosinolates	Ton *et al.*, 2020
	Lipid synthesis and accumulation	Comparative genomics and transcriptome analyses	Identified a candidate gene, *orf188* promoting adenosine triphosphate synthesis in cells	Liu *et al.*, 2019b
	Oil content, Fatty acid composition	Lipidomics, genomics, metabolomics	Identification of *Fad2* and *FAD3* genes improving unsaturated fatty acid levels	Dar *et al.*, 2017
Soybean	Aminoacids, isoflavones, lipids, organic acids	HPLC-MS; phenotypic and metabolic genome-wide association studies and multi-omics analyses	284 and 279 candidate genes were found to be significantly associated with seed oil-related traits and metabolites respectively	Liu *et al.*, 2020
	Oil quality, Protein content	Genomics (QTL mapping)	Identification of fatty acid desaturase genes (*FAD2-1*) for improved oil quality	Kumar *et al.*, 2021
Sesame	Sesame seed allergy	2-DE followed by immuno-labelling	7S vicilin-type globulin, 2S albumin seed maturation protein, and embryogenic abundant protein	Beyer *et al.*, 2002
Groundnut	Oleic/linoleic acid ratio	Genomics, transcriptomics, metabolomics	Identification of *ahFAD2A* and *ahFAD2B* improving oil stability and shelf life	Zhang *et al.*, 2024
Sunflower	Lipid and fatty acid composition (major and minor)	GWAS; UPLC–MS	Genomic regions on chromosomes 3 and 14, explaining up to 34.5% of the total variation of docosanoic acid (22:0) and rare FA in sunflower oil	Chernova *et al.*, 2021
Pulses				
Chickpea	Seed protein, fiber, oil, and starch	GWAS	11 significantly associated SNPs	Mugabe *et al.*, 2023
	High amino acid content	Proteomics (2DE-MS)	Identification of protein for high amino acid	Santos *et al.*, 2017
Lentil	Protein content	Proteomics, genomics	Identification of key proteins including seed hardness and digestability	Lamonaca *et al.*, 2024

Continued

Table 10.2. Continued

Crop	Trait	Technique	Key findings/genes involved	Reference
Pigeonpea	Seed quality (antioxidants, proteins, detoxicants)	RNA-Seq	27,441 expressed genes effecting seed quality directly or indirectly	Pazhamala *et al*., 2016
Common bean	Seed Zn content	GWAS	Gene Phvul001G233500, encoding for an E3 ubiquitin-protein ligase explained the association	Caproni *et al*., 2020
Fiber crops				
Flax	In-planta fiber morphometrics, as well as ex-planta fiber mechanical properties	Genome-wide transcription profiling	Three significantly correlated genes responsible for cell wall dynamics, either directly (Expansin-like protein), or indirectly (NAD(P)-binding Rossmann-fold superfamily protein)	Chabi *et al*., 2023
Cotton	Fiber quality (strength, uniformity and micronaire)	GWAS	Identified three novel QTLs for uniformity, seven for micronaire and two candidate genes for strength, i.e. *Gh_A07G1574* and *Gh_A07G1581*	Mohammed *et al*., 2025

rice, and maize, leading to enhanced selection accuracy and improved performance in traits such as grain yield and quality (Budhlakoti *et al*., 2022).

The integration of advanced statistical models and genomic information further enhances the predictive capabilities of GS, allowing for better decision making in breeding programs. Additionally, GS supports multi-trait selection (Bhat *et al*., 2015), enabling breeders to target several desirable traits simultaneously. This is crucial in developing climate-resilient crops that can adapt to changing environmental conditions while maintaining high nutritional value. The use of genomic selection has been associated with increased genetic gain per unit time, shorter breeding cycles, and improved overall efficiency in crop improvement efforts. Its application is expected to play a vital role in meeting future food demands sustainably while improving the nutritional profile and resilience of crops.

Unlike phenotypic selection, molecular markers are independent of environmental influences, providing a stable and reliable basis for selection across different growth stages (Hasan *et al*., 2021). Their integration into breeding programs significantly enhances efficiency by allowing early selection based on genetic information, eliminating the need to wait for phenotypic expression. Gene editing has facilitated the development of crops with enhanced nutritional profiles, increased resistance to pests and diseases, and improved tolerance to environmental stresses like drought, ultimately enhancing crop quality (Abdallah *et al*., 2015). Furthermore, it significantly accelerates the development process compared to traditional breeding methods, enabling rapid responses to evolving agricultural challenges.

10.2.2 Transcriptomics

Transcriptome profiling provides a comprehensive overview of all RNA molecules, including mRNA, tRNA, rRNA, sRNA, and other non-coding RNAs, within a cell, tissue, organ, or entire organism at a specific time, an analysis not achievable at the genomic level (Weckwerth *et al*., 2020; Chaturvedi *et al*., 2021). Transcriptomics, a powerful subset of functional genomics, has transformed the study of gene expression in agricultural crops. By examining gene expression patterns, transcriptomics offers valuable insights into gene regulation, enhancing our understanding of crop biology (Nejat *et al.*, 2018). Unlike the stable genome, the transcriptome is dynamic, varying under different conditions (such as developmental stage, tissue type, and environmental factors), making it a promising molecular tool for investigating an organism's responses to stress (Escandón *et al.*, 2021). Techniques like RNA sequencing and microarray analysis have become essential in transcriptomics research.

10.2.2.1 Microarray analysis

Microarray analysis is a core technique in transcriptomics which facilitates the parallel assessment of expression levels for several genes. It has significantly advanced gene expression profiling, biomarker identification, and pathway analysis. The technology employs a solid surface, like a glass slide or silicon chip, containing numerous immobilized probes, each representing a specific DNA sequence corresponding to a gene or transcript. The process begins with the preparation of RNA or cDNA samples, labeled using fluorescent dyes like Cy3 or Cy5 to facilitate gene expression quantification. These labeled samples are then hybridized onto the microarray, and fluorescence signals are captured using a scanner. The resulting data undergo background correction, normalization, and statistical analysis, often utilizing bioinformatics tools for interpretation (Kuhn, 2001; Ritchie *et al.*, 2015). This comprehensive workflow has been pivotal in broadening our comprehension of gene expression and associated biological processes.

10.2.2.2 RNA sequencing

RNA sequencing (RNA-Seq), or whole transcriptome shotgun sequencing, is a groundbreaking technology in transcriptomics that enables a thorough analysis of the transcriptome in cells or tissues. This technique provides in-depth knowledge about gene expression, alternative splicing, novel transcript identification, and the study of functional RNAs (Wang *et al.*, 2009). RNA-Seq works by converting RNA into complementary DNA (cDNA) and then performing high-throughput sequencing, offering various advantages compared to traditional microarray methods, having greater sensitivity, a broad dynamic range, and the potential to recognize novel transcripts (Ozsolak and Milos, 2011). The RNA-Seq process usually involves library creation, sequencing, and data analysis. Sequencing is carried out with NGS platforms like Illumina, Ion Torrent, or PacBio, generating large volumes of short- or long-sequence reads (Conesa and Mortazavi, 2014).

In RNA-Seq, data analysis encompasses different key steps, such as quality control, alignment of reads or transcript assembly, quantification of gene expression levels, and differential expression analysis, supported by a range of bioinformatics tools and pipelines. RNA-Seq has significantly advanced transcriptomics by constructing an unbiased and comprehensive view of the transcriptome, making it a key tool in biological research. It has been applied to study gene regulation, disease mechanisms, and developmental processes, facilitating the discovery of new genes and alternative splicing events across a wide array of organisms, from model species to non-model species (Nookaew *et al.*, 2012). RNA-Seq of sorghum grain was performed by Cruet-Burgos and Rhodes (2023) to analyze the transcriptional profiles of four accessions with varying carotenoid levels during grain development. Importantly, geranyl geranyl pyrophosphate synthase (GGPPS), phytoene synthase (PSY), and phytoene desaturase (PDS) were identified as potential markers for pro-vitamin A carotenoid biofortification in sorghum grain.

10.2.3 Metabolomics

Metabolomics is a rapidly advancing and intriguing branch of omics tools that is now widely used for crop improvement. The metabolome encompasses the full array of micromolecule metabolites present in plant systems. In plants, the metabolome is branched into two categories primary and secondary metabolites (Pott *et al.*, 2021). Profiling these metabolites provides in-depth views of the biochemical processes underlying plant metabolism. Understanding the metabolic networks governing complex processes in metabolomics plays an important role in metabolomics-assisted breeding, aiding the development of superior cultivars with enhanced quality and production. Additionally, data from mQTL studies offer deeper insights into quantitative genetics. Metabolic profiling connects genotype and phenotype, facilitating metabolic dissection. The process begins with identifying SNP markers to pinpoint candidate genes (Dawid and Hille, 2018). Metabolic profiling narrows the void between phenotype and genotype, providing different avenues for metabolic dissection. This begins with the identification of genetic markers, coupled with mQTL mapping studies, to locate candidate genes and associated genomic regions (Sharma *et al.*, 2021). Lipidomics is a specialized branch of metabolomics focusing on the large-scale study of lipids within biological systems. It involves the comprehensive analysis of the lipidome, which encompasses all lipid molecular species present in a cell, tissue, organism, or ecosystem.

10.2.3.1 High-throughput methods

Metabolomics research has seen significant advancements in the techniques and tools used, particularly in mass spectrometry (MS) and nuclear magnetic resonance spectroscopy (NMR) (Khakimov *et al.*, 2014). Modern metabolomics platforms enable large-scale surveys that include identified and unidentified metabolites. NMR-based metabolite detection utilizes the magnetic properties of atomic nuclei under a magnetic field. This non-destructive technique is widely used for identifying low-molecular-weight metabolites (<50 kDa) and supports various applications such as metabolite fingerprinting, profiling, metabolic flux analysis along with obtaining atomic structural details of compounds in biological samples.

On the contrary to NMR, MS offers higher sensitivity, enabling broader metabolome coverage. Recent advancements in ionization techniques, like atmospheric pressure chemical ionization (APCI), electrospray ionization (ESI), and matrix-assisted laser desorption/ionization time-of-flight (MALDI-TOF), have further enhanced the precision of MS (Razzaq *et al.*, 2019). To improve throughput, MS is often coupled with chromatography techniques, including gas chromatography (GC), liquid chromatography (LC), capillary electrophoresis (CE), Fourier transform ion-cyclotron resonance (FT-ICR), and field asymmetric waveform ion mobility spectrometry (FAIMS).

10.2.4 Phenomics

Phenomics is an interdisciplinary technique that investigates the interplay between genomics and environmental factors, significantly advancing plant breeding by offering comprehensive insights into plant traits ranging from the molecular to the physiological levels (Mir *et al.*, 2019). It integrates biology, bioinformatics, computer science, engineering, and statistics to develop a holistic understanding of complex plant characteristics. By leveraging high-performance computing and artificial intelligence, phenomics enables the efficient analysis of diverse phenotypic data and physio-biochemical traits of plants. It also facilitates the development of integrative tools to combine phenomics data with other omics datasets for a more comprehensive understanding.

10.2.4.1 High-throughput phenomics for identification of quality traits

The adoption of advanced high-throughput phenotyping techniques enables continuous plant imaging at frequent intervals, facilitating the efficient analysis of various growth processes. These methods employ image processing algorithms to extract traits from high-resolution images of plant samples. High-throughput phenomics platforms integrate cutting-edge non-invasive imaging, spectroscopy, image analysis, robotics, and high-performance computing technologies (Yang *et al.*, 2020).

10.2.5 Proteomics

The term 'proteome' originates from PROTEins expressed by a genOME. Similar to the field of genomics, proteomics is the scientific discipline dedicated to the investigation and characterization of the entire group of proteins present within a cell at a particular moment. While genomic studies give detailed information into the theoretical potential of genetic information, proteomic studies offer a deeper understanding of the functional components responsible for carrying out specific cellular processes. Unlike the genome, which remains relatively stable, the proteome exhibits dynamic properties. The examination of proteins reveals the presence of post-translational modifications (PTMs), which play a crucial role in comprehending biological functions (Nat *et al.*, 2007). Proteomics, a technique for characterizing the whole spectrum of proteins expressed by an organism, is divided into four different branches: structural, functional, sequence, and expression proteomics. Structural proteomics specifically focuses on determining the molecular architecture, including the precise amino acid sequence, of proteins involved in specific biological processes. This information is then cross-referenced with a database of identified genes (Lefkovits, 2003). The methodologies employed in structural proteomics encompass a wide range of techniques, involving computational approaches and instrumental methods such as X-ray crystallography and NMR spectroscopy (Liu and Hsu, 2005). Additionally, computer-aided modeling and electron microscopy can be utilized to analyze structural proteomics data (Maveyraud and Mourey, 2020). The accumulation of a vast number of protein structures will provide invaluable insights into the principles governing protein folding and unfolding mechanisms, as well as a deeper understanding of their biological roles (Liu and Hsu, 2005). While significant advancements in X-ray crystallography (Lamzin and Perrakis, 2000) and multidimensional NMR spectroscopy (Montelione *et al.*, 2000) have enabled structural biologists to elucidate protein structures and their biological functions by determining atomic coordinates, certain experimental constraints persist.

Petersen *et al.* (2006) studied the potential of maize pollen to cause allergies compared to native grass pollen (*Phleum pratense*). They used 2Dgel electrophoresis followed by immunoblotting to analyze the proteins in both types of pollen. Their findings suggest that maize pollen is less likely to trigger allergic reactions than native grass pollen. The proteomic analysis of rice revealed that many allergenic proteins are concentrated in the seeds. This finding suggests that proteomic analysis can be a valuable tool for identifying allergens in food products (Koller *et al.*, 2002).

10.2.6 Ionomics

Ionomics is the study of the presence and distribution of metals, non-metals, and metalloids in living organisms, utilizing high-throughput techniques (Salt *et al.*, 2008). More broadly, ionomics focuses on analyzing an organism's ionome, aiming to quantify as many elements as possible in individual samples. This field seeks to understand how various factors, such as metabolism, genetics, development, and environmental conditions, influence the elemental composition of specific organs, tissues, and cell types (Pita-Barbosa *et al.*, 2019). The thought of the ionome was introduced about a decade ago, and substantial progress has been made in this area, particularly through its integration with genetics and other high-throughput technologies to identify genes regulating the ionome. Deficiency of essential minerals can lead to severe symptoms and potentially plant death under extreme conditions, while excess levels of both essential and non-essential minerals can impair plant growth by causing biochemical, structural, and physiological disruptions (Afshan *et al.*, 2015; Pawlowski *et al.*, 2019). In plants, toxicity from heavy metals can interfere with or alter important biochemical and physiological processes like photosynthesis, chlorosis, nutrient uptake, biomass accumulation, water regulation, and senescence, ultimately leading to plant mortality (Adrees *et al.*, 2015). Plants can typically survive as long as the supply of a specific nutrient meets their requirements, making both nutrient deficiencies and excesses critical factors that influence food production and global food security. Since ionomics allows for the measurement of multiple elements, it is particularly valuable for studying how changes in one element's abundance can affect others in plants. The ZIP transporter family, initially identified in plants, is responsible for transporting a range of cations, including Fe, Zn, Cd, and Mn (Ajeesh Krishna *et al.*, 2020). ZIP transporters play a key role in zinc biofortification of grains and plant adaptation to various soil types. Notably, ZIP transporter genes have been identified in several crops, with significant validation in Arabidopsis (*AtZIP1*, *AtZIP2*, *AtZIP3*, and *AtZIP4*) and rice (*OsZIP1*, *OsZIP3*, *OsZIP4*, *OsZIP5*, *OsZIP7*, and *OsZIP8*).

10.2.7 Nutrigenomics

Nutrigenomics, a burgeoning field at the intersection of nutrition and genomics, holds promise in elucidating the intricate relationship between dietary components and gene expression, paving the way for personalized nutrition strategies tailored to individual genetic makeup. At its core, nutrigenomics investigates how dietary compounds interact with the genome to modulate gene expression, influencing metabolic pathways, physiological responses, and ultimately, health outcomes (Mead, 2007). By deciphering these molecular mechanisms, researchers can identify genetic variations that impact nutrient metabolism and response, shedding light on individual susceptibility to diet-related diseases. Recent studies have highlighted the role of nutrigenomics in addressing the growing epidemic of obesity and metabolic disorders. GWAS has identified genetic variants associated with differential responses to dietary interventions, illustrating the importance of personalized dietary recommendations for optimal health outcomes. Moreover, nutrigenomic approaches have revealed gene–diet interactions influencing nutrient absorption, metabolism, and utilization, offering insights into precision nutrition strategies for disease prevention and management (Mishra *et al.*, 2022).

10.3 Integrating Multiomics Approaches

10.3.1 Data integration

Combining omics big data with advanced computational tools and statistical models has the potential to expedite the identification of complex genetic loci and candidate genes linked to important agronomic traits. Numerous statistical models, including linear and nonlinear models, are widely used in phenotype prediction. In linear models, genomic best linear unbiased prediction (GBLUP) has been extensively used to predict genotypes and phenotypes. In nonlinear models, machine learning (ML) provides a scalable and interpretable framework for multiomics integration and is usually applied to the phenotype prediction of crops. Deep learning (DL) plays a significant role in high-dimension biology because it reduces the number of random variables to consider (Subramanian *et al.*, 2020).

The integration of multiomics datasets with panomics platforms and system biology offers the potential to predict complex crop traits and uncover regulatory networks, paving the way for genetic improvement. Incorporating various scales of trait prediction and decision-making models can make crop breeding processes more efficient and intelligent. Given the massive volume of high-throughput data, tools, databases, and approaches for multiomics system biology are essential for effective data integration. Recently, systemic multiomics integration (MOI) has enabled the comprehensive assimilation, annotation, and modeling of these extensive datasets (Jamil *et al.*, 2020).

Additionally, numerous web-based platforms, including CARMO, OmicsAnalyst, MapMan4, KBCommons, and Plant Regulomics, have been developed to analyze functional information derived from multiomics data (Zhou *et al.*, 2021). Recently, the ENCODE-standardized platform, ChIP-Hub, has emerged as a valuable resource for plant regulome and epigenome data (Fu *et al.*, 2022). Furthermore, integrating regulome data with genetic variations has enabled CRISPR-Cerela to facilitate precise gene editing in wheat, rice, and maize (He *et al.*, 2021).

10.3.2 Challenges in integrating data from multiomics layers

Integrating diverse data from various platforms and formats across the genotype–phenotype spectrum presents a significant challenge, particularly in analyzing and interpreting results. Another key issue lies in enhancing phenotypic prediction using large-scale multiomics data. Developing innovative algorithms or models such as those employing artificial intelligence (AI), ML, and DL, is crucial for predicting heterosis and complex phenotypes (Ballard *et al.*, 2024). Therefore, robust visualization tools, multi-omics analytical approaches, statistical genetic models, bioinformatic tools, and cloud computing must be integrated through interdisciplinary efforts to identify candidate variations associated with agronomic traits.

10.3.3 Tools and methodologies

Plant systems biology offers a comprehensive understanding of the morphological, genetic, physiological, molecular, and biochemical aspects of both dynamic and static biological processes in plants. This systems biology approach integrates big data to identify and characterize plant genes and pathways. Recent advancements in high-throughput technologies across genomics, epigenomics, transcriptomics, proteomics, metabolomics, and phenomics have significantly enhanced the analysis of plant genetic resources, enabling the identification of exotic alleles. Multiomics data undergo preprocessing, feature selection, and dimensionality reduction to ensure data cleaning and the reduction of variables for more effective analysis.

ML is a powerful approach for studying plant systems biology by integrating and analyzing multiomics data. It enables prediction without explicit programming and is categorized into supervised, unsupervised, and reinforcement learning. ML facilitates the identification of regulatory pathways and genes, as demonstrated in potatoes, where Random Forest detected genes for traits like tuber shape and enzymatic discoloration. Tools like MiRduplexSVM and ASRmiRNA identify plant microRNAs, while ML also predicts transcription-factor-binding sites and cis-regulatory elements for abiotic stress (Zhang *et al.*, 2022). Although tools for human epigenome modifications exist, their application to plants is still underexplored.

DL, a subset of ML, uses neural networks for supervised and unsupervised learning. Models like CNN and SVM have been applied to detect biotic and abiotic stresses in plants, such as identifying aphid colonies on *Brassica napus* and predicting nitrogen deficiency in rice, with ResNet-50 + SVM showing superior performance (Iqbal *et al.*, 2021). DL methods have also identified orphan genes in Moso bamboo and are widely used in agriculture for plant disease and plant detection. Furthermore, DL techniques predict plant epigenome modifications using DNA/RNA methylation and histone modification data.

10.4 Application of Multiomics in Quality-Trait Improvement

Multiomics techniques (genomics, transcriptomics, proteomics, and metabolomics) can connect genotype and phenotype, integrating a wealth of information for comprehensive analysis, thereby facilitating the detection of candidate genes governing various important traits. GWAS and transcriptome-wide association studies (TWAS) are complementary methods that link genetic variations and gene expression to phenotypic traits. This combined approach has identified key genes in major cereals, pulses, oilseeds and some fiber crops as described in the following sections.

10.4.1 Cereals

Cereals including rice, wheat, maize, sorghum, millets, oat, and so on serve as staple foods in many diets due to their high carbohydrate and nutritional value. Using an integrative approach combining GWAS and transcriptome profiling for upland rice, natural variation in the promoter region of *DROUGHT1* (DROT1) was identified as a key factor in drought resistance (Sun *et al.*, 2022). Similarly, GWAS and transcriptomic analyses revealed *MADS26* as a gene influencing seed germination in maize (Ma *et al.*, 2022). A non-targeted metabolomics study of leaves from 385 maize lines, coupled with eQTL analysis, identified drought-tolerance hub genes (Zhang *et al.*, 2021). Recently, metagenomic data were used as an external quantitative phenotype alongside genomic and transcriptomic data to identify candidate genes in barley that shape microbiota composition (Escudero-Martinez *et al.*, 2022). Another GWAS involving 325 diverse rice accessions pinpointed loci associated with pericarp color. Additionally, ICP–MS analysis of leaf and grain ionomes in rice enabled the creation of genetic maps, identification of QTLs, and understanding of mineral element genetic diversification (Zhang *et al.*, 2014). Using transcriptomics and proteomics-based screening, small secreted proteins (SSPs) were discovered to regulate rice immunity against rice blast (Wang *et al.*, 2020).

10.4.2 Oilseeds

Soybeans, sunflower, rapeseed, groundnut, sesame, etc. are the oilseeds cultivated primarily for their oil content. With increasing demand for vegetable oils driven by health trends and biofuel production, the importance of oilseeds is growing these days. There have been a few achievements on utilization of more than one omics tools in oilseed crops. By integrating BSA-Seq and RNA-Seq, a *WOX* gene associated with plant architecture in rapeseed was identified, contributing to the compact phenotype (Ye *et al.*, 2022). Integrated QTL and eQTL mapping identified candidate genes linked to fatty acid composition and flowering time in *Brassica napus* (Li *et al.*, 2018). Furthermore, the integration of RAD-Seq and mGWAS with metabolome and transcriptome data revealed novel candidate genes for seed coat color, elucidating the molecular mechanism behind the yellow seed trait in *Brassica rapa* (Zhao *et al.*, 2022). Transcriptomic and metabolic studies have been conducted to understand aflatoxin biosynthesis in groundnut (*Arachis hypogaea*), aiding the development of resistant varieties. Additionally, groundnut cultivars with high oleic acid and low linoleic acid content have been developed through QTL analysis.

10.4.3 Pulses

Pulses are the dry, edible seeds of plants belonging to the legume family, such as chickpeas, lentils, redgram, dry peas, and beans. They are recognized for their high nutritional value, particularly their protein and fiber content, making them a staple in many diets around the world. Combined application of multiomics like DNA methylation, transcriptome, and small-RNA profiling revealed regulatory networks and genes associated with hybrid vigor in pigeonpea (*Cajanus cajan*) (Sinha *et al.*, 2020). Key proteins including seed hardness and digestibility through proteomics and genomics in lentil are identified (Lamonaca *et al.*, 2024).

10.4.4 Fibre crops

Fiber crops are plants specifically cultivated for their fibers, which are used in a variety of products including textiles, ropes, and paper. The integration of proteomic and transcriptomic profiles provided insights into systematic salt-tolerance mechanisms in cotton (Peng *et al.*, 2018). A combination of GWAS, TWAS, and eQTL analysis uncovered the genetic regulatory network governing cell-wall biosynthesis in cotton (Li *et al.*, 2020). Furthermore, combining eQTL and GWAS analyses in cotton identified a KIP-related protein, a key transcriptional regulator of genes involved in initiating secondary cell wall development. Three novel QTLs for UNI (uniformity), seven for MIC (micronaire) and two candidate genes for STR (strength), i.e. *Gh_A07G1574* and *Gh_A07G1581*, were identified through GWAS for fiber quality in cotton (Mohammed *et al.*, 2025).

10.5 Challenges in Multiomics for Quality-Trait Improvement

10.5.1 Data complexity and integration

Multiomics approaches generate large volumes of data from different levels. Integrating data from multiple omics layers to understand the complex interplay between genes, proteins, and metabolites requires advanced computational methods (Li *et al.*, 2023). Managing, processing, and analyzing this vast amount of data efficiently is a significant challenge. Redundancy in data can lead to overfitting models, while missing data complicate the extraction of reliable insights. Proper data preprocessing and statistical techniques are required to handle these issues. The diversity of omics technologies and heterogenous nature of their data types make it difficult to integrate them into a coherent model. Multidisciplinary approaches are needed to align and merge data effectively for meaningful interpretation.

10.5.2 Technological limitations

Current technologies in multiomics can sometimes fail to capture all relevant molecular information, particularly in complex organisms (Zhang *et al.*, 2022). There may be biases or incomplete profiles in the data, especially in non-model species or under specific environmental conditions. There is a lack of standardized protocols and benchmarks for multiomics approaches, which leads to inconsistencies between studies and makes comparisons challenging. While various powerful tools are available, they are often not applied to specific agricultural needs or quality traits. There is a need for more specialized tools for analysis, interpretation, and integration of multiomics data in plant breeding.

10.5.3 Cost and scalability

Sequencing technologies, mass spectrometry, and other omics technologies can be expensive, limiting the widespread adoption in agricultural research. Many regions, especially in developing countries, lack the necessary infrastructure and expertise to implement cutting-edge multiomics approaches. This limits the scalability of these methods to improve quality traits in crops at a global level. Storing large multiomics datasets, especially when dealing with plant breeding programs involving numerous genotypes and environments, presents significant logistical challenges.

10.5.4 Ethical and regulatory concerns

The ownership of multiomics data, especially when collected from multiple sources or countries, raises ethical questions. There is a need to establish clear policies regarding data sharing, access, and use. Genetic modifications or the use of certain technologies derived from multiomics research can raise concerns about the potential environmental and health impacts. Regulation of these technologies needs to balance innovation with safety concerns, ensuring that the benefits of multiomics in crop improvement can be realized responsibly. Regulatory frameworks need to address the ethical implications of using such technologies in agricultural practices. Addressing these challenges requires advances in technology, computational methods, and policy frameworks to make multiomics approaches feasible and effective for quality-trait improvement in agriculture (Scossa *et al.*, 2021).

10.6 Conclusions

The application of omics-based tools, especially genomics, transcriptomics, proteomics, metabolomics, and phenomics, to study the plant responses to various abiotic and biotic stresses has been crucial in uncovering candidate genes, proteins, and metabolic pathways associated with important quality traits in major crop species. Some of the identified candidate genes and metabolic pathways have been incorporated into breeding pipelines via molecular breeding approaches or genetic engineering methods of genomics-assisted or marker-assisted breeding programs. Furthermore, recent advancements in metabolomics and high-throughput phenotyping platforms have enhanced the effectiveness of genomics, transcriptomics, and proteomics in crop improvement. Moreover, genome-editing technologies complement these multiomics strategies by enabling precise modifications to specific genes identified through omics analyses. This synergy not only enhances the ability to introduce beneficial traits but also ensures that these modifications are informed by a thorough understanding of the underlying biological mechanisms. As a result, the combination of multiomics and genome editing is poised to revolutionize crop breeding practices. It promises to deliver crops that are not only higher yielding but also more resilient to climate change and capable of meeting the nutritional needs of a growing global population. The future of agricultural innovation lies in this integrated approach, which will significantly contribute to food security and sustainability efforts worldwide.

References

Abdallah, N.A., Prakash, C.S. and McHughen, A.G. (2015) Genome editing for crop improvement: Challenges and opportunities. *GM Crops & Food* 6(4), 183–205.

Abu, M. and Yimer, O. (2023) Breeding for quality traits and its contribution for the improvement of field crops. *Journal of Plant Biology and Crop Research* 6(2), 1094.

Adrees, M., Ali, S., Rizwan, M., Zia-ur-Rehman, M., Ibrahim, M. *et al.* (2015) Mechanisms of silicon-mediated alleviation of heavy metal toxicity in plants: A review. *Ecotoxicology and Environmental Safety* 119, 186–197.

Afshan, S., Ali, S., Bharwana, S.A., Rizwan, M., Farid, M. *et al.* (2015) Citric acid enhances the phyto-extraction of chromium, plant growth, and photosynthesis by alleviating the oxidative damages in *Brassica napus* L. *Environmental Science and Pollution Research* 22, 11679–11689.

Ajeesh Krishna, T.P., Maharajan, T., Victor Roch, G., Ignacimuthu, S. and Antony Ceasar, S. (2020) Structure, function, regulation and phylogenetic relationship of ZIP family transporters of plants. *Frontiers in Plant Science* 11, 662.

Aleem, M., Razzaq, M.K., Aleem, M., Yan, W., Sharif, I. *et al.* (2024) Genome-wide association study provides new insight into the underlying mechanism of drought tolerance during seed germination stage in soybean. *Scientific Reports* 14(1), 20765.

Anand, A., Subramanian, M. and Kar, D. (2023) Breeding techniques to dispense higher genetic gains. *Frontiers in Plant Science* 13, 1076094.

Ankit Kumar, D.K.J., Deep, H. and Rohila, N. (2022) Quality breeding: Future perspective for nutritional security. *Just Agriculture* 8(2), 1–5.

Arya, S.S., Salve, A.R. and Chauhan, S. (2016) Peanuts as functional food: A review. *Journal of Food Science and Technology* 53, 31–41.

Azmach, G., Menkir, A., Spillane, C. and Gedil, M. (2018) Genetic loci controlling carotenoid biosynthesis in diverse tropical maize lines. *G3: Genes, Genomes, Genetics* 8(3), 1049–1065.

Ballard, J.L., Wang, Z., Li, W., Shen, L. and Long, Q. (2024) Deep learning-based approaches for multi-omics data integration and analysis. *Bio Data Mining* 17(1), 38.

Bentley, A.R., Chen, C. and D'Agostino, N. (2022) Genome wide association studies and genomic selection for crop improvement in the era of Big Data. *Frontiers in Genetics* 13, 873060.

Beyer, K., Bardina, L., Grishina, G. and Sampson, H.A. (2002) Identification of sesame seed allergens by 2- dimensional proteomics and Edman sequencing: Seed storage proteins as common food allergens. *Journal of Allergy and Clinical Immunology* 110, 154–159.

Bhat, J.A., Salgotra, R.K. and Dar, M.Y. (2015) Phenomics: A challenge for crop improvement in genomic era. *Molecular Plant Breeding* 6, 22.

Bradbury, L.M., Fitzgerald, T.L., Henry, R.J., Jin, Q. and Waters, D.L. (2005) The gene for fragrance in rice. *Plant Biotechnology Journal* 3(3), 363–370.

Breseghello, F. and Coelho, A.S.G. (2013) Traditional and modern plant breeding methods with examples in rice (*Oryza sativa* L.). *Journal of Agricultural and Food Chemistry* 61(35), 8277–8286.

Budhlakoti, N., Kushwaha, A.K., Rai, A., Chaturvedi, K.K., Kumar, A. *et al.* (2022) Genomic selection: A tool for accelerating the efficiency of molecular breeding for development of climate-resilient crops. *Frontiers in Genetics* 13, 832153.

Bundo, M., Martin-Cardoso, H., Pesenti, M., Gomez-Ariza, J., Castillo, L. *et al.* (2022) Integrative approach for precise genotyping and transcriptomics of salt tolerant introgression rice lines. *Frontiers in Plant Science* 12, 797141. DOI: 10.3389/fpls.2021.797141.

Cao, P., Zhao, Y., Wu, F., Xin, D., Liu, C. *et al.* (2022) Multi-omics techniques for soybean molecular breeding. *International Journal of Molecular Sciences* 23(9), 4994.

Caproni, L., Raggi, L., Talsma, E.F., Wenzl, P. and Negri, V. (2020) European landrace diversity for common bean biofortification: A genome-wide association study. *Scientific Reports* 10(1), 19775.

Chabi, M., Goulas, E., Galinousky, D., Blervacq, A.S., Lucau-Danila, A. *et al.* (2023) Identification of new potential molecular actors related to fiber quality in flax through Omics. *Frontiers in Plant Science* 14, 1204016.

Chaturvedi, P., Wiese, A.J., Ghatak, A., Záveská Drábková, L., Weckwerth, W. *et al.* (2021) Heat stress response mechanisms in pollen development. *New Phytologist* 231, 571–585. DOI: 10.1111/nph.17380.

Chauhan, H.S., Chhabra, R., Rashmi, T., Muthusamy, V., Zunjare, R.U. *et al.* (2022) Impact of vte4 and crtrb1 genes on composition of vitamin-E and provitamin-A carotenoids during kernel-stages in sweet corn. *Journal of Food Composition and Analysis* 105, 104264. DOI: 10.1016/j.jfca.2021.104264.

Chen, J., Hu, X., Shi, T., Yin, H., Sun, D. *et al.* (2020) Metabolite-based genome-wide association study enables dissection of the flavonoid decoration pathway of wheat kernels. *Plant Biotechnology Journal* 18(8), 1722–1735.

Chernova, A.I., Gubaev, R.F., Singh, A., Sherbina, K., Goryunova, S.V. *et al.* (2021) Genotyping and lipid profiling of 601 cultivated sunflower lines reveals novel genetic determinants of oil fatty acid content. *BMC Genomics* 22, 1–15.

Conesa, A. and Mortazavi, A. (2014) The common ground of genomics and systems biology. *Front Genet* 5, 208.

Cruet-Burgos, C. and Rhodes, D.H. (2023) Unraveling transcriptomics of sorghum grain carotenoids: A step forward for biofortification. *BMC Genomics* 24(1), 233.

Dar, A.A., Choudhury, A.R., Kancharla, P.K. and Arumugam, N. (2017) The FAD2 gene in plants: Occurrence, regulation, and role. *Frontiers in Plant Science* 8, 1789.

Dawid, C. and Hille, K. (2018) Functional metabolomics—a useful tool to characterize stress-induced metabolome alterations opening new avenues towards tailoring food crop quality. *Agronomy* 8(8), 138.

Deshmukh, R., Sonah, H., Patil, G., Chen, W., Prince, S. *et al.* (2014) Integrating omics approaches for abiotic stress tolerance in soybean. *Frontiers in Plant Science* 5, 244.

Dhar, A. (2023) Quality parameters of wheat for industrial uses and its agronomic measures 2, 34–39.

Escandón, M., Castillejo, M.Á., Jorrín-Novo, J.V. and Rey, M.D. (2021) Molecular research on stress responses in Quercus spp.: From classical biochemistry to systems biology through omics analysis. *Forests* 12(3), 364.

Escudero-Martinez, C., Coulter, M., Alegria Terrazas, R., Foito, A., Kapadia, R. *et al.* (2022) Identifying plant genes shaping microbiota composition in the barley rhizosphere. *Nature Communications* 13(1), 3443.

Fu, L.Y., Zhu, T., Zhou, X., Yu, R., He, Z. *et al.* (2022) ChIP-Hub provides an integrative platform for exploring plant regulome. *Nature Communications* 13(1), 3413.

Gaikwad, K.B., Rani, S., Kumar, M., Gupta, V., Babu, P.H. *et al.* (2020) Enhancing the nutritional quality of major food crops through conventional and genomics-assisted breeding. *Frontiers in Nutrition* 7, 533453.

Gali, K.K., Sackville, A., Tafesse, E.G., Lachagari, V.R., McPhee, K. *et al.* (2019) Genome-wide association mapping for agronomic and seed quality traits of field pea (*Pisum sativum* L.). *Frontiers in Plant Science* 10, 1538.

Guo, X., Ge, Z., Wang, M., Zhao, M., Pei, Y. *et al.* (2023) Genome-wide association study of quality traits and starch pasting properties of maize kernels. *BMC Genomics* 24(1), 59.

Hasan, N., Choudhary, S., Naaz, N., Sharma, N. and Laskar, R.A. (2021) Recent advancements in molecular marker-assisted selection and applications in plant breeding programmes. *Journal of Genetic Engineering and Biotechnology* 19(1), 128.

He, C., Liu, H., Chen, D., Xie, W.Z., Wang, M. *et al.* (2021) CRISPR-cereal: A guide RNA design tool integrating regulome and genomic variation for wheat, maize and rice. *Plant Piotechnology Journal* 19(11), 2141–2143. DOI: 10.1111/pbi.13675.

Iqbal, Z., Iqbal, M.S., Khan, M.I.R. and Ansari, M.I. (2021) Toward integrated multi-omics intervention: Rice trait improvement and stress management. *Frontiers in Plant Science* 12, 741419.

Jamil, I.N., Remali, J., Azizan, K.A., Nor Muhammad, N.A., Arita, M. *et al.* (2020) Systematic multi-omics integration (MOI) approach in plant systems biology. *Frontiers in Plant Science* 11, 944.

Jha, U.C., Nayyar, H., Parida, S.K., Deshmukh, R., Wettberg, E.J.B. *et al.* (2022) Ensuring global food security by improving protein content in major grain legumes using breeding and 'omics' tools. *International Journal of Molecular Sciences* 23(14), 7710. DOI: 10.3390/ijms23147710.

Jin, M., Zhang, X., Zhao, M., Deng, M., Du, Y. *et al.* (2017) Integrated genomics-based mapping reveals the genetics underlying maize flavonoid biosynthesis. *BMC Plant Biology* 17, 1–17.

Jukanti, A.K., Gaur, P.M., Gowda, C.L.L. and Chibbar, R.N. (2012) Nutritional quality and health benefits of chickpea (*Cicer arietinum* L.): A review. *British Journal of Nutrition* 108(S1), S11–S26.

Kalyana Babu, B., Sood, S., Gaur, V.S. and Kumar, A. (2022) Comparative genomics of finger millet. In: Kumar, A., Sood, S., Kalyana Babu, B., Gupta, S.M. and Dayakar Rao, B. (eds) *The Finger Millet Genome*. Springer-Nature, Basel, Switzerland, pp. 113–121. ISBN 978-303-100-868-9

Khakimov, B., Bak, S. and Engelsen, S.B. (2014) High-throughput cereal metabolomics: Current analytical technologies, challenges and perspectives. *Journal of Cereal Science* 59(3), 393–418.

Khandagale, K., Krishna, R., Roylawar, P., Ade, A.B., Benke, A. *et al.* (2020) Omics approaches in Allium research: Progress and way ahead. *PeerJ* 8.

Koller, A., Washburn, M.P., Lange, B.M., Andon, N.L., Deciu, C. *et al.* (2002) Proteomic survey of metabolic pathways in rice. *Proceedings of the National Academy of Sciences* 99(18), 11969–11974.

Kuhn, E. (2001) From library screening to microarray technology: Strategies to determine gene expression profiles and to identify differentially regulated genes in plants. *Annals of Botany* 87(2), 139–155.

Kulski, J.K. (2016) Next-generation sequencing—an overview of the history, tools, and "omic" applications. *Next Generation Sequencing-Advances, Applications and Challenges* 10, 61964.

Kumar, V., Vats, S., Kumawat, S., Bisht, A., Bhatt, V. *et al.* (2021) Omics advances and integrative approaches for the simultaneous improvement of seed oil and protein content in soybean (*Glycine max* L.). *Critical Reviews in Plant Sciences* 40(5), 398–421. DOI: 10.1080/07352689.2021.1954778.

Lamichhane, S. and Thapa, S. (2022) Advances from conventional to modern plant breeding methodologies. *Plant Breeding and Biotechnology* 10(1), 1–14.

Lamonaca, A., De Angelis, E., Monaci, L. and Pilolli, R. (2024) One shot-proteomic profiling, allergenicity assessment and varietal discrimination of lentils seeds by discovery high resolution tandem mass spectrometry. *Journal of Food Composition and Analysis* 106456.

Lamzin, V.S. and Perrakis, A. (2000) Current state of automated crystallographic data analysis. *Nature Structural Biology* 7(11), 978–981.

Lefkovits, I. (2003) Quantitative proteomics of lymphocytes. *Comparative and Functional Genomics* 4(5), 531–536.

Li, H., Tahir ul Qamar, M., Yang, L., Liang, J., You, J. *et al.* (2023) Current progress, applications and challenges of multi-omics approaches in sesame genetic improvement. *International Journal of Molecular Sciences* 24(4), 3105.

Li, R., Jeong, K., Davis, J.T., Kim, S., Lee, S. *et al.* (2018) Integrated QTL and eQTL mapping provides insights and candidate genes for fatty acid composition, flowering time, and growth traits in a F2 population of a novel synthetic allopolyploid *Brassica napus*. *Frontiers in Plant Science* 9, 1632.

Li, X., Gao, J., Song, J., Guo, K., Hou, S. *et al.* (2022) Multi-omics analyses of 398 foxtail millet accessions reveal genomic regions associated with domestication, metabolite traits, and anti-inflammatory effects. *Molecular Plant* 15(8), 1367–1383.

Li, Z., Wang, P., You, C., Yu, J., Zhang, X. *et al.* (2020) Combined GWAS and eQTL analysis uncovers a genetic regulatory network orchestrating the initiation of secondary cell wall development in cotton. *New Phytologist* 226, 1738–1752. DOI: 10.1111/nph.16468.

Loni, F., Ismaili, A., Nakhoda, B., Darzi Ramandi, H. and Shobbar, Z.S. (2023) The genomic regions and candidate genes associated with drought tolerance and yield-related traits in foxtail millet: An integrative meta-analysis approach. *Plant Growth Regulation* 101(1), 169–185.

Liu, H.L. and Hsu, J.P. (2005) Recent developments in structural proteomics for protein structure determination. *Proteomics* 5(8), 2056–2068.

Liu, J., Hao, W., Liu, J., Fan, S., Zhao, W. *et al.* (2019b) A novel chimeric mitochondrial gene confers cytoplasmic effects on seed oil content in polyploid rapeseed (*Brassica napus*). *Molecular Plant* 12, 582–596.

Liu, J.Y., Li, P., Zhang, Y.W., Zuo, J.F., Li, G. *et al.* (2020) Three-dimensional genetic networks among seed oil-related traits, metabolites and genes reveal the genetic foundations of oil synthesis in soybean. *The Plant Journal* 103(3), 1103–1124.

Lou, D., Wang, H., Liang, G. and Yu, D. (2017) OsSAPK2 confers abscisic acid sensitivity and tolerance to drought stress in rice. *Frontiers in Plant Science* 8, 993.

Luo, F., Yu, Z., Zhou, Q. and Huang, A. (2022) Multi-omics-based discovery of plant signalling molecules. *Metabolites* 12(1), 76.

Ma, L., Wang, C., Hu, Y., Dai, W., Liang, Z. *et al.* (2022) GWAS and transcriptome analysis reveal MADS26 involved in seed germination ability in maize. *Theoretical and Applied Genetics* 135, 1717–1730. DOI: 10.1007/s00122-022-04065-4.

Mahendrakar, M.D., Parveda, M., Kishor, P.B.K. and Srivastava, R.K. (2020) Discovery and validation of candidate genes for grain iron and zinc metabolism in pearl millet [*Pennisetum glaucum* (L.) r. br.]. *Scientific Reports* 10, 16562.

Matsuda, F., Nakabayashi, R., Yang, Z., Okazaki, Y., Yonemaru, J.I. *et al.* (2015) Metabolome-genome-wide association study dissects genetic architecture for generating natural variation in rice secondary metabolism. *The Plant Journal* 81(1), 13–23.

Maveyraud, L. and Mourey, L. (2020) Protein X-ray crystallography and drug discovery. *Molecules* 25(5), 1030.

Mead, M.N. (2007) Nutrigenomics: The genome-food interface. *Environmental Health Perspectives* 115(12), A582. DOI: 10.1289/ehp.115-a582.

Mir, R.R., Reynolds, M., Pinto, F., Khan, M.A. and Bhat, M.A. (2019) High-throughput phenotyping for crop improvement in the genomics era. *Plant Science* 282, 60–72.

Mishra, U.N., Jena, D., Sahu, C., Devi, R., Kumar, R. *et al.* (2022) Nutrigenomics: An inimitable interaction amid genomics, nutrition and health. *Innovative Food Science & Emerging Technologies* 82, 103196.

Mohammed, J., Thyssen, G.N., Hinze, L., Zhang, J., Zeng, L. *et al.* (2025) A GWAS identified loci and candidate genes associated with fiber quality traits in a new cotton MAGIC population. *Theoretical and Applied Genetics* 138(1), 10.

Montelione, G.T., Zheng, D., Huang, Y.J., Gunsalus, K.C. and Szyperski, T. (2000) Protein NMR spectroscopy in structural genomics. *Nature Structural Biology* 7(11), 982–985.

Mugabe, D., Frieszell, C.M., Warburton, M.L., Coyne, C.J., Sari, H. *et al.* (2023) Kabuli chickpea seed quality diversity and preliminary genome-wide association study identifies markers and potential candidate genes. *Agrosystems, Geosciences & Environment* 6(4), e20437.

Müller, K. and Hippe, J. (1987) Influence of differences in nutrition on important quality characteristics of some agricultural crops. *Plant and Soil* 100, 35–45.

Muthamilarasan, M., Singh, N.K. and Prasad, M. (2019) Multi-omics approaches for strategic improvement of stress tolerance in underutilized crop species: A climate change perspective. *Advances in Genetics* 103, 1–38.

Muthuramalingam, P., Jeyasri, R., Rakkammal, K., Satish, L., Shamili, S. *et al.* (2022) Multi-omics and integrative approach towards understanding salinity tolerance in rice: A review. *Biology* 11, 1022.

Nakano, Y. and Kobayashi, Y. (2020) Genome-wide association studies of agronomic traits consisting of field-and molecular-based phenotypes. *Reviews in Agricultural Science* 8, 28–45.

Nat, R., Nilbratt, M., Narkilahti, S., Winblad, B., Hovatta, O. *et al.* (2007) Neurogenic neuroepithelial and radial glial cells generated from six human embryonic stem cell lines in serum-free suspension and adherent cultures. *Glia* 55(4), 385–399.

Nayak, S.N., Aravind, B., Malavalli, S.S., Sukanth, B.S., Poornima, R. *et al.* (2021) Omics technologies to enhance plant based functional foods: An overview. *Frontiers in Genetics* 12, 742095.

Nejat, N., Ramalingam, A. and Mantri, N. (2018) Advances in transcriptomics of plants. *Advances in Biochemical Engineering/Biotechnology* 161–185.

Nirmal, R.C., Furtado, A., Wrigley, C. and Henry, R.J. (2016) Influence of gene expression on hardness in wheat. *PLoS ONE* 11(10), e0164746.

Nirmal, R.C., Furtado, A., Rangan, P. and Henry, R.J. (2017) Fasciclin-like arabinogalactan protein gene expression is associated with yield of flour in the milling of wheat. *Scientific Reports* 7(1), 12539.

Nookaew, I., Papini, M., Pornputtapong, N., Scalcinati, G., Fagerberg, L. *et al.* (2012) A comprehensive comparison of RNA-Seq-based transcriptomic analysis from reads to differ ential gene expression and cross-comparison with microarrays: A case study in *Saccharomyces cerevisiae. Nucleic Acids Research* 40(20), 10084–10097.

Ortiz, R. (2023) Challenges for crop improvement. *Emerging Topics in Life Sciences* 7(2), 197–205.

Ozsolak, F. and Milos, P.M. (2011) RNA sequencing: Advances, challenges and opportunities. *Nature Reviews Genetics* 12(2), 87–98.

Panahi, B., Gharajeh, N.H., Jalaly, H.M. and Golkari, S. (2024) Leveraging multi-omics and machine learning approaches in malting barley research: From farm cultivation to the final products. *Current Plant Biology* 39, 100362.

Pandey, S. (2019) Review on medicinal importance of *Vigna* genus. *Plant Science Today* 6(4), 450–456.

Pawlowski, M.L., Helfenstein, J., Frossard, E. and Hartman, G.L. (2019) Boron and zinc deficiencies and toxicities and their interactions with other nutrients in soybean roots, leaves, and seeds. *Journal of Plant Nutrition* 42(6), 634–649.

Pazhamala, L.T., Agarwal, G., Bajaj, P., Kumar, V., Kulshreshtha, A. *et al.* (2016) Deciphering transcriptional programming during pod and seed development using RNA-Seq in pigeonpea (*Cajanus cajan*). *PLoS ONE* 11(10), e0164959.

Peng, Z., He, S., Gong, W., Xu, F., Pan, Z. *et al.* (2018) Integration of proteomic and transcriptomic profiles reveals multiple levels of genetic regulation of salt tolerance in cotton. *BMC Plant Biology* 18, 128. DOI: 10.1186/s12870-018-1350-1.

Petersen, A., Dresselhaus, T., Grobe, K. and Becker, W.M. (2006) Proteome analysis of maize pollen for allergy-relevant components. *Proteomics* 6(23), 6317–6325.

Pita-Barbosa, A., Ricachenevsky, F.K. and Flis, P.M. (2019) One "OMICS" to integrate them all: Ionomics as a result of plant genetics, physiology and evolution. *Theoretical and Experimental Plant Physiology* 31(1), 71–89.

Pott, D.M., Durán-Soria, S., Osorio, S. and Vallarino, J.G. (2021) Combining metabolomic and transcriptomic approaches to assess and improve crop quality traits. *CABI Agriculture and Bioscience*. DOI: 10.1186/s43170-020-00021-8.

Raza, A. (2022) Plant biotechnological tools: Solutions for raising climate-resilient crop plants. *Modern Phytomorphology* 15, 132–133.

Razzaq, A., Sadia, B., Raza, A., Khalid Hameed, M. and Saleem, F. (2019) Metabolomics: A way forward for crop improvement. *Metabolites* 9(12), 303.

Rhodes, D.H., Hoffmann Jr, L., Rooney, W.L., Herald, T.J., Bean, S. *et al.* (2017) Genetic architecture of kernel composition in global sorghum germplasm. *BMC Genomics* 18, 15.

Ritchie, M.E., Phipson, B., Wu, D., Hu, Y., Law, C.W. *et al.* (2015) Limma powers differential expression analyses for RNA sequencing and microarray studies. *Nucleic Acids Research* 43(7), 47.

Rooney, L.W. and Murty, D.S. (1982) Evaluation of sorghum food quality.

Roychowdhury, R., Ballén-Taborda, C. and Chaturvedi, P. (2023) Characterizing and improving traits for resilient crop development. *Frontiers in Plant Science* 14, 1307327.

Salt, D.E., Baxter, I. and Lahner, B. (2008) Ionomics and the study of the plant ionome. *Annual Review of Plant Biology* 59, 709–733.

Sansa, O., Abberton, M.T., Ariyo, J., Paliwal, R., Ige, A. *et al.* (2024) Genome-wide association studies of photosynthetic and agronomic traits in cowpea collection. *G3: Genes, Genomes, Genetics* 14(12), jkae233.

Santos, T., Marinho, C., Freitas, M., Santos, H.M., Oppolzer, D. *et al.* (2017) Unravelling the nutriproteomics of chickpea (*Cicer arietinum*) seeds. *Crop and Pasture Science* 1041–1051.

Satyavathi, C.T., Tomar, R.S., Ambawat, S., Kheni, J., Padhiyar, S.M. *et al.* (2022) Stage specific comparative transcriptomic analysis to reveal gene networks regulating iron and zinc content in pearl millet [*Pennisetum glaucum* (L.) r. br.]. *Scientific Reports* 12, 276.

Scossa, F., Alseekh, S. and Fernie, A.R. (2021) Integrating multi-omics data for crop improvement. *Journal of Plant Physiology* 257, 153352.

Sharma, D., Jamra, G., Singh, U.M., Sood, S. and Kumar, A. (2017) Calcium biofortification: Three pronged molecular approaches for dissecting complex trait of calcium nutrition in finger millet (*Eleusine coracana*) for devising strategies of enrichment of food crops. *Frontiers in Plant Science* 7.

Sharma, N., Siddappa, S., Malhotra, N., Thakur, K., Salaria, N. *et al.* (2022) Advances in potato functional genomics: Implications for crop improvement. *Plant Cell, Tissue and Organ Culture (PCTOC)* 148(3), 447–464.

Sharma, V., Gupta, P., Kagolla, P., Hangargi, B., Veershetty, A. *et al.* (2021) Metabolomics intervention towards better understanding of plant traits. *Cells* 10(2), 346.

Shivhare, R., Asif, M.H. and Lata, C. (2020) Comparative transcriptome analysis reveals the genes and pathways involved in terminal drought tolerance in pearl millet. *Plant Molecular Biology* 103, 639–652. DOI: 10.1007/s11103-020-01015-w.

Shobha, D. and Ravishankar, C.R. (2017) Quality assessment and evaluation of ragi for development of multipurpose mix.

Singh, U.M., Chandra, M., Shankhdhar, S.C. and Kumar, A. (2014) Transcriptome wide identification and validation of calcium sensor gene family in the developing spikes of finger millet genotypes for elucidating its role in grain calcium accumulation. *PLoS ONE* 9, e103963.

Sinha, P., Singh, V., Saxena, R., Kale, S., Li, Y. *et al.* (2020) Genomewide analysis of epigenetic and transcriptional changes associated with heterosis in pigeonpea. *Plant Biotechnology Journal* 18, 1697–1710. DOI: 10.1111/pbi.13333.

Subramanian, I., Verma, S., Kumar, S., Jere, A. and Anamika, K. (2020) Multi-omics data integration, interpretation, and its application. *Bioinformatics and Biology Insights* 14, 1177932219899051.

Sultana, S., Faruque, M. and Islam, M.R. (2022) Rice grain quality parameters and determination tools: A review on the current developments and future prospects. *International Journal of Food Properties* 25(1), 1063–1078.

Sun, W., He, Q., Liu, J., Xiao, X., Wu, Y. *et al.* (2023) Dynamic monitoring of maize grain quality based on remote sensing data. *Frontiers in Plant Science* 14, 1177477.

Sun, X., Xiong, H., Jiang, C., Zhang, D., Yang, Z. *et al.* (2022) Natural variation of DROT1 confers drought adaptation in upland rice. *Nature Communication* 13, 4265. DOI: 10.1038/s41467-022-31844-w.

Susmitha, P., Kumar, P., Yadav, P., Sahoo, S., Kaur, G. *et al.* (2023) Genome-wide association study as a powerful tool for dissecting competitive traits in legumes. *Frontiers in Plant Science* 14, 1123631.

Tang, S., Zhao, H., Lu, S., Yu, L., Zhang, G. *et al.* (2021) Genome-and transcriptome-wide association studies provide insights into the genetic basis of natural variation of seed oil content in *Brassica napus*. *Molecular Plant* 14, 470–487.

Ton, L.B., Neik, T.X. and Batley, J. (2020) The use of genetic and gene technologies in shaping modern rapeseed cultivars (*Brassica napus* L.). *Genes* 11(10), 1161.

U.S. Department of Agriculture (n.d.) *Global agricultural productivity must increase to meet food demand*. Economic Research Service, U.S. Department of Agriculture. Available at: https://www.ers.usda.gov/data-products/chart-gallery/gallery/chart-detail/?chartId=108060 (accessed 29 December 2024).

Venu, R.C., Sreerekha, M.V., Nobuta, K., Beló, A., Ning, Y. *et al.* (2011) Deep sequencing reveals the complex and coordinated transcriptional regulation of genes related to grain quality in rice cultivars. *BMC Genomics* 12, 1–13.

Verma, D.K., Thakur, M. and Srivastav, P.P. (2020) Biochemistry and molecular aspects of 2-acetyl-1-pyrroline biosynthesis in rice (*Oryza sativa* L.): A review. *Israel Journal of Plant Sciences* 67(3–4), 129–143. DOI: 10.1163/22238980-bja10012.

Wang, P., Yao, S., Kosami, K., Guo, T., Li, J. *et al.* (2020) Identification of endogenous small peptides involved in rice immunity through transcriptomics and proteomics-based screening. *Plant Biotechnology Journal* 18, 415–428. DOI: 10.1111/pbi.13208.

Wang, Z., Gerstein, M. and Snyder, M. (2009) RNA-Seq: A revolutionary tool for transcriptomics. *Nature Reviews Genetics* 10(1), 57–63. DOI: 10.1038/nrg2484.

Waters, D.L., Henry, R.J., Reinke, R.F. and Fitzgerald, M.A. (2006) Gelatinization temperature of rice explained by polymorphisms in starch synthase. *Plant Biotechnology Journal* 4(1), 115–122.

Weckwerth, W., Ghatak, A., Bellaire, A., Chaturvedi, P. and Varshney, R.K. (2020) PANOMICS meets germplasm. *Plant Biotechnology Journal* 18, 1507–1525. DOI: 10.1111/pbi.13372.

Wei, W., Li, S., Wang, Y., Wang, B., Fan, G. *et al.* (2021) Metabolome-based genome-wide association study provides genetic insights into the natural variation of foxtail millet. *Frontiers in Plant Science* 12, 665530.

Wu, X., Fang, P., Zhang, P., Sun, T., Wang, X. *et al.* (2022) Improvement for quality and safety traits in horticultural plants. *Frontiers in Plant Science* 13, 927779.

Xu, W., Zhang, H., Zhang, Y., Shen, P., Li, X. *et al.* (2022) A paired-end whole-genome sequencing approach enables comprehensive characterization of transgene integration in rice. *Communications Biology* 5, 667.

Yang, W., Feng, H., Zhang, X., Zhang, J., Doonan, J.H. *et al.* (2020) Crop phenomics and high-throughput phenotyping: Past decades, current challenges, and future perspectives. *Molecular Plant* 13(2), 187–214.

Ye, S., Yan, L., Ma, X., Chen, Y., Wu, L. *et al.* (2022) Combined BSA-seq based mapping and RNA-seq profiling reveal candidate genes associated with plant architecture in *Brassica napus*. *International Journal of Molecular Sciences* 23, 2472. DOI: 10.3390/ijms23052472.

Zanini, S.F., Bayer, P.E., Wells, R., Snowdon, R.J., Batley, J. *et al.* (2022) Pangenomics in crop improvement-from coding structural variations to finding regulatory variants with pangenome graphs. *Plant Genome* 15, e20177.

Zenda, T., Liu, S., Dong, A., Li, J., Wang, Y. *et al.* (2021) Omics-facilitated crop improvement for climate resilience and superior nutritive value. *Frontiers in Plant Science* 12, 774994.

Zhang, F., Wu, J., Sade, N., Wu, S., Egbaria, A. *et al.* (2021) Genomic basis underlying the metabolome-mediated drought adaptation of maize. *Genome Biology* 22, 260. DOI: 10.1186/s13059-021-02481-1.

Zhang, H., Tang, Y., Yue, Y. and Chen, Y. (2024) Advances in the evolution research and genetic breeding of peanut. *Gene* 148425.

Zhang, L., Xu, J., Ding, Y., Zhang, S., Liu, W. *et al.* (2023) GWAS of grain color and tannin content in Chinese *sorghum* based on whole-genome sequencing. *Theoretical and Applied Genetics* 136(77). DOI: 10.1007/s00122-023-04307-z.

Zhang, M., Lv, D., Ge, P., Bian, Y., Chen, G. *et al.* (2014) Phosphoproteome analysis reveals new drought response and defense mechanisms of seedling leaves in bread wheat (*Triticum aestivum* L). *Journal of Proteomics* 109, 290–308. DOI: 10.1016/j.jprot.2014.07.010.

Zhang, R., Zhang, C., Yu, C., Dong, J. and Hu, J. (2022) Integration of multi-omics technologies for crop improvement: Status and prospects. *Frontiers in Bioinformatics* 2, 1027457.

Zhao, H., Shang, G., Yin, N., Chen, S., Shen, S. *et al.* (2022) Multi-omics analysis reveals the mechanism of seed coat color formation in *Brassica rapa* L. *Theoretical and Applied Genetics* 135, 2083–2099. DOI: 10.1007/s00122.

Zhou, M. and Zheng, S. (2022) Multi-omics uncover the mechanism of wheat under heavy metal stress. *International Journal of Molecular Sciences* 23(24), 15968.

Zhou, M., Varol, A. and Efferth, T. (2021) Multi-omics approaches to improve malaria therapy. *Pharmacological Research* 167, 105570.

Zhou, R., Jiang, F., Niu, L., Song, X., Yu, L. *et al.* (2022) Increase crop resilience to heat stress using omic strategies. *Frontiers in Plant Science* 13, 891861.

11 Omics Approaches for Climate-resilient Cereal Crops

Banshidhar[1]*, Priyanka Jaiswal[1], Digvijay Singh[1] and Mithilesh Kumar Singh[2]
[1]Department of Genetics and Plant Breeding, ANDUAT, Kumarganj, Ayodhya, Uttar Pradesh, India; [2]Department of Genetics and Plant Breeding, Dr. Rajendra Prasad Central Agricultural University, Pusa, Samastipur, Bihar, India

Abstract

The abrupt and unprecedented change in climatic conditions has posed a serious threat to agriculture and allied sectors particularly in terms of long-term productivity and sustainability. The prevalent situation of global warming as a result of excessive emission of greenhouse gases from various sources including agricultural fields and non-directional change in rainfall pattern has not only led to severe yield losses but has also changed the agroecosystem of crops resulting in the occurrence of new insect-pest and pathogen complexes that will adversely affect the tolerance of crops towards various biotic and abiotic stresses. There is therefore an urgent need to develop improved varieties that have resilience towards climatic changes. Developing such cultivars has been advocated as an innovative sustainable solution and an efficient and economical approach to safeguard agroecosystem productivity. Breeding for climate-smart cereal crop cultivars that have higher yield potential along with resilience will entail exploring crop wild relatives and novel genetic resources from which we could transfer the gene of interest to a suitable background. For searching and evaluating these genetic resources, omics approaches are indispensable and will help in the discovery of even the rarest genes, proteins, and metabolites and to identify their function. Various omics approaches, namely genomics and pangenomics, transciptomics, proteomics, and metabolomics, either alone or in an integrated manner offer new avenues for cereal crop improvement, by facilitating the identification of neglected germplasm resources and gaining a new, clear understanding of underlying factors.

11.1 Introduction

The paradigm shifts of plant breeding from phenotypic selection to genomic-assisted selection has enabled plant breeders to maximize genetic gain for the targeted selection by assembling only positive genetic traits at a much faster rate than ever before, with the promise of higher yields and also various biotic and abiotic stress tolerances to ensure long-term productivity.

Presently, genomics is at the forefront of any sort of crop improvement program by supporting the discovery of novel genetic variation. Various genomic tools such as gene annotation, high-throughput genome sequencing and resequencing, genome mapping, genome editing, genomic selection, etc. have revolutionized crop breeding, particularly against climatic rigors. In recent times, plant breeders have mainly focused on improving the yield and other trait of economic interest of major cereals like rice, wheat,

*Corresponding author: banshidhargpb@nduat.org

DOI: 10.1079/9781800628717.0011

maize (corn), barley, pearl millets, sorghum, etc. by crossing and selection of superior but closely related cultivars that may lack genes for stability on account of rigorous selection in the direction of suitability for intensive cultivation. This has resulted in a boom in yield potential but a bust of genetic diversity. Considering the urgent need for crop diversification, particularly in terms of varietal adoption, plant breeders are exploring the crop wild relative and wild forms, landraces and other genetic stocks that have been designated as the storehouse of a number of valuable genes for climatic resilience.

Newer insights in sequencing technologies, particularly next-generation and third-generation high-throughput sequencing, have facilitated the construction of a comprehensive genome map of a genetic line in less time for an easy and robust comparison with a crop wild relative reference. This allowed for the identification of those valuable genes that have been lost in cultivars during domestication and subsequent improvement but have assembled in the wild over millions of years under the selective pressure of changing climatic factors. Furthermore, such identified genes have been characterized for structure and function via gene annotation followed by their transfer into a compatible host by crossing, or transgenic techniques (Table 11.1). Similarly, genome editing has also facilitated the modification or distortion of candidate genes to achieve the desired response. In recent times various reverse genetic approaches like insertional mutagenesis, TILLING and EcoTILLING have also emerged as important genomic tools for the identification of genes of interest by using engineered sequences or mutational techniques.

11.2 Pangenomics

Pangenomics (study of a whole set of genes present within a species) offers novel insights and tools for developing climate-resilient crop varieties in cereals. It revolutionizes the extent of utilization of genetic diversity in developing superior cultivars. Pangenomic studies encompass the entire genomic variation within a species, including core genes (genes that are present in all the members of a species) and accessory genes (present in some but absent in at least one member). A comprehensive understanding of core genes allows for the identification of genes necessary for survival, while that of accessories genes allows for identification of those genes that confer specific traits,

Table 11.1. Successful example of genomic techniques applied to the improvement of cereals. Author's own table.

Genomic technique	Target trait	Crop	References
Genome mapping and gene annotation	Heat tolerance	Wheat	Paliwal *et al.*, 2012; Bhusal *et al.*, 2017
Genome editing	Drought tolerance	Rice, wheat, maize	Shi *et al.*, 2017; Kim *et al.*, 2018; Santosh Kumar *et al.*, 2020; Qin *et al.*, 2020
	Cold tolerance	Rice	Shen *et al.*, 2017; Zeng *et al.*, 2020
Genome sequencing and resequencing sequencing	Abiotic stress	Cereals	Roy *et al.*, 2024
TILLING and EcoTILLING	Drought tolerance	Rice	Yu *et al.*, 2012
	Heat tolerance	Barley	Xia *et al.*, 2013
Gene silencing mediated by RNAi	Drought tolerance	Rice, maize	Wei *et al.*, 2009; Liu *et al.*, 2009; Zhou *et al.*, 2010; Xia *et al.*, 2012
Pangenomics	Cold tolerance	Rice	Zhao *et al.*, 2018
Epigenomics	Drought tolerance	Barley	Kapazoglou *et al.*, 2013
	Heat tolerance	Wheat	Hu *et al.*, 2015; Liu *et al.*, 2019; Zheng *et al.*, 2019

particularly tolerance towards different climatic rigors in their area of adaptation. Pangenome analysis would be highly beneficial in identifying genes that are related to a response to abiotic stresses in crop wild relatives, landraces, and other germplasm resources. Since the crop wild relatives grow in more diverse environments than the modern crop species, they may host genes that could support the adaptation of crops to new and variable environments. Such insights gained from pangenome analysis could guide breeding programs aimed at developing climate-resilient cultivars or at the re- domestication of crops by advocating the selection of traits for environmental adaptation along with higher crop productivity.

11.3 Epigenomics

Epigenomics refers to the study of the complete set of epigenetic modifications in the genome of an organism. It defines role of external factors and cellular environment in the alteration of gene expression without any change in underlying DNA sequence via epigenetic modifications such as DNA methylation, histone modification, chromatin remodeling, etc. These modifications play crucial roles in responses to environmental changes. Thus, a comprehensive knowledge of such modifications and their roles in the regulation of gene expression in relation to different climatic conditions can be helpful in understanding how plants adapt to their environment and how their traits can be manipulated. Such studies also help in the identification of new genes, and their regulators and modifiers that confer resilience to climatic changes and develop crops that can thrive in challenging conditions. Further, epigenomics also leads to development of epigenetic markers that can be used in marker-assisted breeding programs to select for target traits, allowing new strategies for developing climate-resilient varieties.

11.4 Transcriptomics

Transcriptomics is the study of the complete set of RNA transcripts produced in an individual or population of cells under specific conditions at a certain status. Transcriptomics studies all forms of RNA ranging from mRNA, rRNA, tRNA, and other noncoding RNA. It includes spatio-temporal bioinformation influenced by diverse tissue types, developmental stages, and experimental as well as environmental events. In contrast to the genome, the transcriptome can vary with external environmental conditions. Thus, a comparative study of an RNA transcript from cells under stressed conditions with those that are under normal conditions enables researchers to gain deeper insights into gene expression changes in response to environmental stress. By analyzing gene expression profiles during stress conditions, we can identify genes that confer resilience. This knowledge also helps in identifying key genes associated with abiotic stress. Identifying which genes are activated or suppressed under these conditions provides insights into the plant's adaptive mechanisms. With the recent advances in the application of omics technologies, numerous biological pathways and candidate genes associated with climatic resilience have been identified based on differentially expressed genes in the stress response. Currently, the three main techniques, namely hybridization-based approaches, sequence-based approaches, and RNA sequencing are used for transcriptome analysis. Hybridization-based approaches or microarray technologies are used for gene expression profiling and sequence-based analysis is used to detect the cDNA sequences directly. The most advanced technique is RNA-sequencing (RNA-Seq) methods that are used for high-throughput quantitative analysis of entire transcriptome. Transcriptomic analysis is extensively utilized for comparing wild relatives of crops and then used in the improvement of cultivated crops. Understanding the transcriptome is essential for interpreting the functional elements of the genome and revealing the molecular constituents of cells and tissues, and also for understanding development and adaptation.

Transcriptomics provides insights into how plants adjust their metabolic pathways and physiological processes in response to climate change (Table 11.2). This knowledge can guide breeding efforts to improve water-use efficiency, nutrient uptake, and overall plant health. Transcriptomics is a vital tool in climate-resilient

Table 11.2. Successful example of transcriptomic approaches applied to cereals. Author's own table.

Transcriptomic techniques	Target trait	Crop	References
Transcriptome profiling	Heat stress	Wheat	Rangan *et al.*, 2020
Validation of transcription factor	Heat stress	Wheat	Seni *et al.*, 2021
RNA-Seq analysis	Heat stress	Wheat	Paul *et al.*, 2022
Differential expression of drought-responsive genes	Drought stress	Wheat	Barratt *et al.*, 2023
Transcriptome analysis	High-temperature stress	Barley	Mangelsen *et al.*, 2011
In silico analysis	Heat stress	Barley	Zakavi *et al.*, 2019
Transcriptome analysis	Heat and drought stress	Barley	Mahalingam *et al.*, 2022
Transcriptome profiling	Heat and drought stress	Barley	Mikołajczak *et al.*, 2023
Differential expression of heat-responsive genes	Heat stress	Maize	Jagtap *et al.*, 2023
Transcriptomic analysis	Cold and heat stress	Maize	Qian *et al.*, 2019; Li *et al.*, 2020; Han *et al.*, 2023
Integrated transcriptomics and metabolomics	Heat stress	Maize	Zhao *et al.*, 2024
Transcriptome profiling	Cold stress	Rice	Yan *et al.*, 2023
Transcriptome and physio-biochemical profiling	Drought stress	Rice	Kaur *et al.*, 2023
Comparative transcriptomic analysis	Anaerobic and cold stress	Rice	Thapa *et al.*, 2023
RNA-Seq analysis	Transient heat stress	Rice	Nalini Chandran *et al.*, 2024

breeding, enabling the identification and incorporation of genetic traits that enhance crop tolerance to environmental stresses, thereby supporting sustainable agricultural practices in the face of climate change. Transcriptomic studies encompass various methodologies, each offering distinct insights into gene expression and regulation. For example, bulk transcriptomics involves analyzing gene expression across a population of cells or tissues, providing an overview of average gene activity in a sample, while single-cell transcriptomics is an advanced technique that allows researchers to examine gene expression at the individual cell level within diverse populations. This innovative approach has enhanced our understanding of cellular diversity. Similarly, spatial transcriptomics provides insights into how cells are organized and interact by combining genomic analysis with spatial data to map gene expression patterns within tissues or organisms, while CircRNA transcriptomics examines circular RNAs, a unique class of non-coding RNAs with regulatory roles. More generally, metatranscriptomics focuses on all transcripts within a community, such as a microbial community, illuminating the roles of different microbial species and their interactions. For instance, this approach can reveal how various soil microbes contribute towards climatic resilience. Transcriptome profiling has considerably revolutionized the understanding of the molecular basis of gene regulation during stress. It is also an important tool for the assembly of a crop reference genome, with RNA-Seq being used for the study of gene expression, whole-genome molecular marker development, and the identification of markers in known-function genes, especially the one that confer tolerance mechanisms against climatic stresses. RNA-Seq has been used successfully as an alternative to whole-genome sequencing, unraveling genes and regulatory regions and providing myriad molecular markers that can be used extensively in breeding climate-resilient cereals.

11.5 Proteomics

Proteomics refers to the comprehensive study of complete set of proteins produced by an organism, known as the proteome, at a specific time and under particular conditions involving their functions, structures, interactions, and modifications. Proteomic analysis includes the extraction and separation of a protein sample, and its identification, verification, and comparison with the protein resources in database. Gel-based and chromatography-based approaches are used for the separation and isolation of proteins from the sample. Among a number of alternatives like 1-D electrophoresis, 2-D electrophoresis, size-exclusion chromatography, ion-exchange chromatography and high-performance liquid chromatography (HPLC), 2-D electrophoresis is the most commonly used proteomics tool, in which the protein is separated based on isoelectric points followed by separation based on molecular weights using 2-D SDS-PAGE.

Further analysis techniques like mass spectrometry (MS), X-ray crystallography, nuclear magnetic resonance (NMR), protein microarrays, two-hybrid screening, etc. are used for protein identification. MS is a technique that is useful for determining the size of a protein or protein complex. X-ray crystallography and NMR are techniques useful for determining the 3-D structure of a protein or protein complex, while protein microarrays and two-hybrid screening are useful for determining protein–protein interactions. Proteome study reveals the protein diversity, abundance, isoforms, localization, interactions with other proteins, and post-translational modifications (PTMs). Compared with mRNA that is relatively less stable to environmental fluctuations, the proteins are the direct effectors of the plant responses to developmental or environmental changes. Proteomics offers insights that can enhance crop production, quality and resilience. By studying the protein profiles of plants, we can define the adaptive mechanism triggered in plant in response to a particular stress. Proteomics has novel potential in unfolding the response mechanism of plants towards abiotic stresses like drought, heat and temperature extremes (Table 11.3). Understanding these responses can aid in developing crops that can thrive in challenging environmental conditions. Proteomics reveals information on proteins and their interactions in responses to abiotic stresses (Fig. 11.1). By analyzing protein expression changes and the interactive mechanism of these proteins under different environmental conditions, we can identify the molecular basis of adaptive traits that enable plants to survive and thrive in a changing climate. Proteomics can also help in defining specific proteins as biomarkers that can be used in selective breeding or genetic engineering to develop resilient crop varieties more efficiently. Studying the proteome can provide insights into metabolic pathways that are activated under stress. Understanding these adaptive mechanisms can guide interventions to improve stress tolerance and allows for the breeding of crops that can show enhanced resilience to better withstand climate-related challenges. In addition to the direct application to crops, proteomics can also be applied to understanding the complex interactions between plants and beneficial microbes, as well as the response of entire agroecological systems to climate change.

11.6 Metabolomics

Metabolomics is the comprehensive study of small molecules (metabolites) involved in metabolic processes and their interaction within cells, biofluids, tissues or organisms. The term 'metabolomics' was first proposed in 1999 by Professor Nicholson. The primary objective of metabolomics is to identify the metabolites corresponding to a particular phenotype, and define the involved mechanisms in the development of that particular phenotype, providing insights into the physiological state of a cell, tissue, or organism. Metabolomics play a crucial role in enhancing climate resilience by providing insights into how plants respond to climate change. The metabolite profile of a plant changes when it encounters any stress; by comparing such a profile with the normal profile we can define the mechanism of a plant response towards various climatic stress like drought, heat, chilling injury, floods, etc. This information can guide the breeding of more resilient crops. By optimizing the metabolite profiles of crops for better yields and nutritional quality under climate stress, metabolomics can contribute to

Table 11.3. Successful examples of proteomic approaches applied to cereals. Author's own table.

Proteomic techniques	Target trait	Crop	References
Proteomic analyses using HPLC and LC tandem mass spectrometry	Heat stress	Wheat	Fei *et al.*, 2022
Proteomic analyses using 2-D electrophoresis and matrix-assisted laser desorption/ionization – time of flight (MALDI–TOF)	Heat stress	Wheat	Majoul *et al.*, 2004; Nandha *et al.*, 2018
	Water-deficit stress		Shayan *et al.*, 2020
	Drought stress		Şen *et al.*, 2024
	Drought stress	Barley	Rodziewicz *et al.*, 2019
	Heat	Maize	Abou-Deif *et al.*, 2019
Proteome analysis using tags for relative and absolute quantitation isobaric (iTRAQ)	Post-anthesis heat stress	Wheat	Zhang *et al.*, 2017b
	Cold stress		Xu *et al.*, 2022
	Drought stress	Maize	Li *et al.*, 2021; Gao *et al.*, 2023
	Low light stress	Rice	Liu *et al.*, 2020
	Low temperature stress		Qing *et al.*, 2022
iTRAQ and virus-induced gene silencing	Cold stress	Wheat	Zhang *et al.*, 2017a
Proteome analysis using 2D-difference gel electrophoresis and MALDI–TOF	Cold stress	Wheat	Kosová *et al.*, 2013
	Drought stress		Lu *et al.*, 2022
	Drought and heat stress	Barley	Rollins *et al.*, 2013
Proteome analysis using 2D-PAGE and liquid chromatography-electrospray ionization–mass spectrometry	Drought stress	Wheat	Michaletti *et al.*, 2018
Proteome analysis using 2-D electrophoresis and MALDI–TOF	Heat stress	Barley	Süle *et al.*, 2004;
	Drought stress		Chmielewska *et al.*, 2016;
	Hypoxic stress during waterlogging		Luan *et al.*, 2018
	Chilling stress	Rice	Lee *et al.*, 2009
	Drought stress		Agrawal *et al.*, 2016
Proteome analysis using 2-D electrophoresis and ion-trap mass spectrometry	Ultraviolet-B radiation stress	Rice	Sah *et al.*, 2022
Proteome analysis using liquid chromatography–mass spectrometry	Osmotic stress	Barley	Wang *et al.*, 2020
Proteome analysis using 2D-PAGE and MALDI-TOF	Low-temperature stress	Maize	Ramazan *et al.*, 2023
Proteome analysis using nanoflow liquid chromatography–tandem mass spectrometry	Drought stress	Rice	Hamzelou *et al.*, 2020

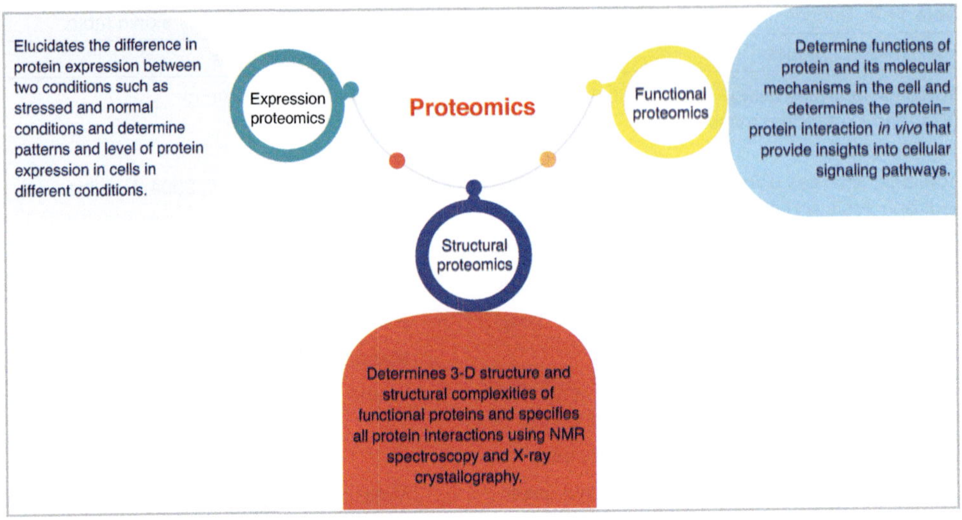

Fig. 11.1. Different proteomics approaches used in cereals. Author's own image.

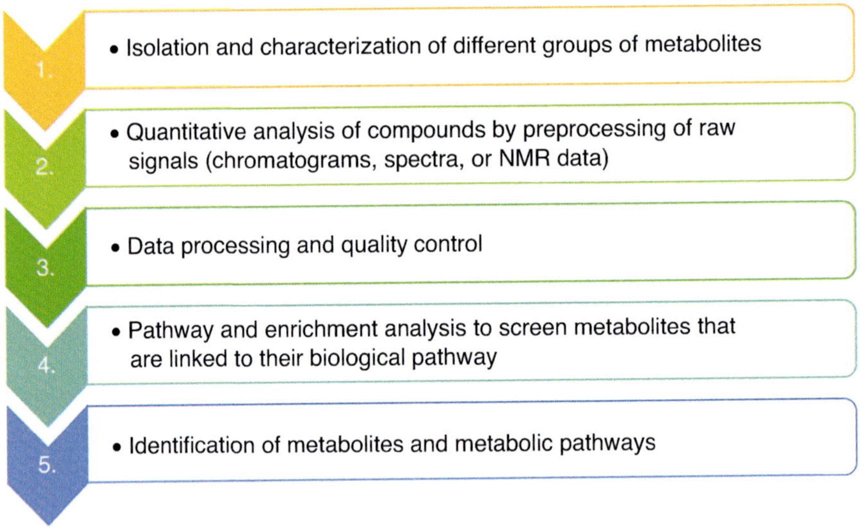

Fig. 11.2. Steps involved in metabolomics. Author's own image.

strategies for ensuring food security in a changing environment (Table 11.4; Fig. 11.2). MS and NMR are commonly employed techniques in the identification and quantification of various classes of metabolites, including sugars, organic acids, amino acids, lipids, and secondary metabolites. MS-based metabolomics typically uses liquid chromatography (LC) or gas chromatography separation to reduce sample complexity before MS analysis. MS allows for the reliable identification of metabolites, especially when it is used in tandem with chromatographic separation methods. It supports selective qualitative and quantitative analyses, making it a widely used technique. The alternative approach, NMR, is a non-destructive and highly reproducible spectroscopic technique to quantify the concentration and characterize the chemical structure of metabolites based on the principle of energy absorption and re-emission by atomic nuclei due

Table 11.4. Successful examples of metabolic approaches applied to cereals. Author's own table.

Metabolomic techniques	Target trait	Crop	References
GC–MS-based metabolite profiling	Submergence	Rice	Fukushima *et al.*, 2020
	High night-time temperature stress	Rice	Dhatt *et al.*, 2019
Non-targeted metabolite profiling was carried out using gas chromatography–mass spectrometry (GC–TOF–MS)	Drought stress	Rice	Ghorbanzadeh *et al.*, 2023
Integrated 16S and metabolomics study using liquid chromatography–mass spectrometry (LC–MS)	Drought stress	Rice	Sun *et al.*, 2023
GC–MS based metabolite profiling	Drought stress	Rice	Casartelli *et al.*, 2018
		Wheat	Bowne *et al.*, 2012; Fadoul *et al.*, 2021
		Barley	Chmielewska *et al.*, 2016
		Triticeae	Ullah *et al.*, 2017
HPLC-based metabolite profiling	High-temperature stress	Rice	Mishra *et al.*, 2024
	Drought stress	Maize	Gaffney *et al.*, 2021
LC–MS/MS based metabolite profiling	Drought and heat stress	Rice	Da Costa *et al.*, 2022
	Cold stress		Gu *et al.*, 2023
	High temperature	Wheat	Matsunaga *et al.*, 2021
	Drought stress		Michaletti *et al.*, 2018; Itam *et al.*, 2020
	Drought stress	Barley	Wang *et al.*, 2024b
Real-time metabolomics using picoliter pressure probe electrospray ionization (picoPPESI)–MS	Heat stress	Rice	Wada *et al.*, 2013
Gas chromatography–electron impact ionization TOF/MS	Drought and heat stress	Rice	Lawas *et al.*, 2019
Extractive electrospray ionization mass spectrometry based metabolomic analysis	Chilling stress	Rice	Du *et al.*, 2020
Ultra performance liquid chromatography (UPLC)/ tandem mass spectrometry	Chilling stress	Rice	Zhang *et al.*, 2016
	Heat tolerance	Wheat	Luo *et al.*, 2022
Untargeted metabolomics liquid chromatography high resolution mass spectrometry	Post anthesis heat stress	Wheat	Thomason *et al.*, 2018
UPLC–MS/MS system	Drought stress	Wheat	Lv *et al.*, 2022
	Heat and drought stress	Maize	Li *et al.*, 2022
	Waterlogging	Barley	Wang *et al.*, 2024a

to variations in an external magnetic field. A quantitative analysis of metabolite composition provides insights into the fundamental aspects of dynamic changes in metabolite profiles under different conditions. It also facilitates the identification of metabolite biomarkers associated with specific physiological processes during stress conditions. Knowledge of these markers supports precise and rapid development of cereal crops with enhanced climatic resilience. In plant metabolomics, researchers utilize both targeted and untargeted approaches to analyze the diverse metabolites in plant tissues. In targeted metabolomics, some specific set of known metabolites, such as flavonoids and alkaloids, that are common in some known metabolic pathways of plant defense against a stress are identified and quantified. In untargeted metabolomics, an exploratory approach is followed to study the entire set of metabolites under a particular condition. This method facilitates the discovery of novel compounds and pathways involved in plant responses to environmental stresses.

11.7 Conclusions

It is evident that elite cereal crop varieties that have resilience to climatic changes are crucially needed to withstand climate change. However, because of the rate at which the climate is changing it is difficult to develop crop varieties with the desired set of characteristics using conventional methods of crop improvement. Thus, for the development of superior cultivars at an accelerated rate, novel technological interventions are required to reinforce the weakened lines of otherwise robust conventional methods of crop improvement. Omics-assisted approaches are the most judicious alternative as they provide new insights into observed phenotypic changes under normal versus stressed environments and the underlying genotypic modifications. The various omics approaches such as genomics, transcriptomics, proteomics, metabolomics, etc. are pivotal in guiding crop improvement programs, especially those that would capitalize on the development of high yielding cereal cultivars with improved adaptation to climatic changes.

References

Abou-Deif, M.H., Rashed, M.A.S., Khalil, K.M. and Mahmoud, F.E.S. (2019) Proteomic analysis of heat shock proteins in maize (*Zea mays* L.). *Bulletin of the National Research Centre* 43, 1–9.

Agrawal, L., Gupta, S., Mishra, S.K., Pandey, G., Kumar, S. *et al.* (2016) Elucidation of complex nature of PEG induced drought-stress response in rice root using comparative proteomics approach. *Frontiers in Plant Science* 7, 1466.

Barratt, L.J., Reynolds, I.J., Franco Ortega, S. and Harper, A.L. (2023) Transcriptomic and co-expression network analyses on diverse wheat landraces identifies candidate master regulators of the response to early drought. *Frontiers in Plant Science* 14, 1212559.

Bhusal, N., Sarial, A.K., Sharma, P. and Sareen, S. (2017) Mapping QTLs for grain yield components in wheat under heat stress. *PLoS ONE* 12(12), e0189594.

Bowne, J.B., Erwin, T.A., Juttner, J., Schnurbusch, T., Langridge, P. *et al.* (2012) Drought responses of leaf tissues from wheat cultivars of differing drought tolerance at the metabolite level. *Molecular Plant* 5(2), 418–429.

Casartelli, A., Riewe, D., Hubberten, H.M., Altmann, T., Hoefgen, R. *et al.* (2018) Exploring traditional aus-type rice for metabolites conferring drought tolerance. *Rice* 11, 1–16.

Chmielewska, K., Rodziewicz, P., Swarcewicz, B., Sawikowska, A., Krajewski, P. *et al.* (2016) Analysis of drought-induced proteomic and metabolomic changes in barley (*Hordeum vulgare* L.) leaves and roots unravels some aspects of biochemical mechanisms involved in drought tolerance. *Frontiers in Plant Science* 7, 1108.

Da Costa, M.V.J., Ramegowda, V., Ramakrishnan, P., Nataraja, K.N. and Sheshshayee, M.S. (2022) Comparative metabolite profiling of rice contrasts reveal combined drought and heat stress signatures in flag leaf and spikelets. *Plant Science* 320, 111262.

Dhatt, B.K., Abshire, N., Paul, P., Hasanthika, K., Sandhu, J. *et al.* (2019) Metabolic dynamics of developing rice seeds under high night-time temperature stress. *Frontiers in Plant Science* 10, 1443.

Du, S., Cui, M., Cai, Y., Xue, A., Hao, Y. *et al.* (2020) Metabolomic analysis of chilling response in rice (*Oryza sativa* L.) seedlings by extractive electrospray ionization mass spectrometry. *Environmental and Experimental Botany* 180, 104231.

Fadoul, H.E., Martínez Rivas, F.J., Neumann, K., Balazadeh, S., Fernie, A.R. *et al.* (2021) Comparative molecular and metabolic profiling of two contrasting wheat cultivars under drought stress. *International Journal of Molecular Sciences* 22(24), 13287.

Fei, L., Chu, J., Zhang, X., Dong, S., Dai, X. *et al.* (2022) Physiological and proteomic analyses indicate delayed sowing improves photosynthetic capacity in wheat flag leaves under heat stress. *Frontiers in Plant Science* 13, 848464.

Fukushima, A., Kuroha, T., Nagai, K., Hattori, Y., Kobayashi, M. *et al.* (2020) Metabolite and phytohormone profiling illustrates metabolic reprogramming as an escape strategy of deepwater rice during partially submerged stress. *Metabolites* 10(2), 68.

Gaffney, I., Sallach, J.B., Wilson, J., Bergström, E. and Thomas-Oates, J. (2021) Metabolomic approaches to studying the response to drought stress in corn (*Zea mays*) cobs. *Metabolites* 11(7), 438.

Gao, C., Peng, X., Zhang, L., Zhao, Q., Ma, L. *et al.* (2023) Proteome and *ubiquitylome* analyses of maize endoplasmic reticulum under heat stress. *Genes* 14(3), 749.

Ghorbanzadeh, Z., Hamid, R., Jacob, F., Zeinalabedini, M., Salekdeh, G.H. *et al.* (2023) Comparative metabolomics of root-tips reveals distinct metabolic pathways conferring drought tolerance in contrasting genotypes of rice. *BMC Genomics* 24(1), 152.

Gu, S., Zhuang, J., Zhang, Z., Chen, W., Xu, H. *et al.* (2023) Multi-omics approach reveals the contribution of OsSEH1 to rice cold tolerance. *Frontiers in Plant Science* 13, 1110724.

Hamzelou, S., Pascovici, D., Kamath, K.S., Amirkhani, A., McKay, M. *et al.* (2020) Proteomic responses to drought vary widely among eight diverse genotypes of rice (*Oryza sativa*). *International Journal of Molecular Sciences* 21(1), 363.

Han, X., Zhang, D., Hao, H., Luo, Y., Zhu, Z. *et al.* (2023) Transcriptomic analysis of three differentially senescing maize (*Zea mays* L.) inbred lines upon heat stress. *International Journal of Molecular Sciences* 24(12), 9782.

Hu, Z., Song, N., Zheng, M., Liu, X., Liu, Z. *et al.* (2015) Histone acetyltransferase GCN5 is essential for heat stress-responsive gene activation and thermotolerance in *Arabidopsis*. *Plant Journal* 84, 1178–1191. DOI: 10.1111/tpj.13076.

Itam, M., Mega, R., Tadano, S., Abdelrahman, M., Alam, M.U. *et al.* (2020) Metabolic and physiological responses to progressive drought stress in bread wheat. *Scientific Reports* 10, 17189. DOI: 10.1038/s41598-020-74303-6.

Jagtap, A.B., Yadav, I.S., Vikal, Y., Praba, U.P., Kaur, N. *et al.* (2023) Transcriptional dynamics of maize leaves, pollens and ovules to gain insights into heat stress-related responses. *Frontiers in Plant Science* 14, 1117136.

Kapazoglou, A., Drosou, V., Argiriou, A. and Tsaftaris, A.S. (2013) The study of a barley epigenetic regulator, HvDME, in seed development and under drought. *BMC Plant Biology* 13, 172. DOI: 10.1186/1471-2229-13-172.

Kaur, S., Seem, K., Duhan, N., Kumar, S., Kaundal, R. *et al.* (2023) Transcriptome and physio-biochemical profiling reveals differential responses of rice cultivars at reproductive-stage drought stress. *International Journal of Molecular Sciences* 24(2), 1002.

Kim D., Alptekin B., Budak H. (2018). CRISPR/Cas9 genome editing in wheat. FunctIntegr Genomics 18, 31–41. DOI: 10.1007/s10142-017-0572-x..

Kosová, K., Vítámvás, P., Planchon, S., Renaut, J., Vanková, R. *et al.* (2013) Proteome analysis of cold response in spring and winter wheat (*Triticum aestivum*) crowns reveals similarities in stress adaptation and differences in regulatory processes between the growth habits. *Journal of Proteome Research* 12(11), 4830–4845.

Lawas, L.M.F., Li, X., Erban, A., Kopka, J., Jagadish, S.K. *et al.* (2019) Metabolic responses of rice cultivars with different tolerance to combined drought and heat stress under field conditions. *GigaScience* 8(5), giz050.

Lee, D.G., Ahsan, N., Lee, S.H., Lee, J.J., Bahk, J.D. *et al.* (2009) Chilling stress-induced proteomic changes in rice roots. *Journal of Plant Physiology* 166(1), 1–11.

Li, H., Yang, M., Zhao, C., Wang, Y. and Zhang, R. (2021) Physiological and proteomic analyses revealed the response mechanisms of two different drought-resistant maize varieties. *BMC Plant Biology* 21, 1–15.

Li, H., Tiwari, M., Tang, Y., Wang, L., Yang, S. *et al.* (2022) Metabolomic and transcriptomic analyses reveal that sucrose synthase regulates maize pollen viability under heat and drought stress. *Ecotoxicology and Environmental Safety* 246, 114191.

Liu, Q., Zhang, Y.C., Wang, C.Y., Luo, Y.C., Huang, Q.J. *et al.* (2009) Expression analysis of phytohormone regulated microRNAs in rice, implying their regulation roles in plant hormone signaling. *FEBS Letters* 583, 723–728.

Liu, Y., Pan, T., Tang, Y., Zhuang, Y., Liu, Z. *et al.* (2020) Proteomic analysis of rice subjected to low light stress and overexpression of OsGAPB increases the stress tolerance. *Rice* 13, 1–10.

Li, Y., Wang, X., Li, Y., Zhang, Y., Gou, Z. *et al.* (2020) Transcriptomic analysis revealed the common and divergent responses of maize seedling leaves to cold and heat stresses. *Genes* 11(8), 881.

Liu, J., Zhi, P., Wang, X., Fan, Q. and Chang, C. (2019) Wheat WD40-repeat protein TaHOS15 functions in a histone deacetylase complex to fine-tune defense responses to *Blumeria graminis* f.sp. *tritici*. *Journal of Experimental Botany* 70, 255–268. DOI: 10.1093/jxb/ery330.

Luan, H., Shen, H., Pan, Y., Guo, B., Lv, C. *et al.* (2018) Elucidating the hypoxic stress response in barley (*Hordeum vulgare* L.) during waterlogging: A proteomics approach. *Scientific Reports* 8(1), 9655.

Lu, F., Duan, W., Cui, Y., Zhang, J., Zhu, D. *et al.* (2022) 2D-DIGE based proteome analysis of wheat-*Thinopyrum* intermedium 7XL/7DS translocation line under drought stress. *BMC Genomics* 23(1), 369.

Luo, Y., Wang, Y., Xie, Y., Gao, Y., Li, W. *et al.* (2022) Transcriptomic and metabolomic analyses of the effects of exogenous trehalose on heat tolerance in wheat. *International Journal of Molecular Sciences* 23(9), 5194.

Lv, L., Chen, X., Li, H., Huang, J., Liu, Y. *et al.* (2022) Different adaptive patterns of wheat with different drought tolerance under drought stresses and rehydration revealed by integrated metabolomic and transcriptomic analysis. *Frontiers in Plant Science* 13, 1008624.

Mahalingam, R., Duhan, N., Kaundal, R., Smertenko, A., Nazarov, T. *et al.* (2022) Heat and drought induced transcriptomic changes in barley varieties with contrasting stress response phenotypes. *Frontiers in Plant Science* 13, 1066421.

Majoul, T., Bancel, E., Triboï, E., Ben Hamida, J. and Branlard, G. (2004) Proteomic analysis of the effect of heat stress on hexaploid wheat grain: Characterization of heat-responsive proteins from non-prolamins fraction. *Proteomics* 4(2), 505–513.

Mangelsen, E., Kilian, J., Harter, K., Jansson, C., Wanke, D. *et al.* (2011) Transcriptome analysis of high-temperature stress in developing barley caryopses: Early stress responses and effects on storage compound biosynthesis. *Molecular Plant* 4(1), 97–115.

Matsunaga, S., Yamasaki, Y., Mega, R., Toda, Y., Akashi, K. *et al.* (2021) Metabolome profiling of heat priming effects, senescence, and acclimation of bread wheat induced by high temperatures at different growth stages. *International Journal of Molecular Sciences* 22(23), 13139.

Michaletti, A., Naghavi, M.R., Toorchi, M., Zolla, L. and Rinalducci, S. (2018) Metabolomics and proteomics reveal drought-stress responses of leaf tissues from spring-wheat. *Scientific Reports* 8(1), 5710.

Mikołajczak, K., Kuczyńska, A., Krajewski, P., Kempa, M. and Nuc, M. (2023) Transcriptome profiling disclosed the effect of single and combined drought and heat stress on reprogramming of genes expression in barley flag leaf. *Frontiers in Plant Science* 13, 1096685.

Mishra, A., Singh, B.B., Shakil, N.A., Shamim, M.D., Homa, F. *et al.* (2024) Effect of high temperature stress on metabolome and aroma in rice grains. *Plant Gene* 38, 100450.

Nalini Chandran, A.K., Paul, P., Dhatt, B.K., Sandhu, J., Irvin, L. *et al.* (2024) Phenotypic and transcriptomic responses of diverse rice accessions to transient heat stress during early grain development. *Frontiers in Plant Science* 15, 1429697.

Nandha, A.K., Mehta, D.R., Tulsani, N.J., Umretiya, N. and Delvadiya, N. (2018) Proteomic analysis in wheat to study the effect of heat stress on flag leaf. *International Journal of Current Microbiology and Applied Sciences* 7(2), 3432–3439.

Paliwal, R., Röder, M.S., Kumar, U., Srivastava, J.P. and Joshi, A.K. (2012) QTL mapping of terminal heat tolerance in hexaploid wheat (*T. aestivum* L.). *Theoretical and Applied Genetics* 125, 561–575.

Paul, S., Duhan, J.S., Jaiswal, S., Angadi, U.B., Sharma, R. *et al.* (2022) RNA-Seq analysis of developing grains of wheat to intrigue into the complex molecular mechanism of the heat stress response. *Frontiers in Plant Science* 13, 904392.

Qian, Y., Ren, Q., Zhang, J. and Chen, L. (2019) Transcriptomic analysis of the maize (*Zea mays* L.) inbred line B73 response to heat stress at the seedling stage. *Gene* 692, 68–78.

Qing, D., Deng, G., Pan, Y., Gao, L., Liang, H. *et al.* (2022) ITRAQ-based quantitative proteomic analysis of japonica rice seedling during cold stress. *Breeding Science* 72(2), 150–168.

Qin, Q., Wang, Y., Huang, L., Du, F., Zhao, X. *et al.* (2020) A U-box E3 ubiquitin ligase OsPUB67 is positively involved in drought tolerance in rice. *Plant Molecular Biology* 102, 89–107. DOI: 10.1007/s11103-019-00933-8.

Ramazan, S., Jan, N. and John, R. (2023) Comparative protein analysis of two maize genotypes with contrasting tolerance to low temperature. *BMC Plant Biology* 23(1), 183.

Rangan, P., Furtado, A. and Henry, R. (2020) Transcriptome profiling of wheat genotypes under heat stress during grain-filling. *Journal of Cereal Science* 91, 102895.

Rodziewicz, P., Chmielewska, K., Sawikowska, A., Marczak, Ł., Łuczak, M. *et al.* (2019) Identification of drought responsive proteins and related proteomic QTLs in barley. *Journal of Experimental Botany* 70(10), 2823–2837.

Rollins, J.A., Habte, E., Templer, S.E., Colby, T., Schmidt, J. *et al.* (2013) Leaf proteome alterations in the context of physiological and morphological responses to drought and heat stress in barley (*Hordeum vulgare* L.). *Journal of Experimental Botany* 64(11), 3201–3212.

Roy, A., Dutta, S., Das, S. and Choudhury, M.R. (2024) Next-generation sequencing in the development of climate-resilient and stress-responsive crops–a review. *The Open Biotechnology Journal* 18(1).

Sah, S.K., Jumaa, S., Li, J. and Reddy, K.R. (2022) Proteomic analysis response of rice (*Oryza sativa*) leaves to ultraviolet-B radiation stress. *Frontiers in Plant Science* 13, 871331.

Santosh Kumar, V.V., Verma, R.K., Yadav, S.K., Yadav, P., Watts, A. *et al.* (2020) CRISPR-Cas9 mediated genome editing of drought and salt tolerance (OsDST) gene in Indica mega rice cultivar MTU101. *Physiology and Molecular Biology of Plants* 26(6), 1099–1110. DOI: 10.1007/s12298-020-00819-w.

Şen, A., Gümüş, T., Temel, A., Öztürk, İ. and Çelik, Ö. (2024) Biochemical and proteomic analyses in drought-tolerant wheat mutants obtained by gamma irradiation. *Plants* 13(19), 2702.

Seni, S., Kaur, S., Malik, P., Yadav, I.S., Sirohi, P. *et al.* (2021) Transcriptome based identification and validation of heat stress transcription factors in wheat progenitor species *Aegilops speltoides*. *Scientific Reports* 11(1), 22049.

Shayan, S., Norouzi, M., Vahed, M.M., Mohammadi, S.A. and Toorchi, M. (2020) Leaf proteome pattern of two bread wheat varieties under water deficit stress conditions. *Current Plant Biology* 23, 100146.

Shen, C., Que, Z., Xia, Y., Tang, N., Li, D. *et al.* (2017) Knock out of the annexin gene OsAnn3 *via* CRISPR/Cas9-mediated genome editing decreased cold tolerance in rice. *Journal of Plant Biology* 60, 539–547. DOI: 10.1007/s12374-016-0400-1.

Shi, J., Gao, H., Wang, H., Lafitte, R., Archibald, L. *et al.* (2017) ARGOS8 variants generated by CRISPR-Cas9 improve maize grain yield under field drought stress conditions. *Plant Biotechnology Journal* 15, 207–216. DOI: 10.1111/pbi.12603.

Süle, A., Vanrobaeys, F., Hajós, G.Y., Van Beeumen, J. and Devreese, B. (2004) Proteomic analysis of small heat shock protein isoforms in barley shoots. *Phytochemistry* 65(12), 1853–1863.

Sun, C., Wang, R., Tang, G., Cai, S., Shi, H. *et al.* (2023) Integrated 16S and metabolomics revealed the mechanism of drought resistance and nitrogen uptake in rice at the heading stage under different nitrogen levels. *Frontiers in Plant Science* 14, 1120584.

Thapa, R., Tabien, R.E., Johnson, C.D. and Septiningsih, E.M. (2023) Comparative transcriptomic analysis of germinating rice seedlings to individual and combined anaerobic and cold stress. *BMC Genomics* 24(1), 185.

Thomason, K., Babar, M.A., Erickson, J.E., Mulvaney, M., Beecher, C. *et al.* (2018) Comparative physiological and metabolomics analysis of wheat (*Triticum aestivum* L.) following post-anthesis heat stress. *PLoS ONE* 13(6), e0197919.

Ullah, N., Yüce, M., Neslihan Öztürk Gökçe, Z. and Budak, H. (2017) Comparative metabolite profiling of drought stress in roots and leaves of seven *Triticeae* species. *BMC Genomics* 18, 1–12.

Wada, H., Hatakeyama, Y., Nakashima, T., Nonami, H., Erra-Balsells, R. *et al.* (2013) On-site single pollen metabolomics reveals varietal differences in phosphatidylinositol synthesis under heat stress conditions in rice. *Scientific Reports* 10(1).

Wang, F., Zhou, Z., Liu, X., Zhu, L., Guo, B. *et al.* (2024a) Transcriptome and metabolome analyses reveal molecular insights into waterlogging tolerance in Barley. *BMC Plant Biology* 24(1), 1–15.

Wang, J., Yao, L., Hao, J., Li, C., Li, B. *et al.* (2024b) Growth properties and metabolomic analysis provide insight into drought tolerance in barley (*Hordeum vulgare* L.). *International Journal of Molecular Sciences* 25(13), 7224.

Wang, Y., Sang, Z., Xu, S., Xu, Q., Zeng, X. *et al.* (2020) Comparative proteomics analysis of Tibetan hull-less barley under osmotic stress via data-independent acquisition mass spectrometry. *GigaScience* 9(3), giaa019.

Wei, L., Zhang, D., Xiang, F. and Zhang, Z. (2009) Differentially expressed miRNAs potentially involved in the regulation of defense mechanism to drought stress in maize seedlings. *International Journal of Plant Sciences* 170, 979–989.

Xia, K., Wang, R., Ou, X., Fang, Z., Tian, C. *et al.* (2012) OsTIR1 and OsAFB2 downregulation via osmiR393 overexpression leads to more tillers, early flowering and less tolerance to salt and drought in rice. *PLoS ONE* 7, e30039.

Xia, Y., Li, R., Ning, Z., Bai, G., Siddique, K.H. *et al.* (2013) Single nucleotide polymorphisms in *HSP17.8* and their association with agronomic traits in Barley. *PLoS ONE* 8, e56816.

Xu, K., Zhao, Y., Gu, J., Zhou, M., Gao, L. *et al.* (2022) Proteomic analysis reveals the molecular mechanism underlying the cold acclimation and freezing tolerance of wheat (*Triticum aestivum* L.). *Plant Science* 318, 111242.

Yan, T., Sun, M., Su, R., Wang, X., Lu, X. *et al.* (2023) Transcriptomic profiling of cold stress-induced differentially expressed genes in seedling stage of Indica rice. *Plants* 12(14), 2675.

Yu, S., Liao, F., Wang, F., Wen, W., Li, J. *et al.* (2012) Identification of rice transcription factors associated with drought tolerance using the EcoTILLING method. *PLoS ONE* 7, e30765.

Zakavi, M., Tohidfar, M. and Javadi, S.M. (2019) In silico analysis and gene expression of heat stress responses genes in *Hordeum vulgare* L. *Biocatalysis and Agricultural Biotechnology* 18, 101061.

Zeng, Y., Wen, J., Zhao, W., Wang, Q. and Huang, W. (2020) Rational improvement of rice yield and cold tolerance by editing the three genes OsPIN5b, GS3, and OsMYB30 with the CRISPR–Cas9 system. *Frontiers in Plant Science* 10, 933–946. DOI: 10.3389/fpls.2019.01663.

Zhang, J., Luo, W., Zhao, Y., Xu, Y., Song, S. *et al.* (2016) Comparative metabolomic analysis reveals a reactive oxygen species-dominated dynamic model underlying chilling environment adaptation and tolerance in rice. *New Phytologist* 211(4), 1295–1310.

Zhang, N., Zhang, L., Zhao, L., Ren, Y., Cui, D. *et al.* (2017a) iTRAQ and virus-induced gene silencing revealed three proteins involved in cold response in bread wheat. *Scientific Reports* 7(1), 7524.

Zhang, Y., Pan, J., Huang, X., Guo, D., Lou, H. *et al.* (2017b) Differential effects of a post-anthesis heat stress on wheat (*Triticum aestivum* L.) grain proteome determined by iTRAQ. *Scientific Reports* 7(1), 3468.

Zhao, P., Sun, L., Zhang, S., Jiao, B., Wang, J, *et al.* (2024) Integrated transcriptomics and metabolomics analysis of two maize hybrids (ZD309 and XY335) under heat stress at the flowering stage. *Genes* 15(2), 189.

Zhao, Q., Feng, Q., Lu, H., Li, Y., Wang, A. *et al.* (2018) Pan-genome analysis highlights the extent of genomic variation in cultivated and wild rice. *Nature Genetics* 50(2), 278–284.

Zheng, M., Liu, X., Lin, J., Liu, X., Wang, Z. *et al.* (2019) Histone acetyltransferase GCN5 contributes to cell wall integrity and salt stress tolerance by altering the expression of cellulose synthesis genes. *The Plant Journal* 97, 587–602.

Zhou, L.G., Liu, Y.H., Liu, Z.C., Kong, D.Y., Duan, M. *et al.* (2010) Genome-wide identification and analysis of drought-responsive microRNAs in *Oryza sativa*. *Journal Experimental Botany* 61, 4157–4168.

12 Multilevel Applications of Omics Technology in Climate-resilient Horticulture

Afiya Rajan Selin[1], Akshara Kampurath[1], Perinba C. Suresh[2], Subramanian Manivannan[1], Senthilkumar Shanmugasundram[1]* and Abinaya Karunakaran[1]
[1]Department of Horticulture, Central University of Tamil Nadu, Thiruvarur, India;
[2]Department of Horticulture, North-Eastern Hill University, Tura, India

Abstract

Global food production has faced previously unheard-of difficulties in recent years due to climate change, especially for horticultural crops that are extremely vulnerable to changes in the environment. Because they may reveal the molecular, physiological, and metabolic reactions to environmental stressors, omics technologies such as proteomics, transcriptomics, ionomics, phenomics, and genomics have become effective instruments for accelerating the conception of environmentally resilient crops. The current chapter presents various analyses with the integration of multi-level omics technologies to increase the resilience of horticulture crops and emphasize how omics has helped discover genes and traits that can withstand stress, clarify intricate biological systems, and aid in the breeding of climate-resilient cultivars. The chapter also explores the applications and examples demonstrating how various omics platforms have been utilized to enhance tolerance to pest infestations, heat, salinity, and drought stressors. It discusses the future prospects and challenges in this field, highlighting the importance of applying systems biology approaches and integrating omics data to develop sustainable horticultural practices that can address the impacts of climate change.

12.1 Introduction

Horticultural crops provide essential food, fibers and ornamental products constituting a significant portion of global agricultural production. As global population numbers increase and environmental pressures mount, the challenge of improving these crops to meet food security and sustainability goals becomes more urgent. While traditional breeding methods have contributed to crop improvement, they are slow and labor intensive. Global horticulture is seriously threatened by climate change, affecting crop productivity, quality, and sustainability (Hatfield and Prueger, 2015). Climate change effects are becoming more and more obvious, and their detrimental effects on horticulture crops are especially worrisome. Because horticulture supplies vital vitamins, minerals, and other nutrients to people all over the world, it plays a major role in ensuring food security globally. However, environmental stressors including pests, salt, temperature fluctuations, and drought can affect horticulture crops. Food production and agricultural livelihoods are at risk because these pressures frequently result in lower crop

*Corresponding author: senthilshanmugam87@gmail.com

 From Gene Discovery to Climate-Resilient Crops: Omics in Crop Improvement (eds S. Pandey *et al.*)
DOI: 10.1079/9781800628717.0012

yields, worse quality, and higher susceptibility to diseases.

Omics technologies, namely genomics, transcriptomics, proteomics, ionomics, metabolomics, phenomics, offer comprehensive tools to understand plant responses and accelerate breeding for stress tolerance. By combining these tools, scientists can better understand the molecular mechanisms behind climate resilience by identifying important genes, proteins, and metabolites in plants' responses to environmental stress. Horticultural breeding might be revolutionized by omics technology, which makes it possible to precisely identify advantageous characteristics for stress tolerance. Additionally, integrating multiomics data provides a broader understanding of plant responses to climate fluctuations. This chapter aims to explore how multilevel omics approaches can be used to improve the climate resistance of horticultural crops.

12.2 Genomics

Genomics refers to the learning of an organism's complete genetic material, encompassing the genes, regulatory regions, and non-coding sequences within its genome. It plays a crucial role in the identification of genetic variation and the discovery of functional genes that contribute to key traits in horticultural crops. Horticultural plants, including fruits, vegetables, ornamentals, and medicinal herbs, play crucial roles in human life by providing food, aesthetic value, and ecological benefits. The study of horticulture genomics has significantly improved with the development of sequencing tools, which have shown the genetic complexity of these plants. Many horticultural species exhibit high heterozygosity and polyploidy due to their long evolutionary and domestication histories. A deeper understanding of their genetic diversity has been made possible by recent advances in genome sequencing and pangenome reconstruction, which have helped identify crucial features for breeding and conservation. Furthermore, the creation of horticulture genome databases has made it easier to store, share, and visualize data effectively, assisting with international research initiatives.

Domestication and crop improvement have been central themes in horticultural genomics, with genome-wide association studies (GWAS), quantitative trait locus (QTL) mapping, and pangenomics playing key roles. Studying domestication syndromes in crops like tomato and grape has provided valuable insights into genetic changes associated with human selection. Advances in molecular techniques have enabled researchers to identify genes responsible for desirable traits, such as enhanced yield, disease resistance, and stress tolerance. Understanding the genetic basis of domestication allows for the targeted improvement of horticultural crops while preserving genetic diversity. Functional genomics studies have further expanded knowledge of specific gene families that influence important traits in horticultural plants. For example, the NRAMP gene family in *Kandelia obovata* has been linked to metal transport and stress resilience, while the PEBP gene family in pineapple regulates flowering time. Growth-regulating factor (GRF) proteins in pineapple also play a role in floral organ development and gibberellin response. Additionally, transcriptomic studies on tea plants have uncovered molecular pathways responsible for selenium accumulation, highlighting the potential health benefits of certain horticultural crops. Proteomics and metabolomics approaches have also been applied to crops like kale, revealing how light conditions influence its nutritional composition and categorization into distinct metabolic groups.

Genome editing technologies, including TALENs, ZFNs and CRISPR-Cas9, have revolutionized horticultural crop improvement by enabling precise modifications of plant genomes. These tools offer the potential to enhance disease resistance, improve nutritional content, and increase stress tolerance in a more efficient manner than traditional breeding methods. CRISPR-Cas9, in particular, has been widely used to introduce beneficial traits with high precision. However, challenges like off-target effects, regulatory considerations, and delivery methods remain obstacles to widespread adoption. Despite these challenges, genome editing presents a capable avenue for progressing horticulture in a sustainable manner. Future directions in horticultural genomics will focus on large-scale sequencing projects, the integration of multiomics approaches, and the development of advanced breeding strategies. Expanding genome databases and bioinformatics tools will be essential for improving data accessibility and analysis. Additionally, genome editing and synthetic biology approaches may further enhance crop resilience, quality, and nutritional value. By leveraging genomic research, scientists can develop innovative solutions to address global food security challenges while promoting sustainable agricultural practices. Through the integration of cutting-edge molecular technology and traditional wisdom, horticultural genomics is positioned to play a significant role in determining the direction of agriculture in the future.

Table 12.1. Targeted genes in fruit crops and their traits through genomics (Wang *et al.*, 2023).

Crops	Target gene	Character
Apple	*MdMYB1*	Anthocyanin biosynthesis
Pear	*PyMYB114* interacts with other transcription factors (*PybHLH3 and PyERF3*)	Anthocyanin biosynthesis
Peach	BLOOD (BL) with PpNAC1 to activate transcription of *PpMYB10.1*	Anthocyanin accumulation
	PpOFP1	Flat-fruit shape formation
Citrus	Ruby2-Ruby1 cluster	Anthocyanin biosynthesis
	CitRWP	Polyembryony

12.3 Applications of Genomics in Horticulture

12.3.1 Genome sequencing

Recent advancements in sequencing technologies, particularly next-generation sequencing (NGS), have facilitated comprehensive genome sequencing across various horticultural crops (Tables 12.1 and 12.2). These genomic sequences

Table 12.2. Targeted genes in vegetable crops and their traits through genomics (adapted from Devi *et al.*, 2022 CC BY 4.0).

Crops	Target genes	Characters
Tomato	SlHyPRP1	Salt tolerance
	K transporter HKT 1,2	Salt tolerance
	BZR 1	Heat stress tolerance
	SlMAPK3	Drought stress response
	SlCBF	Decrease stress in chilling tolerance
	SlPelo and SlMlo1	Tomato yellow leaf curl virus (TYLCV) and powdery mildew fungus
	Coat protein-replicase	Resistance yellow leaf curl virus
	SlMlo1	Resistance to powdery mildew
	SlGAI	Gibberellin responsive dwarf mutant
	SGR1, Blc, LCY-E, LCY-B1, LCY-B2	Increase lycopene content
	SlMYB12	Pink fruit color
Potato	Coilin gene	Biotic (PVY) and abiotic stress resistance
	StDNDI,StCHL 1 and DMG400000582(StDMRR6-1)	Late blight resistance
Pumpkin	GRF12, AHA1 and HAK5	Salt sensitivity
Carrot	DcMYB113-Like	Anthocyanin biosynthesis

are instrumental in uncovering the genetic foundations of traits associated with climate resilience, including drought tolerance, heat resistance, pest resistance, and salinity tolerance. For example, genomic insights have been gained in crops such as tomatoes (*Solanum lycopersicum*) (Schwarz *et al.*, 2014), apples (*Malus domestica*) and strawberries (*Fragaria* ×*ananassa*).

According to the Tomato Genome Consortium (2012), the genome of tomato was among the first horticultural crops to be sequenced. Afterwards, genes linked to drought and heat tolerance were identified. For example, the SlDREB2 gene, a member of the dehydration-responsive element-binding (DREB) family, plays a critical role in the tomato plant's response to drought. The genomes of several citrus species have been sequenced, making it possible to identify genes linked to stress tolerance. *Citrus sinensis* has shown a high genetic response to salinity stress and has been found to contain key genes that regulate ion homeostasis and osmotic balance. The identification of genes linked to stress tolerance, such as those encoding heat shock proteins and transcription factors triggered by stress, has been made easier by the sequencing of the grapevine genome. The creation of citrus cultivars with improved resistance to salinity and drought depends on these realizations. As a result, genome sequencing is essential for the development of molecular markers, gene discovery, and molecular breeding methods.

12.3.2 Genetic mapping and marker development

High-density genetic maps for a variety of horticulture crops have been made possible by genome sequencing data, and these maps are crucial for locating QTLs linked to important characteristics including fruit quality, yield, and disease and pest resistance. In breeding operations, genomic resources like as single nucleotide polymorphisms (SNPs) help marker-assisted selection (MAS) by making it easier to identify QTLs associated with desirable traits. For example, QTLs for disease resistance, fruit size, and color have been found in apples, enhancing marker-assisted breeding efforts.

12.3.3 CRISPR-Cas9 gene editing

The ability to precisely alter specific genes has revolutionized crop breeding by CRISPR-Cas9 and other gene-editing technologies. This method has been used to alter traits such as fruit ripening, pest resistance, and stress tolerance in horticultural crops. To enhance fruit quality attributes like shelf life and pest resistance, for instance, CRISPR has been utilized to modify the genes of tomatoes.

12.3.4 Comparative genomics for crop improvement (pangenomics and epigenomics)

Research on gene variations across cultivars and epigenetic changes under stress conditions has provided valuable insights (Springer and Schmitz, 2017). By comparing the genomes of different crop species, scientists are able to pinpoint conserved genetic regions and gene families that are essential for the growth and development of plants. The identification of genes linked to important horticultural characteristics such as fruit size, flavor, and ripening in crops like tomato and pepper has been made easier by this comparative research.

12.4 Transcriptomics: Decoding Gene Expression Under Stress

Transcriptomics is the learning of the complete set of RNA molecules produced by a plant under specific conditions. Transcriptomic studies help decode how plants respond to environmental stresses at the RNA level. Transcriptomic data can help to recognize stress-responsive genes, pathways, and regulatory networks that contribute to resilience. The genetic improvement of fruit and vegetable crops is crucial for acknowledging emerging challenges such as climate change, biotic and abiotic stresses, and increasing consumer demand for quality produce. Traditional breeding methods, though effective, are time-consuming due to factors like long juvenile phases, polyploidy, and self-incompatibility. The advent of biotechnological tools, particularly transcriptomics, has revolutionized crop improvement by enabling the study of gene expression patterns in response to developmental stages and conditions of environment. RNA sequencing (RNA-Seq) provides a high-resolution, comprehensive understanding of transcriptome dynamics, offering valuable insights for crop enhancement.

Transcriptomics has been widely applied to major fruit crops such as apple, banana, citrus, and grape to study metabolic changes during fruit development, ripening, and stress responses. This approach has also been instrumental in identifying genes linked to resistance mechanisms against pests and diseases. Similarly, in vegetable crops like tomato, cucumber, and *Brassica* spp., transcriptomic studies have facilitated the discovery of genes responsible for disease resistance, stress tolerance, and nutritional quality. The application of transcriptomics to underutilized crops can further contribute to food security by unlocking their genetic potential and enhancing their adaptability to various climatic conditions. The rapid advancements in RNA-Seq technology have significantly improved the accuracy and efficiency of transcriptomic studies. Single-cell RNA sequencing (scRNA-Seq) has emerged as a powerful tool to analyze gene expression variability at the cellular level, offering deeper insights into plant developmental biology. Moreover, bioinformatics tools play a crucial role in processing and analyzing vast transcriptomic datasets, helping researchers identify differentially expressed genes and functional pathways. Despite these advancements, challenges remain in terms of computational complexity, cost, and the need for well-annotated reference genomes for non-model horticultural crops.

Understanding complicated biological processes in plants is improved by combining transcriptomics with other omics technologies including proteomics, metabolomics, and genomics. This multiomics method makes it possible to investigate metabolic pathways in more detail, which results in focused breeding plans and the creation of hardy horticulture crops. The combination of transcriptomics with genome editing tools like CRISPR-Cas9 holds great potential for accelerating the breeding of high-yielding, stress-tolerant, and nutritionally superior crop varieties. As the demand for sustainable and resilient agricultural practices grows, transcriptomics will continue to be a vital tool in horticultural research. Future studies should focus on expanding transcriptomic analyses to a wider range of crops, especially those with economic and nutritional importance. Additionally, advancements in artificial intelligence (AI) and machine learning (ML) can aid in deciphering complex gene regulatory networks, making transcriptomic data more actionable for breeders and farmers. Horticultural transcriptomics will contribute significantly to agricultural sustainability and worldwide food security by utilizing these state-of-the-art technologies.

12.5 Key Implications for Transcriptomics in Horticultural Crops

12.5.1 Gene expression profiling

Identifying genes involved in stress responses, growth regulation, and fruit development is possible through the analysis of gene expression profiles. For example, genes linked to citrus greening disease and the reaction to cold stress have been found by RNA-Seq investigations in citrus. RNA-Seq has also been used in tomatoes to profile gene expression changes during ripening, helping to identify key genes involved in color change, texture, and flavor formation.

12.5.2 Developmental and environmental regulation of gene expression

Additionally, transcriptomic research is essential for comprehending how fruit development and ripening are regulated. The transcriptional regulation of fruit ripening and the way that particular genes govern the synthesis of chemicals that affect flavor and aroma in tomatoes have been clarified by RNA-Seq. Also in apple, the studies have provided insights into how genes involved in cell-wall metabolism and hormone regulation contribute to fruit ripening and softening. Similarly, transcriptomic studies in leafy vegetables like lettuce help identify genes responsible for the regulation of leaf morphology, which is critical for improving crop yields.

12.5.3 Non-coding RNAs (miRNAs and ncRNAs)

In horticultural crops, non-coding RNAs (ncRNAs), such as long lncRNAs and micro-RNAs (miRNAs), are essential for controlling a number of biological processes. By altering transcriptional and post-transcriptional gene expression, these ncRNAs affect stress responses, growth, development, and fruit ripening. For instance, ncRNAs control plant development processes including flowering and fruit maturity, whereas miRNAs regulate plant stress tolerance by targeting genes linked to abiotic stress responses. Better varieties with increased stress tolerance and higher quality features may result from an understanding of the activities of ncRNAs in horticulture crops.

12.5.4 Biotic and abiotic stress response

The study of transcriptomics examines the responses of horticulture crops to biotic (such as pests and diseases) and abiotic (such as temperature and drought) challenges (Tables 12.3 and 12.4). In order to enhance crops, stress-responsive genes and transcription factors can be found by gene expression profiling under stress. Key genes implicated in resistance to bacterial wilt, a significant disease in cucurbit crops, have been identified through investigations conducted on cucumbers. RNA-Seq research on cold stress in strawberries has identified important genes that respond to cold and can be targeted to improve cold tolerance.

Transcriptome analyses of lettuce (*Lactuca sativa*) have identified genes that are crucial

Table 12.3. Transcriptome analysis in various fruit crop genotypes (adapted from Singh and Pradhan, 2020).

Crops	Genotypes	Characters
Apple	Transgenic Gala apple	Transcriptome of Hcrvf12-transformed resistant line to Venturia inaequalis
	Mondial Gala	Changes during fruit maturation and ripening
	Golden Delicious	Response to inoculation to *Penicillium expansum* and *P. digitatum*
	Fuji	Resistance responses against Valsamali
	Dwarfing apple rootstock 'SH6'	Response to drought, cold and high salinity
Banana	Dwarf Cavendish	Fruit ripening
	Saba and Grand Naine	Response to drought stress
	Minhou wild banana (*M. itinerans*)	Anthocyanin biosynthesis genes
Citrus	*Citrus sp.*	Citrus bark cracking viroid (CBCVd)
Grape	*V.vinifera cv.* Cabernet Sauvignon and *V. aestivalis cv.* Norton	Berry peel development and flavonoid biosynthesis
	Muscat of Hamburg and *V. amurensis*	Genome wide transcriptional profile analysis in response to cold stress
	Thompson Seedless	Transcriptome analysis of vine under salinity and identification of key genes
Mango	Kent	Fruit ripening
	Dashehari	Ripening associated genes
	Neelam and Dashehari	Alternate bearing
	Amrapali	Floral malformation
Litchi	*Litchi chinensis* Sonn. cv. Nuomici	Fruit maturation
Kiwi	*Actinidia chinensis Planch.*	Anthocyanin accumulation
Pineapple	*Ananas comosus L. cv.* Smooth Cayenne	Response to ethephon induction

Table 12.4. Transcriptomic analysis for abiotic stress tolerance in vegetable crops (Mangal *et al*., 2023).

Crops	Target genes	Characters
Radish	NAU-RG	Lead (Pb) stress
Potato	Longshu 3	Drought
Onion	Acc. 1656 (T), Acc. 1627 (S)	Drought
Pepper	Bukang, ECW	Cold, drought, and salinity
	Cuneo CCu07	Drought
	A188 (T), A122 (S)	Cold
Potato	DM1-3 516 R44 clone	Heat, drought, and salinity
	Desiree (T), Unica (T), Agria (S), Russett Burbank (S)	Heat, drought
Garlic	T36 and T141	Salt
	Purple Glazer	Drought
Onion	36,122 (T), 36,101 (S)	Cold
Tomato	MicroTom	Salt and oxidative
Lettuce	KNV482 (T), KNV664 (T), KNV242 (S), KNV302 (S)	Heat

S, susceptible; T, tolerant.

for drought resistance mechanisms, including those that encode late embryogenesis abundant (LEA) proteins and drought-responsive transcription factors. Scientists can identify genes by looking at patterns of gene expression to create drought-resistant lettuce types. Moreover, genes linked to the plant's resistance to downy mildew have been found using transcriptome analysis in lettuce. Transcriptome investigations of citrus species have also shown genes associated with responses to drought and salt stress. For instance, in *Citrus sinensis*, the transcript levels of antioxidant enzymes such as catalase (CAT) and superoxide dismutase (SOD) are significantly upregulated under drought. Certain findings suggest that citrus may be more resilient to salinity and water deprivation if certain stress-related genes are expressed more effectively.

12.6 Proteomics: Identifying Stress-Responsive Proteins

The extensive study of proteins, their roles, and interactions is known as proteomics. Stress-responsive proteins, enzymes, and signaling molecules that mediate plant responses to environmental stresses can be found using proteomic analysis. These proteins play a key role in controlling stress tolerance, antioxidant defense systems, and metabolic activities. In horticulture research, proteomics is a potent technique that allows for a better knowledge of how plants react to abiotic stressors such salt, drought, severe temperatures, and nutritional imbalances. Through modifications to molecular and physiological processes, these environmental conditions have a substantial impact on crop output and quality. Researchers can find biomarkers and important metabolic processes involved in plant resilience by using proteomic studies, which concentrate on differentially expressed proteins that are essential for stress adaption. Advanced proteomic techniques, including mass spectrometry (MS) and bioinformatics, are now widely used to analyze stress-responsive proteins, providing valuable insights for improving crop performance under adverse conditions.

Horticultural crops, such as fruits, vegetables, and ornamentals, often experience stress-related metabolic disruptions that affect their growth and postharvest quality. Proteomics helps in identifying proteins linked to stress tolerance, nutrient absorption, and antioxidant mechanisms that safeguard plant health. Understanding these molecular changes allows researchers to develop biotechnological strategies, such as breeding stress-resistant varieties or engineering plants with enhanced tolerance. Additionally, post-translational modifications of proteins, crucial for stress signaling and metabolic regulation, are being extensively studied to optimize crop resilience and productivity. Recent advancements in quantitative proteomics, including techniques like isobaric tagging—isobaric tag for relative and absolute quantitation (iTRAQ)—and inductively coupled plasma–MS (ICP–MS), have accelerated the identification of novel proteins and metabolic pathways in horticultural crops. These findings contribute to improving agronomic traits, including disease resistance, shelf life, and stress adaptability. Proteomics also aids in understanding plant–pathogen interactions, enabling the development of disease-resistant cultivars by targeting key defense-related proteins. This approach is crucial for safeguarding horticultural crops from emerging threats and ensuring sustainable production.

In light of the growing need for crops that are both more adaptable to climate change and more nutritious, proteomics is essential to the advancement of horticulture science. Researchers are able to create thorough models of plant stress responses by combining proteomic data with other omics techniques like genomics and metabolomics. These discoveries aid in the creation of horticultural crops of the future that are more productive and more tolerant of the environment. The use of proteomic technologies in horticulture will greatly enhance sustainable agriculture methods and global food security as they develop further.

12.7 Key Contributions of Proteomics in Horticultural Crops

12.7.1 Protein identification

To find proteins linked to significant characteristics including yield, quality, and pest resistance, MS-based proteomics can be utilized (Table 12.5). For example, proteomic research on apples has found proteins associated with resistance to fungi like *Venturia inaequalis*. Proteomic analysis of tomatoes revealed a number of important proteins associated with controlling fruit ripening, providing information on how to increase fruit quality or shelf life after harvest.

12.7.2 Understanding disease-resistance mechanisms

Proteins that are essential for plant immune responses have been identified by proteomic research. Proteomic analysis, for instance, has

Table 12.5. Applications of proteomics in horticultural crops. Author's own table.

Crop	Characteristic	Achievement
Tomato, grapevine, strawberry	Stress tolerance (abiotic and biotic)	Identified stress responsive proteins involved in drought, heat, salt stress and resistance to pathogen (Kosová *et al.*, 2018)
Banana, apple, tomato	Fruit ripening and quality	Helped in understanding of fruit ripening and development mechanisms and postharvest physiology (Palma *et al.*, 2011)
Mango, peach, citrus	Postharvest shelf life and storage	Proteins related to ethylene signaling, senescence and oxidative stress regulation were identified
Tomato, grape and tea	Plant–microbe interactions	Plant defense and beneficial microbial associations that enhance growth and immunity were identified
Tea, ginseng, basil	Secondary metabolism and phytochemicals	Proteins involved in biosynthesis of flavonoids, anthocyanins and terpenoids for flavor and medicinal properties were identified (Yonekura-Sakakibara *et al.*, 2019)
Rose, orchid cucumber	Growth development regulation	Analysis of hormonal and signaling proteins regulating shoot, root and flower development were identified

been used to discover proteins in peppers that have been involved in the plant's defensive mechanism against bacterial wilt. These proteins include transcription factors that trigger defense genes and enzymes that break down pathogen cell walls.

12.7.3 Metabolic pathways and fruit quality

Fruit color, flavor, and nutritional value are influenced by secondary metabolites such as flavonoids, terpenes, and carotenoids, which been biosynthesized by proteins and studied using proteomics. According to proteomic research, proteins in tomatoes were involved in the synthesis of volatile chemicals that give the fruit its scent.

12.7.4 Protein–protein interactions (PPIs)

PPI includes understanding signaling pathways involved in heat- and drought-stress responses. In grapes, proteomics has been used to identify protein networks involved in the parameter of fruit development and ripening.

12.7.5 Stress response proteins

Proteomics is used to analyze stress-related proteins in horticultural crops. In citrus, proteomic studies have identified proteins involved in resistance to drought and salinity stress. For example, heat shock proteins (HSPs) are involved in stress responses that help maintain fruit quality under high temperatures.

HSP70 is essential for preserving fruit set and quality in fruit crops under heat stress, such as tomatoes. HSP70 and HSP90 in grapes are essential for shielding cellular structures from heat-induced damage. These HSPs contribute to the resistance of grapevine (*Vitis vinifera*) to heat stress by stabilizing enzymes and cellular membranes in high-temperature environments. HSPs are crucial for crops like lettuce and cabbage to withstand heat as well as other stresses like salinity and drought. For example, HSP101 has been found to be a crucial component of the ability of lettuce to withstand heat. Dehydration-responsive proteins and antioxidant enzymes are among the proteins in lettuce (*Lactuca sativa*) that are involved in osmotic regulation and defense against oxidative stress. Because they preserve cellular integrity and guard against oxidative damage, these proteins are essential in helping the plant deal with water scarcity.

12.8 Metabolomics

Metabolomics involves the learning of metabolites or small molecules such as lipids, amino acids, sugars, and secondary metabolites. Metabolomic analyses help in understanding the biochemical processes underlying plant growth, development, and stress responses. Metabolomics is a cutting-edge approach in horticultural research that involves the inclusive analysis of metabolites, the small molecules that drive plant growth, development, and stress responses. By studying these biochemical compounds, researchers can gain deeper insights into the metabolic pathways that regulate essential traits in fruits, vegetables, and ornamental plants. Metabolomics employs advanced analytical techniques such as MS and nuclear magnetic resonance (NMR) spectroscopy to profile metabolites, helping scientists understand plant physiology at a molecular level. This knowledge is crucial for improving crop quality, stress resilience, and postharvest management.

In horticulture, one of the main uses of metabolomics is enhancing crop quality and nutritional value. By identifying bioactive compounds responsible for flavor, aroma, color, and texture, researchers can breed or engineer crops with improved sensory attributes. Additionally, metabolomics helps in fortifying crops with essential nutrients and health-promoting compounds, making them more beneficial for human consumption. This approach has a crucial role in postharvest research, where metabolite profiling helps optimize storage conditions, prolong shelf life, and minimize losses due to spoilage. Understanding how horticulture crops react to environmental stressors like drought, salt, severe temperatures, and disease infections is another important application of metabolomics. Researchers can find biomarkers linked to stress tolerance and create more resilient crop varieties by examining metabolic changes under stress. Additionally, by identifying advantageous bacteria that improve nutrient uptake, disease resistance, and general plant health, metabolomics helps researchers understand plant–microbe interactions in the rhizosphere and phyllosphere. This information aids in the creation of environmentally friendly farming methods that lessen reliance on chemical inputs.

In modern breeding programs, metabolomics is increasingly integrated with genomics and transcriptomics to accelerate the development of superior horticultural crops. By mapping metabolite QTL and conducting metabolic GWAS, researchers can pinpoint genetic markers linked to desirable characters like enhanced yield, disease resistance, and improved nutritional content. The application of metabolomics in functional genomics and metabolic engineering further enables the targeted modification of metabolic pathways to produce high-value compounds. With its wide-ranging applications, metabolomics is transforming horticultural science, contributing to more sustainable and efficient crop production.

12.9 Metabolomic Contributions in Horticultural Crops

12.9.1 Metabolic profiling

The volatile organic molecules (VOCs) that give food its flavor and scent are examined using metabolomics. The production of sweeter, more delicious melons has resulted from the use of metabolomic analysis to pinpoint the components that give them their distinct flavor characteristics. Key metabolites that contribute to sweetness, acidity, and smell have been identified with the aid of strawberries and citrus.

12.9.2 Secondary metabolites

Understanding how secondary metabolites like flavonoids, phenolic chemicals, and terpenes are biosynthesized and contribute to the flavor and health benefits of fruits and vegetables depends on metabolomics. Metabolomic profiling has shown particular polyphenolic chemicals in berries that are associated with antioxidant qualities. Metabolomics has been utilized in tomatoes to find cultivars with greater levels of lycopene, a carotenoid containing antioxidants.

12.9.3 Stress-related metabolites

In reaction to stress, plants generate certain metabolites. For instance, plants frequently collect osmolytes like proline in response to drought, which aid in preventing cell dehydration. These compounds may be identified and measured thanks to metabolomics, which also sheds light on how plants handle stress (Hosseini *et al.*, 2020). Breeders can select for tomato cultivars with superior flavor by using metabolomics to determine the metabolites that contribute to the fragrance profile. Polyphenols, which are antioxidants that aid in shielding plants from oxidative stress, were found to accumulate in grapevines under heat stress. Likewise, it was discovered that the amino acid proline accumulated in response to heat and dehydration, indicating that it serves as a protective osmolyte.

12.9.4 Liquid chromatography–mass spectrometry (LC–MS) and gas chromatography–mass spectrometry (GC–MS)

Advanced analytical technologies like these are frequently employed to analyze agricultural crops under stress. With the use of these methods, thorough metabolite profiling is made possible, offering important new information about how crops react biochemically to physiological or environmental stress. Polar metabolites are most suited for LC–MS, whereas volatile and semi-volatile substances are best suited for GC–MS. When combined, they provide a potent method for researching metabolic alterations brought on by stress and aid in the discovery of yield resilience biomarkers. Increasingly, these approaches are being used to improve crop management techniques and comprehend the mechanisms underlying plant stress.

12.10 Phenomics in Horticulture for High-Throughput Phenotyping

The integration of high-throughput phenotyping (HTP) in plant research has revolutionized the study of crop traits, particularly in horticulture, by enabling rapid, non-destructive, and large-scale data collection. Phenomics bridges the gap between genotypic and phenotypic data, enhancing the proficiency of breeding programs and accelerating genetic improvement. Horticultural phenomics plays a central role in optimizing yield, quality, and stress resilience in fruits and vegetables, which are highly perishable and sensitive to environmental factors. Advanced imaging technologies such as RGB imaging, hyperspectral imaging, fluorescence imaging, LiDAR, thermal imaging, and X-ray CT have been widely employed to assess morphological, physiological, biochemical, and postharvest traits. These tools improve the accuracy and speed of trait evaluation, aiding in selection for desirable genotypes and enhancing decision-making in breeding programs.

In breeding, HTP increases selection intensity and accuracy by screening a large number of genotypes for desirable traits. In cultivation, it helps monitor plant stress, nutrient deficiencies, and disease outbreaks in real time, allowing for precision agriculture interventions. In post-harvest research, phenomics supports quality assessment by evaluating external traits like size, shape, and color and internal attributes such as firmness and soluble solid content. Despite advancements, challenges remain, including data management, the high cost of phenotyping platforms, and the need for standardization across studies. However, integrating artificial intelligence and machine learning with HTP data has further refined phenotypic analysis, improving predictive modeling and trait selection. The use of UAV-based phenotyping has also expanded data collection across large fields, making phenomics more accessible for real-world applications. Future research should focus on refining multi-omics approaches, improving data integration, and enhancing automation in phenotypic analysis. With further development, horticulture phenomics will be essential in tackling the problems of global food security by facilitating effective crop breeding, better post-harvest handling, and optimised production.

12.11 Applications of Phenomics in Horticulture

12.11.1 High-throughput phenotyping

High-throughput phenotyping is a cutting-edge technique used in horticultural crops to rapidly assess a large number of plant traits. Using cutting-edge technologies like robots, sensors, and unmanned aerial vehicles (UAVs), this method measures important metrics including growth rate, disease resistance, and stress tolerance in real time. It makes the process of identifying desirable features for breeding programs easier by gathering a large amount of data from many plant phenotypes. High -throughput phenotyping accelerates crop improvement, allowing for more efficient selection and enhanced productivity in horticulture. High-throughput phenotyping integrates omics data with plant morphology and physiology

12.11.2 Precision agriculture

Phenomics also plays a critical role in precision agriculture, where detailed phenotypic data is used to optimize crop management practices. Automated phenotyping systems provide real-time data that can be integrated into decision-support systems for better resource allocation and improved crop management.

12.11.3 Automated greenhouse systems

Automated greenhouse systems for phenomics in horticultural crops combine advanced technologies like environmental sensors, imaging systems, and data analytics platforms to monitor and optimize plant growth. These systems enable continuous, high-throughput phenotyping, collecting detailed data on plant morphology, growth patterns, and responses to environmental stressors. By automating routine tasks such as irrigation, temperature control, and light regulation, these systems allow researchers to create more accurate and repeatable growing conditions. The integration of such systems is transforming the way horticultural crops are bred and managed, leading to improved crop resilience and yield.

12.11.4 AI and machine learning in phenomics

AI and machine learning (ML) are revolutionizing phenomics in horticultural crops by enabling efficient data analysis and interpretation. These technologies leverage large datasets from sensors, imaging systems, and environmental conditions to predict plant traits, growth patterns, and responses to stress. Machine learning algorithms can identify subtle patterns in complex phenotypic data, accelerating crop breeding programs and improving yield predictions. AI-based systems also assist in automating tasks such as trait classification and disease detection, making phenotyping faster and more precise. By integrating AI and ML, researchers can enhance crop resilience and optimize management strategies for horticultural crops.

12.11.5 Integration of omics with phenotypic data

The integration of phenotypic data with genomic, transcriptomic, and metabolomic data is a prevailing tool for identifying the genetic basis of complex traits. This approach has been used in citrus to identify traits linked to fruit quality and disease resistance. The use of automated phenotyping platforms in grapes, for example, has allowed researchers to rapidly assess key traits such as fruit size, color, and yield, which are crucial for quality and production (Table 12.6).

12.12 Ionomics

Ionomics, the study of elemental accumulation in plants, plays a key role in understanding uptake of nutrients, transport, and storage across different plant species. This field relies on high-throughput analytical techniques such as inductively coupled plasma (ICP)–MS and atomic emission spectrometry (ICP–AES) to quantify various essential and nonessential elements in plant tissues. In horticultural crops, ionomic studies reveal how different species, cultivars, and plant organs accumulate minerals, impacting their nutritional value and safety. The findings from such research contribute to plant breeding programs aimed at enhancing mineral content in edible parts while minimizing the accumulation of toxic elements. One of the key applications of ionomics in horticulture is biofortification, a strategy to boost the nutritive value of crops by increasing their essential mineral content. Deficiencies of micronutrients like iron (Fe) and zinc (Zn) affect large populations worldwide, particularly in developing countries. Ionomic studies help identify plant varieties that naturally accumulate higher levels of these nutrients, enabling targeted breeding efforts. Similarly, reducing toxic element accumulation, such as cadmium (Cd) in leafy vegetables, is critical for food safety. By understanding how different crops absorb and transport minerals, researchers can develop strategies to produce safer and more nutritious food.

The ionomic profile of a plant is influenced by multiple factors, including species, cultivar, environmental conditions, and organ type. For instance, calcium (Ca) accumulates more in leaves and roots compared to seeds and fruits due to its xylem-dependent transport. In contrast, elements like cadmium and boron show higher accumulation in eggplant (*Solanum melongena*) fruit than in tomato (*S. lycopersicum*) fruit, likely due to their effective phloem transport. In order to maximize crop yield and guarantee appropriate nutrient distribution in edible portions, it is imperative to understand these variances. Additionally, because some minerals help plants tolerate biotic and abiotic stress, ionomics sheds light on the biology of plant stress. Genomic methods like QTL mapping and GWAS are increasingly using ionomics advancements. These methods support

Table 12.6. Applications of phenotypic techniques in fruit trees (Adapted from Huang *et al.*, 2020 CC BY 4.0).

Applications	Crop	Device	Detected parameters	Advantages
Pigment and nutrient composition	Apple	Ocean Optics Fiber Spec	Chlorophyll	Easy to operate Non-destructive method
		Field Spec	Chlorophyll	
	Vineyard	Ocean Optics	Chlorophyll, carotenoid	
		Field Spec	Moisture, nitrogen, ,agnesium	
Water stress	Citrus	Field Spec	Water status	Obtain the spectral value directly, not affected by sunlight conditions
	Olive	Field Spec	Water potential	
	Vineyard	Field Spec	Equivalent water thickness	
		Field spec handheld	Predrawn leaf water potential	
		Micro PHAZIR	Solute potential, relative water content	
		NIR Spec	Solute potential	
		PSS	Solute potential	
Biochemical parameters	Mango	Field Spec	Total soluble solids, titrable acidity	The method of data processing is simple
	Vineyard	PSS	Total soluble solids, anthocyanin, polyphenols	
Architecture parameters	Apple	Nikon Fisheye lens	Leaf area index	Low cost, high accuracy especially suitable for the determination of leaf area index
		CID Fisheye lens	Leaf area index	
		Digimax Samsung	Leaf area index	
	Almond	Nikon Fisheye lens	Leaf area index	
	Vineyard	Digimax Samsung	Leaf area Index	
		Canon EOS	Leaf area	
	Citrus	Canon EOS	Leaf area	
Biochemical Parameters	Mango	Kodak D	Chlorophyll – a, chlorophyll – b, chlorophyll – t, carotenoids, total soluble solids, titrable acidity	RGB images can reflect color information well
Architecture parameters	Olive	Tetracam mini	Canopyy area, tree height	Suitable for the information acquisition of the whole orchard canopy
		Panasonic lumix	Tree height, crown diameter	
	Vineyard	ADC-Snap	Tree height	

Continued

Table 12.6. Continued

Applications	Crop	Device	Detected parameters	Advantages
Pigment and nutrient contents	Apple	Multispectral imager	Chlorophyll	Wide spectral band range real time monitoring of a large area
	Citrus	Mini-MCA	Total nitrogen, total soluble sugar, starch	
		Micro hyperspec VNIR model	Chlorophyll fluorescence	
	Pear	Tetracam Micro-MCA	Leaf percentage	
	Vineyard	Multispectral sensor	Carotenoid content	
		Micro-hyperspec VNIR Model	Carotenoid content	
Biochemical parameters	Mango	Resonon Pika	Yield	Quick detection, time saving, no need for chemical treatment
	Vineyard	Resonon Pika	Total soluble solids, anthocyanin concentration	
Diseases detection	Avocado	Modified Canon	Distinguish laurel wilt disease	Suitable for disease detection over large scales, rot influenced by the variation in agronomic characteristics
		Tetracam mini MCA	Distinguish laurel disease	
		ADC Micro	Identify white root rot disease	
	Olive	Micro-hyperspec VNR	Severity levels	
	Almond	Micro-hyperspec VNIR	Red leaf blotch development	
Water Stress	Citrus	IR Thermal camera	Solute potential	Suitable for canopy temperature detection at the orchard level, reducing the deployment of a large number of temperature sensors
	Pear	FLIR	Solute potential	
	Vineyard	FLIR	Solute potential	
		FLIR	Solute potential	
	Olive	Tau	Solute potential	
		Thermal CAM	Water status	
		FLIR One	Leaf water potential	
	Almond	Miricle	Solute potential	
	Peach	Miricle	Solute potential	
	Apricot	Miricle	Solute potential	
Disease detection	Apple	Varloscan	Infected area of scab disease	
	Olive	Miricle	Severity levels of VW	
		FLIR	Severity level of VW	
	Almond	FLIR	Severity levels of red leaf blotch	

VW, Verticulum wilt.

functional genomics research by assisting in the identification of genes involved in ion uptake, transport, and storage in plants. Large-scale ionomic databases such as the Purdue Ionomics Information Management System (PiiMS) store elemental composition data, facilitating the study of plant mineral interactions at a genetic level. This integration allows scientists to link ionomic traits with genetic variations, aiding the development of stress-tolerant and nutrient-rich crops.

In conclusion, horticultural ionomics provides a comprehensive framework for improving crop nutrition, food safety, and stress resilience (Table 12.7). By combining high-throughput analytical techniques with genetic studies, researchers can enhance the understanding of mineral dynamics in plants. This knowledge is instrumental in addressing global challenges like hidden hunger, soil contamination, and climate-induced stress. As ionomics continues to evolve, its applications will be critical for developing sustainable agricultural practices and improving human nutrition through biofortified and resilient horticultural crops.

12.13 Integration of Multiomics for Climate Resilience

A comprehensive knowledge of plant biology is made possible by the integration of data from several omics layers, including transcriptomics, proteomics, ionomics, metabolomics, phenomics, and genomes. Researchers may build thorough models of crop development, stress reactions, and trait regulation by combining various technologies. Integrating multiomics data for a comprehensive knowledge of plant adaptation is key to the future of climate-resilient horticulture. Systems biology, which simulates the relationships between genes, proteins, metabolites, and phenotypes, is based on omics integration. In tomatoes, systems biology approaches have been used to integrate omics data and model fruit ripening processes.

Precision breeding, also known as 'genome editing' or 'gene editing,' utilizes cutting-edge genomic methods to make highly accurate modifications to specific genes. Precision DNA modifications in horticultural crops are made possible by methods like CRISPR/Cas9, which enhance characteristics like yield, fruit quality, disease resistance, and drought tolerance. This method has the advantage of significantly speeding up the breeding process by removing the necessity for traditional hybridization. Additionally, high-throughput genotyping facilitates quick screening of genetic diversity, helping to identify key genes linked to desirable traits. Advanced computational models can be used to predict crop performance. Combining omics data allows for a more accurate identification of molecular markers for breeding. For example, multiomics approaches in cabbage have revealed complex interactions between the genome, transcriptome, and metabolome that influence stress tolerance.

12.14 Achievements in Horticulture

12.14.1 Enhancement of drought tolerance in tomato through genomic and metabolomic approaches

Researchers used a multiomics strategy to study the genetic and metabolic mechanisms that impart drought tolerance in tomatoes, including transcriptomics, metabolomics, and genomics (Taheri *et al.*, 2022). Important genes involved in water retention and osmotic regulation were found by means of genome sequencing and examination of metabolic changes under drought stress. As a result of this strategy, tomato cultivars with enhanced drought resistance were created.

12.14.2 Improving heat tolerance in cucumber via genomic and transcriptomic integration

A combination of genomic and transcriptomic data was used to explore heat-tolerance mechanisms in cucumber (Yu *et al.*, 2022). By analyzing gene expression patterns under heat stress, scientists identified HSPs and heat-responsive genes. In cucumber breeding projects, these genes were utilised for marker-assisted selection in order to create heat-resistant cultivars because they were associated with enhanced heat tolerance.

Table 12.7. Applications implied in horticultural crops through ionomics. Author's own table.

Crop	Achievement	References
Coconut	Metals such as iron, copper, manganese and zinc with its complexes in varying concentrations ranging from 0.2 to 24 ppm	Alchoubassi *et al.*, 2021
	Sugar content in coconut has a high elemental mass fraction compared to cane and beetroot sugars	Zdiniakova and de la Calle, 2020
	In coconut meat, concentration of lead was high	Zhang and Rui, 2013
	In virgin coconut oil, elemental profiling of 17 elements namely Na, Mg, Al, P, Ca, Cr, Mn, Fe, Ni, Cu, Zn, SE, Rb, Sr, Mo, Cs, Pb were identified	Jamwal *et al.*, 2022
Amla	Amla powder has both transition (V, Cr, Ni, Cu, Zn) and non-transition elements (Rb, Sr,Ba)	Reddy *et al.*, 2010
	Emblica officinalis has the level of metal in the range of Cu >Cr>Mn>Zn.	Verma, *et al.*, 2013
	The freeze-dried Amla powder showed the highest retention of minerals, including calcium (79.6 mg/100 g), phosphorus (12.38 mg/100 g), and iron (88.03 mg/100 g), highlighting its superior nutrient preservation	Mishra *et al.*, 2009
Mandarin	Geographical origin of mandarin juices were identified using ionomics	Hidalgo *et al.*, 2023
	Nutritional optimum value for Khasi mandarin: 2.26–2.83%N, 0.11–0.16% P, 1.86–2.07% K, 1.85–2.12% Ca, 0.33–0.48% Mg, 170.10–225.10 mg/kg Fe, 74.03–83.43 mg/kg Mn, 1.24–2.45 mg/kg Cu, and 19.84–21.28 mg/kg Zn	Rymbai *et al.*, 2024
	Leaf analysis of kinnow mandarin revealed Mn (52.2%) as most dominant constraint Zn (61.2%) followed by Mn (48.3%), N (41.2%), and P (35.6%)	Srivastava and Patil, 2016
Large cardamom	235.03 ± 21.49 and 227.49 ± 25.82 mg/g dry weight were the total protein contents for *A. subulatum,* whereas 201.9 ± 37.79 and 294.99 ± 37.93 mg/g dry weight were the total protein contents for *A. xanthioides* seeds and fruit, respectively. Twenty times as much carvacrol is present in *A. subulatum* as in *A. xanthioides*. Both species also showed lower amounts of a-thujene, phyllanderenes, ascaridole, and pinocarvone. Both species are regarded as excellent providers of antimicrobials, such as thyme oil, gallic acid, and apinene	Alruhaili *et al.*, 2023
Moringa	Thirteen essential minerals were identified by the mineral analysis: Al, Ca, Co, Cr, Cu, Fe, K, Mg, Mn, Ni, P, SE, and Zn. Chrome (919.99 ppm) and selenium (1097.84 ppm) have the highest mineral content, respectively, at the second highest level. The minerals magnesium (60.84 ppm) and sodium (68.83 ppm) are present in trace amounts	Natsir *et al.*, 2019
	The average concentrations of heavy metals for different *M. oleifera* products drop in the following order: Al (255.718 mg/kg) > Fe (247.608 mg/kg) > Zn (78.373 mg/kg) > Mn (73.135 mg/kg) > Ni (26.408 mg/kg) > Cu (8.54 mg/kg) > Pb (2.449 mg/kg) > Cr (1.164 mg/kg)>As (0.362 mg/kg) > Cd (0.122 mg/kg) > Hg (0.087 mg/kg). The highest average Zn concentration was found in leaf powders (92.778 mg/kg)	Limmatvapirat *et al.*, 2015

12.14.3 Development of cold-resistant apple varieties using omics approaches

In apple breeding, multiomics tools were used to identify genes involved in cold resistance (Saini *et al.*, 2025). By integrating genomic, transcriptomic, and proteomic data, researchers identified pathways that help apple trees survive frost damage. These findings were used to breed apple cultivars that could withstand colder climates, aiding in the expansion of apple cultivation to regions previously unsuitable due to frost.

12.14.4 Improvement of pest resistance in strawberries through genomic and proteomic integration

The development of pest-resistant strawberry cultivars has benefited from a multiomics strategy that combines genomic sequencing and proteomic profiling (Kumar *et al.*, 2021). This integrated approach identified key genes and proteins associated with resistance to pests, leading to the identification of candidate genes for breeding pest-resistant strawberries.

12.14.5 Nutritional quality enhancement in spinach using metabolomics and transcriptomics

Multiomics approaches were applied to spinach to understand how environmental factors affect its nutritional composition. By analyzing transcriptomic and metabolomic data, researchers were able to identify key metabolites responsible for nutrient content in spinach (Han *et al.*, 2025). These findings allowed the development of spinach varieties with improved levels of vitamins and minerals, enhancing their nutritional quality.

12.15 Constraints in the Application of Omics in Horticultural Crops

12.15.1 Genomic complexity

Large repetitive sequences, high degrees of heterozygosity, or polyploidy (the presence of multiple sets of chromosomes) are characteristics of the complex genomes of many horticultural crops. Accurate gene identification and genome assembly are made more difficult by this. For instance, crops with such complexity such as current rose varieties or potatoes like *Solanum tuberosum* make annotation and sequencing difficult.

12.15.2 High costs and technical barriers

Omics technologies, including high-throughput sequencing genomics, transcriptomics, proteomics, and metabolomics, often involve considerable financial costs. These approaches are resource demanding, requiring expensive equipment, reagents, and computational support. This can pose a significant obstacle, especially for researchers and institutions with constrained funding.

12.15.3 Environmental and phenotypic variability

Horticultural crops are typically cultivated in environments with fluctuating conditions, and phenotypic plasticity makes it challenging to identify consistent genetic links. Elements such as soil, water availability, temperature, and pests influence trait expression, which means that results from omics studies may not be universally applicable across diverse growing environments.

12.15.4 Limited functional annotation

Despite the abundance of data provided by omics technologies, the functional annotation of genes and proteins is frequently lacking. A large portion of the genome is frequently left unannotated, particularly for non-model species, which makes it more difficult to associate particular genetic markers with desirable characteristics. This problem may make it more difficult to apply omics effectively in agricultural improvement initiatives.

12.16 Future Prospects for Omics in Horticultural Crops

12.16.1 Precision breeding and crop improvement

By locating genes and biomarkers linked to desired characteristics like yield, disease resistance, and stress tolerance, omics technology can greatly improve plant breeding. These insights can speed up and enhance the accuracy of developing superior crop varieties when paired with genomic selection or CRISPR/Cas9-based genome editing. For example, combining omics and CRISPR-Cas9 can hasten the development of crops with particular characteristics, including enhanced nutritional profiles or disease resistance.

12.16.2 Climate resilience and stress tolerance

The search for the genetic components that provide resistance to environmental stressors like drought, excessive salinity, or harsh temperatures is a promising application of omics techniques. More resilient horticulture crops may result from a better understanding of the molecular mechanisms underpinning stress tolerance. As climate change presents new difficulties for horticulture, this becomes more and more crucial. Crops can be modified to tolerate unfavorable climatic circumstances while retaining productivity using transcriptomics and genomes.

12.16.3 Nutritional enhancement

Omics technologies can be applied to enhance the nourishing profiles of horticultural crops by identifying key metabolic pathways and genes involved in the synthesis of beneficial compounds (e.g. antioxidants, polyphenols and vitamins). Personalized nutrition is another avenue where crops could be tailored to meet specific dietary needs, improving human health on a global scale. The future of functional foods may heavily rely on such omics insights.

12.16.4 Automated and smart farming practices

The integration of omics data with AI and ML models offers opportunities for precision agriculture. Through real-time data collection from field sensors and omics analyses, by optimizing crop management techniques like fertilization, irrigation, and pest control, AI systems can eventually boost crop yield while reducing their negative effects on the environment. This technology-driven approach will be crucial in enhancing resource use efficiency in agriculture.

12.16.5 Microbiome research

Plant health, nutrient uptake, and stress tolerance are all significantly influenced by the plant microbiome. Metagenomics and metabolomics are two examples of omics methods that can be used to find helpful microbes that suppress diseases or encourage plant growth. By lowering the need for chemical pesticides and fertilizers, research on the plant microbiome combined with genetics has the potential to completely transform sustainable crop production methods.

12.17 Conclusions

The integration of omics technologies, including transcriptomics, proteomics, metabolomics, ionomics and genomes, with horticulture research provides a comprehensive approach to creating crops that can withstand climate change. These technologies provide a better understanding of how plants react to environmental stressors by discovering important genes, proteins, and metabolites that help plants tolerate harsh environments including heat, salinity, and drought. By integrating data across multiple levels, omics approaches help pinpoint biomarkers and new biological pathways, which can be utilized in breeding programs to accelerate the creation of crops with enhanced stress resistance. Furthermore, interpreting the intricate relationships between plants and their surroundings through the application of omics technology can help develop better farming techniques and climate-adaptive production systems. The ability to monitor and modify the molecular networks governing stress resilience opens up new possibilities for sustainable horticultural practices that maintain high productivity and food security amid climate change. The multilevel use of omics technology in horticulture is essential for increasing crop resilience and furthering global sustainability goals. In the face of a rapidly changing environment, researchers can provide creative solutions to foster agricultural biodiversity and safeguard future food systems by combining state-of-the-art molecular tools with complex data analysis.

References

Alchoubassi, G., Kinska, K., Bierla, K., Lobinski, R. and Szpunar, J. (2021) Speciation of essential nutrient trace elements in coconut water. *Food Chemistry* 339, 127680.

Alruhaili, M.H., Almuhayawi, M.S., Gattan, H.S., Alharbi, M.T., Nagshabandi, M.K. *et al.* (2023) Insight into the phytochemical profile and antimicrobial activities of *Amomum subulatum* and *Amomum xanthioides*: An *in vitro* and *in silico* study. *Frontiers in Plant Science* 14, 1136961.

Devi, R., Chauhan, S. and Dhillon, T.S. (2022) Genome editing for vegetable crop improvement: Challenges and future prospects. *Frontiers in Genetics* 13, 1037091.

Han, K., Zhang, J., Wang, C. and Xie, J. (2025) Integrated metabolome and transcriptome analysis of fulvic acid relieves nitrate stress-induced damage in spinach (*Spinacia oleracea* L.) by regulating multiple defense pathways. *Plant Stress* 16, 100808.

Hatfield, J.L. and Prueger, J.H. (2015) Challenge for future agriculture. In: Redden, R., Yadav, S.S., Maxted, N., Dulloo, M.E. and Guarino, L. (eds) *Crop Wild Relatives and Climate Change*. John Wiley & Sons, pp. 24–43.

Hidalgo, M.J., Gaiad, J.E., Goicoechea, H.C., Mendoza, A., Pérez-Rodríguez, M. *et al.* (2023) Geographical origin identification of mandarin fruits by analyzing fingerprint signatures based on multielemental composition. *Food Chemistry X* 20, 101040.

Hosseini, R., Hassanpour, N., Liu, L.-P. and Hassoun, S. (2020) Pathway-activity likelihood analysis and metabolite annotation for untargeted metabolomics using probabilistic modeling. *Metabolites* 10(5), 183. DOI: 10.3390/metabo10050183.

Huang, Y., Ren, Z., Li, D. and Liu, X. (2020) Phenotypic techniques and applications in fruit trees: A review. *Plant Methods* 16, 1–22.

Jamwal, R., Kumari, S., Kelly, S., Cannavan, A. and Singh, D.K. (2022) Assessment of geographical origin of virgin coconut oil using inductively coupled plasma mass spectrometry along with multivariate chemometrics. *Current Research in Food Science* 5, 545–552.

Kosová, K., Vítámvás, P., Urban, M.O., Prášil, I.T. and Renaut, J. (2018) Plant Abiotic Stress Proteomics: The Major Factors Determining Alterations in Cellular Proteome. *Frontiers in Plant Science* 9, 122. DOI: 10.3389/fpls.2018.00122.

Kumar, A., Anju, T., Kumar, S., Chhapekar, S.S., Sreedharan, S. *et al.* (2021) Integrating omics and gene editing tools for rapid improvement of traditional food plants for diversified and sustainable food security. *International Journal of Molecular Sciences* 22(15), 8093.

Limmatvapirat, C., Limmatvapirat, S., Charoenteeraboon, J., Wessapan, C., Kumsum, A. *et al.* (2015) Comparison of eleven heavy metals in *Moringa oleifera* Lam. products. *Indian Journal of Pharmaceutical Sciences* 77(4), 485.

Mangal, V., Lal, M.K., Tiwari, R.K., Altaf, M.A., Sood, S. *et al.* (2023) A comprehensive and conceptual overview of omics-based approaches for enhancing the resilience of vegetable crops against abiotic stresses. *Planta* 257(4), 80.

Mishra, P., Srivastava, V., Verma, D., Chauhan, O.P. and Rai, G.K. (2009) Physico-chemical properties of Chakiya variety of Amla (*Emblica officinalis*) and effect of different dehydration methods on quality of powder. *African Journal of Food Science* 3(10), 303–306.

Natsir, H., Wahab, A.W., Budi, P., Dali, S. and Arif, A.R. (2019) Amino acid and mineral composition of *Moringa oleifera* leaves extract and its bioactivity as antioxidant. *Journal of Physics: Conference Series* 1317(1), 012030.

Palma, J.M., Corpas, F.J. and del Río, L.A. (2011) Proteomics as an approach to the understanding of the molecular physiology of fruit development and ripening. *Journal of Proteomics* 74(8), 1230–1243.

Reddy, N.C.G., Reddy, K.N.M., Reddy, S.L. and Endo, T. (2010) Spectroscopic studies on *Satavari root* and *Amla*. *Transactions of the Materials Research Society of Japan* 35(3), 519–527.

Rymbai, H., Deshmukh, N., Ramesh, T., Verma, V., Baiswar, P. *et al.* (2024) DRIS standards for identifying yield-limiting nutrients in Khasi Mandarin and their relationship with fruit yield on acidic soil of the eastern Himalayas, India. *Journal of Plant Nutrition* 47(8), 1331–1346.

Saini, A.K., Jahed, K.R., Neres, D.F., Wright, R.C. and Sherif, S.M. (2025) Investigating frost response, rootstock-dependent cold tolerance, and floral bud mortality in apple cultivars through transcriptomic insights. *Plant Stress* 16(10082), 9.

Schwarz, D., Thompson, A.J. and Kläring, H.P. (2014) Guidelines to use tomato in experiments with a controlled environment. *Frontiers in Plant Science* 5, 625.

Singh, S.K. and Pradhan, S. (2020) Transcriptomics in fruit crops: Present status and future prospects. *Indian Journal of Horticulture* 77(4), 563–581.

Springer, N.M. and Schmitz, R.J. (2017) Exploiting induced and natural epigenetic variation for crop improvement. *Nature Reviews Genetics* 18(9), 563–575.

Srivastava, A.K. and Patil, P. (2016) Nutrient indexing in "'Kinnow" Mandarin (*Citrus deliciosia* Lour. × *Citrus nobilis* Tanaka) grown in Indogangetic plains. *Communications in Soil Science and Plant Analysis* 47(18), 2115–2125.

Taheri, S., Gantait, S., Azizi, P. and Mazumdar, P. (2022) Drought tolerance improvement in *Solanum lycopersicum*: An insight into "OMICS" approaches and genome editing. *Biotech* 12(3), 63.

Tomato Genome Consortium (2012) The tomato genome sequence provides insights into fleshy fruit evolution. *Nature* 485(7400), 635–641.

Verma, N.K., Vibha, K. and Kumar, A. (2013) A comparative study of heavy metals in *Emblica officinalis*, *Phyllanthus emblica*, and *Azadirachta indica*. *International Research Journal of Biological Sciences* 2(8), 16–19.

Wang, R., Li, X., Sun, M., Xue, C., Korban, S.S. *et al.* (2023) Genomic insights into domestication and genetic improvement of fruit crops. *Plant Physiology* 192(4), 2604–2627.

Yonekura-Sakakibara, K., Higashi, Y. and Nakabayashi, R. (2019) The origin and evolution of plant flavonoid metabolism. *Frontiers in Plant Science* 10, 943.

Yu, B., Ming, F., Liang, Y., Wang, Y., Gan, Y. *et al.* (2022) Heat stress resistance mechanisms of two cucumber varieties from different regions. *International Journal of Molecular Sciences* 23(3), 1817.

Zdiniakova, T. and de la Calle, M.B. (2020) Feasibility study about the use of element profiles determined by ED-XRF as screening method to authenticate coconut sugar commercially available. *European Food Research and Technology* 246(10), 2101–2109.

Zhang, Y. and Rui, Y. (2013) Determination of trace elements and heavy metals in coconut meat simultaneously by ICP-MS. *Asian Journal of Chemistry* 25(9), 5105.

13 Omics Approaches for Climate-resilient Horticulture Crops

Mayank Tiwari[1], Kajal Mehra[1*], Garima Aggarwal[1], Bishawajit Kumar[1], Anand Singh Jeena[1], Namata Kumari[1] and Shalu Kumari[2]

[1]Department of Genetics and Plant Breeding, College of Agriculture, G.B. Pant University of Agriculture and Technology, Pantnagar, Uttarakhand, India; [2]Department of Genetics and Plant Breeding, I.A.S, BHU, Varanasi, India

Abstract

Horticultural crops, including fruits, vegetables, ornamental plants, aromatic herbs, plantation crops, and medicinal plants, are crucial for agricultural success and national economies. These crops are vital sources of nutrients essential for human health. However, climate change and global warming are causing environmental (abiotic) stresses such as drought, heat, cold, and salinity, which negatively impact plant growth. These abiotic stresses are major constraints on crop productivity. Plants must adapt to external environmental and soil conditions through their biological mechanisms, or their growth, development, and yield will be severely affected. Enhancing traits related to increased crop production and climate resilience in crop breeding is a key approach to overcome these challenges. Recent innovations in high-throughput sequencing and bioinformatics have significantly accelerated research in genomics, transcriptomics, proteomics, metabolomics, ionomics, phenomics, and methylomics. These advancements are improving our understanding of plant adaptation to changing environments. Recently, these omics technologies have been applied to horticultural crops, enabling more effective Genomics selection, genome-wide association studies, and quantitative trait loci mapping. By integrating multiomics approaches with modern plant breeding and genetic engineering techniques, we can develop crops that are nutritionally rich, climate-resilient, and capable of meeting growing demands for food, nutrition, and energy in a sustainable manner.

13.1 Introduction

The rapidly expanding global population, coupled with the adverse effects of climate change, presents significant challenges to food security worldwide (McCouch, 2013). To overcome these challenges, crop improvement strategies must prioritize enhancing critical agronomic traits beyond mere yield, such as adaptability, resistance, and nutritional quality. These strategies require both direct and indirect selection techniques (Varshney *et al.*, 2021). Horticulture, a crucial aspect of agriculture, plays an important role in human development by facilitating the transition from nomadic lifestyles to established agricultural communities. It encompasses the cultivation, propagation, processing, and marketing of a diverse range of plants, including fruits, vegetables, nuts, flowers, herbs, and ornamental species (Tang, 2017).

As stationary organisms, plants are highly sensitive to fluctuations in their environment,

*Corresponding author: mehra9505@gmail.com

 From Gene Discovery to Climate-Resilient Crops: Omics in Crop Improvement (eds S. Pandey *et al.*)
DOI: 10.1079/9781800628717.0013

especially abiotic stresses. These stresses, which include temperature extremes, drought, salinity, and waterlogging, can significantly impair plant functions, leading to reduced yields, particularly in horticultural crops (Rao *et al.*, 2016). The ability of plants to withstand such stresses depends on their adaptive biological mechanisms, which are often inadequate under severe climatic shifts. Therefore, developing climate-resilient varieties through advanced breeding techniques is paramount.

The introduction of omics technology has transformed plant science by providing comprehensive tools to explore biological systems at multiple levels. The term 'omics' refers to the extensive study of biological molecules, including genomics, proteomics transcriptomics, phenomics, and metabolomics (Khalid *et al.*, 2019). These strategies allow scholars to investigate the complex molecular interactions that underpin plant responses to abiotic stresses. Omics-based research has been especially useful in understanding how various biological molecules, such as genes, proteins, and metabolites, interact under stress conditions to regulate plant growth and development (Devireddy *et al.*, 2021).

Single-cell omics, a more recent advancement, has further expanded our understanding of how plants cope with abiotic stresses. By analyzing gene expression at the single-cell level, researchers can identify the precise cellular mechanisms responsible for stress tolerance. This is particularly valuable for transcriptomics, which reveals how environmental stresses induce changes in gene expression, leading to shifts in metabolic, proteomic, and phenomic profiles (Zhu, 2016). Integrating these multiomics approaches is essential for gaining a comprehensive analysis of the regulatory system underlying in plant stress responses, allowing for more targeted and efficient breeding strategies.

In this context, leveraging multiomics tools for horticultural crops has the capacity to increase their resistance to climatic variations, while also improving their nutritional value and productivity. To fully understand crop tolerance and the intricate regulatory processes that propel responses to abiotic stresses, integrating various omics approaches will be essential. This holistic approach will enable more precise interventions, leading to the development of climate-smart crops that can thrive in an unpredictable environment.

13.2 Climate and Abiotic Stresses in Horticultural Crops

Climate plays an important part in determining the growth and productivity of horticultural crops, as each species has specific requirements for temperature, radiation, and water availability. Temperature extremes, whether high or low, are among the most impactful abiotic stresses. High temperatures can cause cellular damage, including membrane lysis and protein denaturation, which disrupts physiological processes vital to development and growth (Arias *et al.*, 2022). Conversely, cold stress includes not only freezing temperatures but also dehydration, reduced light availability, and fluctuating soil conditions. Plants subjected to cold stress may suffer from frost damage, leading to reduced metabolic activity and inhibited growth (Su *et al.*, 2022). Adaptation to these extremes is crucial for survival, particularly for crops grown in regions with significant seasonal temperature variations (Zapata *et al.*, 2017).

Solar radiation is another pivotal factor in crop development. Moderate levels of radiation enhance photosynthesis and metabolic processes, but excessive radiation—particularly when coupled with elevated temperatures—can impair photosynthetic efficiency, leading to reduced crop yields (Romero *et al.*, 2016). This is especially critical for crops grown in regions prone to high solar intensity, where managing radiation levels is key to maintaining productivity (Arias *et al.*, 2022).

Water availability, or the lack thereof, also poses significant challenges. On the one hand, drought stress, which is becoming more prevalent due to climate change, can reduce photosynthetic activity by limiting CO_2 uptake and causing stomatal closure. This leads to altered leaf morphology, impaired root and shoot growth, and reduced crop yields (Akinci, 1997; Liliane and Charles, 2020). On the other hand, waterlogging—often a result of excessive rainfall—can lead to hypoxic conditions in the soil. Under these conditions, oxygen levels decrease rapidly, impairing root respiration and

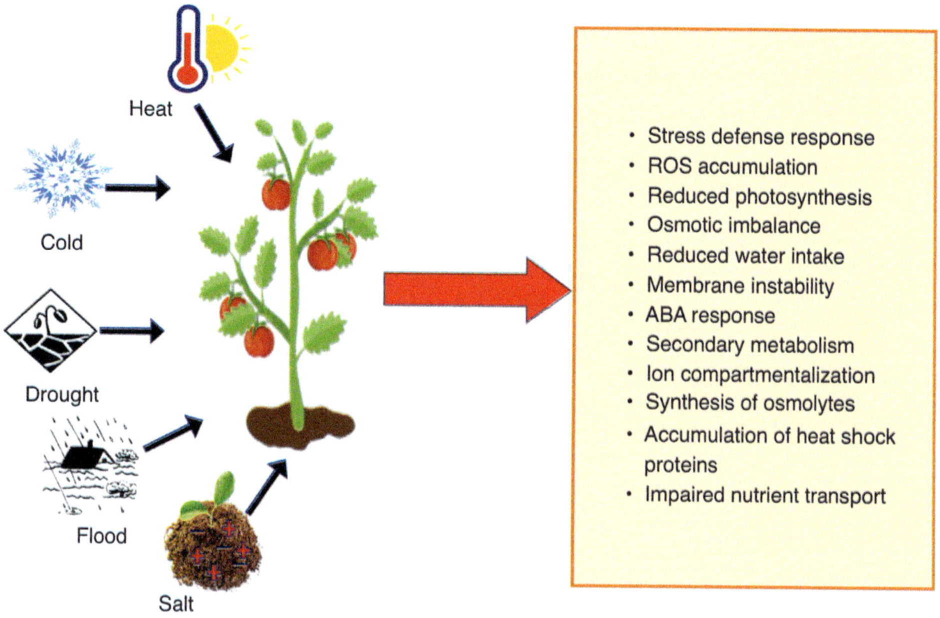

Fig. 13.1. Different type of climatic stress and their impact on plants. ABA, abscisic acid. Author's own image.

nutrient absorption, ultimately resulting in root decay and plant death (Cassia *et al.*, 2018). Both drought and waterlogging present unique challenges that require specific management practices to mitigate their effects.

In addition to temperature, radiation, and water stress, rising levels of greenhouse gases such as CH_4, CO_2 and N_2O exacerbate abiotic stresses by contributing to global warming. Elevated CO_2 levels, for instance, can lead to reduced nitrogen availability in plants, lowering their protein content and overall nutritional value (Cassia *et al.*, 2018). The combination of these elements impacts not just the physiological health of plants but also threatens long-term agricultural productivity, particularly for crops that are sensitive to environmental fluctuations.

Overall, the combined effects of temperature fluctuations, excessive solar radiation, water stress, and increasing greenhouse gas concentrations create complex challenges for horticultural crops (Fig. 13.1). Addressing these abiotic stresses requires innovative agronomic practices and technological interventions aimed at enhancing crop resilience to sustain food production in an increasingly volatile climate (Sahu *et al.*, 2014; Zhu, 2016).

13.3 Integrated Omics Approaches for Climate-resilient Horticulture Crops

Omics is a rapidly advancing field that offers the methodologies, technological expertise and interdisciplinary insight required to enhance our understanding of all genomic and transcriptomic processes in plants. It is particularly useful for studying how plant metabolism changes in response to environmental interactions (Kumar *et al.*, 2021a). Multiomics integration methods aim to link genotypes with phenotypes, providing a comprehensive view of biological processes, such as how plants respond to abiotic stresses. By integrating information from many high-throughput omics technologies, multiomics integration serves as a robust approach for gaining deeper insights into these complex interactions.

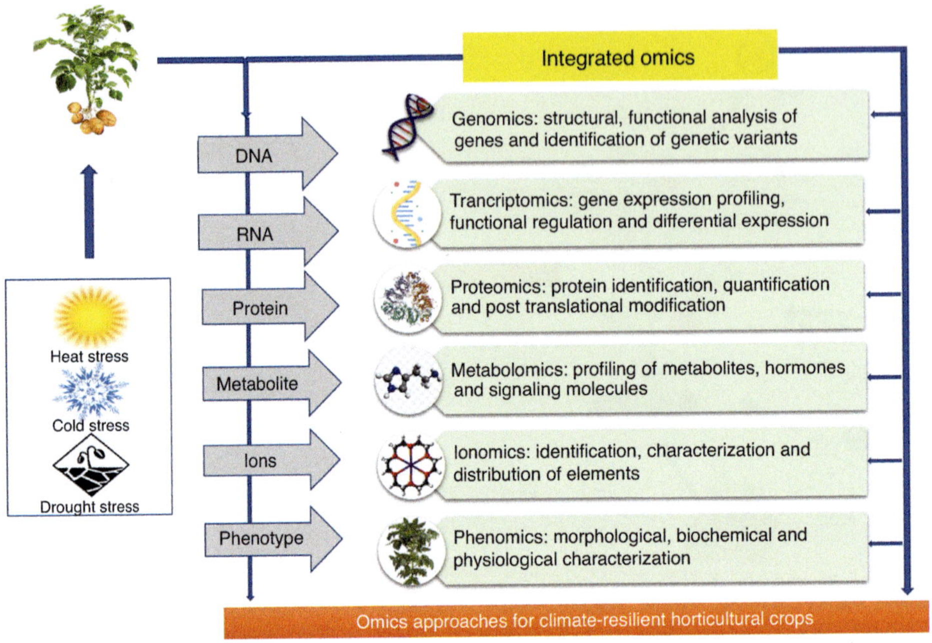

Fig. 13.2. Multiomics approaches for climate-resilience horticulture crops. Author's own image.

Key components of omics encompass several distinct areas: genomics involves creating genetic and genomic resources, functional genomics, gene mapping, and genomic selection; transcriptomics focuses on expression profiling and gene regulation; proteomics deals with protein identification and their effects; metabolomics includes profiling metabolites, studying their regulation, pathway, and intermediates; phenomics covers the automated analysis of phenotypic and physiological traits; and ionomics pertains to the identification, composition, effects, and interaction of elements (Fig. 13.2) (Pazhamala *et al.*, 2021).

13.3.1 Genomics

A genome represents the complete set of nucleic acids (DNA or RNA) within an organism, encompassing all of its genes. Genomics, or genomic science, involves examining the genome's structure, mapping, evolution, function, and alterations. To respond to abiotic stresses, plants must undergo regulatory changes that activate various genes and pathways (Bohnert *et al.*, 2006). Advances in next-generation sequencing technologies have significantly advanced research in crop functional genomics (Kersey, 2019).

13.3.1.1 Structural genomics

The area of structural genomics determines the 3-D structures of genes, aiming to identify, position, and organize them along chromosomes. This approach provides critical insights into the spatial configuration of genes and their associated proteins, contributing to our understanding of how genetic information translates into functional traits. The advent of high-throughput next-generation sequencing (HT-NGS) technologies, such as Roche's GS FLX Titanium and GS Junior, has revolutionized plant genomic research, facilitating breeding improvements and enabling the efficient sequencing and re-sequencing of large genomes (Table 13.1).

These advanced platforms have greatly enhanced our ability to investigate complex genomic structures (Takagi *et al.*, 2013).

Table 13.1. List of horticulture crops sequenced. Author's own table.

Crop	Method used	Year	References
Grapevine (Vitis vinifera L.)	Whole-genome shotgun sequencing	2007	(Velasco et al., 2007)
Cucumber (Cucumis sativus L.)	Whole-genome shotgun sequencing	2009	Huang et al., 2009
Potato (Solanumtuberosum L.)	Next-generation sequencing (NGS)	2011	Potato Genome Sequencing Consortium, 2011
Chinese cabbage (Brassica rapa L. ssp. pekinensis)	Whole genome sequencing	2011	Wang et al., 2011
Tomato (Solanum lycopersicum L.)	Whole-genome shotgun sequencing	2012	Tomato Genome Consortium, 2012
Pepper (Capsicum annuum)	Whole genome sequencing	2014	Kim et al. 2014
Spinach (Spinacia oleracea)	Whole-genome shotgun sequencing	2017	Xu et al., 2017
Apple (Malus domestica)	Whole-genome sequencing, high-throughput sequencing	2010	Velasco et al., 2010
Strawberry (Fragaria vesca)	Roche/454 Illumina/Solexa and Life Technologies	2011	Shulaev et al., 2011
Peach (Prunus persica)	Sanger whole genome short gun method	2013	Verde et al., 2013
Banana (Musa acuminata)	Sanger (ABI 3730xl sequencers) and Roche/454 (GSFLX pyrosequencing platform) method	2012	D'Hont et al., 2012
Pear (Pyrus bretschneideri)	Bacterial artificial chromosome and NGS	2013	Wu et al., 2013
Mango (Mangifera indica)	Whole-genome shotgun sequencing	2020	Wang et al., 2020
Carrot (Daucus carota)	Illumina paired-end and mate-paired libraries, BAC end sequences	2016	Iorizzo et al., 2016
Lettuce (Lactuca sativa)	Illumina HiSeq 2000	2017	Reyes-Chin-Wo et al., 2017
Coffee (Coffea canephora)	Whole-genome shotgun and transcriptome sequencing	2014	Denoeud et al., 2014
Olive (Olea europaea)	Whole-genome shotgun and transcriptome sequencing	2016	Unver et al., 2017
Pomegranate (Punica granatum)	Illumina HiSeq 2000 platform	2017	Qin et al., 2017
Transgenic papaya (Carica papaya)	Whole-genome shotgun sequencing	2008	Ming et al. 2008
Ginseng (Panax ginseng)	Whole-genome sequencing, Illumina HiSeq 2500	2017	Jayakodi et al., 2018

A notable advancement in structural genomics is the application of whole-genome shotgun sequencing, which aids in identifying quantitative trait loci (QTLs) by sequencing and comparing bulked samples from populations exhibiting specific traits. This approach allows for the detection of potential QTLs that contribute to phenotypic variation in traits like abiotic stress tolerance or disease resistance (Table 13.2). Furthermore, genotyping-by-sequencing (GBS) is an efficient technique used to identify single nucleotide polymorphisms (SNPs) and small

Table 13.2. QTLs and indicators for abiotic stress (drought, waterlogging, heat, cold tolerances) response in horticulture crops. Author's own table.

Crop	No. of QTLs identified	Chromosome location	Methods used	Abiotic stresses	References
Tomato (S. lycopersicum)	54	Chr 1, 2, 3, 10, and 11	Multi-parent advanced generation intercross (MAGIC) derived from eight parental lines	Drought	Diouf et al. 2018
Cucumber (Cucumis sativus)	5	Chr 2, 3, and 6	165 F8 RILs from CG104 × CG37 hybrid via single-seed descent.	Salt	Liu et al. 2021
	1	Chr 6	186 F8 RILs from CG104 × CG37 hybrid via single-seed descent.	Cold	Li et al. 2022
		Chr 3, 4, and 6	88 diverse accessions	Heat	Wang et al. 2023
	1	Chr 1	RILs (86) derived from heat-sensitive	Heat	Liu et al.,2021
Pepper	64	-	F2 population crossed between heat-tolerant (DLS-161–10) and sensitive (DChBL-240) genotypes.	Heat	Ts et al. 2023
Broccoli (Brassica oleracea var. italica)	5	-	Doubled haploid	Heat	Branham et al. 2017
Cauliflower	18	Chr 1, 2, 3, 4, 6, 8, and 9	Diversity panel	Heat	Matschegewski et al. 2015
Potato	45	Chr 3, 5, 8, 10, and 12	Diploid backcross progeny (BC1)	Drought	Khan et al. (2015)
	3	-	Diploid mapping population	Heat	Trapero-Mozos et al. 2018
Pumpkin	3	Chr 1, 4, and 10	F2 population crossed between cold stress tolerant and sensitive genotypes	Cold	Xu et al. 2017

insertions or deletions (INDELs), which serve as valuable markers for genetic diversity and trait analysis in plants.

SNPs, in particular, are widely used in marker-assisted breeding, the construction of high-density genetic maps, QTL mapping, and association studies. They are essential to comprehending linkage disequilibrium (LD) and conducting population genetic studies. Additionally, SNPs underpin many horticultural traits, such as fruit size, color, flavor, and resistance to diseases or pests, making them essential tools in advancing agricultural research and improving crop varieties.

Structural genomics thus fills the void between gene sequence data and functional genomics, providing an in-depth comprehension of how genes, proteins, and their 3D structures interact to regulate biological processes. The insights derived from structural genomics not only improve the annotation of genomes but also support the development of innovative breeding approaches aimed at improving crop resilience and productivity.

13.3.1.2 Functional genomics

Functional genomics examines data from complete or partial genome sequencing to understand gene functions and interactions, utilizing two main strategies: reverse and forward genetics. The forward genetics approach involves studying a mutant with an interesting phenotype to identify the genes responsible for that trait. Conversely, reverse genetics involves analyzing an organism's phenotype by disrupting a known gene (Singh *et al.*, 2022). Functional genomics aids in uncovering gene interactions and regulatory networks through various methodologies.

1. Transcriptome assembly: transcriptome or EST sequencing is a crucial initial step in exploring the gene content of a species and has proven to be an effective method for generating functional genomic data for non-model organisms. Expressed sequence tags (ESTs) are fragments of expressed genes or partial gene sequences obtained from cDNA clones. They help in determining the corresponding genes and are valuable for understanding the functional roles of expressed genes. EST databases offer resources for comparing interspecies genetic and physical markers, as well as gene expression profiles for various clones, particularly through comparative genomics approaches.
2. Serial analysis of gene expression (SAGE): To measure a significant quantity of transcripts simultaneously, the technique of SAGE has been developed. SAGE is a high-throughput sequencing-base method that enables the identification and quantification of tissue-specific gene expression by cloning short sequence tags (13–26 bp) derived from expressed poly (A) transcripts (Velculescu *et al.*, 1997).

13.3.1.3 Comparative genomics

Comparative genomics involves comparing entire genomes or segments of genomes to identify fundamental biological similarities and differences and to explore evolutionary relationships between organisms (Avni *et al.*, 2022). This field compares biological sequences by aligning them to detect conserved regions, revealing significant synteny among related species. Additionally, comparative genomics can identify small-scale variations within different genomes, offering insights into the impact of these changes on protein structure and function, and uncovering important regulatory elements in DNA (Fang *et al.*, 2021). The concept of the 'genome zipper' emerged from comparative genomics, aiding in the determination of the virtual gene order within partially sequenced genomes.

13.3.2 Transcriptomics

Crops face increasing environmental pressure, which adds complexity to the already challenging breeding process. Understanding the molecular-level biological processes throughout the plant's life cycle and within its cells is crucial for grasping how various factors impact plant development. Advances in sequencing technologies have enabled high-throughput transcriptomics analysis, allowing the deciphering of intricate transcriptional modifications related to phenotype development, particularly in the context of genotype×environment interactions. Transcriptome data facilitate the identification of transcriptional regulatory elements and mechanisms essential for cellular function. The significance of transcriptomic analysis is underscored by its ability to provide detailed insights into the molecular mechanisms and genetic foundations of plant disease resistance, offering valuable information that can be utilized to develop resistant genotypes through breeding (Guo *et al.*, 2017).

Transcriptomics is the study of the 'transcriptome,' which encompasses the complete set of RNA transcripts produced by an organism's genome in a particular cell or tissue (Zhou *et al.*, 2022). This profiling provides a comprehensive view of all RNA molecules, including mRNA, tRNA, rRNA, small RNAs (sRNAs), and other non-coding RNAs, such as microRNAs (miRNAs), circular RNAs (circRNAs), long non-coding RNAs (lncRNAs), and small interfering RNAs (siRNAs), within a cell, tissue, organ, or entire organism at a specific time point—something that cannot be achieved through genomic analysis alone

(Chaturvedi *et al.*, 2021). Transcriptomics analysis has identified numerous stress-responsive genes and their expression patterns under abiotic stress conditions. For example, transcriptome studies in tomatoes have uncovered regulators of secondary growth pathways, such as GLYCOALKALOID METABOLISM (GAME) 9, also known as JRE4, which is an AP2/ERF transcription factor that responds to various abiotic stresses (Guo *et al.*, 2022).

Non-coding RNAs play various crucial roles; for instance, miRNAs are involved in the cleavage and post-transcriptional modification of target mRNAs. Different miRNAs have regulatory functions under abiotic stresses such as drought, salinity, and cold. These miRNAs, which are 19–23 nucleotides long, regulate several biological processes. In tomatoes, the overexpression of miR169C, which controls genes related to stomatal activity, enhances drought tolerance (Zhang *et al.*, 2011). Similarly, Buv-miR13 regulates WD repeat proteins that are crucial for stress tolerance in cucumbers (Li *et al.*, 2014). Beyond regulating transcription factors (TFs), miRNAs also target stress signaling pathways involved in root development, leaf morphogenesis, and stress responses (Curaba *et al.*, 2014). In addition, lncRNAs are important regulators of various biological processes, while circRNAs act as miRNA sponges and play roles in transcriptional and splicing regulation, as well as the modulation of primary gene expression in plants. Whole-transcriptome sequencing has been employed to map the global landscape of these mRNAs in various plants, including Chinese cabbage (Shi *et al.*, 2022) and citrus (Fu *et al.*, 2019) (Table 13.3).

Early transcriptomic platforms were based on the idea of immobilizing transcripts on an array to capture the mRNA content of a sample through nucleotide sequence complementarity. These were later advanced by replacing the transcripts with 20–70 nucleotide-long oligonucleotides, specifically designed to target known gene sequences, resulting in more refined microarray techniques (Knudsen, 2004). Transcriptomics offers valuable insights into gene expression patterns helping researchers determine which genes are activated or suppressed at various developmental stages or under different environmental conditions. This information reveals how genetic data are converted into functional

Table 13.3. Transcriptomic analysis for climate-resilience horticulture crops. Author's own table.

Abiotic stress	Tissue	Crops	Genotype	Differentially expressed genes (DEGs)	Sequencing platform	References
Drought	Leaf	Potato	Longshu3	1336	Illumina HiSeq™ 2000	Zhang *et al.* (2014)
		Onion	Acc. 1656 (T), Acc. 1627 (S)	>1100	Illumina Next 500	Ghodke *et al.* (2020)
		Tomato	M82	966	Illumina Hiseq 1500	Iovieno *et al.* (2016)
Cold	Leaf	Onion	36,122 (T), 36,101	491	Illumina HiSeq 2000	Han *et al.* (2016)
		Pepper	A188 (T), A122 (S)		BGISEQ-500	Miao *et al.* (2021)
Heat	Leaf	Potato	Desiree (T), Unica (T), Agria (S), Rus Sett Burbank (S)		Microarray	Demirel *et al.* (2020)
		Lettuce	KNV482 (T), KNV664 (T)	31 Transcription factors	Illumina Truseq™	Wei *et al.*, 2021

S, susceptible; T, tolerant.

outcomes and can pinpoint genes crucial for specific plant traits. For instance, in horticultural research, RNA sequencing (RNA-Seq) can be employed to analyze transcriptomics changes occurring during fruit ripening or in response to disease (Chen *et al.*, 2019).

Techniques for analyzing the transcriptome primarily focus on quantifying mRNA levels, either relative or absolute. These techniques fall into two main categories: those based on hybridization and those based on amplification. Microarrays are a prominent example of the hybridization-based approach and were among the first platforms developed for studying genome-wide expression. DNA microarrays offer a systematic and comprehensive way to explore the transcriptome (Brown and Botstein, 1999). This method involves immobilizing specific DNA sequences (probes) on an array, which are then used to detect mRNA in a sample (target). The construction of probes is a critical aspect of microarray design as they must hybridize with very specific sequences. Thus, probe design necessitates a fundamental understanding of the genomic sequence of the species, Consequently, microarray technology has been predominantly applied to model organisms for many years (Witzel *et al.*, 2015).

13.3.3 Proteomics

The ability of crop to adapt to various environmental changes relies on a series of protein modifications resulting from shifts in gene expression. Even proteins that are consistently expressed under normal conditions often exhibit altered expression levels under stress (Hakeem *et al.*, 2012). Proteomics involves the comprehensive analysis of the proteome within an organism, tissue, or cell whether during regular growth and development or in response to environmental changes. The primary goal of proteomics is to uncover protein diversity, abundance, isoforms, localization, interactions with other proteins, and post-translation modifications (PTMs) (Labuschagne, 2018).

While complete genome sequences for many organisms are now accessible, they alone are insufficient to fully understand the biological functions of gene products and validate their predicated roles. Proteomics is a cutting-edge and dynamic field of biology that focuses on studying proteomes—the comprehensive set of proteins present in a cell of an organism at a specific moment. Proteomics is categorized into four main areas: sequence, structure, function, and expression proteomics (Aizat and Hassan, 2018). By examining protein structures, function, and interaction, proteomics provides valuable insights into the cellular mechanisms that underlie various plant traits. Furthermore, it helps identify PTMs, protein–protein interactions, and the effects of environmental factors on protein function. The methodologies employed in proteomics include mass spectrometry (MS), ELISA, western blotting, and matrix-assisted laser desorption ionization-time of flight (MALDI-TOF), along with various bioinformatics tools (Chaudhary *et al.*, 2019).

The 'omics' strategies enhance genomics and transcriptomics by offering a more direct connection to cellular function and phenotypes, as proteins are the primary executors of cellular processes. In horticulture, proteomics is valuable for identifying important proteins that are involved in biological functions such as photosynthesis, respiration, signaling pathways, and stress responses (Table 13.4). For instance, comparing proteomic profiles linked to disease-resistant and disease-susceptible plant varieties can uncover proteins associated with disease resistance. Additionally, studying protein–protein interactions can reveal complex networks that control plant development and responses to environmental factors. Understanding these interactions, particularly those involved in fruit ripening, can lead to improved strategies for enhancing fruit quality and extending shelf life (Wang *et al.*, 2021).

Various strategies are employed to analyze different types of proteomics. For instance, sequence proteomics is typically analyzed using high-performance liquid chromatography (HPLC). Structural proteomics involves techniques such as nuclear magnetic resonance (NMR), electron microscopy, X-ray diffraction of protein crystals, and crystallization. Conversely, functional proteomics is assessed through methods like yeast one-hybrid (Y1H) and two-hybrid (Y2H) systems, as well as protein microarray profiling (Mao *et al.*, 2020).

Table 13.4. An overview of the proteomics research on horticulture crops under various climate conditions. Author's own table.

Crop/tissue	Proteomics approach used	No. of identified proteins	Biological phenomenon	References
Solanum lycopersicon (Tomato)				
Cell culture	2-DE N-terminal sequencing	NS	Identification of heat shock proteins	Nover and Scharf (1984)
Roots	Comparative proteomics approach	~100	Proteins involved in drought tolerance	Zhou *et al.* (2013)
Solanum tuberosum (Potato)				
Shoots	2DE/MALDI-TOF-TOF-MS	94	Cold stress response	Folgado *et al.* (2013)
Shoot tips	2DE–i.e. F/SDS-PAGE MALDI- TOF–MS/MS	100	Drought stress response	Bündig *et al.* (2016)
Daucus carota (Carrot)				
Fruit	NS	3	Cold storage response	Louarn *et al.* (2013)

NS, not specified.

In proteomic studies, two widely used laboratory techniques are protein electrophoresis and MS. Conventional gel-based methods include two-dimensional electrophoresis (2-DE) and difference in-gel electrophoresis (DIGE). These techniques are valued for their simplicity, reproducibility, extensive molecular mass range, and sensitivity to PTMs. However, 2-DE has limitations, including issues with reproducibility, difficulties in detecting less abundant or hydrophobic proteins, challenges in identifying basic proteins, and the potential for co-migration of proteins or issues with very large or small proteins. To address these challenges, multidimensional protein identification technology (MudPIT) has emerged as a non-gel method that allows for both qualitative and quantitative analysis (Lee *et al.*, 2020)

MS techniques, such as liquid chromatography–tandem MS (LC-MS/MS), ion trap–MS (IT–MS), and MALDI–MS, are also widely used. Additionally, advanced methods like fluorophore-tagged protein immunoprecipitation and label-free MS-based quantification have been developed to enhance the precision of identifying low-abundance signaling and regulatory protein complexes (Pittenauer and Allmaier, 2009).

Investigating changes in plant proteomes is essential for understanding how proteins regulate stress responses throughout development and environmental challenges. Variations in protein expression can arise at different development stages and under abiotic stresses, leading to the production of distinct protein sets tailored to specific environmental contexts. Proteomics research plays a pivotal role in identifying these protein markers and exploring how their abundance relates to physiological indicators of stress resilience. By analyzing protein profiles and interactions, proteomics offers valuable insights into plant adaptation mechanisms under adverse conditions.

13.3.4 Metabolomics

Metabolomics focuses on the extensive chemical diversity of plant metabolites, with each species showcasing a unique set that poses analytical challenges in simultaneous profiling and quantification. This advanced omics technology has become increasingly vital for investigating metabolites and their role in crop resilience, particularly with the rise of untargeted metabolome detection (Kumar *et al.*, 2021b). While genomics, transcriptomics, and proteomics elucidate the genes and proteins involved in plant stress responses, a thorough understanding of metabolic pathways is crucial for comprehending how plants react to stressors (Weckwerth and Kahl, 2013).

Metabolomics systematically analyzes all metabolites within an organism, emphasizing how their levels fluctuate in response to environmental stimuli. This approach reveals how gene functions translate into metabolite production throughout different developmental stages and conditions. Unlike other omics disciplines, which focus on upstream processes, metabolomics highlights the biochemical outcomes and physiological effects, as metabolites serve as the end products of gene and protein activity (Zhou *et al.*, 2022).

The metabolome encompasses primary metabolites—essential for cellular functions—and secondary metabolites, which are species specific and play roles in environmental interactions, such as plant defense against biotic and abiotic stresses (Verpoorte *et al.*, 2008). Understanding the metabolome's response to environmental stressors is essential for functional gene tagging and mining, positioning metabolomics as a crucial technology for studying stress adaptation mechanisms.

Environmental changes significantly influence the metabolome, leading to physiological alterations in organisms (Bundy *et al.*, 2005). Stress-tolerant organisms typically exhibit distinct physiological states, making it vital to understand the metabolome under both normal and stress conditions. This understanding allows for the identification of characteristic metabolites linked to stress responses and facilitates the investigation of metabolic pathway alterations and the activation of stress-responsive genes. Thus, metabolomics increasingly provides insights into how organisms adapt to abiotic stress (Jia *et al.*, 2016).

To analyze complex metabolite mixtures, metabolomics employs various separation and detection techniques, including HPLC, gas chromatography (GC), and MS. Advanced methods such as ultraviolet light spectroscopy (UV/VIS), high-resolution mass spectrometry (HRMS), Fourier transform infrared spectroscopy (FTIR), and NMR are also utilized for specialized analysis. Given the dynamic nature of plant metabolism, it is essential to consider all metabolites simultaneously, as individual metabolites can interact and vary in expression (Fischbach and Clardy, 2007). For instance, low oxygen stress during storage can significantly alter metabolic profiles, as demonstrated in a study of cultured tomato cells, which showed altered profiles, including the accumulation of glycolysis intermediates and increased levels of lactate and sugar alcohols (Ampofo-Asiama *et al.*, 2014).

Key methods in plant metabolomics include:

1. Metabolite profiling: this approach focuses on metabolites with similar chemical properties and typically requires separation techniques. It aims to identify and quantify a broad range of metabolites.
2. Metabolic fingerprinting: this method does not involve separation techniques but uses various analyzers to compare spectral data. It provides a general profile of the metabolome and is useful for distinguishing between samples based on their overall metabolic patterns.
3. Targeted analysis: this approach involves the identification and quantitative analysis of specific metabolic compounds of interest. It is used to provide detailed information on particular metabolites.

These techniques can be employed individually or in combination, depending on the specific objectives of the research (Ramalingam *et al.*, 2015).

13.3.5 Ionomics

The term 'ionome' describes the complete set of minerals, nutrients, and trace elements within an organism, representing the inorganic component of cellular and systemic functions. Ionomics, often likened to the functional genomics of elements, involves the quantitative and simultaneous assessment of the elemental makeup of living organisms, including metals, non-metals, and metalloids. This approach measures how the ionome changes in response to physiological conditions, developmental stages, and genetic alterations (Salt *et al.*, 2008).

The primary focus of ionomic studies is the quantitative evaluation of ions within an organism, assessing how their concentrations fluctuate in response to external stimuli. These ions are regulated by various biochemical pathways that facilitate essential processes such as mineral transport and enzymatic activity, both of which are critical for maintaining cellular function. Any disruption in ion regulation can lead to significant shifts in metabolic activity. Consequently, analyzing the plant ionome is vital for comprehending the role of ions in supporting key biological functions. Ionomic research also sheds light on gene regulation, enabling the distinction between natural alleles and mutants that display variations in specific elemental profiles (Chen *et al.*, 2009).

Ions play a pivotal role in maintaining plant homeostasis under diverse environmental conditions. Ion transporters are integral to both metabolism and stress response, governing the composition of elements within the plant and influencing its biochemistry, physiology, and nutrition. Plants have evolved a variety of mechanisms for nutrient uptake, tailored to the specific soil types and environmental conditions that shape their growth and development (Fujita *et al.*, 2013).

Research has demonstrated that certain genes have a direct impact on the ionome and elemental composition in plant tissues. Variations in these genes can lead to shifts in gene expression and alterations in the multi-elemental profile, often due to changes in water and ion transport. The ionomics approach is particularly effective in distinguishing between mutants with distinct elemental profiles and natural genetic variations. Advanced high-throughput technologies, such as inductively coupled plasma–atomic emission spectrometry (ICP–AES) and ICP–MS have proven invaluable in analyzing the ionomic profiles of various plant species (Salt *et al.*, 2008).

13.3.6 Phenomics

The study of phenomics focuses on a wide range of phenotypes, particularly emphasizing genotype–environment–phenotype (G × P × E) interactions in specific conditions, often through high-throughput analyses (Ninomiya, 2022). The term phenotype refers to observable traits such as appearance, behavior, and development, which are influenced by interactions between genetics, the environment, and management. This makes phenomics central to understanding the complex dynamics of genotype–phenotype–environotype relationships (Song *et al.*, 2021).

Phenotype variations arise from a complex interplay between genotypes and a range of external factors. To gain a deeper understanding of the genetic underpinnings of complex traits it is essential to conduct a thorough and accurate assessment of phenotypes, Phenotypes exhibiting desirable traits can serve as valuable sources of key alleles, which can be identified through high-throughput sequencing. While significant advancements have been made in sequencing technologies to enhance genotypic information, progress in phenotyping has been relatively slower, resulting in limited characterization of complex quantitative traits (Zhao *et al.*, 2019).

Traditional phenotyping methods, which include visual trait assessments and physiological measurements, are often slow and labor intensive. The challenge in plant breeding lies in the lack of efficient high-throughput technologies capable of rapidly assessing phenotypes under field conditions. This makes integrating phenomics into breeding programs crucial for selecting high-performing crops. Modern phenomics, which combines genomics, transcriptomics, and metabolomics, is emerging as a key strategy to enhance crop traits and resilience (Biswas *et al.*, 2021). Technological advances, such as near-infrared spectroscopy (NIRS), are enhancing the precision of phenotyping, allowing real-time imaging and analysis of plant traits. Moreover, decreasing costs in genotyping technologies are expected to make genome resequencing a standard practice in identifying allelic variations (Furbank and Tester, 2011). However, while omics technologies have advanced, the precision of these molecular approaches still depends on accurate phenotypic data, which remains a significant challenge for improving crop traits in diverse environmental conditions (Stamp and Visser, 2012).

13.3.6.1 Challenges and advances in phenomics

One of the significant challenges in phenomics is the difficulty of phenotyping large mapping populations, which is essential for identifying QTLs or key genes. This process is often labor intensive and time consuming, particularly when attempting to clone major QTLs from these populations. Additionally, conducting replicated trials across multiple environments and over several years is a resource-intensive and logistically demanding task. Another bottleneck is the reliance on destructive phenotyping methods, where plants are sampled at specific growth stages, preventing continuous monitoring throughout their development. The slow pace and high costs associated with traditional phenotyping further limit the efficiency of evaluating large sample sets. Furthermore, inaccuracies in phenotypic data, especially when assessing allelic variation in candidate genes, can undermine the reliability of these evaluations across different germplasm collections (Furbank and Tester, 2011).

The gap between genotypic and phenotypic data, commonly known as the 'GP gap,' underscores the need for a transformative approach in plant phenotyping. Bridging this divide requires the development of advanced phenomics technologies that can automate and accelerate the phenotyping process, facilitating data collection on a much larger scale.

13.3.6.2 Phenomics revolution

To address the GP gap, recent advancements in phenomics focus on several key innovations. Non-invasive imaging techniques, including high-resolution cameras and scanners, enable real-time monitoring of plant growth and development without damaging the plants. Coupled with this, advanced image analysis software and algorithms allow for the extraction of both quantitative and qualitative phenotypic data, greatly enhancing throughput and accuracy. Spectroscopy methods, particularly NIRS, provide detailed insights into a plant's biochemical composition and physiological state. High-performance computing resources are also crucial, enabling the processing and analysis of the vast datasets generated by high-throughput phenotyping systems. Additionally, robotic systems are increasingly employed to automate the handling and processing of large sample sets, significantly reducing labor costs and time.

These technological advancements are poised to revolutionize plant breeding by integrating with other omics approaches (Table 13.5). The synergy between genomics, transcriptomics, proteomics, metabolomics, ionomics, and phenomics will deepen our understanding of plant traits, leading to more effective breeding strategies. The ability to phenotype plants efficiently and accurately on a large scale will accelerate the development of crops with desirable traits, reducing both time and costs.

Global efforts are underway to refine and implement these cutting-edge technologies. Initiatives like the Plant Phenomics Institute (http://www.plantphenomics.org.au/) are at the forefront of this revolution. As phenomics technologies continue to evolve, they hold great promise for closing the GP gap and advancing crop improvement research.

13.3.7 Metagenomics

Metagenomics has transformed our capacity for learning about microbial communities without the need for culturing, providing insights into the diverse, often unseen microbial life present in various environments, including the plant rhizosphere. This approach is instrumental in exploring microbial diversity, particularly for communities that are difficult or impossible to culture using traditional methods (Chen and Pachter, 2005).

Applications in plant rhizosphere research include:

1. Microbial diversity and distribution: metagenomics allows researchers to analyze the complete microbial community present in the rhizosphere of plants, revealing the distribution and diversity of microorganisms within this critical zone of plant growth. This comprehensive view includes both culturable and non-culturable microbes, providing a more accurate picture of the microbial ecosystem.
2. Plant growth promotion: Metagenomic analysis can identify microbial communities

Table 13.5. A summary of multiomics methods. Author's own table.

Type of omics	Definition	Common methods used	Application in horticulture
Genomics	The investigation of a species' whole gene pool, or genome, and its roles.	Whole-genome sequencing (WGS) genotyping by sequencing (GBS)	Pangenome analysis, plant breeding, genetic diversity resistance research
Transcriptomics	The investigation of the entire collection of RNA transcripts generated by the genome in particular conditions.	Single-cell RNA sequencing (scRNA-Seq), RNA-Seq, microarray analysis	Understanding plant response to stress, gene expression studies, identification of key regulatory genes
Proteomics	The investigation of the entire protein family as gene expressions and their roles	Two-dimensional gel electrophoresis mass spectrometry	Studying protein interaction networks, protein expression analysis, discovering disease resistance proteins
Metabolomics	The examination of all the small-molecule chemicals in a biological sample	Liquid chromatography-mass spectrometry (LC–MS), gas chromatography–mass spectrometry (GC–MS)	Profiling of plant-targeted and untargeted metabolites understanding plant metabolic pathways, flavour and fragrance research
Ionomics	Refers to the quantitative assessment of all ions present in a living organism and examines how ion production varies in response to various external stimuli.	ICP-MS and ICP–AES are two examples of high-throughput elemental analysis	Element identification plant morpho-physiology
Phenomics	Investigation of the entire collection of phenotypes involving interactions between genotype and environment (G × P×E) under certain environmental conditions	NIRS (near-infrared spectroscopy)	Genotype–phenotype–environotype interactions

with plant growth-promoting traits. These microbes can enhance plant growth through various mechanisms such as nitrogen fixation, phosphorus solubilization, and hormone production.

3. Biocontrol: metagenomics helps in identifying microbes with biocontrol capabilities, which can suppress plant pathogens through competition, production of antimicrobial compounds, or inducing systemic resistance in plants.
4. Antibiotic production: by studying the genomic content of microbial communities, researchers can discover novel antibiotics produced by microbes in the rhizosphere, which could be valuable for agricultural and medical applications.
5. Xenobiotic degradation: metagenomic approaches can identify microbes capable of degrading xenobiotics (e.g. pesticides, pollutants) in the soil, which can be crucial for bioremediation efforts and soil health management (Handelsman *et al.*, 2007).

Advancements in metagenomics include:

1. High-throughput sequencing: sequencing technology advancements have significantly

increased the throughput and resolution of metagenomic analyses. This enables the detailed characterization of microbial communities at a much higher depth, uncovering previously hidden microbial diversity and functional potential.

2. Bioinformatics tools: sophisticated bioinformatics tools and software are used to analyze metagenomic data, allowing for the identification of microbial taxa, functional genes, and metabolic pathways. This facilitates the understanding of complex interactions within microbial communities and their roles in plant health.

The integration of metagenomics with other omics approaches, such as genomics and metabolomics, will further enhance our understanding of microbial interactions in the plant rhizosphere. As metagenomics technologies continue to evolve, they hold great promise for improving plant health, optimizing crop yields, and developing sustainable agricultural practices.

By exploring the microbial communities present in the plant rhizosphere, metagenomics provides valuable insights into the complex interactions between plants and their associated microbes, paving the way for innovative solutions in agriculture and environmental management.

13.3.7.1 Advanced omics approaches for microbiome analysis

To understand the composition, diversity, and functional potential of microbiomes, especially in contexts like plant rhizospheres, several advanced omics approaches are employed:

1. Metagenomics: this offers a thorough perspective on the microbial community by sequencing all the DNA present in a sample. It aids in determining the diversity of microorganisms and their potential functions according to the existence of particular genes.
2. Metatranscriptomics: this focuses on the RNA transcript present in a microbial community at a given time, providing insights into gene expression and functional activity.
3. Metaproteomics: this analyzes the entire set of proteins expressed by a microbial community, reflecting the functional state of the microbiome.

The combined application of metagenomics, metatranscriptomics, and metaproteomics is expected to provide a more complete understanding of microbiome dynamics and interactions in various environments. Advancements in sequencing technologies, bioinformatics, and high-throughput analyses will continue to enhance the resolution and scope of microbiome studies, paving the way for improved stress management strategies and microbial application in agriculture (Turner *et al.*, 2013).

13.3.8 Methylomics

One heritable epigenetic process is DNA methylation. The five common sequencing techniques for DNA methylation are: HpaIItiny fragment enrichment by ligation-mediated PCR (HELP-seq) (Oda *et al.*, 2009), oxidative bisulfite sequencing (OxBS-seq), reduced representation bisulfite sequencing (RRBS), methylation DNA immune precipitation sequencing (MeDIP-seq), and whole genome bisulfite methylation sequencing (WGBS). These techniques can be applied in a variety of research directions.

13.3.9 Interactomics

Interactomics explores how biomolecules interact within a cell and the outcomes of these interactions. It examines the intricate physical, biochemical, and functional relationships between DNA, RNA, proteins, lipids, and small metabolites that drive cellular activities. While the term ‘interactome’ often refers to networks of protein–protein interactions, another crucial aspect is the protein–DNA interactome or gene-regulatory network. In plants, this interactome includes transcription factors (TFs) and chromatin-regulatory proteins interacting with their target genes. By studying these protein–protein and protein–DNA interactions under both normal and stress conditions, we can better understand the regulatory mechanisms underlying stress tolerance.

Recent advancements in interactomics have led to the development of various technologies, which are broadly classified into three main categories:

1. *In silico* methods involve computational techniques such as simulation, text mining, and data analysis to study biomolecular interactions.
2. *In vivo* methods are conducted on living organisms and include techniques like protein-fragment complementation assay (PFCA), yeast two-hybrid (Y2H), and protein–protein interaction trap (MAPPIT).
3. *In vitro* methods are conducted outside living organisms in controlled lab environments, utilizing techniques such as tandem affinity purification–MS(TAP-MS), protein microarrays, and luminescence-based mammalian interactome (LUMIER) tools.

While interactomics has started to be applied to plant studies, particularly regarding abiotic stress tolerance, the high-throughput application of these methods in plants is still developing (Low *et al.*, 2019).

13.3.10 Proteogenomics

Proteogenomics is a cutting-edge, integrative approach that merges advancements in genomics, transcriptomics, and proteomics. This method aims to catalog the proteins expressed within a cell by integrating high-throughput NGS data with MS-based proteomics. Specifically, proteogenomics excels in identifying proteins by combining genomic, transcriptomic, and proteomic data from the same crop species or sample. This integrative approach has led to the discovery of novel proteins and has provided deep insights into gene regulatory mechanisms and cellular signaling related to abiotic stress tolerance, Proteogenomics has illuminated the ways plants respond to abiotic stress and adapt to environmental changes, with a particular focus on the role of protein phosphorylation in these processes (Hajheidari and Huang, 2022).

13.4 Challenges in Multiomics Approaches

One of the significant hurdles facing agricultural systems today is the looming environmental crisis. With the global population expected to reach 10 billion in the next few decades, the strain on food production systems is becoming more pronounced. This stress is compounded by projections of global temperature increases, estimated to rise by 1.1 to 5.4°C by the century's end (Tollefson, 2020). Such climate shifts are anticipated to cause crops to endure more frequent and severe bouts of heat stress, drought, and other abiotic pressures during critical growth periods. Furthermore, the intensifying effects of climate change are likely to exacerbate a range of abiotic stresses, including drought, salinity, heat, cold and submergence.

To mitigate these challenges, recent advances in molecular plant breeding are employing multiomics strategies. These approaches allow plant breeders to incorporate desirable genetic traits into crop genomes with increased precision and speed, utilizing an expanded gene pool. By integrating multiomics technologies, crop improvement efforts are becoming more targeted and effective, blending modern molecular techniques with traditional breeding practices for superior outcomes (Rio *et al.*, 2022).

Despite the promising advances, multiomics research also faces several key challenges, particularly in the integration and management of large datasets. First, data archiving and sharing pose a significant barrier. As the volume and complexity of omics data grow, ensuring reproducibility and accessibility of these datasets becomes crucial. The principles of findability, accessibility, interoperability, and reusability (FAIR) guide the development of data-sharing frameworks, enabling researchers to access and build on existing data while maintaining transparency and reliability.

Another critical challenge is data interpretation. Multiomics research generates vast amounts of complex data, making it difficult to analyze and derive meaningful conclusions. A central goal of these studies is the discovery of biomarkers—molecular indicators that can predict specific traits or responses in crops. For these biomarkers to be useful in breeding programs, they must be identified with high specificity and sensitivity, followed by rigorous validation. Accurate biomarker discovery is essential for translating multiomics data into practical improvements in crop resilience and performance.

As multiomics technologies continue to advance, addressing these integration challenges will be crucial for achieving breakthroughs in crop improvement. By refining data-sharing mechanisms and improving biomarker identification processes, researchers can better harness the power of these technologies to overcome the pressing environmental challenges of modern agriculture.

13.5 Future Directions in Omics Research for Crop Improvement

Single-cell transcriptomics and ionomics have emerged as promising fields with the potential to address complex biological questions. However, these areas face several technical and experimental design challenges that need to be addressed. Despite these hurdles, advancements in gene editing, bioinformatics tools, and machine learning have made significant strides, providing robust support for these omics technologies.

By integrating multiomics approaches with modern plant breeding techniques, high-throughput experimental methods, advanced bioinformatics, and computational modeling, researchers can expect to develop crops that are more productive, nutritionally rich, and resilient to climate change. This integration not only deepens our understanding of plant biology but also helps identify key traits that contribute to crop improvement.

Omics-based research has proven to be an effective platform for linking extensive genomic knowledge with the diverse improvement needs of various crops. Techniques such as proteomics and transcriptomics are invaluable for studying organ development, non-coding RNAs, and responses to abiotic and biotic stresses, as well as other critical agronomic traits. Consequently, there is an urgent need for further omics-based research, particularly in proteomics and metabolomics, to advance molecular breeding in horticultural crops and enhance their resilience and productivity.

13.6 Conclusions

Agriculture's reliance on climate makes it particularly vulnerable to stresses caused by changing weather patterns, resulting in substantial annual losses. During the last 20 years, significant advancements have been made using cutting-edge omics technologies to develop climate-smart cultivars. Given the rapid increase in the global population and the ongoing impacts of climate change, there is an urgent need for innovative biotechnological and breeding tools to create crop varieties that are resistant or tolerant to various stresses.

In this context, the field of horticulture is increasingly integrating omics technologies—such as genomics, transcriptomics, proteomics, and metabolomics—to advance research and development. Significant efforts have been directed toward developing genetic maps using diverse techniques and mapping populations. These efforts play a crucial role in improving yield-related traits through approaches like comparative genomics, map-based cloning, QTL/gene mapping, and marker-assisted selection.

In conclusion, omics-based breeding research and advanced technologies are vital for improving crop yield, nutritional content, and stress tolerance to address the growing global food demand. To fully realize these advancements, it is important for plant scientists, computational biologists, breeding companies, and farmers to collaborate across disciplines. Building a cohesive community that shares knowledge and explores innovative multiomics approaches will drive ground-breaking discoveries in crop breeding.

References

Aizat, W.M. and Hassan, M. (2018) "Proteomics in systems biology," in omics applications for systems biology. In: Aizat, W., Goh, H.H. and Baharum, S. (eds) *Omics Applications for Systems Biology*. Advances in Experimental Medicine and Biology, Springer, Cham, pp. 31–49. DOI: 10.1007/978-3-319-98758-3_3.

Akinci, S. (1997) Physiological responses to water stress by *Cucumis sativus* L. and related species. PhD thesis, University of Sheffield. Department of Animal and Plant Sciences, England.

Ampofo-Asiama, J., Baiye, V., Hertog, M., Waelkens, E., Geeraerd, A. *et al.* (2014) The metabolic response of cultured tomato cells to low oxygen stress. *Plant Biology* 16, 594–606.

Arias, L.A., Berli, F., Fontana, A., Bottini, R. and Piccoli, P. (2022) Climate change effects on grapevine physiology and biochemistry: Benefits and challenges of high altitude as an adaptation strategy. *Frontiers in Plant Science* 13. DOI: 10.3389/fpls.2022.835425.

Avni, R., Lux, T., Minz-Dub, A., Millet, E., Sela, H. *et al.* (2022) Genome sequences of three *Aegilops* species of the section Sitopsis reveal phylogenetic relationships and provide resources for wheat improvement. *The Plant Journal* 110, 179–192.

Biswas, A., Andrade, M.H.M.L., Acharya, J.P., Souza, C.L., Lopez, Y. *et al.* (2021) Phenomics-assisted selection for herbage accumulation in Alfalfa (*Medicago sativa* L.). *Frontiers in Plant Science* 12, 756768–756768. DOI: 10.3389/fpls.2021.756768.

Bohnert, H.J., Gong, Q., Li, P. and Ma, S. (2006) Unraveling abiotic stress tolerance mechanisms–getting genomics going. *Current Opinion in Plant Biology* 9, 180. DOI: 10.1016/j.pbi.2006.01.003.

Branham, S.E., Stansell, Z.J., Couillard, D.M. and Farnham, M.W. (2017) Quantitative trait loci mapping of heat tolerance in broccoli (*Brassica oleracea* var. *italica*) using genotyping-by-sequencing. *Theoretical and Applied Genetics* 130(3), 529–538.

Brown, P.O. and Botstein, D. (1999) Exploring the new world of the genome with DNA microarrays. *Nature Genetics* 21(1), 33–37. DOI: 10.1038/4462.

Bündig, C., Jozefowicz, A.M., Mock, H.P. and Winkelmann, T. (2016) Proteomic analysis of two divergently responding potato genotypes (*Sola num tuberosum* L.) following osmotic stress treatment *in vitro*. *Journal of Proteomics* 143(227), 241. DOI: 10.1016/j.jprot.2016.04.048.

Bundy, J.G., Willey, T.L., Castell, R.S., Ellar, D.J. and Brindle, K.M. (2005) Discrimination of pathogenic clinical isolates and laboratory strains of *Bacillus cereus* by NMR-based metabolomic profiling. *FEMS Microbiology Letters* 242, 127–136. DOI: 10.1016/j.femsle.2004.10.048.

Cassia, R., Nocioni, M., Correa-Aragunde, N. and Lamattina, L. (2018) Climate change and the impact of greenhouse gasses: CO2 and NO, friends and foes of plant oxidative stress. *Frontiers in Plant Science* 9, 273.

Chaturvedi, P., Wiese, A.J., Ghatak, A., Záveská Drábková, L. and Weckwerth, W. (2021) Heatstress response mechanisms in pollen development. *NewPhytologist* 231, 571–585. DOI: 10.1111/nph.17380.

Chaudhary, J., Khatri, P., Singla, P., Kumawat, S., Kumari, A. *et al.* (2019) Advances in omics approaches for abiotic stress tolerance in tomato. *Biology* 8(4), 90.

Chen, K. and Pachter, L. (2005) Bioinformatics for whole genome shotgun sequencing of microbial communities. *PLoS Computational Biology* 1, 106–112. DOI: 10.1371/journal.pcbi.0010024.

Chen, G., Ning, B. and Shi, T. (2019) Single-cell RNA-seq technologies and related computational data analysis. *Frontiers in Genetics*10, 317.

Chen, H., Liu, N., Xu, R., Chen, X., Zhang, Y. *et al.* (2021) Quantitative proteomics analysis reveals the response mechanism of peanut (*Arachis hypogaea* L.) to imbibitional chilling stress. *Plant Biology* 23(3), 517–527.

Chen, Z., Shinano, T., Ezawa, T., Wasaki, J., Kimura, K. *et al.* (2009) Elemental interconnections in *Lotus japonicus*: A systematic study of the effects of elements additions on different natural variants. *Soil Science and Plant Nutrition* 55, 91–101. DOI: 10.1111/j.1747-0765.2008.00311.x.

Curaba, J., Singh, M.B. and Bhalla, P.L. (2014) MiRNAs in the crosstalk between phytohormone signalling pathways. *Journal of Experimental Botany*65, 1425–1438. DOI: 10.1093/jxb/eru002.

Demirel, U., Morris, W.L. and Ducreux, L.J.M. (2020) Physiological, bio chemical, and transcriptional responses to single and combined abiotic stress in stress-tolerant and stress-sensitive potato geno types. *Frontiers Plant Science* 11(169). DOI: 10.3389/fpls.2020.00169.

Denoeud, F., Carretero-Paulet, L., Dereeper, A., Droc, G., Guyot, R. *et al.* (2014) The coffee genome provides insight into the convergent evolution of caffeine biosynthesis. *Science* 345(6201), 1181–1184. DOI: 10.1126/science.1255274.

Devireddy, A.R., Zandalinas, S.I., Fichman, Y. and Mittler, R. (2021) Integration of reactive oxygen species and hormone signaling during abiotic stress. *The Plant Journal* 105, 459–476. DOI: 10.1111/tpj.15010.

D'hont, A., Denoeud, F., Aury, J.M., Baurens, F.C., Carreel, F. *et al.* (2012) The banana (*Musa acuminata*) genome and the evolution of monocotyledonous plants. *Nature* 488(7410), 213–217.

Diouf, I.A., Derivot, L., Bitton, F., Pascual, L. and Causse, M. (2018) Water deficit and salinity stress reveal many specific QTL for plant growth and fruit quality traits in tomato. *Frontiers in Plant Science* 9, 279.

Fang, C., Hamilton, J.P., Vaillancourt, B., Wang, Y.W., Wood, J.C. *et al.* (2023) Cold stress induces differential gene expression of retained homeologs in *Camelina sativa* cv Suneson. *Frontiers in Plant Science* 14, 1271625. DOI: 10.3389/fpls.2023.1271625.

Fang, Y., Qin, X., Liao, Q., Du, R., Luo, X. *et al.* (2021) The genome of homosporous maidenhair fern sheds light on the euphyllophyte evolution and defences. *Nature Plants* 8, 1024–1037.

Fischbach, M.A. and Clardy, J. (2007) One pathway, many products. *Nature Chemical Biology* 3, 353–355. DOI: 10.1038/nchembio0707-353.

Folgado, R., Panis, B. and Sergeant, K. (2013) Differential protein expression in response to abiotic stress in two potato species: *Solanum commersonii* Dun and *Solanum tuberosum* L. *International Journal of Molecular Sciences* 14, 4912–4933. DOI: 10.3390/ijms14034912.

Fujita, Y., Venterink, H.O., Van Bodegom, P.M., Douma, J.C., Heil, G.W. *et al.* (2013) Low investment in sexual reproduction threatens plants adapted to phosphorus limitation. *Nature* 505, 82–86. DOI: 10.1038/nature12733.

Furbank, R.T. and Tester, M. (2011) Phenomics-technologies to relieve the phenotyping bottleneck. *Trends Plant Science* 16, 635–644.

Fu, X.Z., Zhang, X.Y., Qiu, J.Y., Zhou, X., Yuan, M. *et al.* (2019) Whole-transcriptome RNA sequencing reveals the global molecular responses and ceRNA regulatory network of mRNAs, lncRNAs, miRNAs and circRNAs in response to copper toxicity in Ziyang Xiangcheng (*Citrus junos* Sieb. Ex Tanaka). *BMC Plant Biology* 19, 509. DOI: 10.1186/s12870-019-2087-1.

Ghodke, P., Khandagale, K. and Thangasamy, A. (2020) Comparative transcriptome analyses in contrasting onion (*Allium cepa* L.) genotypes for drought stress. *PLoS ONE 15* e0237457. DOI: 10.1371/journal.pone.0237457.

Guo, H., Mao, M., Deng, Y., Sun, L., Chen, R. *et al.* (2022) Multi-omics analysis reveals that SlERF. D6 synergistically regulates SGAs and fruit development. *Frontiers in Plant Science* 13, 860577.

Guo, S., Zuo, Y., Zang, Y., Wu, C., Su, W. *et al.* (2017) Large-scale transcriptome comparison of sunflower genes responsive to *Verticillium dahliae*. *BMC Genomics* 18, 42.

Hajheidari, M. and Huang, S.C. (2022) Elucidating the biology of transcription factor–DNA interaction for accurate identification of Cis-regulatory elements. *Current Opinion in Plant Biology* 68, 102232.

Hakeem, K.R., Chandna, R., Ahmad, P., Iqbal, M. and Ozturk, M. (2012) Relevance of proteomic investigations in plant abiotic stress physiology. *Omics* 16, 621–635. DOI: 10.1089/omi.2012.0041.

Han, J., Thamilarasan, S.K. and Natarajan, S. (2016) De novo assembly and transcriptome analysis of bulb onion (*Allium cepa* L.) dur ing cold acclimation using contrasting genotypes. *PLoS ONE 11* e0161987. DOI: 10.1371/journal.pone.0161987.

Handelsman, J., Tiedje, J., Alvarez-Cohen, L., Ashburner, M., Cann, I.K.O. *et al.* (2007) *The New Science of Metagenomics: Revealing the Secrets of Our Microbial Planet*. National Academy of Science, Washington, DC.

Huang, S., Li, R., Zhang, Z., Li, L.I., Gu, X. *et al.* (2009) The genome of the cucumber, *Cucumis sativus* L. *Nature Genetics* 41(12), 1275–1281.

Iorizzo, M., Ellison, S., Senalik, D., Zeng, P., Satapoomin, P. *et al.* (2016) A high-quality carrot genome assembly provides new insights into carotenoid accumulation and asterid genome evolution. *Nature Genetics* 48(6), 657–666.

Iovieno, P., Punzo, P. and Guida, G. (2016) Transcriptomic changes drive physiological responses to progressive drought stress and rehydration in tomato. *Frontiers Plant Science* 7, 371. DOI: 10.3389/fpls.2016.00371.

Jayakodi, M., Choi, BS. and Lee, S.C. (2018) Ginseng Genome Database: An open-access platform for genomics of *Panax ginseng*. *BMC Plant Biology* 18, 62.

Jia, X., Sun, C., Zuo, Y., Li, G., Li, G. *et al.* (2016) Integrating transcriptomics and metabolomics to characterise the response of *Astragalus membranaceus* Bge. var. *mongolicus* (Bge.) to progressive drought stress. *BMC Genomics* 17, 188. DOI: 10.1186/s12864-016-2554-0.

Kersey, P.J. (2019) Plant genome sequences: Past, present, future. *Current Opinion Plant Biology* 48, 1–8. DOI: 10.1016/j.pbi.2018.11.001.

Khalid, N., Aqeel, M. and Noman, A. (2019) System biology of metal tolerance in plants: an integrated view of genomics, transcriptomics, metabolomics, and phenomics. In: Sablok, G. (ed.) *Plant Metallomics and Functional Omics*. Cham: Springer. DOI: 10.1007/978-3-030-19103-0_6.

Khan, M.A., Saravia, D., Munive, S., Lozano, F., Farfan, E. *et al.* (2015) Multiple QTLs linked to agro-morphological and physiological traits related to drought tolerance in potato. *Plant Molecular Biology Reporter* 33(5), 1286–1298. DOI: 10.1007/s11105-014-0824-z.

Kim, S., Park, M., Yeom, S.I., Kim, Y.M., Lee, J.M. *et al.* (2014) Genome sequence of the hot pepper provides insights into the evolution of pungency in *Capsicum species*. *Nature Genetics* 46(3), 270–278.

Knudsen, S. (2004) *Guide to Analysis of DNA Microarray Data*, 2nd edn. Wiley-Liss, Hoboken, N.J. DOI: 10.1002/0471670278.

Kumar, J., Sen Gupta, D., Djalovic, I., Kumar, S. and Siddique, K.H. (2021a) Root-omics for drought tolerance in cool-season grain legumes. *Physiologia Plantarum* 172, 629–644. DOI: 10.1111/ppl.13313.

Kumar, R., Sharma, V., Suresh, S., Ramrao, D.P., Veershetty, A. *et al.* (2021b) Understanding omics driven plant improvement and de novo crop domestication: Some examples. *Frontiers Genetics* 12. DOI: 10.3389/fgene.2021.637141.

Labuschagne, M.T. (2018) Review of cereal grain proteomics and its potential for sorghum improvement. *Journal of Cereal Science* 84, 151–158. DOI: 10.1016/j.jcs.2018.10.010.

Lee, P.Y., Saraygord-Afshari, N. and Low, T.Y. (2020) The evolution of two-dimensional gel electrophoresis-from proteomics to emerging alternative applications. *Journal of Chromatography A, 1615* 460763.

Li, Y., Chen, J., Li, X., Jiang, H., Guo, D. *et al.* (2022) Adaptive response and transcriptomic analysis of flax (*Linum usitatissimum* L.) seedlings to salt stress. *Genes* 13(10). DOI: 10.3390/genes13101904.

Liliane, T.N. and Charles, M.S. (2020) Factors affecting yield of crops. In: *Agronomy-Climate Change & Food Security*. IntechOpen, London, UK, p. 9.

Li, Q., Zhao, P., Li, J., Zhang, C., Wang, L. *et al.* (2014) Genome wide analysis of the WD-repeat protein family in cucumber and *Arabidopsis*. *Molecular Genetics and Genomics* 289, 103–124. DOI: 10.1007/s00438-013.

Louarn, S., Nawrocki, A. and Thorup-Kristensen, K. (2013) Proteomic changes and endophytic micromycota during storage of organically and conventionally grown carrots. *Postharvest Biology and Technology* 76, 26–33. DOI: 10.1016/j.postharvbio.2012.08.011.

Low, T.Y., Mohtar, M.A., Ang, M.Y. and Jamal, R. (2019) Connecting proteomics to next-generation sequencing: Proteogenomics and its current applications in biology. *Proteomics* 19, e1800235.

Mao, C., He, J., Liu, L., Deng, Q., Yao, X. *et al.* (2020) OsNAC2 integrates auxin and cytokinin pathways to modulate rice root development. *Plant Biotechnology Journal* 18, 429–442. DOI: 10.1111/pbi.13209.

Matschegewski, C., Zetzsche, H., Hasan, Y., Leibeguth, L., Briggs, W. *et al.* (2015) Genetic variation of temperature-regulated curd induction in cauliflower: Elucidation of floral transition by genome-wide association mapping and gene expression analysis. *Frontiers in Plant Science* 6, 720. DOI: 10.3389/fpls.2015.00720.

McCouch, S. (2013) Feeding the future. *Nature* 499, 23–24.

Miao, W., Song, J. and Huang, Y. (2021) Comparative transcriptomics for pepper (*Capsicum annuum* L.) under cold stress and after rewarming. *Applied Science* 11(10204). DOI: 10.3390/app112110204.

Ming, R., Hou, S., Feng, Y., Yu, Q., Dionne-Laporte, A. *et al.* (2008) The draft genome of the transgenic tropical fruit tree papaya (*Carica papaya* Linnaeus). *Nature* 452(7190), 991–996.

Ninomiya, S. (2022) High-throughput field crop phenotyping: Current status and challenges. *Breeding Science* 72, 3–18. DOI: 10.1270/jsbbs.21069.

Nover, L. and Scharf, K.-D. (1984) Synthesis, modification and structural binding of heat-shock proteins in tomato cell cultures. *European Journal of Biochemistry* 139(303–313), 1432–1033. DOI: 10.1111/j.1432-1033.1984.tb08008.x.

Oda, M., Glass, J.L., Thompson, R.F., Mo, Y., Olivier, E.N. *et al.* (2009) High-resolution genome-wide cytosine methylation profiling with simultaneous copy number analysis and optimization for limited cell numbers. *Nucleic Acids Research* 37, 3829–3839.

Pazhamala, L.T., Kudapa, H., Weckwerth, W., Millar, A.H. and Varshney, R.K. (2021) Systems biology for crop improvement. *Plant Genome* 14, e20098. DOI: 10.1002/tpg2.20098.

Pittenauer, E. and Allmaier, G. (2009) High-energy collision induced dissociation of biomolecules: MALDITOF/RTOF mass spectrometry in comparison to tandem sector mass spectrometry. *Combinatorial Chemistry & High Throughput Screening* 12(2), 137–155.

Qin, G., Xu, C., Ming, R., Tang, H., Guyot, R. *et al.* (2017) The pomegranate (*Punica granatum* L.) genome and the genomics of punicalagin biosynthesis. *The Plant Journal* 91(6), 1108–1128.

Ramalingam, A., Kudapa, H., Pazhamala, L.T., Weckwerth, W. and Varshney, R.K. (2015) Proteomics and metabolomics: Two emerging areas for legume improvement. *Frontiers in Plant Science* 6, 1116. DOI: 10.3389/fpls.2015.01116.

Rao, N.K.S., Laxman, R.H. and Shivashankara, K.S. (2016) Physiological and morphological responses of horticultural crops to abiotic stresses. *Abiotic stress physiology of horticultural crops*10, 1007. DOI: 10.1007/978-81-322-2725-0_1.

Reyes-Chin-Wo, S., Wang, Z., Yang, X., Kozik, A., Arikit, S. *et al.* (2017) Genome assembly with in vitro proximity ligation data and whole-genome triplication in lettuce. *Nature Communications* 8(1), 14953.

Rio, S., Akdemir, D., Carvalho, T. and Sánchez, J.I.Y. (2022) Assessment of genomic prediction reliability and optimization of experimental designs in multi-environment trials. *Theoretical and Applied Genetics* 135, 405–419.

Romero, P., Fernandez-Fernandez, J.I. and Botia, P. (2016) Interannual climatic variability effects on yield, berry and wine quality indices in long-term deficit irrigated grapevines, determined by multivariate analysis. *International Journal of Wine Research* 8, 3–17. DOI: 10.2147/IJWR.S107312.

Sahu, M., Dehury, B., Modi, M.K. and Barooah, M. (2014) Functional genomics and bioinformatics approach to understand regulation of abiotic stress in cereal crops. In: *Crop Improvement in the Era of Climate*. I.K.International Publishing House Pvt. Ltd, Delhi, India, p. 205.

Salt, D.E., Baxter, I. and Lahner, B. (2008) Ionomics and the study of the plant ionome. *Annual Review of Plant Biology* 59, 709–733. DOI: 10.1146/annurev.arplant.59.032607.092942.

Shi, F., Pang, Z., Liu, C., Zhou, L., Tan, C. *et al.* (2022) Whole transcriptome analysis and construction of an anther development-related ceRNA network in Chinese cabbage (*Brassica campestris* L. ssp. *pekinensis*). *Scientific Reports* 12, 2667. DOI: 10.1038/s41598-022-06556-2.

Shulaev, V., Sargent, D.J., Crowhurst, R.N., Mockler, T.C., Folkerts, O. *et al.* (2011) The genome of woodland strawberry (*Fragaria vesca*). *Nature Genetics* 43(2), 109–116.

Singh, R., Kumar, K., Bharadwaj, C. and Verma, P.K. (2022) Broadening the horizon of crop research: A decade of advancements in plant molecular genetics to divulge phenotype governing genes. *Planta* 255, 46.

Song, P., Wang, J., Guo, X., Yang, W. and Zhao, C. (2021) High-throughput phenotyping: Breaking through the bottleneck in future crop breeding. *The Crop Journal* 9(S9).

Stamp, P. and Visser, R. (2012) The twenty-first century, the century of plant breeding. *Euphytica* 186, 585–591.

Su, X., Gao, T., Zhang, P., Li, F., Wang, D. *et al.* (2022) Comparative physiological and transcriptomic analysis of sesame cultivars with different tolerance responses to heat stress. *Physiology and Molecular Biology of Plants* 28, 1131–1146.

Takagi, H., Abe, A. and Yoshida, K. (2013) QTL-seq: Rapid mapping of quantitative trait loci in rice by whole genome resequencing of DNA from two bulked populations. *The Plant Journal* 74, 174–183. DOI: 10.1111/tpj.12105.

Tang, L. (2017) Cultural changes and food production. *International Journal of Business and Social Research* 7, 19–33.

The Potato Genome Sequencing Consortium (2011) Genome sequence and analysis of the tuber crop potato. *Nature* 475, 189–195.

The Tomato Genome Consortium (2012) The tomato genome sequence provides insights into fleshy fruit evolution. *Nature* 485, 635–641.

Tollefson, J. (2020) How hot will earth get by 2100. *Nature* 580(7804), 443–445.

Trapero-Mozos, A., Morris, W.L., Ducreux, L.J., McLean, K., Stephens, J. *et al.* (2018) Engineering heat tolerance in potato by temperature-dependent expression of a specific allele of HEAT-SHOCK COGNATE 70. *Plant Biotechnology Journal* 16(1), 197–207.

Ts, A., Srivastava, A., Tomar, B.S., Behera, T.K., Krishna, H. *et al.* (2023) Genetic analysis of heat tolerance in hot pepper: Insights from comprehensive phenotyping and QTL mapping. *Frontiers in Plant Science* 14, 1232800. DOI: 10.3389/fpls.2023.1232800.

Turner, T.R., Ramakrishnan, K., Walshaw, J., Heavens, D., Alston, M. *et al.* (2013) Comparative metatranscriptomics reveals kingdom level changes in the rhizosphere microbiome of plants. *The ISME Journal* 7, 2248–2258. DOI: 10.1038/ismej.2013.119.

Unver, T., Wu, Z., Sterck, L., Turktas, M., Lohaus, R. *et al.* (2017) Genome of wild olive and the evolution of oil biosynthesis. *Proceedings of the National Academy of Sciences* 114(44), E9413–E9422.

Varshney, R.K., Bohra, A., Roorkiwal, M., Barmukh, R., Cowling, W.A. *et al.* (2021) Fast-forward breeding for a food-secure world. *Trends in Genetics* 37, 1124–1136.

Velasco, R., Zharkikh, A., Troggio, M., Cartwright, D.A., Cestaro, A. *et al.* (2007) A high quality draft consensus sequence of the genome of a heterozygous grapevine variety. *PloS One* 2(12), e1326.

Velasco, R., Zharkikh, A., Affourtit, J., Dhingra, A., Cestaro, A. *et al.* (2010) The genome of the domesticated apple (*Malus*× *domestica* Borkh). *Nature Genetics* 42(10), 833–839.

Velculescu, V.E., Zhang, L., Vogelstein, B. and Kinzler, W.K. (1997) Serial analysis of gene expression. *Science* 270, 484–487.

Verde, I., Abbott, A.G., Scalabrin, S., Jung, S., Shu, S. *et al.* (2013) The high-quality draft genome of peach (*Prunus persica*) identifies unique patterns of genetic diversity, domestication and genome evolution. *Nature Genetics* 45(5).

Verpoorte, R., Choi, Y.H., Mustafa, N.R. and Kim, H.K. (2008) Metabolomics: Back to basics. *Phytochem Rev* 7, 525–537. DOI: 10.1007/s11101-008-9091-7.

Wang, P., Luo, Y., Huang, J., Gao, S., Zhu, G. *et al.* (2020) The genome evolution and domestication of tropical fruit mango. *Genome Biology* 21(1), 60.

Wang, Y., Guo, L., Zhao, X., Zhao, Y., Hao, Z. *et al.* (2021) Advances in mechanisms and omics pertaining to fruit cracking in horticultural plants. *Agronomy* 11, 1045.

Weckwerth, W. and Kahl, G. (2013) *The Handbook of Plant Metabolomics*. JohnWiley and Sons, New York, NY. DOI: 10.1002/9783527669882.

Wei, S., Zhang, L. and Huo, G. (2021) Comparative transcriptomics and metabolomics analyses provide insights into thermal resistance in lettuce (*Lactuca sativa* L.). *Scientia Horticulturae* 289, 110423. DOI: 10.1016/j.scienta.2021.110423.

Witzel, K., Neugart, S., Ruppel, S., Schreiner, M., Wiesner, M. *et al.* (2015) Recent progress in the use of "omics" technologies in Brassicaceous vegetables. *Frontiers in Plant Science* 6, 244.

Wu, J., Wang, Z., Shi, Z., Zhang, S. and Ming, R. (2013) The genome of the pear (*Pyrus bretschneideri* Rehd). *Genome Research* 23(2), 396–408.

Xu, C., Jiao, C., Sun, H., Cai, X., Wang, X. *et al.* (2017) Draft genome of spinach and transcriptome diversity of 120 Spinacia accessions. *Nature Communications* 8(1), 15275.

Xu, Y., Guo, S.R., Shu, S., Ren, Y. and Sun, J. (2017) Construction of a genetic linkage map of rootstock-used pumpkin using SSR markers and QTL analysis for cold tolerance. *Scientia Horticulturae* 220, 107–113.

Zapata, D., Salazar-Gutierrez, M., Chaves, B., Keller, M. and Hoogenboom, G. (2017) Predicting key phenological stages for 17 grapevine cultivars (*Vitis vinifera* L.). *American Journal of Enology and Viticulture*68, 60–72. DOI: 10.5344/ajev.2016.15077.

Zhang, N., Liu, B. and Ma, C. (2014) Transcriptome characterization and sequencing-based identification of drought-responsive genes in potato. *Molecular Biology Reports* 41, 505–517. DOI: 10.1007/s11033-013-2886-7.

Zhang, X., Zou, Z., Gong, P., Zhang, J., Ziaf, K. *et al.* (2011) Over-expression of microRNA169 confers enhanced drought tolerance to tomato. *Biotechnology Letters* 33, 403–409. DOI: 10.1007/s10529-010-0436-0.

Zhao, C., Zhang, Y., Du, J., Guo, X., Wen, W. *et al.* (2019) Crop henomics: Current status and perspectives. *Frontiers in Plant Science* 10, 714.

Zhou, R., Jiang, F., Niu, L., Song, X., Yu, L. *et al.* (2022) Increase crop resilience to heat stress using omic strategies. *Frontiers in Plant Science* 13. DOI: 10.3389/fpls.2022.891861.

Zhou, S., Palmer, M. and Zhou, J. (2013) Differential root proteome expression in tomato genotypes with contrasting drought tolerance exposed to dehydration. *Journal of the American Society for Horticultural Science* 138, 131–141. DOI: 10.21273/jashs.138.2.131.

Zhu, J.K. (2016) Abiotic stress signaling and responses in plants. *Cell* 167, 313–324.

14 Omics Approaches for Climate-resilient Oilseed Crops

Namata Kumari[1], Bishawajit Kumar[1], Mithilesh Kumar Singh[2], Kajal Mehra[1], Mayank Tiwari[1], Shalu Kumari[3]* and Garima Aggarwal[1]

[1]Department of Genetics and Plant Breeding, G. B. Pant University of Agriculture & Technology, Pantnagar (Udham Singh Nagar), India; [2]Department of Genetics and Plant Breeding, Dr. Rajendra Prasad Central Agricultural, University, Pusa (Samastipur), India; [3]Department of Genetics and Plant Breeding, I.A.S, BHU, Varanasi, India

Abstract

The increasing challenges posed by climate change, including rising global temperatures, erratic rainfall patterns, and intensifying abiotic and biotic stresses, threaten the productivity of oilseed crops critical to global food, nutritional, and energy security. Abiotic stresses such as drought, heat, salinity, and cold severely impair physiological processes like photosynthesis, nutrient uptake, and water-use efficiency, reducing yield and quality. Biotic stresses, exacerbated by climate change, further compound these issues by promoting pests, diseases, and weed proliferation. Developing climate-resilient oilseed crops through advanced breeding strategies and omics technologies presents a promising solution. Omics approaches—genomics, transcriptomics, proteomics, metabolomics, phenomics, and ionomics—enable the identification and manipulation of stress-responsive genes, proteins, and metabolites, enhancing crop resilience. Integrated omics-assisted breeding, combined with the exploration of genetic diversity from crop wild relatives and underutilized species, offers a sustainable path to improving oilseed crop adaptation. This chapter delves into the potential of multiomics technologies to address abiotic and biotic stress in oilseed crops, fostering the development of climate-resilient cultivars and ensuring sustainable oil production amid global climate change.

14.1 Introduction

Optimizing climate change adaptation, agricultural productivity, food security, and environmental protection stands as one of the most pressing challenges for scientists in the 21st century. The undeniable shifts in climate—marked by rising global temperatures, unpredictable rainfall patterns, and expanding regions affected by floods or water shortages—are driving the growth of drought- and salinity-prone areas, which stifle plant growth and reduce oil seed crop productivity (Lamaoui *et al.*, 2018). Furthermore, these climatic changes are likely to intensify both individual and combined abiotic stresses, including drought, heat, salinity, cold, and submergence (Pandey *et al.*, 2017; Anwar *et al.*, 2021). In addition to abiotic challenges, climate change is also exacerbating biotic stresses by increasing populations of insects, pests, and pathogens, worsening disease outbreaks, promoting the spread of weed species, reducing

*Corresponding author: shaluprasad14@bhu.ac.in

 From Gene Discovery to Climate-Resilient Crops: Omics in Crop Improvement (eds S. Pandey *et al.*)
DOI: 10.1079/9781800628717.0014

beneficial soil microbes, and threatening key plant pollinators (Kole *et al.*, 2015; Raza *et al.*, 2019; Shahzad *et al.*, 2021). These effects carry significant implications for global food security, as they negatively impact plant growth, development, and yield, ultimately affecting worldwide agricultural production (Dhankher and Foyer, 2018; Nhamo *et al.*, 2019) and oil production.

At the same time, the global human population continues to rise, driven by high fertility rates in developing regions (UN, 2017), with projections estimating a population of 9.15 billion by 2050 (Alexandratos and Bruinsma, 2012). This demographic surge is further straining agricultural production and food supply systems, as it is estimated that 56% more food will need to be produced to feed an additional 3 billion people using the same or fewer resources compared to 2010 (Ranganathan *et al.*, 2018). The situation is similarly grim for leading oil crops (oil palm, soybean, sunflower, rapeseed, etc.), which have shown a global annual plateau in yield during the past several years (Food and Agriculture Organization of the United Nations, 2017; Ray *et al.*, 2019). Oil-producing crops, including oilseeds and oil-bearing varieties, occupy about 10% of global arable land and are essential sources of calories for humans and livestock, as well as raw materials for various industrial products, including oleochemicals. They are increasingly being utilized for biofuels, particularly biodiesel. The five primary oil crops—oil palm (*Elaeis guineensis*), soybean (*Glycine max*), oilseed rape (*Brassica napus*), sunflower (*Helianthus annuus*), and olives (*Olea europaea*)—together account for approximately 80% of global oil production (FAOSTAT, 2014). Oil palm is a productive perennial tree that grows mainly in South-east Asia and yields oil rich in C16:0 and C18:1 fatty acids, although it is challenging to genetically manipulate. Soybean serves primarily as a protein crop, with transgenic varieties developed for high oleate content for biodiesel production. Oilseed rape is the leading temperate oilseed crop, producing C18:1-rich oil, while sunflowers yield oil high in C18:2. Olives, which are drought-resistant Mediterranean fruit trees, produce C18:1-rich oil but are also difficult to genetically modify. Most oilseed crops are annuals grown in temperate regions, including soybean, corn (maize), rapeseed, and sunflower. Key oil-rich fruit crops include *E. guineensis*, *Cocos nucifera*, *O. europaea*, and *Arachis hypogaea* (groundnut). Oil-bearing crops consist of annual oilseeds and perennial plants, with seeds, fruits (mainly mesocarp), and nuts valued for the edible or industrial oils extracted from them. Oil content varies significantly, ranging from 10–15% in coconuts to over 50% in *Sesamum indicum* (sesame) seeds and *E. guineensis* kernels. While soybeans have a high protein content of around 40%, most other oilseeds contain lower protein levels (15–25%), and some oil-bearing crops have even less (Nogueira, 2011). Non-edible plant oils, such as those derived from *Jatropha curcas* (30–50% oil) and soapnut (*Sapindus mukorossi* and *Sapindus trifoliatus*; 52% oil), offer promising new sources for biodiesel production without competing with edible oils from traditional crop plants. Other potential oil crops include *Azadirachta indica*, *Calophyllum inophyllum*, and *Pongamia pinnata*, among 75 species known to contain 30% or more oil in their seeds, fruits, or nuts. Notable plant families that encompass oil-producing species include Amaryllidaceae, Apocynaceae, Asclepiadaceae, Compositae, Convolvulaceae, Cruciferae, Euphorbiaceae, Flacourtiaceae, Lauraceae, Leguminosae, Malvaceae, Moraceae, Myrcaceae, and Palmae. A small percentage (5–6%) of global oil crop production is utilized for seeds (oilseeds) and animal feed, with roughly 8% used for food, while the remaining 86% is processed into oil. Economically, oilseed crops are among the most diverse and valuable, serving crucial roles in both developed and developing nations for food, feed, and industrial purposes. The complexity of the agroecosystems in which these crops are cultivated worldwide indicates that the effects of global climate change (GCC) on these systems will vary significantly across different regions and over time. Industrial applications rely on the specific properties of saturated fatty acids (SFAs) found in oils, while the protein-rich byproduct left after oil extraction from oilseeds serves as a valuable nutrient source for livestock. Oilseed meals from crops such as *G. max*, *A. hypogaea*, *B. napus*, and *Linum usitatissimum* (flax) are particularly high in protein and, when combined with cereal grains, offer nutritionally balanced feeds for livestock (Colombini *et al.*, 2014). In recent years, the oilseed value chain has seen a significant increase in market value, driven by several key stages of processing—from seeds to oils, meals, and biodiesel. Market dynamics, especially in Europe and the United States, have favored oilseed production, notably for biodiesel, which has led to rapid growth in oilseed exports and related products, increasing from 5% in 1961 to over 12% by 2011. Currently, more than 50% of vegetable oils, primarily from *E. guineensis*, *G. max*, and *Brassica* spp., are traded globally (Tubiello *et al.*, 2013). To meet rising demand, efforts to boost oil crop productivity will need to tackle the growing challenges of abiotic and biotic stresses driven by global climate change. Given the diverse oil content and economic significance of these crops, they are vital for food and industrial uses globally. Oilseed crops are sensitive to these climatic changes and face considerable challenges under climate change, particularly from abiotic stresses such as drought, high temperatures, and soil salinity. These stresses impair critical physiological functions, including photosynthesis, nutrient uptake, and water-use efficiency, leading to reduced yield and crop quality.

For example, soybean is particularly susceptible to drought during flowering and pod formation, resulting in significant yield losses. At higher latitudes (longer day length), there is a delay in the occurrence of subsequent developmental stages and problems with plant maturation before the onset of autumn frost. Global warming allows the cultivation range of warm-season species (including soybean) to be shifted; however, periodic droughts and very high temperatures limit crop production. Adverse weather events result in a reduction in soybean seed yield of around 30%. Environmental stresses related to day length, high and low temperatures and water shortage or excess have the greatest impact on soybean yields, as we have no influence on them and can only, to a very limited extent, offset their negative effects. In the case of canola (rapeseed), abiotic stresses are projected to reduce global crop production by nearly 70% (Lohani *et al.*, 2020). Climate change has intensified the frequency of such events, often leading to the simultaneous occurrence of multiple abiotic stresses, which worsens their impact. Drought affects around 40% of the global land area, while salinity impacts 7%, making them the most critical environmental factors for crop yield (Trenberth *et al.*, 2014). Plant vulnerability to these stresses depends on factors like the timing, duration, and intensity of the stress. In real-world conditions, plants often face multiple stresses simultaneously, making it difficult to develop varieties that are tolerant to multiple stressors. As a rainfed spring-sown crop, sunflower (*H. annuus* L.) is increasingly exposed to negative impacts of climate change, especially to high temperatures and drought stress. It suffer from reduced seed set and oil content when exposed to high temperatures and salinity. Although mustard is cultivated over a large area, its productivity has declined in recent years due to the negative impact of various abiotic stress factors on plant growth and development. Drought, salt, frost and heat are the major abiotic stresses that determine the productivity of rapeseed mustard.

These crops are essential not only for human nutrition, in which a large body of research has shown that consumption of unsaturated fats leads to better health outcomes than consumption of saturated fats (Williams, 2000, May; Ursin, 2003; Calder and Yaqoob, 2009; Calder, 2015; Kris-Etherton and Krauss, 2020), but also for the production of edible oils, biofuels, and other industrial applications, making them crucial for global food and nutritional security. Oil and fats typically account for 20–35% of the calorie intake of an average, healthy adult (Mayo Clinic). As a result of climate change, productivity has been reduced, particularly in regions like the Indian subcontinent, Eastern Europe, Russia, China, and Canada, where these crops are extensively cultivated. Furthermore, climate change exacerbates biotic stresses by increasing the proliferation of pests, diseases, and weeds, further diminishing oilseed crop yields.

To address these challenges, it is critical to develop climate-resilient oilseed crops through advanced breeding strategies and technological interventions (Vaughan *et al.*, 2018). Proposals to tackle these challenges include developing climate-resilient crops, improving the efficiency of natural resource use, integrating agricultural intensification with the protection of natural ecosystems, and diversifying agricultural systems (Gil *et al.*, 2017; Dhankher and Foyer, 2018; Evans and Lawson, 2020). Specifically, the development of climate-resilient crop cultivars with desirable agronomic traits is considered one of the most feasible, cost-effective, sustainable, and efficient strategies for adapting our agricultural systems to climate change (Mba *et al.*, 2012; Kumari *et al.*, 2020; Kim *et al.*, 2021). Breeding climate-smart crops will involve exploring crop wild relatives and revisiting neglected and underutilized species to tap into the novel allelic variations they offer, thus expanding the genetic diversity available to crop breeders (Brozynska *et al.*, 2016; Gupta *et al.*, 2017; Ananda *et al.*, 2020; Klein *et al.*, 2020; Kamenya *et al.*, 2021). Furthermore, it will be essential to utilize advanced crop breeding techniques and methodologies, integrated with both conventional practices and enhanced data analysis pipelines (Qian *et al.*, 2019; Steinwand and Ronald, 2020; Pourkheirandish *et al.*, 2020; Bohra *et al.*, 2020; Ahmar *et al.*, 2020). Fortunately, the rapid advancements in omics technologies have transformed our approaches to crop improvement by enhancing the tools available to breeders and promoting omics-assisted breeding programs aimed at various agronomic traits (Langridge and Fleury, 2011; Li and Yan, 2020). Omics technology serves as a contemporary molecular tool for understanding the functional genomic systems of organisms (Hu *et al.*, 2018; Banerjee *et al.*, 2019), involving processes like DNA sequencing and the profiling of expressed transcripts and translated proteins (Missanga *et al.*, 2021). The term 'omics' derives from the Greek word '-ome,' which means 'whole,' and refers to scientific disciplines that analyze the complete array of biological molecules within biological systems (SETAC, 2019). These fields include genomics, transcriptomics, proteomics, metabolomics, phenomics (Hasin *et al.*, 2017) and epigenomics.

Omics approaches, including genomics, transcriptomics, proteomics, metabolomics, phenomics, ionomics, and epigenomics, offer powerful tools to dissect the complex molecular mechanisms that underlie crop responses to environmental stressors. Furthermore, omics technologies provide powerful tools for enhancing the resilience of oilseed crops. Genomics enables the identification of key stress-tolerance genes, while transcriptomics reveals changes in gene expression in response to environmental stressors. Proteomics and metabolomics offer insights into the functional dynamics of proteins and metabolites during stress responses. Phenomics examines the morphological structural traits and physiological functional traits, as well as component content traits of cells, tissues, organs, canopy, whole plants, or even populations. Furthermore, epigenomics examines heritable changes in gene regulation that contribute to long-term adaptation to abiotic stress. These technologies, when integrated through systems biology approaches, provide a comprehensive understanding of how oilseed crops respond to climate stress. Such knowledge can inform breeding programs aimed at developing climate-resilient varieties, offering a path toward sustainable oilseed crop production in the face of ongoing climate change. This chapter aims to explore the role and use of multiomics technologies in addressing mainly abiotic stress of oilseed crops. By leveraging omics data, it is possible to develop climate-resilient oilseed crops, ensuring global oil productivity in the face of a growing population.

14.2 Integrated Omics Approaches for Enhancing Climate Resilience in Oilseed Crop Improvement

In modern molecular biology, the suffix omics specially refers to a collection of technologies applied to the analysis of a huge and complete dataset of a particular class or type of biological molecule in a cell, tissue, organ, or whole organism. In other words, plant molecular biology focuses on studying cellular processes, their genetic foundations, and how they interact with environmental changes. This multifaceted and in-depth exploration involves large-scale experiments that target entire genetic, structural, or functional components. These extensive studies are collectively referred to as omics (Deshmukh *et al.*, 2014). The key omics sub-disciplines driving foundational systems biology research and modern crop improvement strategies include genomics (generation of genetic and genomic resources, gene mapping, functional genomics, and genomic selection), transcriptomics (gene regulation and expression profiling), proteomics (protein identification and effects), metabolomics (metabolite profiling, regulation, pathway and intermediates), phenomics (automated study and analysis of phenotypic and physiological effects), and ionomics (elemental identification, composition, effects, and interactions). Over the past few decades, several promising omics technologies have emerged. These omics approaches have uncovered each corresponding molecular biological aspect integrated within plant systems (Salt *et al.*, 2008) .These omics-based approaches have proven invaluable for uncovering the genetic and molecular mechanisms underlying crop development by analyzing changes in DNA, transcript levels, proteins, metabolites, and mineral nutrients in response to environmental and physiological stress in oilseed crops. These fields primarily focus on the comprehensive analysis of the genome, transcriptome, proteome, metabolome and phenotypes, respectively. All of these omics branches are closely linked to bioinformatics). Overall, analyses in the omics fields are structured according to Francis Crick'sclassical central dogma of molecular biology (Crick, 1970), involving targeted investigations of each molecule at specific levels. However, the complexity of biological systems indicates that dynamic environmental and spatiotemporal molecular interactions do not adhere to this straightforward path of reductionism and cannot be examined from a static topological perspective (Franklin and Vondriska, 2011; Wolkenhauer and Muir, 2011). Comprehensive multiomics approaches, employing sophisticated techniques, have been utilized to identify and elucidate the critical components of stress responses, senescence, and yield in a variety of economically significant) oilseed crops. The introduction of next-generation sequencing (NGS) technologies has enabled high-throughput and rapid data generation across genomes, epigenomes, transcriptomes, proteomes, metabolomes, phenomes, and ionomics. The emergence of omics strategies, alongside technological advancements like gene sequencing and mutagenesis, has opened new pathways for improving climate-resilient oilseed crops. These methods facilitate better predictions of gene functions and a deeper understanding of the molecular mechanisms underlying essential traits such as stress tolerance and yield stability (Kumar *et al.*, 2024). This is vital for developing superior oilseed cultivars that can withstand abiotic and biotic stresses while maintaining high nutritional quality. By introgressing genes or quantitative trait loci (QTLs) from identified donor genotypes through forward or reverse genetic approaches, we can enhance the resilience of oilseed crops to climate challenges (Bahuguna *et al.*, 2018). In this chapter, we delve into multiomics approaches, their applications, and potential implementations in crop science, with a focus on enhancing crop yields and improving biotic and abiotic stress tolerance in climate-resilient oilseed crops. In subsequent sections, we elaborate on various omics strategies that can form the basis for advancing genetic development, increasing yields, improving breeding practices, and strengthening the resilience of oilseed crops to both biotic and abiotic stresses.

14.2.1 Genomics

Before the 1980s, conventional plant breeding methods played a crucial role in improving oilseed crops, but progress in enhancing complex traits such as crop yield, oil content, certain quality characteristics, and resistance to biotic and abiotic stresses, including those related to climate, was relatively slow. Additionally, traditional approaches were limited in their ability to precisely identify the genes responsible for these complex traits. However, as climate change increases the frequency and intensity of stresses like drought, heat, and salinity, the need for rapid crop improvement becomes even more critical. Genomics-assisted breeding has revolutionized crop improvement by providing the tools to address these challenges. Genomics-assisted crop improvement offers significant potential for identifying major loci linked to key traits, including climate resilience, and selecting plants with favorable allele combinations that enhance their adaptability to changing environmental conditions. Techniques such as marker-assisted selection (MAS), marker-assisted backcrossing (MABC), or marker-assisted recurrent selection (MARS) enable

Table 14.1. List of oilseed crops sequenced. Author's own table.

Crop	Method used	Year	References
Soybean (*G. max*)	Whole-genome shotgun sequencing (WGS)	2010	Schmutz *et al.*, 2010
Castor bean (*Ricinus communis*)	WGS	2010	Chan *et al.*, 2010
Flax (*L. usitatissimum*)	WGS	2012	Wang *et al.*, 2012
Oil palm (*E.s guineensis*)	Roche/454 and Sanger BAC end sequencing	2013	Singh *et al.*, 2013
Camelina (*Camelina sativa*)	Illumina and Roche 454 next-generation sequencing (NGS) approach	2014	Kagale *et al.*, 2014
Canola (rapeseed; *B. napus*)	Whole-genome resequencing	2014	Chalhoub *et al.*, 2014
Sesame (*S. indicum*)	Whole genome shotgun sequencing	2014	Wang *et al.*, 2014
Mustard (*B. juncea*)	Illumina-based resequencing	2016	Yang *et al.*, 2016
Sunflower (*H.s annuus*)	PacBio sequencing.	2017	Badouin *et al.*, 2017
Groundnut (*A. hypogaea*)	SMRT and Illumina HiSeq 2000	2019	Zhuang *et al.*, 2019

the development of oilseed crops that are not only high-yielding and nutritionally rich but also better equipped to withstand the stresses brought on by climate change (Varshney *et al.*, 2010). Significant advancements have been made in the genomics of major oilseed crops such as soybean, Brassica, groundnut, and sunflower, although further biotechnological interventions are needed to improve other oilseed crops. These breakthroughs include the development of molecular markers, advanced genotyping platforms, enhanced genetic mapping techniques, and the identification of marker-trait associations, which have led to the discovery of genes and QTLs. Moreover, the reduction in sequencing costs and the availability of crop genome sequences (Table 14.1) have driven significant progress in this area. These innovations have enabled gene discovery and a deeper understanding of gene networks and complex quantitative traits, crucial for developing superior plant varieties.

In the context of climate change, these genomic tools are especially important for breeding crops that are resilient to increasingly severe and unpredictable environmental stresses, such as heat, drought, and salinity. Genomics-assisted breeding has improved the precision of conventional breeding programs, making it possible to develop cultivars with enhanced resistance or tolerance to both biotic and abiotic stresses, which are becoming more prevalent with climate change. The identification of key genes and loci associated with stress tolerance can accelerate the development of climate-resilient oilseed crops, ensuring stable yields under adverse conditions. In this chapter, we will discuss essential requirements for genomics-assisted crop improvement and provide examples that highlight how these tools can contribute to developing oilseed crops capable of withstanding climate-related challenges.

Many economically important traits in oilseed crops, such as yield, oil content, quality traits, and tolerance to biotic and abiotic stresses, are genetically complex and controlled by polygenes scattered across the genome. These polygenic traits were difficult to map using classical or Mendelian genetics due to their small individual effects on phenotypic expression. Limited information was available about these genes, including their chromosomal locations, the impact of individual genes or QTLs, and how they influence specific traits (Tuberosa and Salvi, 2004). However, advancements in genomic tools and the development of molecular markers have now made it possible to locate and map the positions of these genes or QTLs on chromosomes, allowing researchers to study their individual effects. In the context of climate resilience, this capability is particularly valuable for identifying genes that confer tolerance to stresses such as heat, drought, and salinity, which are becoming more critical due to climate change. Two major genomic approaches—QTL or linkage mapping and association or linkage disequilibrium (LD) mapping—are commonly used to identify marker-trait associations in oilseed crops. Each approach has its strengths and limitations (Neale and Savolainen, 2004). By employing these techniques, researchers can more precisely target and improve traits related to climate resilience, enabling the development of oilseed crops that are better equipped to thrive in increasingly harsh environments. For linkage mapping, we need to have genetic resources in the form of a biparental mapping population and genomic resources in the form of molecular markers and genome maps. A number of molecular markers and genome maps have been developed for most of the major oilseed crops such as soybean (Lark *et al.*, 1993), groundnut (Garcia *et al.*, 1995; Varshney *et al.*, 2009), *B. napus* (Landry *et al.*, 1991), sunflower (Berry *et al.*, 1995), sesamum (Dixit *et al.*, 2005), *B. juncea* (Sharma *et al.*, 1994), linseed (Oh *et al.*, 2000), and safflower (Amiri *et al.*, 2001). For mapping of QTLs/genes of complex traits the QTL mapping approach has been adopted in various oilseed crops. QTL mapping has been done for seed and oil yield (Diers *et al.*, 1992), abiotic stress tolerance (Lee *et al.*, 2004), biotic stress tolerance (Diers *et al.*, 1991), and oil quality (Bachlava *et al.*, 2008).

Soybean is the most successful food legume where molecular markers have been routinely integrated into breeding programs. This has resulted in the development of numerous improved lines and varieties with resistance to various soybean cyst nematode (SCN) races (Arelli *et al.*, 2010), Phytophthora root rot, brown stem rot, insect resistance (Walker *et al.*, 2004), low linolenic acid content (Sauer *et al.*, 2008), yield enhancement (Concibido *et al.*, 2003), and resistance to soybean mosaic virus (Saghai Maroof *et al.*, 2008). Additionally, several varieties, such as JTN-5503, JTN-5303, DS-880, and JTN-5109, have been released, particularly in the United States, with disease and SCN resistance (Arelli and Young, 2009). The identification of molecular markers linked to traits of interest has advanced significantly in soybean. Linkage mapping has been extensively used to identify genes conferring resistance to biotic stresses such as Sclerotinia stem rot (Guo *et al.*, 2008), brown stem rot (Patzoldt *et al.*, 2005), Phytophthora stem rot (Han *et al.*, 2008), Asian soybean rust (Murithi *et al.*, 2016), soybean mosaic virus (Shi *et al.*, 2009), sudden death syndrome (Kazi *et al.*, 2008), and SCN (Wu *et al.*, 2009). Likewise, QTL mapping has been effective for abiotic stresses such as waterlogging (Githiri *et al.*, 2006), salt stress (Lee *et al.*, 2004), manganese toxicity (Kassem *et al.*, 2004), and aluminum tolerance (Qi *et al.*, 2008).

As in soybean, in the initial years of the century, many molecular markers such as rapid amplification of polymorphic DNA (RAPD), amplified fragment length polymorphism (AFLP), inter-simple sequence repeat (ISSR), and expressed sequence tags–simple sequence repeats (EST-SSR) were used to assess the genetic diversity in flax (Oh *et al.*, 2000).

In groundnut, eight QTLs observed for resistance to late leaf spot (LLS), two (*Qlf2* and *Qlf7*) explained 32.7% of total phenotypic variation for lesion frequency per square centimeter of leaf, three (*Qlc4*, *Qlc7*, and *Qlc8*) observed 40.3% of total phenotypic variation for lesion number per leaf, and three QTLs (*Qlad3*, *Qlad6*, and *Qlad8*) explained 31.2% of total phenotypic variation for leaf area damage (Cuc and Quang, 2008). Khedikar *et al.* (2010) reported that one major QTL linked with rust resistance explained 6.90–55.20% variation. They observed that linking the simple sequence repeat (SSR) marker IPAHM 103 to this QTL could be used for marker-assisted backcrossing. Shoba *et al.* (2012) reported 32–52% phenotypic variation for LLS resistance and suggested using the primer PM 384100 for marker-assisted breeding. Sujay *et al.* (2012) identified 28 QTLs for LLS and 15 QTLs for rust. A major QTL for LLS, namely QTLLLS01, explained 10.27–62.34% of phenotypic variance. Four new markers (GM2009, GM1536, GM2301, and GM2079) for rust resistance have been identified as explaining 82.96% of phenotypic variance. Rust and LLS-linked markers were found in linkage group

03 (eight SSRs) and linkage group 04 (three SSRs) (Gajjar *et al.*, 2014). On the other hand, Varshney *et al.* (2014) observed one dominant (IPAHM 103) and three codominant (GM 2079, GM 1536, and GM 2301) markers for rust resistance in one QTL region in linkage group AhXV, which explained phenotypic variation of 82.66%. Three markers—namely, RKN410, RKN440, and RKN229—were linked to an identified resistance gene from *Arachis cardenasii* or *Arachis diogoi* for root-knot nematode (Burow *et al.*, 1996). The dominant polymerase chain reaction (PCR)-based marker 197/909 has been reported to be linked to nematode resistance (Chu *et al.*, 2007). Five SSR markers (PM375, PM36, PM45, pPGPseq8D9, and Ah-041) have been associated with seed length, pod length, number of pods per plant, 100-seed weight, maturity, and oil content (Gomez Selvaraj *et al.*, 2009).

Researchers from Oil Crops Research Institute of the Chinese Academy of Agricultural Sciences, BGI and other institutes have successfully cracked the nuclear genome of sesame, generating 54.5 Gb of high-quality data from the elite cultivar 'Zhongzhi No.13' using the Illumina Hiseq2000 platform. This has been the biggest breakthrough in sesame research for decades (Wang *et al.*, 2014). The highquality draft genome encompassing 27,148 genes distributed on 16 linkage groups (LGs) with 274 Mb of size, has become the reference genome for biology study in sesame. This genome with a contig N50 of 52.2 kb and a scaffold N50 of 2.1 Mb has been recently upgraded to reach 13 pseudochromosomes, 94.3% of the estimated genome size and 97.2% of the predicted gene models in sesame (Wang *et al.*, 2016). In parallel, another genome sequencing project was initiated under the auspices of the Sesame Genome Working Group (SGWG). By 2013, they assembled from the variety 'Yuwhi 11' a genome size of 293.7 Mb out of the 354 Mb estimated in sesame and predicted the function of 23,713 genes. More recently, two new genome sequences from sesame landraces ('Baizhima' and 'Mishuozhima') have also been released, increasing the genome sequence resources available for this crop (Wei *et al.*, 2016b).

14.2.2 Transcriptomics

Transcriptomics deals with the transcriptome, which refers to the complete set of RNA transcripts that are produced by genome of an organism in a cell or tissue (Raza *et al.*, 2021). Transcriptome profiling is dynamic and has emerged as a promising technique to analyze gene expression in response to any stimuli over a certain time period (Duque *et al.*, 2013; El-Metwally *et al.*, 2014). This approach enables researchers to study the differential expression of genes *in vitro*, providing insights into the initial functions of specific genes. Initially, transcriptome dynamics were explored using traditional methods such as cDNAs-AFLP, differential display-PCR (DD-PCR), and suppression subtractive hybridization (SSH), though these techniques had limitations in resolution (Nataraja *et al.*, 2017). With the advent of more advanced technologies, RNA expression profiling has become possible through techniques like microarrays, digital gene expression profiling, NGS, RNA sequencing (RNA-Seq), and serial analysis of gene expression (SAGE) (Kawahara *et al.*, 2012; Duque *et al.*, 2013). Transcriptome or EST sequencing serves as the initial step in exploring the gene content of a species and has proven to be an effective approach for generating functional genomic data in non-model organisms. Transcriptome analysis, which includes protein-coding genes, non-coding genes, and regulatory genes, is crucial for understanding the intricate molecular mechanisms and gene regulatory networks involved in resistance to various biotic stresses. Protein-coding genes play a role in downstream signaling related to defense pathways, while non-coding genes, such as microRNAs (miRNAs), long non-coding RNAs (lncRNAs), and transcription factors (TFs), regulate the expression of functional genes. Methods like northern blotting, microarrays, and RNA-Seq are commonly used to conduct transcriptomic studies in oilseed crops. As a result, microarray profiling and deep sequencing are being utilized to identify numerous biotic and abiotic stress-responsive miRNAs. These miRNAs hold potential for engineering resistance to both biotic and abiotic stresses in oilseed crops.

The availability of reliable tools for routine gene expression analysis in oilseed crops has enabled researchers to assess key traits, particularly those related to oil biosynthesis and resistance to abiotic (and biotic) stresses such as drought, salinity, extreme temperatures, and nutrient deficiencies (Table 14.2). However, linking specific changes in plant metabolism to a single stress factor is challenging, as multiple abiotic and biotic stresses often occur simultaneously, potentially triggering antagonistic responses that lead to the accumulation, degradation, or consumption of both primary and secondary metabolites. Genome-wide transcription analysis in response to abiotic stresses is a significant development in providing a basis for effective genetic manipulation in order to improve stress tolerance in oilseed crops (Jung *et al.*, 2014). Genes underlying environmental adaptability tend to be overretained in polyploid oilseed crop species (Li *et al.*, 2014), and functionally redundant multiple-copy genes related to abiotic stress have been found in several oilseed crops.

The genetic and molecular mechanisms underlying the early osmotic adjustment response to drought in drought-resistant and drought-sensitive *B. napus* remain unclear. However, heterosis may enhance overall drought stress tolerance in this crop (Hatzig *et al.*, 2014). Heterozygous F1 hybrids of winter *B. napus* exhibited improved vigor and greater resilience to drought stress compared to homozygous inbred lines. Interestingly, drought-resistant genotypes did not show typical hormonal adjustments or osmolyte accumulation, suggesting that alternative physiological mechanisms may be at play in their drought resistance. For example, the aquaporin CsPIP2;1 in *Camelina sativa*, which plays a role in salt and drought stress responses, is expressed in both roots and leaves (Jang *et al.*, 2014).

Although *Jatropha curcas* is highly drought-tolerant, only a limited number of drought-inducible genes have been identified, including ethylene-responsive element binding factor (JcERF), betaine aldehyde dehydrogenase 1 (JcBD1), and plasma membrane intrinsic protein 2 (JcPIP2). However, a substantial number of differentially expressed genes have been observed in *J. curcas* under drought stress, with 332 genes being upregulated during drought stress, 585 during the recovery stage, and 374 genes involved in both stages (Cartagena *et al.*, 2015).

Tropical oil-producing crops, such as *E. guineensis*, are highly sensitive to low temperatures (Lei *et al.*, 2014). RNA-Seq analysis of *E. guineensis* under cold stress revealed a gene expression pattern mediated by C-repeat binding factors, which differs from the pattern observed in *Arabidopsis thaliana*. The C-repeat binding factor–mediated signal transduction pathway plays a key role in cold tolerance. In *Plukenetia volubilis*, a potential oilseed perennial tree, the expression of genes involved in polyunsaturated fatty acid (PUFA) biosynthesis and oil deposition in developing seeds demonstrated that low temperatures promote PUFA biosynthesis and deposition in its seeds. This effect was also observed in *G. max* and *L. usitatissimum*.

In *Brassica rapa* and other polyploids, genes associated with adaptability to nutrient deficiencies (e.g. Zn, Fe) or exposure to stress (e.g. Cd) are often over-retained at the gene dosage level and functionally redundant at the transcriptional level (Li *et al.*, 2014). Most of these multiple-copy genes exhibit similar expression changes under stress conditions. However, little is known about their differential expression in response to such stresses. The redundancy of these genes enables polyploids to better adapt to abiotic stress, but it also complicates the regulatory network in *B. rapa* compared to the diploid *A. thaliana*. The diversified expression patterns between over-retained and redundant genes likely reflect an evolutionary mechanism that supports the increased functional demands of polyploids, including a broader range of ontogenetic functions and specialized cell types within plant tissues.

Successful use of RNA-Seq data for detecting and identifying soybean viruses was demonstrated by Jo *et al.* (2017) for five viruses, including SMV, for marker-based diagnosis of crop pathogens. Similarly, Díaz-Cruz *et al.* (2019) harnessed the potential of the RNA-Seq approach for identifying major and residual pathogens, including bacteria, fungi, and viruses. For SCN resistance, differentially expressed genes (DEGs) associated with hormone signal transduction pathways, *MAPK* signaling, and *WRKY*-controlled transcriptional regulation were found to be

Table 14.2. Transcriptomic analysis for climate-resilience oilseed crops. Author's own table.

Abiotic stress	Tissue	Crops	Genotype	Differentially expressed genes (DEGs)	Sequencing platform	References
Drought	Leaf	Soybean	Liaodou 15	2985	Illumina HiSeq 2500	Li *et al.*, 2022a
	Leaf	Soybean	-	2771	Illumina HiSeq 4000 platform	Xu *et al.*, 2018a
Heat	Seed coat	Soybean	Pioneer 93B15	1576	Illumina Genome HiSeq 2000	Leisner *et al.*, 2017
Waterlogging	Stem, leaf	Oil palm	-	1090 in stem, 13 in leaves	Illumina NovaSeq	Lim *et al.*, 2023
Drought	Root	Sesame	Drought-tolerant vs sensitive genotypes	-	Illumina HiSeq4000 platform	Dossa *et al.*, 2017
Heat	Seed	*Brassica napus*	Huyou 2	442	Illumina Hiseq 2500 platform	Gao *et al.*, 2021
Drought	Seed	*B. napus*	Drought-resistant vs sensitive genotypes	169	Illumina HiSeq 2000	Wang *et al.*, 2017
Cold	Seed	*B. napus*	Cold-tolerant vs sensitive genotypes	138	RNA-seq	Mehmood *et al.*, 2021
Cold	Immature leaf	Camelina	*Camelina sativa* cv Suneson	4467	Illumina NovaSeq 6000	Fang *et al.*, 2023
Salt	Seed	Flax	Fiber flax (Heiya No. 19)	7736, 1566 and 454 in alkaline-salt stress, neutral salt stress and alkaline stress, respectively	Illumina HiSeq 2000	Yu *et al.*, 2014
	Seed	Flax	Salt sensitive vs salt tolerant genotypes	7855 in Fanni and 8054 in Longya10	RNA-Seq	Li *et al.*, 2022b

important (Zhang *et al.*, 2017). The identified disease resistance genes could be used as molecular markers for breeding SCN-resistant soybean. RNA-Seq also helped to characterize three-way interactions between soybean, soybean aphids (SBA), and SCN (Neupane *et al.*, 2019), which could be used to mine disease resistance genes involved in complex regulatory pathways. Moreover, the intricate process behind the non-host resistance (NHR) response to SCN was also understood using this approach (Song *et al.*, 2019). This laid the foundation for NHR research that could lead to development of disease-resistant soybean cultivars. *Psuedomonas syringae* is a non-host-specific pathogen causing chlorosis by producing coronatine (COR) toxin. Analysis of COR-treated and control soybean plants revealed DEGs belonging to jasmonic acid production, ethylene synthesis, phenylpropane metabolism, photosynthesis, and salicylic acid signaling pathways (Beyer *et al.*, 2019).

Using the Illumina 2500 platform, a time-course transcriptome profiling of two sesame genotypes with contrasting tolerance levels revealed significant gene expression data on sesame's response to waterlogging stress (Wang *et al.*, 2016). Another key abiotic stress affecting sesame productivity is drought, which remains insufficiently understood at the molecular level. To address this, Dossa *et al.* (2017) investigated gene expression changes in drought-tolerant and sensitive sesame genotypes using the Illumina HiSeq 4000 sequencing platform, under both progressive drought conditions and after re-watering, in order to identify potential genes linked to drought tolerance. Currently, research is focused on understanding sesame's molecular responses to salt stress, an important abiotic challenge. However, relatively little research has been conducted on biotic stress at the molecular level, highlighting the need for further studies to develop disease-resistant crops. An RNA-Seq study has been conducted on both resistant and susceptible sesame cultivars inoculated with *Fusarium oxysporum* f. sp. *sesami* to investigate the molecular mechanisms underlying sesame's resistance to Fusarium Wilt, a major disease that causes 15–30% yield loss worldwide (Wei *et al.*, 2016a).

14.2.3 Proteomics

An organism's genome, inherited from its parents, remains unchanged and consistent across all cells throughout its life cycle. In contrast, proteins are dynamic and undergo processes such as alternative splicing and post-transcriptional and translational modifications (PTMs). As a result, the number of proteins exceeds the number of coding sequences, creating a more complex process for identifying proteins compared to genes. Biotic stresses can significantly affect the proteome by altering protein quantity, cellular localization, PTMs, protein–protein interactions, and their overall biological functions. Proteomics is crucial for providing comprehensive information on protein categories in the cell/tissue/plant under stress conditions as it regulates the plant epigenome, transcriptome and metabolome. Proteins play a direct role in signaling and activating immune responses to external pathogens and pest attacks. Soybean proteomics primarily involves comparative analysis of protein abundance between tolerant and susceptible cultivars, species, and cultivar-specific metabolic processes and pathways, helping to identify differentially expressed proteins (DEPs). Proteomics is a technique used to profile the total expressed proteins in an organism and is divided into four main categories: sequence proteomics, structural proteomics, functional proteomics, and expression proteomics (Mosa *et al.*, 2017; Aizat and Hassan, 2018). Sequence proteomics focuses on determining amino acid sequences, typically identified sequentially using high-performance liquid chromatography (HPLC; Twyman, 2013). Structural proteomics, on the other hand, examines the structure of proteins to infer their potential functions. This can be analyzed through various methods, including computational modeling and experimental techniques such as nuclear magnetic resonance (NMR), crystallization, electron microscopy, and X-ray diffraction of protein crystals (Sali *et al.*, 2003; Woolfson, 2018). Functional proteomics focuses on identifying the functions of proteins, often through methods like yeast one- or two-hybrid assays and protein microarray profiling (Lueong *et al.*, 2014). Advances in protein extraction and separation have significantly accelerated plant proteomic research, both at the sample and genome-wide levels (Nakagami *et al.*, 2012).

Traditional proteomics techniques, such as ion-exchange chromatography (IEC), size-exclusion chromatography (SEC), and affinity chromatography, are commonly used. Additionally, techniques like western blotting and ELISA allow for selective protein analysis. Subsequently, advanced techniques like SDS-PAGE, two-dimensional gel electrophoresis (2-DE), and two-dimensional differential gel electrophoresis (2D-DIGE) were introduced, enabling protein separation through gel-based methods. To enable rapid protein expression analysis, protein microarrays and chips have been developed for detecting small amounts of protein samples. Additionally, advanced quantitative proteomic techniques, such as stable isotope labeling with amino acids in cell culture (SILAC), isotope-coded affinity tag (ICAT) labeling, and isobaric tag for relative and absolute quantification (iTRAQ), have been developed for more accurate protein quantification. Recently, two significant high-throughput techniques, X-ray crystallography and NMR spectroscopy, have been developed for determining the three-dimensional structure of proteins, which can provide valuable insights into their biological functions (Aslam *et al.*, 2017). The applications and significance of both traditional and modern proteomic methods have been explored. Two-dimensional gel electrophoresis and SDS-PAGE are essential for identifying proteins and quantifying their content, respectively (Eldakak *et al.*, 2013). The identified proteins are then analyzed for their molecular mass using techniques like mass spectrometry (MS), ion trap-MS (IT-MS), or liquid chromatography (LC; Fournier *et al.*, 2007). Additionally, methods such as MALDI-TOF, electrospray ionization (ESI), and collision-induced dissociation (CID) are employed to determine the molecular weights of proteins (Tanaka *et al.*, 1988; McLuckey and Stephenson, 1998; Baggerman *et al.*, 2005).

In soybean, researchers studied the interaction between the host and the rust-causing pathogen (*Phakopsora pachyrhizi*) using 2D-GE and LC–MS/MS, identifying two key proteins: pathogenesis-related protein 10 and chalcone isomerase 1. Non-host resistance (NHR) serves as a highly effective defense mechanism in plants. Dong *et al.* (2015) explored the NHR response in soybean to *Bipolaris maydis*, a fungal pathogen commonly affecting maize, which triggered significant protein changes in plant cells during its interaction with fungal spores. The study highlighted the complex interactions between the plant and the pathogen. Various proteomic studies on soybean have employed techniques such as 2D-GE and MALDI-TOF to analyze responses to biotic stresses, including soybean mosaic virus (SMV), root and stem rot, soybean rust, cottonworm (*Prodenia litura*), and MYMIV and MYMV viruses. Functional proteomics approaches have identified reactive oxygen species (ROS) scavengers such as quinone reductase, γ-glutamylcysteine synthetase, dehydrins, and dehydroascorbate reductase in sunflower crops (Shalata *et al.*, 2001; Di-Baccio *et al.*, 2004; Mittova *et al.*, 2004). These proteins play critical roles in protecting the plant from oxidative stress. Recent iTRAQ-based comparative proteomic studies have shed light on various stress-responsive DEPs in different coconut varieties, offering new insights into stress tolerance mechanisms (Yang *et al.*, 2020). Similarly, iTRAQ-based proteomic analysis has enhanced our understanding of somatic embryogenesis in cotton crops (Zhu *et al.*, 2018), demonstrating its value in studying oilseed crops under abiotic stresses. For example, Zeng *et al.* (2017) used iTRAQ proteomics to analyze resistance to the soybean webworm (*Lamprosema indicata*), discovering that resistant plants produced proteins that hindered pest growth. Additionally, Bai *et al.* (2019) employed this technique to assess the effectiveness of *Funneliformis mosseae* (arbuscular mycorrhizal fungus; AMF) in combating root rot disease in soybean. Another study highlighted the identification of *Sinorhizobium fredii* strain Sneb183 as a potential biological control agent against SCN (Wang *et al.*, 2020).

Post-iTRAQ-based proteomic analysis has been particularly useful in identifying translational modifications such as phosphorylation and ubiquitination, along with protein–protein interactions, which are crucial for understanding cellular biological mechanisms. Phosphoproteomics focuses on analyzing protein phosphorylation by detecting phosphoproteins and their phosphorylated amino acid residues, both qualitatively and quantitatively (Mosa *et al.*, 2017). This technique is crucial for understanding protein functions and identifying pathways involved in various cellular processes (Mosa *et al.*, 2017). Moreover, combining proteomics and phosphoproteomics has proven valuable for studying diverse crop functions in response to pathogens, such as phytoplasma and the *Septoria tritici* fungal pathogen in wheat and grapevine (Margaria *et al.*, 2013; Yang *et al.*, 2013). Proteomics plays an essential role in deciphering the functional mechanisms in oilseed crops, especially under stress conditions, thereby contributing to improved crop yields. Several proteomic techniques, such as MALDI-TOF, SDS-PAGE, MS) 2-DE, and peptide-mass fingerprinting (PMF), have been used in crops like rapeseed, soybean, and cotton to understand stress-response pathways (Demirevska *et al.*, 2008; Toorchi *et al.*, 2009; Jangpromma *et al.*, 2010; Nouri and Komatsu, 2010; Deeba *et al.*, 2012; Mohammadi *et al.*, 2012). These methods have been particularly effective for studying drought stress responses in plants (Ghosh and Xu, 2014). Overall, proteomic studies have significantly contributed to understanding how

Table 14.3. Proteomics study carried out on oilseed crops for various climatic stresses. Author's own table.

Crop	Proteomics approach used	No. of identified proteins	Biological phenomenon	References
Canola (*B. napus*)	iTRAQ LC–MS/MS	1976	Proteins involved in drought stress	Koh *et al.*, 2015
	iTRAQ	7736	Proteins involved in waterlogging stress	Xu *et al.*, 2018b
Peanut (*A. hypogaea*)	MALDI-TOF/TOF mass spectrometry	96	Proteins involved in drought stress	Katam *et al.*, 2016
	Tandem Mass Tag (TMT)-based quantitative analysis	5029	Proteins involved in chilling stress	Chen *et al.*, 2021
Camelina (*C. sativa*)	iTRAQ, LC-MS/MS	1532	Proteins involved in various stress tolerance	Alvarez *et al.*, 2015
Mustard (*B. juncea*)	iTRAQ, LC-MS/MS	172	Proteins involved in heat stress	Cheng *et al.*, 2020
Cotton (*Gossypium herbaceum* L.)	2-DE, MALDI-TOF/TOF-MS	550	Proteins involved in drought stress	Deeba *et al.*, 2012
Soybean (*G. max*)	MALDI-TOF/MS, 2-DE	89	Proteins involved in chilling stress	Cheng *et al.*, 2010
	MALDI-TOF/MS	28	Proteins involved in short term drought stress	Alam *et al.*, 2010
	Two-dimensional polyacrylamide gel electrophoresis	321	Proteins involved in salt stress	
Sunflower (*H. annuus*)	Nano-LC–MS/MS	42	Proteins involved in drought stress	Ghaffari *et al.*, 2017
Safflower (*Carthamus tinctorius*)	Two-dimensional polyacrylamide gel electrophoresis	17	Proteins involved in salt stress	Shaki *et al.*, 2020
Castor (*R. communis* L.)	iTRAQ	1670	Proteins involved in cold stress	Wang *et al.*, 2019

crops respond to both abiotic and biotic stress conditions (Table 14.3).

14.2.4 Metabolomics

Metabolomics is defined as the comprehensive study of metabolites that participate in different cellular events in a biological system. However, metabolome denotes the complete set of metabolites synthesized via metabolic pathways in plant system (Fiehn, 2002; Baharum and Azizan, 2018). NGS technologies have emerged as promising tools in order to understand the regulation of gene expression and molecular basis of cellular responses that occur in crops in response to biotic and abiotic stresses (Abdelrahman *et al.*, 2018). However, metabolomics, when combined with NGS, has provided a foundation for predicting an organism's metabolic network from its genome sequence (Weckwerth, 2011). By integrating NGS with metabolite quantification via MS, this approach enables the development of strategies for crop improvement (Pandey *et al.*, 2016). Metabolites can be viewed as the final products of gene expression, reflecting the biochemical phenotype of cells (Weckwerth and Fiehn, 2002). While proteomics focuses on gene products, metabolomics offers insights into the metabolic expression of proteins, helping identify the biochemical processes essential for gene function (Weckwerth and Morgenthal, 2005; Lindon and Nicholson, 2008). Due to their varying chemical and physical properties, metabolites require specialized separation and analytical techniques to generate comprehensive metabolic profiles of plant samples (Jogaiah *et al.*, 2013). Various analytical techniques have been employed in plant systems to quantify metabolites, including thin-layer chromatography (TLC), gas/liquid chromatography–MS (GC/LC–MS), LC-electrochemistry (EC)–MS, NMR, direct infusion MS (DIMS), Fourier-transform infrared (FT-IR), and capillary electrophoresis (CE)—MS (Fiehn *et al.*, 2000; Weckwerth, 2003; Moco *et al.*, 2007; Allwood and Goodacre, 2010; Saito and Matsuda, 2010; Duque *et al.*, 2013; Jogaiah *et al.*, 2013). Among these, CE–MS, GC–MS, LC–MS, and NMR are the most commonly used techniques in plant metabolomics (Fiehn, 2002; Kikuchi and Hirayama, 2007; Moco *et al.*, 2007; Allwood and Goodacre, 2010; Weckwerth, 2010; Kim *et al.*, 2011). These techniques are selected based on factors like selectivity, sensitivity, speed, and accuracy. While NMR is fast and selective, mass spectrometry techniques (CE–MS, GC–MS, and LC–MS) offer high sensitivity and selectivity but are generally more time consuming (Sauter *et al.*, 1991; Sumner *et al.*, 2003).

Proper annotation and reporting of metabolomics data are essential for the accurate identification and analysis of metabolites. Recently, Alseekh *et al.* (2021) proposed guidelines for the annotation and quantification of LC/GC–MS-based metabolomics data. These recommendations include aspects such as sample preparation, sample replication and randomization, quantification, recovery and recombination, ion suppression, and peak misidentification. Following these guidelines can significantly enhance the reliability of metabolite data acquisition and reporting. In addition, the authors provided a detailed workflow for sampling, quenching, metabolite extraction, and storage, emphasizing that adherence to these protocols can help prevent data misinterpretation and improve transparency in the reporting of LC/GC–MS-based metabolomics data (Alseekh *et al.*, 2021). A key goal of metabolic research in oilseed crops is to explore the connections between genotypes and metabolic outputs, particularly focusing on plant growth and chemical composition. This includes a specific emphasis on fatty acid (FA) profiles and how these are impacted by environmental stresses. Understanding these relationships can help optimize crop performance and improve the quality of oilseed production under varying conditions. According to Nikoloski *et al.* (2015), the supply and demand dynamics in central metabolism are likely a more significant driver of fluxes than environmental variations. Various methods have been employed to understand the regulatory mechanisms behind oil accumulation in oilseed crops, including conventional breeding, QTL mapping, transcription factors, biochemistry, genetic manipulation (such as gene overexpression or silencing), and metabolic control analysis. For instance, differences in metabolic and physiological responses between drought-resistant and drought-sensitive *B. napus* genotypes may not be solely due to typical hormonal changes or osmolyte accumulation, but rather to alternative mechanisms used by drought-resistant varieties (Hatzig *et al.*, 2014). A deeper understanding of these differences may emerge by integrating metabolic performance with overall plant physiology. Metabolic control analyses have been utilized to enhance yield in several oilseed crops, including *B. napus*, and the insights gained have been applied to refine genetic manipulation strategies aimed at improving other oilseed crops. In addition, metabolic biosynthesis pathways are being engineered into transgenic oilseed crops to produce seed oils rich in γ-linolenic acid, omega-3, and very long-chain PUFAs that are not typically found in plant sources but offer enhanced health benefits (Ruiz-López *et al.*, 2012). For instance, in metabolically engineered *Crambe abyssinica* lines, the levels of C18:1, C18:2, and C18:3 were significantly reduced, while the level of C22:1 (erucic acid, a key feedstock for the oleochemical industry) increased from 60% in the wild-type to 73% in the top transgenic line. Functional genomics is increasingly applied to understand metabolic pathways and improve genetic breeding in oilseed crops (Maghuly and Laimer, 2013). Through metabolic engineering, plant genomes are manipulated to produce genetically modified oilseed crops with optimized oil quality, content, and composition, making them more suitable as industrial feedstocks. Comparative functional genomic analyses of crops exposed to abiotic stresses, such as extreme temperatures, have led to the discovery of numerous genes and gene products involved in plant adaptation to these stresses. While the biochemical and molecular mechanisms behind crop tolerance to specific abiotic stresses are not yet fully understood, various tolerance indicators at different growth stages (germination, seedling, flowering, and maturation) have been developed. These indicators are used for germplasm screening and selection in breeding programs. Further advancements in metabolic engineering are anticipated to improve crop tolerance to a range of abiotic stresses. These innovations may enable crops to adapt to and thrive in drier regions, where temperature extremes and limited moisture often pose significant challenges to plant growth and productivity.

14.2.5 Phenomics

A phenotype results from complex interactions among environmental factors, genetics, and crop management practices. For crop improvement programs, accurate and rapid phenotyping of thousands of germplasm and breeding lines across various conditions is essential. This can be efficiently achieved using both low- and high-throughput phenomics tools. Precise phenotyping plays a crucial role in enabling plant breeders to achieve selection gains. Over the past two decades, significant advancements in the genomics and transcriptomics of major food crops, such as soybean, have substantially contributed to improving crop productivity. The integration of molecular genetics with accurate phenotyping enhances our understanding of plant genetic architecture. However, advancements in high-throughput phenotyping have lagged behind those in high-throughput genotyping, despite their interdependence in developing stress-resilient genotypes. Therefore, improving phenotyping methods is essential to fully harness the potential of high-throughput genomics. Phenomics combines multiple disciplines, including physics, mathematics, life sciences, and earth sciences, along with artificial intelligence, to process and analyze complex, high-throughput phenotypic data of crops in diverse environments. Innovative phenomics techniques are essential to closing the gap in high-throughput phenotyping. These advanced tools can be utilized to analyze phenomes that emerge in response to biotic stresses and challenging environmental conditions, aiding in the development of sustainable crop varieties. Cutting-edge phenomics platforms, such as the Australian Plant Phenomics Facility (APPF), multispectral and fluorescence imaging for physiological phenotyping, and others spanning ground-based proximal methods to large-scale aerial remote sensing, have been established to support a wider range of crop plants.

Phenomics refers to the process of characterizing phenotypes by gathering high-dimensional data at the whole-organism level

(Houle *et al.*, 2010). The *phenome* represents the complete set of an organism's phenotypic traits, with the plant phenome shaped by interactions between the genome (G), environment (E), and management (M) factors, commonly referred to as G–P–E interactions (Gjuvsland *et al.*, 2013; Großkinsky *et al.*, 2018; Zhao *et al.*, 2019). Precise phenotyping is essential for gene and QTL mapping of specific traits, enabling their identification and functional analysis through both forward and reverse phenomics methods, which support genetic improvements in crops (Kumar *et al.*, 2015). Automated high-throughput imaging technologies are essential for identifying superior germplasm lines and varieties. These non-invasive techniques include color imaging for biomass evaluation, far-infrared imaging for canopy analysis, lidar (light detection and ranging) to measure growth parameters, and magnetic resonance imaging (MRI) to study root systems in crops (Finkel, 2009; Berger *et al.*, 2010; Furbank and Tester, 2011). Roots can also be imaged in laboratory and greenhouse environments without damaging the plant samples. Techniques such as electrical resistance tomography, electrical capacitance, X-ray computed tomography, and positron emission tomography have been used to visualize root systems in soil-grown plants and crops (McGrail *et al.*, 2020). Red, green, blue (RGB) imaging, which utilizes visible light, is another effective tool for phenotyping canopy and root systems (Großkinsky *et al.*, 2018). RGB imaging has been applied in various studies to assess crop traits, such as scanning the root systems of *Triticum durum* in soil-filled rhizoboxes (Bodner *et al.*, 2017), evaluating plant pathosystems and disease symptoms (Mahlein, 2016), and analyzing shoot phenotypes under different stress conditions (Humplík *et al.*, 2015). Infrared thermography has been used to study stomatal conductance under salinity stress in young barley and wheat seedlings (Sirault *et al.*, 2009). Additionally, chlorophyll fluorescence imaging with pulse amplitude modulated (PAM) instruments has been employed to assess abiotic stress responses in crops like tobacco, canola, and cotton (Saranga *et al.*, 2004; Baker, 2008).

Digital imaging techniques have also been employed to quantify boron toxicity as an abiotic stress by mapping wheat and barley populations (Schnurbusch *et al.*, 2010). In terms of biotic stress responses, similar methods have been used to detect and quantify disease symptoms caused by pathogens in barley (Chaerle *et al.*, 2009). Consequently, phenomics applications are crucial for evaluating phenotypic traits in crops subjected to both biotic and abiotic stresses. Field-based phenotyping techniques are particularly significant as they provide more relevant insights into crop performance compared to laboratory or greenhouse studies. Multispectral and hyperspectral technologies are useful for assessing various agronomic traits (Rascher and Pieruschka, 2008). Among these, the laser-induced fluorescence transient (LIFT) method is particularly effective for evaluating the photosynthetic efficiency of crops in the field (Pieruschka *et al.*, 2010). The integration of wireless sensor networks facilitates crop trait phenotyping and the collection of valuable data for breeding purposes (Ruiz-Garcia *et al.*, 2009). Phenomics also provides platforms linked to computational systems for analyzing phenotypic data, including support vector machines (SVMs), artificial neural networks (ANNs), and principal component analysis (Karkee *et al.*, 2009). A major challenge in plant phenomics remains the organization of information systems to store phenotyping datasets, allowing for future reanalysis and the generation of new insights (Cabrera-Bosquet *et al.*, 2012). The integration of GWAS with a high-throughput rice phenotyping facility (HRPF) led to the identification of 15 agronomic traits and 25 associated loci, including the Green Revolution semi-dwarf gene (SD1) in rice (Yang *et al.*, 2014). This advanced phenotyping method, combining rice automatic phenotyping (RAP) and the yield traits scorer (YTS), has facilitated the shift from traditional manual phenotyping to high-throughput phenotyping (HTP) in crop breeding (Yang *et al.*, 2014). More recently, an ontology-driven phenotyping hybrid information system (PHIS) has been developed to integrate and share multi-scale data and metadata (Neveu *et al.*, 2019). The ontology-driven PHIS information system is a robust tool for integrating and managing data, as well as sharing multi-source and multi-scale data from both greenhouse and field conditions. Its ontology-driven architecture establishes relationships between objects and enhances datasets with knowledge and metadata (Neveu *et al.*, 2019). Recent research has introduced the Internet of Things (IoT)-based CropSight system, designed to scale and assess crop phenotyping and genotype–environment (G × E) interactions. The IoT is a global network that uses information and communication technologies to connect sensing and actuating devices, enabling the sharing of information across various platforms. This system is capable of performing high-quality crop phenotyping and monitoring microclimate dynamics, and has been applied in field experiments with wheat crops (Reynolds *et al.*, 2019). Roitsch *et al.* (2019) proposed the use of HTP applications with next-generation sensors, which are expected to enhance crop yields, stress tolerance, and management in the near future. A wide range of techniques, combined with machine learning and advanced computer algorithms, is being explored for the accurate and timely detection of diseases caused by biotic factors in soybean. Advances in phenotypic technologies have greatly improved the early detection, differentiation, and quantification of diseases. Early and reliable disease identification is crucial, particularly for soybean, which is a major commercial crop facing annual yield losses of around 40% due to various biotic agents. Various phenotyping platforms, including unmanned aerial vehicles (UAVs), satellite imagery, aerial systems, and ground-based sensing robots, are equipped with imaging technologies such as optical imaging, chlorophyll fluorescence imaging, spectroscopy, RGB, multispectral, hyperspectral, and thermal imaging to detect biotic stresses in crops. In general, phenomics plays a vital role in advancing crop breeding strategies by integrating phenomic data with other omics fields like genomics, proteomics, and metabolomics. This integration helps to better understand the complex interactions between phenomes, genomes, and environmental factors, ultimately aiding in the improvement of oil crop management.

14.2.6 Ionomics

Ionomics is the study of the elemental composition of metals, metalloids, and non-metals in various plant species, with an emphasis on high-throughput detection and analysis. Technologies such as inductively coupled plasma-MS (ICP–MS) and ICP–atomic emission spectrometry (ICP–AES) are commonly used to analyze the ionomic profile of plants (Salt *et al.*, 2008). This approach plays a crucial role in understanding the elemental composition of plants and its implications for biochemistry, physiology, and nutrition. The ability of plants to absorb and utilize different elements has evolved in response to diverse soil types and other environmental factors that influence growth and development (Fujita *et al.*, 2013). The ionomic profile of crops is influenced by factors such as element availability, root uptake efficiency, transport mechanisms, and environmental stress. Extensive research has been conducted in the field of ionomics. For instance, studies on the ionome of wild and cultivated barley under varying levels of salt tolerance revealed a significant negative correlation between accumulated Na^+ levels and metabolites associated with glycolysis and the tricarboxylic acid (TCA) cycle (Wu *et al.*, 2013). Research on wheat (Guo *et al.*, 2009) and other grasses like *Aneurolepidium chinense* (Shi and Wang, 2005), *Setaria viridis* (Guo *et al.*, 2011), and flax (Guo *et al.*, 2014) demonstrated that Na^+ concentration increases with rising alkalinity stress, as plants store substantial amounts of Na^+ in vacuoles to lower cell water potential. In flax subjected to PEG-induced water stress, key inorganic ions such as K^+, Na^+, Ca2$^+$, and Cl^- were found to play a vital role in osmotic adjustment, enhancing drought tolerance (Guo *et al.*, 2012). In flax, no notable differences were observed in the concentrations of Na^+ and K^+ in shoots when comparing the effects of salinity and alkalinity stress (Guo *et al.*, 2014). This indicates that the mechanism of adaptation in flax shoots to alkaline stress may differ from that of other species like barley (Guo *et al.*, 2009) and *Chloris virgata* (Yang *et al.*, 2010), where shoot K^+ levels were significantly reduced under alkaline stress. Another study on flax reported a marked increase in the uptake of Cl^-, H_2PO^{4-}, and SO_4^{2-}, while NO^{3-} levels decreased in shoots under salt stress. These findings suggest that these anions accumulate in vacuoles to offset the influx of Na^+, maintaining cellular hydration (Parida and Das, 2005). Additionally, inorganic anion concentrations were significantly lower under alkali stress compared to salt stress at the same osmotic potential. This implies that the high pH associated with alkali stress may hinder the absorption of anions like NO^{3-}, H_2PO^{4-}, and SO_4^{2-} in flax (Guo *et al.*, 2013, 2014). The application of various inorganic elements has been shown to enhance tolerance to abiotic stresses. For instance, silicon has demonstrated effectiveness in mitigating the effects of drought, salinity, heat, heavy metals, and UV-B radiation (Liang *et al.*, 2007; Deshmukh *et al.*, 2014). In grapevines, hyperaccumulation of manganese (Mn) in the leaves has been linked to delayed pathogen spread, contributing to resistance against powdery mildew (Yao *et al.*, 2012). Thus, analyzing the elemental composition of plants can provide valuable insights into stress tolerance mechanisms. However, ionomic research in oilseed crops remains relatively underexplored.

14.3 Conclusions

The challenges posed by climate change demand a comprehensive approach to safeguarding oilseed crop productivity and global food security. Multiomics technologies offer transformative potential in understanding and addressing the complex molecular mechanisms underlying oilseed crop responses to abiotic stresses. By integrating genomics, transcriptomics, proteomics, metabolomics, phenomics, and ionomics, researchers can identify key genetic, molecular, and biochemical pathways that confer stress resilience. This knowledge empowers the development of climate-resilient oilseed crops with enhanced tolerance to drought, salinity, heat, and other environmental challenges. Furthermore, the application of omics data enables precision breeding strategies, facilitating the identification and utilization of novel alleles from crop wild relatives and underutilized species. These approaches, combined with advanced bioinformatics and systems biology tools, ensure a holistic understanding of plant stress responses, paving the way for sustainable agricultural practices. As the global demand for oilseed crops continues to rise, integrating omics-assisted breeding into agricultural systems represents a cost-effective, scalable, and sustainable solution to mitigate the impacts of climate change. By fostering innovation and collaboration in the field, we can not only secure oilseed crop yields but also contribute to broader goals of food security, environmental conservation, and economic stability in the face of an uncertain climate future.

References

Abdelrahman, M., El-Sayed, M.A., Hashem, A., Abd-Allah, E.F., Alqarawi, A.A. *et al.* (2018) Metabolomics and transcriptomics in legumes under phosphate deficiency in relation to nitrogen fixation by root nodules. *Frontiers in Plant Science* 9, 922. DOI: 10.3389/fpls.2018.00922.

Ahmar, S., Gill, R.A., Jung, K.-H., Faheem, A., Qasim, M.U. *et al.* (2020) Conventional and molecular techniques from simple breeding to speed breeding in crop plants: Recent advances and future outlook. *International Journal in Molecular Science* 21, 2590. DOI: 10.3390/ijms21072590.

Aizat, W.M. and Hassan, M. (2018) Proteomics in systems biology. In: Aizat, W., Goh, H.H. and Baharum, S. (eds) *Omics Applications for Systems Biology. Advances in Experimental Medicine and Biology*. Springer, Cham, pp. 31–49. DOI: 10.1007/978-3-319-98758-3_3.

Alam, I., Sharmin, S.A., Kim, K.H., Yang, K.J., Choi, M.S. *et al.* (2010) Proteome analysis of soybean roots subjected to short-term drought stress. *Plant and Soil* 333, 491–505. DOI: 10.1007/s11104-010-0365-7.

Alexandratos, N. and Bruinsma, J. (2012) World agriculture towards 2030/2050: The 2012 revision. ESA Working paper No. 12-03. Rome, FAO.

Allwood, J.W. and Goodacre, R. (2010) An introduction to liquid chromatography-mass spectrometry instrumentation applied in plant metabolomic analyses. *Phytochemical Analysis* 21, 33–47. DOI: 10.1002/pca.1187.

Alseekh, S., Aharoni, A., Brotman, Y., Contrepois, K., D'Auria, J. *et al.* (2021) Mass spectrometry-based metabolomics: A guide for annotation, quantification and best reporting practices. *Nature Methods* 18, 747–756. DOI: 10.1038/s41592-021-01197-1.

Alvarez, S., Roy Choudhury, S., Sivagnanam, K., Hicks, L.M. and Pandey, S. (2015) Quantitative proteomics analysis of *Camelina sativa* seeds overexpressing the AGG3 gene to identify the proteomic basis of increased yield and stress tolerance. *Journal of Proteome Research* 14(6), 2606–2616.
Amiri, R.M., Azdi, S.B., Ghanadha, M.R. and Abd, M.C. (2001) Detection of DNA polymorphism in landrace populations of safflower in Iran using RAPD-PCR technique. *Iranian Journal of Agricultural Sciences* 32, 737–745.
Ananda, G.K., Myrans, H., Norton, S.L., Gleadow, R., Furtado, A. *et al.* (2020) Wild sorghum as a promising resource for crop improvement. *Frontiers in Plant Science* 11, 1108. DOI: 10.3389/fpls.2020.01108.
Anwar, K., Joshi, R., Dhankher, O.P., Singla-Pareek, S.L. and Pareek, A. (2021) Elucidating the response of crop plants towards individual, combined and sequentially occurring abiotic stresses. *International Journal of Molecular Sciences* 22(6119), ijms221161. DOI: 10.3390/ijms221161.
Arelli, P. and Young, L. (2009) JTN-5109 soybean germplasm resistant to nematode population infecting cv. hartwig. In: *In 2009 Annual Meeting (November 1-5, 2009). ASA-CSSA-SSSA*.
Arelli, P.R., Concibido, V.C. and Young, L.D. (2010) QTLs associated with resistance in soybean PI567516C to synthetic nematode population infecting cv. Hartwig. *Journal of Crop Science and Biotechnology* 13, 163–167.
Aslam, B., Basit, M., Atif, N.M., Khurshid, M. and Rasool, M.H. (2017) Proteomics: Technologies and their applications. *Journal of Chromatography Science* 55, 182–196. DOI: 10.1093/chromsci/bmw167.
Bachlava, E., Dewey, R.E., Auclair, J., Wang, S., Burton, J.W. *et al.* (2008) Mapping genes encoding microsomal w-6 desaturase enzymes and their cosegregation with QTL affecting oleate content in soybean. *Crop Science* 48, 640–650.
Badouin, H., Gouzy, J., Grassa, C., Murat, F., Staton, S.E. *et al.* (2017) The sunflower genome provides insights into oil metabolism, flowering and Asterid evolution. *Nature* 546, 148–152. DOI: 10.1038/nature22380.
Baggerman, G., Vierstraete, E., De, L.A. and Schoofs, L. (2005) Gel-based versus gel-free proteomics: A review. Comb. Chem. *High Throughput Screen* 8, 669–677. DOI: 10.2174/1386207057749 62490.
Baharum, S.N. and Azizan, K.A. (2018) Metabolomics in systems biology. *Advances in Experimental Medicine and Biology* 1102, 51–68. DOI: 10.1007/978-3-319-98758-3_4.
Bahuguna, R.N., Tamilselvan, A., Muthurajan, R., Solis, C.A. and Jagadish, S.V.K. (2018) Mild preflowering drought priming improves stress defences, assimilation and sink strength in rice under severe terminal drought. *Functional Plant Biology* 45(8), 827–839.
Bai, L., Sun, H.B., Liang, R.T. and Cai, B.Y. (2019) ITRAQ proteomic analysis of continuously cropped soybean root inoculated with *Funneliformis mosseae*. *Front Microbiol* 10, 1–13. DOI: 10.3389/fmicb.2019.00061.
Baker, N.R. (2008) Chlorophyll fluorescence: A probe of photosynthesis *in vivo*. *Annual Review of Plant Biology* 59, 89–113. DOI: 10.1146/annurev.arplant.59.032607.092759.
Banerjee, R., Kumar, G.V. and Kumar, S.P.J. (eds) (2019) *OMICS-based approaches in plant biotechnology*. John Wiley & Sons, Hoboken, NJ.
Berger, B., Parent, B. and Tester, M. (2010) High-throughput shoot imaging to study drought responses. *Journal of Experimental Botany* 61, 3519–3528. DOI: 10.1093/jxb/erq201.
Berry, S.T., Leon, A.J., Hanfrey, C.C., Challis, P., Burkholz, A. *et al.* (1995) Molecular markers analysis of *Helianthus annuus* L.2. Construction of an RFLP map for cultivated sunflower. *Theoretical and Applied Genetics* 91, 195–199.
Beyer, S., Daba, S., Tyagi, P., Bockelma, H.E., Brown-Guedira, G.L. *et al.* (2019) Loci and candidate genes controlling root traits in wheat seedlings—a wheat root GWAS. *Functional & Integrative Genomics* 19(91–107), s10142–018. DOI: 10.1007/s10142-018-0630-z.
Bodner, G., Alsalem, M., Nakhforoosh, A., Arnold, T. and Leitner, D. (2017) RGB and spectral root imaging for plant phenotyping and physiological research: Experimental setup and imaging protocols. *Journal of Visualized Experiments* 126, 56251. DOI: 10.3791/56251.
Bohra, A., Jha, U.C., Godwin, I.D. and Kumar Varshney, R. (2020) Genomic interventions for sustainable agriculture. *Plant Biotechnology Journal* 18, 2388–2405. DOI: 10.1111/pbi.13472.
Brozynska, M., Furtado, A. and Henry, R.J. (2016) Genomics of crop wild relatives: Expanding the gene pool for crop improvement. *Plant Biotechnology Journal* 14, 1070–1085. DOI: 10.1111/pbi.12454.
Burow, M.D., Simpson, C.E., Paterson, A.H. and Starr, J.L. (1996) Identification of peanut (*Arachis hypogaea* L.) RAPD markers diagnostic of root-knot nematode (*Meloidogyne arenaria* (Neal) Chitwood) resistance. *Molecular Breeding* 2, 369–379.

Cabrera-Bosquet, L., Crossa, J., von Zitzewitz, J., Serret, M.D. and Araus, J.L. (2012) High-throughput phenotyping and genomic selection: The frontiers of crop breeding converge. *Journal of Integrative Plant Biology* 54, 312–320. DOI: 10.1111/j.1744-7909.2012.01116.x.

Calder, P.C. and Yaqoob, P. (2009) Omega-3 polyunsaturated fatty acids and human health outcomes. *BioFactors* 35(3), 266–272.

Calder, P.C. (2015) Functional roles of fatty acids and their effects on human health. *Journal of Parenteral and Enteral Nutrition* 39, 18S–32S.

Cartagena, J.A., Seki, M., Tanaka, M., Yamauchi, T., Sato, S. *et al.* (2015) Gene expression profiles in Jatropha under drought stress and during recovery. *Plant Molecular Biology Reporter* 33, 1075–1087.

Chaerle, L., Lenk, S., Leinonen, I., Jones, H.G., Van Der Straeten, D. *et al.* (2009) Multi-sensor plant imaging: Towards the development of a stress-catalogue. *Biotechnology Journal* 4, 1152–1167. DOI: 10.1002/biot.200800242.

Chalhoub, B., Denoeud, F., Liu, S., Parkin, I.A.P., Tang, H. *et al.* (2014) Plant genetics. *early* allopolyploid evolution in the post-neolithic *Brassica napus* oilseed genome. *Science* 345(6199), 950–953. DOI: 10.1126/science.1253435.

Chen, H., Liu, N., Xu, R., Chen, X., Zhang, Y. *et al.* (2021) Quantitative proteomics analysis reveals the response mechanism of peanut (*Arachishypogaea* L.) to imbibitional chilling stress. *Plant Biology* 23(3), 517–527.

Chan, A.P., Crabtree, J., Zhao, Q., Lorenzi, H., Orvis, J. *et al.* (2010) Draft genome sequence of the oilseed species *Ricinus communis*. *Nature Biotechnology* 28(9), 951–956. DOI: 10.1038/nbt.1674.

Cheng, C., Liu, Y., Fang, W., Tao, J., Yang, Z. *et al.* (2020) iTRAQ-based proteomic and physiological analyses of mustard sprouts in response to heat stress. *RSC Advances* 10(10), 6052–6062. DOI: 10.1039/c9ra10089j.

Cheng, L., Gao, X., Li, S., Shi, M., Javeed, H. *et al.* (2010) Proteomic analysis of soybean [*Glycine max* (L.) Meer.] seeds during imbibition at chilling temperature. *Molecular Breeding* 26, 1–17. DOI: 10.1007/s11032-009-9371-y.

Chu, Y., Holbrook, C.C., Timper, P. and Ozias-Akins, P. (2007) Development of a PCR-based molecular marker to select for nematode resistance in peanut. *Crop Science* 47(2), 841–845.

Colombini, S., Broderick, G.A., Galasso, I., Martinelli, T., Rapetti, L. *et al.* (2014) Evaluation of *Camelinasativa* (L.) Crantz meal as an alternative protein source in ruminant rations. *Journal of the Science of Food and Agriculture* 94(4), 736–743.

Concibido, V., La Vallee, B., Mclaird, P., Pineda, N., Meyer, J. *et al.* (2003) Introgression of a quantitative trait locus for yield from *Glycine soja* into commercial soybean cultivars. *Theoretical and Applied Genetics* 106, 575–582.

Crick, F. (1970) Central dogma of molecular biology. *Nature* 227(5258), 561–563.

Cuc, L.M. and Quang, V.D. (2008) QTL mapping of resistance to late leaf spot in groundnut (*Arachis hypogaea* L.). *Academia Journal of Biology* 30(3), 136–140.

Deeba, F., Pandey, A.K., Ranjan, S., Mishra, A., Singh, R. *et al.* (2012) Physiological and proteomic responses of cotton (*Gossypium herbaceum* L.) to drought stress. *Plant Physiology and Biochemistry* 53, 6–18. DOI: 10.1016/j.plaphy.2012.01.002.

Demirevska, K., Simova-Stoilova, L., Vassileva, V., Vaseva, I., Grigorova, B. *et al.* (2008) Drought-induced leaf protein alterations in sensitive and tolerant wheat varieties. *General and Applied Plant Physiology* 34, 79–102. DOI: 10.7892/boris.110728.

Deshmukh, R., Sonah, H., Patil, G., Chen, W., Prince, S. *et al.* (2014) Integrating omic approaches for abiotic stress tolerance in soybean. *Frontiers in Plant Science* 5, 244. DOI: 10.3389/fpls.2014.00244.

Díaz-Cruz, G.A., Smith, C.M., Wiebe, K.F., Villanueva, S.M., Klonowski, A.R. *et al.* (2019) Applications of next-generation sequencing for large-scale pathogen diagnoses in soybean. *Plant Disease* 103(6), 1075–1083. DOI: 10.1094/pdis-05-18-0905-re.

Di-Baccio, D., Navari-Izzo, F. and Izzo, R. (2004) Seawater irrigation: Antioxidant defence responses in leaves and roots of a sunflower (*Helianthus annuus* L.) ecotype. *Journal of Plant Physiology* 161, 1359–1366. DOI: 10.1016/j.jplph.2003.07.001.

Diers, B.W., Mansur, L., Imsande, J. and Shoemaker, R. (1991) Mapping *Phytophthora* resistance loci in soybean with restriction fragment length polymorphism markers. *Crop Science* 32, 377–383.

Diers, B.W., Fehr, W., Keim, P. and Shoemaker, R.C. (1992) RFLP analysis of soybean seed protein and oil content. *Theoretical and Applied Genetics* 83, 608–612.

Dixit, A., Jin, M.H., Chung, J.W., Yu, J.W., Chung, H.K. *et al.* (2005) Development of polymorphic microsatellite markers in sesame (*Sesamum indicum* L.). *Molecular Ecology Notes* 5(4), 736–738.

Dong, Y., Su, Y., Yu, P., Yang, M., Zhu, S. *et al*. (2015) Proteomic analysis of the relationship between metabolism and nonhost resistance in soybean exposed to *Bipolaris maydis*. *PLoS One* 10(10), e0141264.

Dossa, K., Donghua, L., Linhai, W., Xiaomin, Z., Jingyin, Y. *et al*. (2017) Dynamic transcriptome landscape of sesame (*Sesamum indicum* L.) under progressive drought and after rewatering. *Genomics Data* 11, 122–124.

Duque, A.S., Almeida, A.M., Bernardes, A., Marques, J., Farinha, A.P. *et al*. (2013) Chapter 3: Abiotic stress responses in plants: Unraveling the complexity of genes and networks to survive. In: Vahdati, K. and Leslie, C. (eds) *Abiotic Stress: Plant Responses and Applications in Agriculture*. INTECH Open, Rijeka, pp. 49–102. DOI: 10.5772/45842.

Dhankher, O.P. and Foyer, C.H. (2018) Climate resilient crops for improving global food security and safety. *Plant, Cell & Environment* 41, 877–884. DOI: 10.1111/pce.13207.

Eldakak, M., Milad, S.I., Nawar, A.I. and Rohila, J.S. (2013) Proteomics: A biotechnology tool for crop improvement. *Frontiers in Plant Science* 4, 35. DOI: 10.3389/fpls.2013.00035.

El-Metwally, S., Ouda, O.M. and Helmy, M. (2014) *Next Generation Sequencing Technologies and Challenges in Sequence Assembly*, 1st edn. Springer, New York, NY. DOI: 10.1007/978-1-4939-0715-1.

Evans, J.R. and Lawson, T. (2020) From green to gold: Agricultural revolution for food security. *Journal of Experimental Botany* 71, 2211–2215.

Fang, C., Hamilton, J.P., Vaillancourt, B., Wang, Y.W., Wood, J.C. *et al*. (2023) Cold stress induces differential gene expression of retained homeologs in *Camelina sativa* cv Suneson. *Frontiers in Plant Science* 14, 1271625. DOI: 10.3389/fpls.2023.1271625.

FAOSTAT (2014) Food and Agriculture Organization of the United Nations Statistics Division. 2014, 2015. Available at: http://faostat3.fao.org/home/E (accessed 20 July 2025).

Fiehn, O. (2002) Metabolomics – the link between genotypes and phenotypes. *Plant Molecular Biology* 48, 155–171. DOI: 10.1023/A:1013713905833.

Fiehn, O., Kopka, J., Dormann, P., Altmann, T., Trethewey, R.N. *et al*. (2000) Metabolite profiling for plant functional genomics. *Nature Biotechnology* 18, 1157–1161. DOI: 10.1038/81137.

Finkel, E. (2009) With 'phenomics' plant scientists hope to shift breeding into overdrive. *Science* 325, 380–381. DOI: 10.1126/science.325_380.

Food and Agriculture Organization of the United Nations (2017) FAO cereal supply and demand brief.

Fournier, M.L., Gilmore, J.M., Martin-Brown, S.A. and Washburn, M.P. (2007) Multidimensional separations-based shotgun proteomics. *Chemical Reviews* 107, 3654–3686. DOI: 10.1021/cr068279a.

Franklin, S. and Vondriska, T.M. (2011) Genomes, proteomes, and the central dogma. *Circulation: Cardiovascular Genetics* 4(5), 576–576.

Fujita, Y., Venterink, H.O., Van Bodegom, P.M., Douma, J.C., Heil, G.W. *et al*. (2013) Low investment in sexual reproduction threatens plants adapted to phosphorus limitation. *Nature* 505, 82–86. DOI: 10.1038/nature12733.

Furbank, R.T. and Tester, M. (2011) Phenomics—technologies to relieve the phenotyping bottleneck. *Trends in Plant Science* 16, 635–644. DOI: 10.1016/j.tplants.2011.09.005.

Gajjar, K.N., Mishra, G.P., Radhakrishnan, T., Dodia, S.M., Rathnakumar, A.L. *et al*. (2014) Validation of SSR markers linked to the rust and late leaf spot diseases resistance in diverse peanut genotypes. *Australian Journal of Crop Science* 8(6), 927–936.

Gao, G., Hu, J., Zhang, X., Zhang, F., Li, M. *et al*. (2021) Transcriptome analyses reveals genes expression pattern of seed response to heat stress in *Brassica napus* L. *Oil Crop Science* 6. DOI: 10.1016/j.ocsci.2021.04.005.

Garcia, G.M., Stalker, H.T. and Kochert, G. (1995) Introgression analysis of an interspecific hybrid population in peanut (*Arachis hypogaea* L.) using RFLP and RAPD markers. *Genome* 38, 166–176.

Ghaffari, M., Toorchi, M., Valizadeh, M. and Shakiba, M. (2017) Proteomic prospects for tolerance of sunflower (*Helianthus annuus*) to drought stress during the flowering stage. *Crop and Pasture Science* 68(5), 457–465.

Ghosh, D. and Xu, J. (2014) Abiotic stress responses in plant roots: A proteomics perspective. *Frontiers in Plant Science* 5, 6. DOI: 10.3389/fpls.2014.00006.

Gil, J.D., Cohn, A.S., Duncan, J., Newton, P. and Vermeulen, S. (2017) The resilience of integrated agricultural systems to climate change. *Wiley Interdisciplinary Reviews: Climate Change* 8, e461.

Githiri, S.M., Watanabe, S., Harada, K. and Takahashi, R. (2006) QTL analysis of flooding tolerance in soybean at an early vegetative growth stage. *Plant Breeding* 125(6), 613–618.

Gjuvsland, A.B., Vik, J.O., Beard, D.A., Hunter, P.J. and Omholt, S.W. (2013) Bridging the genotype-phenotype gap: What does it take? *The Journal of Physiology* 591, 2055–2066. DOI: 10.1113/jphysiol.2012.248864.

Gomez Selvaraj, M., Narayana, M., Schubert, A.M., Ayers, J.L., Baring, M.R. *et al.* (2009) Identification of QTLs for pod and kernel traits in cultivated peanut by bulked segregant analysis. *Electronic Journal of Biotechnology* 12(2), 3–4.

Großkinsky, D.K., Syaifullah, S.J. and Roitsch, T. (2018) Integration of multiomics techniques and physiological phenotyping within a holistic phenomics approach to study senescence in model and crop plants. *Journal of Experimental Botany* 69, 825–844. DOI: 10.1093/jxb/erx333.

Guo, B., Fedorova, N.D., Chen, X., Wan, C.H., Wang, W. *et al.* (2011) Gene expression profiling and identification of resistance genes to *Aspergillus flavus* infection in peanut through EST and microarray strategies. *Toxins* 3, 737–753. DOI: 10.3390/toxins3070737.

Guo, P., Baum, M., Grando, S., Ceccarelli, S., Bai, G. *et al.* (2009) Differentially expressed genes between drought-tolerant and drought-sensitive barley genotypes in response to drought stress during the reproductive stage. *Journal of Experimental Botany* 60, 3531–3544. DOI: 10.1093/jxb/erp194.

Guo, R., Hao, W. and Gong, D.Z. (2012) Effects of water stress on germination and growth of linseed seedlings (*Linum usitatissimum* L.), photosynthetic efficiency and accumulation of metabolites. *Journal of Agricultural Science* 4, 253–265. DOI: 10.5539/jas.v4n10p253.

Guo, R., Zhou, J., Ren, G. and Hao, W. (2013) Physiological responses of linseed seedlings to iso osmotic polyethylene glycol, salt, and alkali stresses. *Agronomy Journal* 105, 764–772. DOI: 10.2134/agronj2012.0442.

Guo, R., Zhou, J., Hao, W., Gu, F., Liu, Q. *et al.* (2014) Germination, growth, chlorophyll fluorescence and ionic balance in linseed seedlings subjected to saline and alkaline stresses. *Plant Production Science* 17, 20–31. DOI: 10.1626/pps.17.20.

Guo, X., Wang, D., Gordon, S.G., Helliwell, E., Smith, T. *et al.* (2008) Genetic mapping of QTLs underlying partial resistance to *Sclerotinia sclerotiorum* in soybean PI 391589A and PI 391589B. *Crop Science* 48(3), 1129–1139.

Gupta, S.M., Arora, S., Mirza, N., Pande, A., Lata, C. *et al.* (2017) Finger millet: A “certain” crop for an “uncertain” future and a solution to food insecurity and hidden hunger under stressful environments. *Frontiers in Plant Science* 8, 643. DOI: 10.3389/fpls.2017.00643.

Han, Y., Teng, W., Yu, K., Poysa, V., Anderson, T. *et al.* (2008) Mapping QTL tolerance to phytophthora root rot in soybean using microsatellite and RAPD/SCAR derived markers. *Euphytica* 162, 231–239.

Hasin, Y., Seldin, M. and Lusis, A. (2017) Multi-omics approaches to disease. *Genome Biology* 18, 1–15.

Hatzig, S., Zaharia, L.I., Abrams, S., Hohmann, M., Legoahec, L. *et al.* (2014) Early osmotic adjustment responses in drought-resistant and drought-sensitive oilseed rape. *Journal of Integrative Plant Biology* 56(8), 797–809.

Houle, D., Govindaraju, D.R. and Omholt, S. (2010) Phenomics: The next challenge. *Nature Reviews Genetics* 11, 855–866. DOI: 10.1038/nrg2897.

Hu, H., Scheben, A. and Edwards, D. (2018) Advances in integrating genomics and bioinformatics in the plant breeding pipeline. *Agriculture* 8(75). DOI: 10.3390/agriculture8060075.

Humplík, J.F., Lazár, D., Husi̇čková, A. and Spíchal., L. (2015) Automated phenotyping of plant shoots using imaging methods for analysis of plant stress responses – a review. *Plant Methods* 11, 29. DOI: 10.1186/s13007-015-0072-8.

Jang, H.Y., Rhee, J., Carlson, J.E. and Ahn, S.J. (2014) The Camelina aquaporin CsPIP2; 1 is regulated by phosphorylation at Ser273, but not at Ser277, of the C-terminus and is involved in salt-and drought-stress responses. *Journal of Plant Physiology* 171(15), 1401–1412.

Jangpromma, N., Songsri, P., Thammasirirak, S. and Jaisil, P. (2010) Rapid assessment of chlorophyll content in sugarcane using a SPAD chlorophyll meter across different water stress conditions. *Asian Journal of Plant Sciences* 9, 368–374. DOI: 10.3923/ajps.2010.368.374.

Jo, Y., Choi, H., Bae, M., Kim, S.-M., Lee, B.C. *et al.* (2017) De novo Genome assembly and single nucleotide variations for soybean Mosaic Virus using soybean seed Transcriptome data. *Plant Pathology Journal* 33, 478. DOI: 10.5423/PPJ.OA.03.2017.0060.

Jung, H.J., Dong, X., Park, J.I., Thamilarasan, S.K., Lee, S.S. *et al.* (2014) Genome-wide transcriptome analysis of two contrasting *Brassica rapa* doubled haploid lines under cold-stresses using Br135K oligomeric chip. *PLoS ONE* 9(8), e106069.

Jogaiah, S., Govind, S.R. and Tran, L.-S.P. (2013) Systems biology-based approaches toward understanding drought tolerance in food crops. *Critical Reviews in Biotechnology* 33, 23–39. DOI: 10.3109/07388551.2012.659174.

Kagale, S., Koh, C., Nixon, J., Bollina, V., Clarke, W.E. *et al.* (2014) The emerging biofuel crop *Camelina sativa* retains a highly undifferentiated hexaploid genome structure. *Nature Communications* 5, 3706. DOI: 10.1038/ncomms4706.

Kamenya, S.N., Mikwa, E.O., Song, B. and Odeny, D.A. (2021) Genetics and breeding for climate change in Orphan crops. *Theoretical and Applied Genetics* 1–29. DOI: 10.1007/s00122-020-03755-1.

Karkee, M., Steward, B.L., Tang, L. and Aziz, S.A. (2009) Quantifying sub-pixel signature of paddy rice field using an artificial neural network. *Computers and Electronics in Agriculture* 65, 65–76. DOI: 10.1016/j.compag.2008.07.009.

Kassem, M.A., Meksem, K., Kang, C.H., Njiti, V.N., Kilo, V. *et al.* (2004) Loci underlying resistance to manganese toxicity mapped in a soybean recombinant inbred line population of 'Essex2019; x 'Forrest'. *Plant and Soil* 260, 197–204.

Katam, R., Sakata, K., Suravajhala, P., Pechan, T., Kambiranda, D.M. *et al.* (2016) Comparative leaf proteomics of drought-tolerant and-susceptible peanut in response to water stress. *Journal of Proteomics* 143, 209–226.

Kawahara, Y., Oono, Y., Kanamori, H., Matsumoto, T., Itoh, T. *et al.* (2012) Simultaneous RNA-seq analysis of a mixed transcriptome of rice and blast fungus interaction. *PLoS ONE* 7, e49423. DOI: 10.1371/journal.pone.0049423.

Kazi, S., Shultz, J., Afzal, J., Johnson, J., Njiti, V.N. *et al.* (2008) Separate loci underlie resistance to root infection and leaf scorch during soybean sudden death syndrome. *Theoretical and Applied Genetics* 116, 967–977.

Khedikar, Y.P., Gowda, M.V.C., Sarvamangala, C., Patgar, K.V., Upadhyaya, H.D. *et al.* (2010) A QTL study on late leaf spot and rust revealed one major QTL for molecular breeding for rust resistance in groundnut (*Arachis hypogaea* L.). *Theoretical and Applied Genetics* 121, 971–984.

Kikuchi, J. and Hirayama, T. (2007) Practical aspects of uniform stable isotope labeling of higher plants for heteronuclear NMR-based metabolomics. In: Weckwerth, W. (ed.) *Metabolomics, Methods in Molecular Biology*. Humana Press, New York, NY, pp. 273–286. DOI: 10.1007/978-1-59745-244-1_15.

Kim, H.K., Choi, Y.H. and Verpoorte, R. (2011) Plant metabolomics: Where do we stand, where do we go? *Trends Biotechnol* 29, 267–275. DOI: 10.1016/j.tibtech.2011.02.001.

Kim, S.B., Van den Broeck, L., Karre, S., Choi, H., Christensen, S.A. *et al.* (2021) Analysis of the transcriptomic, metabolomic, and gene regulatory responses to *Pucciniasorghi* in maize. *Molecular Plant Pathology* 22, 465–479. DOI: 10.1111/mpp.13040.

Klein, A., Houtin, H., Rond-Coissieux, C., Naudet-Huart, M., Touratier, M. *et al.* (2020) Meta-analysis of QTL reveals the genetic control of yield-related traits and seed protein content in pea. *Scientific Reports* 10, 1–11. DOI: 10.1038/s41598-020-72548-9.

Koh, J., Chen, G., Yoo, M.J., Mudge, J., Bhattacharyya, M.K. *et al.* (2015) Comparative proteomic analysis of *Brassica napus* in response to drought stress. *Journal of Proteome Research* 14(8), 3068–3081. DOI: 10.1021/pr501323d.

Kole, C., Muthamilarasan, M., Henry, R., Edwards, D., Sharma, R. *et al.* (2015) Application of genomics-assisted breeding for generation of climate resilient crops: Progress and prospects. *Front. Plant Sci* 6, 563. DOI: 10.3389/fpls.2015.00563.

Kris-Etherton, P.M. and Krauss, R.M. (2020) Public health guidelines should recommend reducing saturated fat consumption as much as possible: YES. *The American Journal of Clinical Nutrition*.

Kumar, J., Pratap, A. and Kumar, S. (2015) Plant phenomics: An overview. In: Kumar, J., Pratap, A. and Kumar, S. (eds) *Phenomics in Crop Plants: Trends, Options and Limitations*. Springer, New Delhi, pp. 1–10. DOI: 10.1007/978-81-322-2226-2_1.

Kumar, R., Das, S.P., Choudhury, B.U., Kumar, A., Prakash, N.R. *et al.* (2024) Advances in genomic tools for plant breeding: Harnessing DNA molecular markers, genomic selection, and genome editing. *Biological Research* 57(1), 80.

Kumari, P., Rastogi, A. and Yadav, S. (2020) Effects of heat stress and molecular mitigation approaches in orphan legume, Chickpea. *Molecular Biology Reports* 47, 4659–4670.

Lamaoui, M., Jemo, M., Datla, R. and Bekkaoui, F. (2018) Heat and drought stresses in crops and approaches for their mitigation. *Frontiers in Chemistry* 6, 26. DOI: 10.3389/fchem.2018.00026.

Landry, B.S., Hubert, N., Etoh, T., Harada, J.J. and Lincoln, S.E. (1991) A genetic map for *Brassica napus* based on restriction fragment length polymorphism detected with expressed DNA sequences. *Genome* 34, 543–552.

Langridge, P. and Fleury, D. (2011) Making the most of 'omics' for crop breeding. *Trends in Biotechnology* 29, 33–40.

Lark, K.G., Weisemann, J.M., Matthews, B.F., Palmer, R., Chase, K. *et al.* (1993) A genetic map of soybean (*Glycine max* L.) using an intraspecific cross of two cultivars: 'Minosy' and 'Noir 1'. Theor. *Theoretical and Applied Genetics* 86(8), 901–906.

Lee, G.J., Boerma, H.R., Villagarcia, M.R., Zhou, X., Carter, T.E. *et al.* (2004) A major QTL conditioning salt tolerance in S-100 soybean and descendent cultivars. *Theoretical and Applied Genetics* 109(8), 1610–1619.

Lei, X., Xiao, Y., Xia, W., Mason, A.S., Yang, Y. *et al.* (2014) RNA-seq analysis of oil palm under cold stress reveals a different C-repeat binding factor (CBF) mediated gene expression pattern in *Elaeis guineensis* compared to other species. *PLoS ONE* 9(12), e114482.

Leisner, C.P., Yendrek, C.R. and Ainsworth, E.A. (2017) Physiological and transcriptomic responses in the seed coat of field-grown soybean (*Glycine max* L. Merr.) to abiotic stress. *BMC Plant Biology* 17, 242. DOI: 10.1186/s12870-017-1188-y.

Li, Q. and Yan, J. (2020) Sustainable agriculture in the era of omics: Knowledge-driven crop breeding. *Genome Biology* 21, 1–5. DOI: 10.1186/s13059-020-02073.

Li, G.Y., Zhou, J.X., Ma, W.K., Jiang, L.G., Jin, Z.H. *et al.* (2014) De novo assembly of soybean wild relatives for pan-genome analysis of diversity and agronomic traits. *Nature Biotechnology* 32, 1045–1052. DOI: 10.1038/nbt.2979.

Li, M., Li, H., Sun, A., Wang, L., Ren, C. *et al.* (2022a) Transcriptome analysis reveals key drought-stress-responsive genes in soybean. *Frontiers in Genetics* 13, 1060529. DOI: 10.3389/fgene.2022.1060529.

Li, Y., Chen, J., Li, X., Jiang, H., Guo, D. *et al.* (2022b) Adaptive response and transcriptomic analysis of flax (*Linum usitatissimum* L.) seedlings to salt stress. *Genes* 13(10). DOI: 10.3390/genes13101904.

Liang, Y.C., Sun, W.C., Zhu, Y.G. and Christie, P. (2007) Mechanisms of silicon-mediated alleviation of abiotic stresses in higher plants: A review. *Environmental Pollution* 147, 422–428. DOI: 10.1016/j.envpol.2006.06.008.

Lim, H., Kobayashi, M.J., Marsoem, S.N., Irawati, D., Kosugi, A. *et al.* (2023) Transcriptomic responses of oil palm (*Elaeisguineensis*) stem to waterlogging at plantation in relation to precipitation seasonality. *Frontiers in Plant Science* 14, 1213496. DOI: 10.3389/fpls.2023.1213496.

Lindon, J.C. and Nicholson, J.K. (2008) Analytical technologies for metabonomics and metabolomics, and multi-omic information recovery. *Trends in Analytical Chemistry* 27, 194–204. DOI: 10.1016/j.trac.2007.08.009.

Lohani, N., Jain, D., Singh, M.B. and Bhalla, P.L. (2020) Engineering multiple abiotic stress tolerance in canola, *Brassica napus*. *Frontiers in Plant Science* 11, 3.

Lueong, S.S., Hoheisel, J.D. and Alhamdani, M.S.S. (2014) Protein microarrays as tools for functional proteomics: Achievements, promises and challenges. *Journal of Proteomics & Bioinformatics* 7, 1–10. DOI: 10.4172/jpb.S7-004.

Maghuly, F. and Laimer, M. (2013) *Jatropha curcas*, a biofuel crop: Functional genomics for understanding metabolic pathways and genetic improvement. *Biotechnology Journal* 8(10), 1172–1182.

Mahlein, A.K. (2016) Plant disease detection by imaging sensors–parallels and specific demands for precision agriculture and plant phenotyping. *Plant Disease* 100, 241–251. DOI: 10.1094/PDIS-03-15-0340-FE.

Margaria, P., Abbà, S. and Palmano, S. (2013) Novel aspects of grapevine response to phytoplasma infection investigated by a proteomic and phosphoproteomic approach with data integration into functional networks. *BMC Genomics* 14, 38. DOI: 10.1186/1471-2164-14-38.

Mba, C., Guimaraes, E.P. and Ghosh, K. (2012) Re-orienting crop improvement for the changing climatic conditions of the 21st century. *Agriculture & Food Security* 1, 1–17. DOI: 10.1186/2048-7010-1-7.

McGrail, R.K., Van Sanford, D.A. and McNear, D.H. Jr. (2020) Trait-based root phenotyping as a necessary tool for crop selection and improvement. *Agronomy* 10, 1328. DOI: 10.3390/agronomy10091328.

McLuckey, S.A. and Stephenson, J.L. Jr. (1998) Ion/ion chemistry of high-mass multiply charged ions. *Mass Spectrometry Review* 17, 369–407. DOI: 10.1002/(SICI)1098-2787(1998)17:63.0.CO;2-J.

Mehmood, S.S., Lu, G., Luo, D., Hussain, M.A., Raza, A. *et al.* (2021) Integrated analysis of transcriptomics and proteomics provides insights into the molecular regulation of cold response in *Brassica napus*. *Environmental and Experimental Botany* 187, 104480.

Missanga, J.S., Venkataramana, P.B. and Ndakidemi, P.A. (2021) Recent developments in *Lablab purpureus* genomics. A focus on drought stress tolerance and use of genomic resources to develop stress-resilient varieties. *Legume Science*. DOI: 10.1002/leg3.99.
Mittova, V., Guy, M., Tal, M. and Volokita, M. (2004) Salinity up-regulates the antioxidative system in root mitochondria and peroxisomes of the wild salt-tolerant tomato species *Lycopersicon*. *Journal of Experimental Botany* 55, 1105–1113. DOI: 10.1093/jxb/erh113.
Moco, S., Bino, R.J., De Vos, R.C.H. and Vervoort, J. (2007) Metabolomics technologies and metabolite identification. *Trends in Analytical Chemistry* 26, 855–866. DOI: 10.1016/j.trac.2007.08.003.
Mohammadi, P.P., Moieni, A. and Komatsu, S. (2012) Comparative proteome analysis of drought-sensitive and drought-tolerant rapeseed roots and their hybrid F1 line under drought stress. *Amino Acids* 43, 2137–2152. DOI: 10.1007/s00726-012-1299-6.
Mosa, K.A., Ismail, A. and Helmy, M. (2017) Omics and system biology approaches in plant stress research. In: Ismail, K.A., Ismail, A. and Helmy, M. (eds) *Plant Stress Tolerance: An Integrated Omics Approach*. Springer, Cham, pp. 21–34. DOI: 10.1007/978-3-319-59379-1_2.
Murithi, H.M., Beed, F., Tukamuhabwa, P., Thomma, B.P. and Joosten, M.H. (2016) Soybean production in eastern and southern Africa and threat of yield loss due to soybean rust caused by *Phakopsora pachyrhizi*. *Plant Pathology* 65(2), 176–188.
Nakagami, H., Sugiyama, N., Ishihama, Y. and Shirasu, K. (2012) Shotguns in the front line: Phosphoproteomics in plants. *Plant and Cell Physiology* 53, 118–124. DOI: 10.1093/pcp/pcr148.
Nataraja, K.N., Madhura, B.G. and Parvathi, S.M. (2017) Omics: Modern tools for precise understanding of drought adaptation in plants. In: Zargar, S.M. and Rai, V. (eds) *Plant OMICS and Crop Breeding*. Apple Academic Press, Palm Bay, FL, pp. 289–320.
Neale, D.B. and Savolainen, O. (2004) Association genetics of complex traits in conifers. *Trends in Plant Science* 9(7), 325–330.
Neupane, S., Varenhorst, A.J. and Nepal, M.P. (2019) Transcriptome profiling of induced susceptibility effects on soybean-soybean aphid (*Hemiptera*: *Aphididae*) interaction. *BMC Research Notes* 12, 1–4. DOI: 10.1186/S13104-019-4372-3/TABLES/1.
Neveu, P., Tireau, A., Hilgert, N., Nègre, V., Mineau-Cesari, J. *et al*. (2019) Dealing with multi-source and multi-scale information in plant phenomics: The ontology-driven phenotyping hybrid information system. *New Phytologist* 221, 588–601. DOI: 10.1111/nph.15385.
Nhamo, L., Matchaya, G., Mabhaudhi, T., Nhlengethwa, S., Nhemachena, C. *et al*. (2019) Cereal production trends under climate change: Impacts and adaptation strategies in Southern Africa. *Agriculture* 9(30). DOI: 10.3390/agriculture9020030.
Nikoloski, Z., Perez-Storey, R. and Sweetlove, L.J. (2015) Inference and prediction of metabolic network fluxes. *Plant Physiology* 169(3), 1443–1455.
Nogueira, L.A. (2011) Does biodiesel make sense? *Energy* 36(6), 3659–3666.
Nouri, M.Z. and Komatsu, S. (2010) Comparative analysis of soybean plasma membrane proteins under osmotic stress using gel-based and LC MS/MS-based proteomics approaches. *Proteomics* 10, 1930–1945. DOI: 10.1002/pmic.200900632.
Oh, T., Gorman, M. and Cullis, C. (2000) RFLP and RAPD mapping in flax (*Linum usitatissimum*). *Theoretical and Applied Genetics* 101, 590–593.
Pandey, M.K., Roorkiwal, M., Singh, V.K., Ramalingam, A., Kudapa, H. *et al*. (2016) Emerging genomic tools for legume breeding: Current status and future perspectives. *Frontiers in Plant Science* 7, 455. DOI: 10.3389/fpls.2016.00455.
Pandey, P., Irulappan, V., Bagavathiannan, M.V. and Senthil-Kumar, M. (2017) Impact of combined abiotic and biotic stresses on plant growth and avenues for crop improvement by exploiting physio-morphological traits. *Frontiers in Plant Science* 8, 537. DOI: 10.3389/flps.2017.00537.
Parida, A.K. and Das, A.B. (2005) Salt tolerance and salinity effects on plants: A review. *Ecotoxicology and Environmental Safety* 60, 324–349. DOI: 10.1016/j.ecoenv.2004.06.010.
Patzoldt, M.E., Grau, C.R., Stephens, P.A., Kurtzweil, N.C., Carlson, S.R. *et al*. (2005) Localization of a quantitative trait locus providing brown stem rot resistance in the soybean cultivar Bell. *Crop Science* 45(4), 1241–1248.
Pieruschka, R., Klimov, D., Kolber, Z., and Berry, J. A. (2010). Monitoring of cold and light stress impact on photosynthesis by using the laser induced fluorescence transient (LIFT) approach. *Functional Plant Biology*. 37, 395–402. DOI: 10.1071/FP09266.
Pourkheirandish, M., Golicz, A.A., Bhalla, P.L. and Singh, M.B. (2020) Global role of crop genomics in the face of climate change. *Frontiers in Plant Science* 11, 922. DOI: 10.3389/fpls.2020.00922.

Qi, B., Korir, P., Zhao, T., Yu, D., Chen, S. *et al.* (2008) Mapping quantitative trait loci associated with aluminum toxin tolerance in NJRIKY recombinant inbred line population of soybean (*Glycine* max). *Journal of Integrative Plant Biology* 50(9), 1089–1095.

Qian, Y., Ren, Q., Zhang, J., and Chen, L. (2019). Transcriptomic analysis of the maize (*Zea mays* L.) inbred line B73 response to heat stress at the seedling stage. Gene 692, 68–78. DOI: 10.1016/j.gene.2018.12.062.

Ranganathan, J., Waite, R., Searchinger, T. and Hanson, C. (2018) *How to Sustainably Feed 10 Billion People by 2050, in 21 Charts*. World Resource Institute, Washington, D.C.

Rascher, U. and Pieruschka, R. (2008) Spatio-temporal variations of photosynthesis: The potential of optical remote sensing to better understand and scale light use efficiency and stresses of plant ecosystems. *Precision Agriculture* 9, 355–366. DOI: 10.1007/s11119-008-9074-0.

Ray, D.K., West, P.C., Clark, M., Gerber, J.S., Prishchepov, A.V. *et al.* (2019) Climate change has likely already affected global food production. *PLoS ONE* 14(5), e0217148.

Raza, A., Tabassum, J., Kudapa, H. and Varshney, R.K. (2021) Can omics deliver temperature resilient ready-to-grow crops? *Critical Reviews in Biotechnology* 1–24. DOI: 10.1080/07388551.2021.1898332.

Raza, A., Razzaq, A., Mehmood, S.S., Zou, X., Zhang, X. *et al.* (2019) Impact of climate change on crops adaptation and strategies to tackle its outcome: A review. *Plants* 8, 34. DOI: 10.3390/plants8020034.

Reynolds, D., Ball, J., Bauer, A., Davey, R., Griffiths, S. *et al.* (2019) CropSight: A scalable and open-source information management system for distributed plant phenotyping and IoT-based crop management. *Gigascience* 8, 1–35. DOI: 10.1093/gigascience/giz009.

Roitsch, T., Cabrera-Bosquet, L., Fournier, A., Ghamkhar, K., Jiménez- Berni, J. *et al.* (2019) Review: New sensors and data-driven approaches—a path to next generation phenomics. *Plant Science* 282, 2–10. DOI: 10.1016/j.plantsci.2019.01.011.

Ruiz-Garcia, L., Lunadei, L., Barreiro, P. and Robla, J.I. (2009) A review of wireless sensor technologies and applications in agriculture and food industry: State of the art and current trends. *Sensors* 9, 4728–4750. DOI: 10.3390/s90604728.

Ruiz-López, N., Sayanova, O., Napier, J.A. and Haslam, R.P. (2012) Metabolic engineering of the omega-3 long chain polyunsaturated fatty acid biosynthetic pathway into transgenic plants. *Journal of Experimental Botany* 63(7), 2397–2410.

Saghai Maroof, M.A., Jeong, S.C., Gunduz, I., Tucker, D.M., Buss, G.R. *et al.* (2008) Pyramiding of soybean mosaic virus resistance genes by marker-assisted selection. *Crop Science* 48(2), 517–526.

Saito, K. and Matsuda, F. (2010) Metabolomics for functional genomics, systems biology, and biotechnology. *Annual Review of Plant Biology* 61, 463–448. DOI: 10.1146/annurev.arplant.043008.092035.

Sali, A., Glaeser, R., Earnest, T. and Baumeister, W. (2003) From words to literature in structural proteomics. *Nature* 422, 216–225. DOI: 10.1038/nature01513.

Salt, D.E., Baxter, I. and Lahner, B. (2008) Ionomics and the study of the plant ionome. *Annual Review of Plant Biology* 59, 709–733. DOI: 10.1146/annurev.arplant.59.032607.092942.

Saranga, Y., Jiang, C.X., Wright, R.J., Yakir, D. and Paterson, A.H. (2004) Genetic dissection of cotton physiological responses to arid conditions and their inter-relationships with productivity. *Plant, Cell & Environment* 27, 263–277. DOI: 10.1111/j.1365-3040.2003.01134.x.

Sauer, M.L.A., Scott, R.A. and Cheesbrough, T.M. (2008) Marker-assisted selection for low linolenic acid content in soybean. *Journal of Crop Improvement* 21(2), 139–155.

Sauter, H., Lauer, M. and Fritsch, H. (1991) Metabolic profiling of plants – a new diagnostic technique. In: Baker, D.R., Fenyes, J.G. and Moberg, W.K. (eds) *Synthesis and Chemistry of Agrochemicals I*. American Chemical Society, Washington, DC, pp. 288–299. DOI: 10.1021/bk-1991-0443.ch024.

Schmutz, J., Cannon, S.B., Schlueter, J., Ma, J., Mitros, T. *et al.* (2010) Genome sequence of the *palaeopolyploid* soybean. *Nature* 463, 178–183. DOI: 10.1038/nature08670.

Schnurbusch, T., Hayes, J. and Sutton, T. (2010) Boron toxicity tolerance in wheat and barley: Australian perspectives. *Breeding Science* 60, 297–304. DOI: 10.1270/jsbbs.60.297.

SETAC (2019) Technical Issue Paper: OMICS: Complete Systems and Complete Analyses. Pensacola, FL: SETAC, 4.

Shahzad, A., Ullah, S., Dar, A.A., Sardar, M.F., Mehmood, T. *et al.* (2021) Nexus on climate change: Agriculture and possible solution to cope future climate change stresses. *Environmental Science and Pollution Research* 26(93), 1–22.

Shaki, F., Maboud, H.E. and Niknam, V. (2020) Differential proteomics: Effect of growth regulators on salt stress responses in safflower seedlings. *Pesticide Biochemistry and Physiology* 164, 149–155.

Shalata, A., Mittova, V., Volokita, M., Guy, M. and Tal, M. (2001) Response of the cultivated tomato and its wild salt-tolerant relative *Lycopersicon pennellii* to salt-dependent oxidative stress: The root anti-oxidative system. *Physiologia Plantarum* 112, 487–494. DOI: 10.1034/j.1399-3054.2001.1120405.x.

Sharma, A., Mohapatra, T. and Sharma, R.P. (1994) Molecular mapping and character tagging in *Brassica juncea*–I. Degree, nature and linkage relationship of RFLPs and their association with quantitative traits. *Journal of Plant Biochemistry and Biotechnology* 3(2), 85–89.

Shi, D.C. and Wang, D. (2005) Effects of various salt-alkaline mixed stresses on *Aneurolepidium chinense* (Trin.) Kitag. *Plant and Soil* 271, 15–26. DOI: 10.1007/s11104-004-1307-z.

Shi, A., Chen, P., Li, D., Zheng, C., Zhang, B. *et al.* (2009) Pyramiding multiple genes for resistance to soybean mosaic virus in soybean using molecular markers. *Molecular Breeding* 23, 113–124.

Shoba, D., Manivannan, N., Vindhiyavarman, P. and Nigam, S.N. (2012) SSR markers associated for late leaf spot disease resistance by bulked segregant analysis in groundnut (*Arachis hypogaea* L.). *Euphytica* 188, 265–272.

Singh, R., Ong-Abdullah, M., Low, E.T.L., Manaf, M.A.A., Rosli, R. *et al.* (2013) Oil palm genome sequence reveals divergence of interfertile species in Old and New worlds. *Nature* 500, 335–339. DOI: 10.1038/nature12309.

Sirault, X.R.R., James, R.A. and Furbank, R.T. (2009) A new screening method for osmotic component of salinity tolerance in cereals using infrared thermography. *Functional Plant Biology* 36, 970–977. DOI: 10.1071/FP09182.

Song, W., Qi, N., Liang, C., Li, Y., Zhang, Y. *et al.* (2019) Soybean root transcriptome profiling reveals a nonhost resistant response during *Heterodera glycines* infection. *PLoS ONE* 14, e0217130. DOI: 10.1371/JOURNAL.PONE.0217130.

Steinwand, M.A. and Ronald, P.C. (2020) Crop biotechnology and the future of food. *Nature Food* 1, 273–283. DOI: 10.1038/s43016-020-0072-3.

Sujay, V., Gowda, M.V.C., Pandey, M.K., Bhat, R.S., Khedikar, Y.P. *et al.* (2012) Quantitative trait locus analysis and construction of consensus genetic map for foliar disease resistance based on two recombinant inbred line populations in cultivated groundnut (*Arachis hypogaea* L.). *Molecular Breeding* 30, 773–788.

Sumner, L.W., Mendes, P. and Dixon, R.A. (2003) Plant metabolomics: Large-scale phytochemistry in the functional genomics era. *Phytochemistry* 62, 817–836. DOI: 10.1016/s0031-9422(02)00708-2.

Tanaka, K., Waki, H., Ido, Y., Akita, S., Yoshida, Y. *et al.* (1988) Protein and polymer analyses up to *m/z* 100000 by laser ionization time-of-flight mass spectrometry. *Rapid Communications in Mass Spectrometry* 2, 151–153. DOI: 10.1002/rcm.1290020802.

Toorchi, M., Yukawa, K., Nouri, M.Z. and Komatsu, S. (2009) Proteomics approach for identifying osmotic-stress-related proteins in soybean roots. *Peptides* 30, 2108–2117. DOI: 10.1016/j.peptides.2009.09.006.

Trenberth, K.E., Dai, A., Van Der Schrier, G., Jones, P.D., Barichivich, J. *et al.* (2014) Global warming and changes in drought. *Nature Climate Change* 4(1), 17–22.

Tuberosa, R. and Salvi, S. (2004) QTLs and genes for tolerance to abiotic stress in cereals. *Cereal Genomics* 253–315.

Tubiello, F.N., Salvatore, M., Rossi, S., Ferrara, A., Fitton, N. *et al.* (2013) The FAOSTAT database of greenhouse gas emissions from agriculture. *Environmental Research Letters* 8(1), 015009.

Twyman, R.M. (2013) *Principles of Proteomics*. Abingdon, Garland Science Press.

UN (2017) "World Population Prospects: The 2017 Revision, Key Findings and Advance Tables," in Working Paper No. ESA/P/WP/248. New York, NY: United Nations, 46.

Ursin, V.M. (2003) Modification of plant lipids for human health: Development of functional land-based omega-3 fatty acids. *Journal of Nutrition* 133(12), 4271–4274.

Varshney, R.K., Bertioli, D.J., Moretzsohn, M.C., Vadez, V., Krishnamurthy, L. *et al.* (2009) The first SSR-based genetic linkage map for cultivated groundnut (*Arachis hypogaea* L.). *Theoretical and Applied Genetics* 118, 729–739.

Varshney, R.K., Thudi, M., May, G.D. and Jackson, S.A. (2010) Legume genomics and breeding. *Plant Breeding Review* 33, 257–304. DOI: 10.1002/9780470535486.ch6.

Varshney, R.K., Terauchi, R. and McCouch, S.R. (2014) Harvesting the promising fruits of genomics: Applying genome sequencing technologies to crop breeding. *PLoS Biology* 12(6), e1001883.

Vaughan, M.M., Block, A., Christensen, S.A., Allen, L.H. and Schmelz, E.A. (2018) The effects of climate change associated abiotic stresses on maize phytochemical defenses. *Phytochem. Rev* 17, 37–49.

Walker, D.R., Narvel, J.M., Boerma, H.R., All, J.N. and Parrott, W.A. (2004) A QTL that enhances and broadens Bt insect resistance in soybean. *Theoretical and Applied Genetics* 109, 1051–1057.

Wang, L., Yu, S., Tong, C., Zhao, Y., Liu, Y. *et al.* (2014) Genome sequencing of the high oil crop sesame provides insight into oil biosynthesis. *Genome Biology* 15, 1–13. DOI: 10.1186/gb-2014-15-2-r39.

Wang, M., Wang, P., Tu, L., Zhu, S., Zhang, L. *et al.* (2016) Multi-omics maps of cotton fibre reveal epigenetic basis for staged single-cell differentiation. *Nucleic Acids Research* 44, 4067–4079. DOI: 10.1093/nar/gkw238.

Wang, P., Yang, C., Chen, H., Song, C., Zhang, X. *et al.* (2017) Transcriptomic basis for drought-resistance in *Brassica napus* L. *Scientific Reports* 7, 40532. DOI: 10.1038/srep40532.

Wang, X., Li, M., Liu, X., Zhang, L., Duan, Q. *et al.* (2019) Quantitative proteomic analysis of castor (*Ricinus communis* L.) seeds during early imbibition provided novel insights into cold stress response. *International Journal of Molecular Sciences* 20(2), 355. DOI: 10.3390/ijms20020355.

Wang, Y., Yang, R., Feng, Y., Sikandar, A., Zhu, X. *et al.* (2020) iTRAQ-based proteomic analysis reveals the role of the biological control agent, *Sinorhizobium fredii* strain Sneb183, in enhancing soybean resistance against the soybean cyst nematode. *Frontiers in Plant Science* 11, 597819.

Wang, Z., Hobson, N., Galindo, L., Zhu, S., Shi, D. *et al.* (2012) The genome of flax (*Linum usitatissimum*) assembled de novo from short shotgun sequence reads. *The Plant Journal* 72(3), 461–473. DOI: 10.1111/j.1365-313X.2012.05093.x.

Weckwerth, W. (2003) Metabolomics in systems biology. *Annual Review of Plant Biology* 54, 669–689. DOI: 10.1146/annurev.arplant.54.031902.135014.

Weckwerth, W. (2010) Metabolomics: An integral technique in systems biology. *Bioanalysis* 2, 829–836. DOI: 10.4155/bio.09.192.

Weckwerth, W. (2011) Unpredictability of metabolism—the key role of metabolomics science in combination with next-generation genome sequencing. *Analytical and Bioanalytical Chemistry* 400, 1967. DOI: 10.1007/s00216-011-4948-9.

Wei, L., Zhang, H., Duan, Y., Li, C., Chang, S. *et al.* (2016a) Transcriptome comparison of resistant and susceptible sesame (*Sesamum indicum* L.) varieties inoculated with *Fusarium oxysporum* f. sp. *sesami*. *Plant Breeding* 135(5), 627–635. 2016a.

Weckwerth, W. and Fiehn, O. (2002) Can we discover novel pathways using metabolomic analysis? *Current Opinion in Biotechnology* 13, 156–160. DOI: 10.1016/s0958-1669(02)00299-9.

Weckwerth, W. and Morgenthal, K. (2005) Metabolomics: From pattern recognition to biological interpretation. *Drug Discovery Today* 10, 1551–1155. DOI: 10.1016/S1359-6446(05)03609-3.

Wei, X., Zhu, X., Yu, J., Wang, L., Zhang, Y. *et al.* (2016b) Identification of sesame genomic variations from genome comparison of landrace and variety. *Frontiers in Plant Science* 7, 1169.

Williams, C.M. (2000, May) Dietary fatty acids and human health. In: *Annales de Zootechnie*, Vol. 49, No. 3. EDP Sciences, pp. 165–180.

Wolkenhauer, O. and Muir, A. (2011) The complexity of cell-biological systems. In: *Philosophy of Complex Systems*. North-Holland, pp. 355–385.

Woolfson, M. (2018) The development of structural x-ray crystallography. *Physica Scripta* 93, 1–32. DOI: 10.1088/1402-4896/aa9c30.

Wu, J.W., Shi, Y., Zhu, Y.X., Wang, Y.C. and Gong, H.J. (2013) Mechanisms of enhanced heavy metal tolerance in plants by silicon: a review. *Pedosphere* 23, 815–825.

Wu, X., Blake, S., Sleper, D.A., Shannon, J.G., Cregan, P. *et al.* (2009) QTL, additive and epistatic effects for SCN resistance in PI 437654. *Theoretical and Applied Genetics* 118, 1093–1105.

Xu, C., Xia, C., Xia, Z., Zhou, X., Huang, J. *et al.* (2018a) Physiological and transcriptomic responses of reproductive stage soybean to drought stress. *Plant Cell Reports* 37, 1611–1624. DOI: 10.1007/s00299-018-2332-3.

Xu, J., Qiao, X., Tian, Z., Zhang, X., Zou, X. *et al.* (2018b) Proteomic analysis of rapeseed root response to waterlogging stress. *Plants (Basel, Switzerland)* 7(3), 71. DOI: 10.3390/plants7030071.

Yang, C., Guo, W. and Shi, D. (2010) Physiological roles of organic acids in alkali-tolerance of alkali-tolerant halophyte *Chloris virgata*. *Agronomy Journal* 102, 1081–1089. DOI: 10.2134/agronj2009.0471.

Yang, F., Melo-Braga, M.N., Larsen, M.R., Jørgensen, H.J.L. and Palmisano, G. (2013) Battle through signaling between wheat and the fungal pathogen *Septoria tritici* revealed by proteomics and phosphoproteomics. *Molecular & Cellular Proteomics* 12, 2497–2508. DOI: 10.1074/mcp.M113.027532.

Yang, J., Liu, D., Wang, X., Ji, C., Cheng, F. *et al.* (2016) The genome sequence of allopolyploid *Brassica juncea* and analysis of differential homoeolog gene expression influencing selection. *Nature Genetics* 48, 1225–1232. DOI: 10.1038/ng.3657.

Yang, W., Guo, Z., Huang, C., Duan, L., Chen, G. *et al.* (2014) Combining high-throughput phenotyping and genome-wide association studies to reveal natural genetic variation in rice. *Nature Communications* 5, 5087. DOI: 10.1038/ncomms6087.

Yang, Y., Saand, M.A., Abdelaal, W.B., Zhang, J., Wu, Y. *et al.* (2020) TRAQ-based comparative proteomic analysis of two coconut varieties reveals aromatic coconut cold-sensitive in response to low temperature. *Journal of Proteomics* 220(103766), 103766. DOI: 10.1016/j.jprot.2020.

Yao, Y.A., Wang, J.R., Ma, X.M., Lutts, S., Sun, C. *et al.* (2012) Proteomic analysis of Mn-induced resistance to powdery mildew in grapevine. *Journal of Experimental Botany* 63, 5155–5170. DOI: 10.1093/jxb/ers175.

Yu, Y., Huang, W., Chen, H., Yu, L., Lin, Y. *et al.* (2014) Identification of differentially expressed genes in flax (*Linum usitatissimum* l.) under saline-alkaline stress by digital gene expression. *Gene* 549(1), 113–122. DOI: 10.1016/j.gene.2014.07.053.

Zeng, W., Sun, Z., Cai, Z., Tang, M., Lv, Y. *et al.* (2017) Proteomic analysis by iTRAQ-MRM of soybean resistance to *Lamprosema Indicate*. *BMC Genomics* 18, 1–22. DOI: 10.1186/S12864-017-3825-0/FIGURES/7.

Zhang, J., Wen, Z., Li, W., Zhang, S., Chen, J. *et al.* (2017) Genome-wide association study for soybean cyst nematode resistance in Chinese elite soybean cultivars. *Molecular Breeding* 37, 1–10. DOI: 10.1007/S11032-017-0665-1/FIGURES/6.

Zhao, C., Zhang, Y., Du, J., Guo, X., Wen, W. *et al.* (2019) Crop phenomics: Current status and perspectives. *Frontiers in Plant Science* 10, 714. DOI: 10.3389/fpls.2019.00714.

Zhu, H.G., Cheng, W.H., Tian, W.G., Li, Y.J., Liu, F. *et al.* (2018) iTRAQ-based comparative proteomic analysis provides insights into somatic embryogenesis in *Gossypium hirsutum* L. *Plant Molecular Biology* 96, 89–102. DOI: 10.1007/s11103-017-0681-x.

Zhuang, W., Chen, H., Yang, M., Wang, J., Pandey, M.K. *et al.* (2019) The genome of cultivated peanut provides insight into legume karyotypes, polyploid evolution and crop domestication. *Nature Genetics* 51, 865–876. DOI: 10.1038/s41588-019-0402-2.

Index

Printed and bound by CPI Group (UK) Ltd, Croydon, CR0 4YY

13/09/2026

14961088-0002